Food Science and Technology
International Series

Series Editor

Steve L. Taylor
University of Nebraska – Lincoln, USA

Advisory Board

Ken Buckle
The University of New South Wales, Australia

Mary Ellen Camire
University of Maine, USA

Bruce Chassy
University of Illinois, USA

Patrick Fox
University College Cork, Republic of Ireland

Dennis Gordon
North Dakota State University, USA

Robert Hutkins
University of Nebraska, USA

Ronald S. Jackson
Quebec, Canada

Daryl B. Lund
University of Wisconsin, USA

Connie Weaver
Purdue University, USA

Ronald Wrolstad
Oregon State University, USA

Howard Zhang
Ohio State University, USA

A complete list of books in this series appears at the end of this volume.

INNOVATIONS IN FOOD PACKAGING

Innovations in Food Packaging

Edited by
Jung H. Han
Department of Food Science
University of Manitoba
Winnipeg, Manitoba
Canada

ELSEVIER
ACADEMIC
PRESS

Amsterdam · Boston · Heidelberg · London · New York · Oxford
Paris · San Diego · San Francisco · Singapore · Sydney · Tokyo

Permissions may be sought directly from Elsevier's Science & Technology Rights
Department in Oxford, UK: phone: (+44) 1865 843830, fax: (+44) 1865 853333,
e-mail: permissions@elsevier.co.uk. You may also complete your request on-line via
the Elsevier homepage (http://www.elsevier.com), by selecting 'Customer Support'
and then 'Obtaining Permissions'

Elsevier Academic Press
525 B Street, Suite 1900, San Diego, California 92101-4495, USA
http://www.elsevier.com

Elsevier Academic Press
84 Theobald's Road, London WC1X 8RR, UK
http://www.elsevier.com

British Library Cataloguing in Publication Data
A catalogue record for this book is available from the British Library

Library of Congress Catalog Control Number: 2005923849

ISBN 0-12-311632-5

Typeset by Charon Tec Pvt. Ltd, Chennai, India
website: http://www.charontec.com
Printed and bound in Holland
05 06 07 08 09 9 8 7 6 5 4 3 2 1

Contents

Contributors

Aaron L. Brody (Ch. 25), Packaging/Brody, Inc., PO Box 956187, Duluth, GA 30095, USA

Roberto A. Buffo (Ch. 4, 14, 17), Department of Food Science, University of Manitoba, Winnipeg, Manitoba R3T 2N2, Canada

Luis Cisneros-Zevallos (Ch. 11), Department of Horticultural Sciences, Texas A&M University, College Station, TX 77843, USA

Kay Cooksey (Ch. 18), Department of Packaging Science, Clemson University, 228 Poole Agricultural Center, Box 340320, Clemson, SC 29634, USA

John D. Floros (Ch. 10), The Pennsylvania State University, Department of Food Science, 111 Borland Laboratory, University Park, PA 16802, USA

Aristippos Gennadios (Ch. 15), Materials Science Group, Research and Development, Banner Pharmacaps Inc., 4125 Premier Drive, High Point, NC 27265, USA

Alexander O. Gill (Ch. 13), Department of Food Science, University of Manitoba, Winnipeg, Manitoba R3T 2N2, Canada

Colin O. Gill (Ch. 13), Agriculture and Agri-Food Canada, Lacombe Research Centre, 6000 C & E Trail, Lacombe, Alberta T4L 1W1, Canada

Nathalie Gontard (Ch. 16), Joint Research Unit "Agro-polymers Engineering and Emerging Technologies", ENSA.M, INRA, UM II, CIRAD, 2 Place P. Viala, F. 34060, Montpellier, France

Stéphane Guilbert (Ch. 16), Joint Research Unit "Agro-polymers Engineering and Emerging Technologies", ENSA.M, INRA, UM II, CIRAD, 2 Place P. Viala, F. 34060, Montpellier, France

Jung H. Han (Ch. 1, 2, 4, 6, 9, 15 and 17), Department of Food Science, University of Manitoba, Winnipeg, Manitoba R3T 2N2, Canada

Mark A. Hepp (Ch. 26), US Food and Drug Administration, Office of Food Additive Safety, HFS-275, 5100 Paint Branch Parkway, College Park, MD 20740, USA

Colin H. L. Ho (Ch. 9) Department of Food Science, University of Manitoba, Winnipeg, Manitoba R3T 2N2, Canada

Richard A. Holley (Ch. 14), Department of Food Science, University of Manitoba, Winnipeg, Manitoba R3T 2N2, Canada

John M. Krochta (Ch. 11, 22, 23), Department of Food Science & Technology, University of California, One Shields Avenue, Davis, CA 95616, USA

Monique Lacroix (Ch. 18, 20), INRS – Institut Armand-Frappier, Research Laboratories in Sciences Applied to Food, Canadian Irradiation Center, 531 des Prairies boulevard, Laval, Quebec, N7V 1B7, Canada

Dong Sun Lee (Ch. 7), Division and Food Science and Biotechnology, Kyungnam University, Masan 631–701, South Korea

Canh Le Tien (Ch. 20), Université du Qubec à Montréal, Department of Chemistry and Biochemistry, C. P. 8888, Succ. Centre Ville, Montréal, Quebec H3C 3P8, Canada

Zhiqiang Liu (Ch. 19), Department of Food Science, University of Manitoba, Winnipeg, Manitoba R3T 2N2, Canada

Konstantinos I. Matsos (Ch. 10), The Pennsylvania State University, Department of Food Science, 6 Borland Laboratory, University Park, PA 16802, USA

Mina McDaniel (Ch. 24), Department of Food Science & Technology, 100 Wiegand Hall, Oregon State University, Corvalis, OR 97331, USA

Seacheol Min (Ch. 27), Department of Food Science and Technology, University of California, One Shields Avenue, Davis, CA 95616, USA

María B. Pérez-Gago (Ch. 22) Department of Postharvest, Instituto Valenciano de Investigaciones Agrarias, Mondada, Spain 46113

Jong Whan Rhim (Ch. 21), Department of Food Engineering, Mokpo Natl. University, Chonnam 534–729, South Korea

Evangelina T. Rodrigues (Ch. 9), Department of Food Science, University of Manitoba, Winnipeg, Manitoba R3T 2N2, Canada

Michael L. Rooney (Ch. 5, 8), Food Science Australia, CSIRO, North Ryde, NSW 1670, Australia

Martin G. Scanlon (Ch. 2), Department of Food Science, University of Manitoba, Winnipeg, Manitoba R3T 2N2, Canada

Thomas H. Shellhammer (Ch. 21), Department of Food Science & Technology, Oregon State University, Corvalis, OR 97331, USA

Nepal Singh (Ch. 3), Department of Food Science & Technology, University of Georgia, Athens, GA 30602, USA

Rakesh K. Singh (Ch. 3), Department of Food Science & Technology, University of Georgia, Athens, GA 30602, USA

Yoon S. Song (Ch. 26), US Food and Drug Administration, Division of Food Processing and Packaging, HFH-450, National Center for Food Safety and Technology, 6502 South Archer Road, Summit-Argo, IL 60501, USA

Rungsinee Sothornvit (Ch. 23), Department of Food Engineering, Faculty of Engineering at Kamphaengsaen, Kasetsart University, Kamphengsaen Campus, Nakhonpathom 73140, Thailand

Kevin C. Spencer (Ch. 12), Spencer Consulting, 424 Selborne Road, Riverside, IL 60546, USA

Q. Howard Zhang (Ch. 27), Department of Food Science and Technology, The Ohio State University, 2015 Fyffe Road, 110 Parker Food Science and Technology Building, Columbus, OH 43210, USA

Yachuan Zhang (Ch. 4), Department of Food Science, University of Manitoba, Winnipeg, Manitoba, R3T 2N2, Canada

Yanyun Zhao (Ch. 24), Department of Food Science & Technology, 100 Wiegand Hall, Oregon State University, Corvalis, OR 97331, USA

Preface

For two years in a row in 2001 and 2002, I was privileged to host symposia at the annual meetings of Canadian Institute of Food Science and Technology in Toronto and Edmonton, under the title of "New Developments in Food Packaging" and "Active Food Packaging," respectively. Both symposia were quite successful as they provided such invaluable opportunities for the general participants and invited speakers to share with one another their scholarly interests and practical experiences. So I have imposed upon myself an obligation to put the shiny fruits borne in the course of these events in a single basket with a view to extending the network of active packaging specialists. All of the invited speakers and other experts in food packaging have contributed to this book.

Food packaging was originally conceived to secure proper preservation of food in a container by protecting it from the influence of the environmental elements. Current innovations of food packaging technology are geared to ensuring better protection, more efficient quality preservation, and enhanced safety. In addition, active packaging facilitates increased functionality of a package: now a food package is expected to be informative, easy-to-handle and disposal-friendly. By addressing selective topics of these extended functions of food packaging, this book as a whole attempts to modify the traditional notion of food packaging.

This volume was designed to benefit those who are interested in innovative technology of food packaging in general, and experienced field packaging specialists and graduate-level food scientists in particular. This book will be useful as a textbook not only for extension programs of food packaging development in food industry but also for advanced graduate-level food packaging courses.

The text consists of five parts: (1) fundamental theories covering physical chemistry background and quality preservation of foods; (2) active packaging research and development; (3) modified atmosphere packaging of fresh produce, meats and ready-to-eat products; (4) edible and biodegradable coatings and films; and (5) commercialization aspects of these new packaging technologies. Each part is divided into chapters of subject review and detailed technical information. All chapters have been peer-reviewed by responsible professionals to ensure scientific rigor.

This book would not have been possible without the contribution of the chapter authors. They are my mentors, colleagues and academic family members. My special thanks go to Elsevier Academic Press staff in London for initiating the publication process and to Elsevier Academic Press staff in Amsterdam for completing it.

Jung H. Han, Ph.D.

Fundamental theories regarding the physical and chemical background and quality preservation of foods

PART

1

New technologies in food packaging: overview

1

Jung H. Han

Introduction

Packaging is one of the most important processes to maintain the quality of food products for storage, transportation and end-use (Kelsey, 1985). It prevents quality deterioration and facilitates distribution and marketing. The basic functions of packaging are protection, containment, information and convenience (Kelsey, 1985). A good package can not only preserve the food quality but also significantly contribute to a business profit. Beyond the major function of preservation, packaging also has secondary functions – such as selling and sales promotion. However, the main function of food packaging is to achieve preservation and the safe delivery of food products until consumption. During distribution, the quality of the food product can deteriorate biologically and chemically as well as physically. Therefore, food packaging contributes to extending the shelf life and maintaining the quality and the safety of the food products.

Yokoyama (1985) suggested the conditions necessary to produce appropriate packaging:

1. Mass production
2. Reasonable and efficient packaging material
3. Suitable structure and form

Innovations in Food Packaging
ISBN: 0-12-311632-5

4. Convenience
5. Consideration of disposal.

Therefore, according to these conditions, packaging design and development includes not only the industrial design fields, creativity and marketing tools, but also the areas of engineering and environmental science. Preservation, convenience and the other basic functions of packaging are certainly important, but its disposal should be treated as an important aspect of packaging development. This is a problem in package development that may confront us in the future.

The food industry uses a lot of packaging materials, and thus even a small reduction in the amount of materials used for each package would result in a significant cost reduction, and may improve solid waste problems. Packaging technology has attempted to reduce the volume and/or weight of materials in efforts to minimize resources and costs. Several trends in the food packaging evolution have been remarkable (Testin and Vergano, 1990), including source reduction, design improvement for convenience and handling, and environmental concerns regarding packaging materials and processes. Food packaging has evolved from simple preservation methods to convenience, point-on-purchase (POP) marketing, material reduction, safety, tamper-proofing, and environmental issues (Stilwell *et al.*, 1991; see Table 1.1). Since the World Trade Center tragedy in 2001, food technologists have focused their attention on revising packaging systems and package designs to increase food safety and security. The level of concern regarding the use of food and water supplies as a form of bioterrorism has increased (Nestle, 2003). Therefore, many applications of active packaging will be commercially developed for the security and safety enhancement of food products.

Although food packaging has evolved in its various functions, every package still has to meet the basic functions. Food packaging reduces food waste and spoilage during distribution, and decreases the cost of preservation facilities. It extends the shelf life of foods, and provides safe foods to the consumers. A good package has to maintain the safety and quality of foods as well as being convenient, allowing sales promotion, and addressing environmental issues.

The quality of the packaged food is directly related to the food and packaging material attributes. Most food products deteriorate in quality due to mass transfer phenomena, such as moisture absorption, oxygen invasion, flavor loss, undesirable odor absorption, and the migration of packaging components into the food (Kester

Table 1.1	Trends in the evolution of food packaging*
Period	**Functions and issues**
1960s	Convenience, POP marketing
1970s	Lightweight, source reduction, energy saving
1980s	Safety, tamper-evidence
1990s	Environmental impact
2000s	Safety and security

*1960s–1990s data are from Stilwell *et al.* (1991).

and Fennema, 1986; Debeaufort *et al.*, 1998). These phenomena can occur between the food product and the atmospheric environment, between the food and the packaging materials, or among the heterogeneous ingredients in the food product itself (Krochta, 1997). Therefore, mass transfer studies on the migration of package components and food ingredients, on the absorption and desorption of volatile ingredients, flavors and moisture, on gas permeation, and on the reaction kinetics of oxidation and ingredient degradation are essential for food packaging system designs.

Developments in food processing and packaging

Year after year, technology becomes better. Most developments in the field of food technology have been oriented towards processing food products more conveniently, more efficiently, at less cost, and with higher quality and safety levels. Traditional thermal processes have offered tremendous developments in the food processing industry; these include commercial sterilization, quality preservation, shelf-life extension and safety enhancement. Extended shelf-stable products manufactured by retorting or aseptic processing are available in any grocery store and do not require refrigeration. These types of products are very convenient at any place or time and are easy to handle, therefore benefiting producers, processors, distributors, retailers and consumers. The major function of extended shelf-stable food packaging is barrier protection against the invasion of micro-organisms.

Beyond this simple barrier function, there has been more research and development regarding the introduction of new purposes to food packaging systems. Among these, significant new functional packaging systems include active packaging, modified atmosphere packaging (MAP) and edible films/coatings. Figure 1.1 shows the number of publications relating to these subjects in the last 14 years; the data were collected from *Food Science and Technology Abstract*™ using these three topics as key words. Many articles on MAP and edible films/coatings have been published, particularly since the early 1990s, while articles on active packaging have increased only as of the late 1990s. Articles on MAP reached 100 publications per year, prior to articles on edible films/coatings. It is assumed that the reason why the number of articles on active packaging remained low was because articles in this area used key words other than "active packaging" – such as oxygen-scavenging packaging, antimicrobial packaging etc.

The development of new packaging functionalities has been possible because of technological advances in food processing, packaging material science and machinery. Among the many new technologies the development in processing and packaging machinery is notable, leading to higher standards of regulation, hygiene, health and safety. New software and part installations in unit operations have increasingly been introduced, and high-speed automation has been achieved by using new servomotors and software technologies such as the machine vision system (Tucker, 2003). The processing and packaging equipment has new functions that have increased safety, quality and productivity, and therefore it seems that the development of new packaging functions may go hand in hand with the development of new processes, materials and

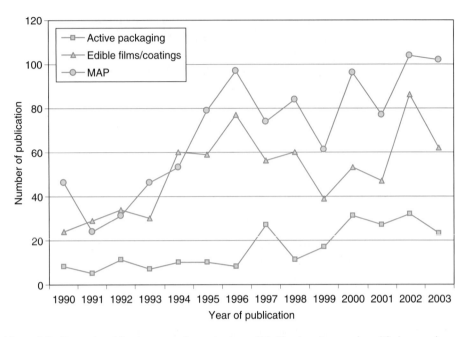

Figure 1.1   Research publications on active packaging, edible films/coatings, and modified atmosphere packaging. Data collected from Food Science and Technology Abstract™.

equipments. Packages may have new purposes if new functional packaging materials and/or materials containing functional inserts are used. Developing new packaging technologies not only implies new materials but also new packaging design systems.

Food packaging technologies

Extra active functions of packaging systems

Many new "extra" functions have been introduced in active packaging technologies, including oxygen-scavenging and intelligent functions, antimicrobial activity, atmosphere control, edibility, biodegradability etc. Food packaging performs beyond its conventional protective barrier property. The new active packaging systems increase security, safety, protection, convenience, and information delivery. Active packaging systems extend the shelf life of food products by maintaining their quality longer, increase their safety by securing foods from pathogens and bioterrorism, and enhance the convenience of food processing, distribution, retailing and consumption.

There are many applications of active packaging technologies, several of which have been commercialized and are used in the food industry; these include oxygen-scavenging, carbon dioxide-absorbing, moisture-scavenging (desiccation) and antimicrobial systems. Oxygen-scavenging systems have been commercialized in the form of a sachet that removes oxygen (Figure 1.2). An oxygen-free environment can prevent food

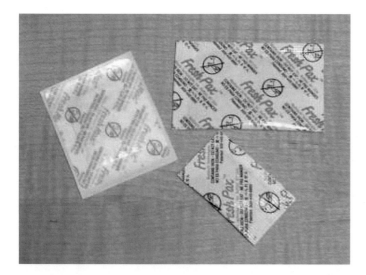

Figure 1.2 Oxygen-scavenging sachets.

oxidation and rancidity, and the growth of aerobic bacteria and moulds. Carbon dioxide-scavenging packaging systems can prevent packages from inflating due to the carbon dioxide formed after the packaging process – for example, packaged coffee beans may produce carbon dioxide during storage as a result of non-enzymatic browning reactions. Fermented products such as kimchi (lactic acid fermented vegetables), pickles, sauces, and some dairy products can produce carbon dioxide after the packaging process. Carbon dioxide-scavenging systems are also quite useful for the products that require fermentation and aging processes after they have been packed. Moisture-scavenging systems have been used for a very long time for packaging dried foods, moisture-sensitive foods, pharmaceuticals and electronic devices; in these systems, desiccant materials are included in the package in the form of a sachet. Recently, the sachets have contained humectants as well as desiccants to control the humidity inside the package more specifically. Moisture-scavenging systems that are based on desiccation are evolving to control the moisture by maintaining a specific relative humidity inside the package by absorbing or releasing the moisture.

Antimicrobial packaging applications are directly related to food microbial safety and bioterrorism, as well as to shelf-life extension by preventing the growth of spoilage and/or pathogenic micro-organisms. The growth of spoilage micro-organisms reduces the food shelf life, while the growth of pathogenic micro-organisms endangers public health. Antimicrobial packaging systems consist of packaging materials, in-package atmospheres and packaged foods, and are able to kill or inhibit micro-organisms that cause food-borne illnesses (Han, 2000, 2003a, 2003b; see Figure 1.3).

Intelligent packaging has been categorized both as a part of active packaging and as a separate entity, depending on different viewpoints. It contains intelligent functions which have been researched to enhance convenience for food manufacturing and distribution and, increasingly, to improve food security and safety verification (Rodrigues and Han, 2003).

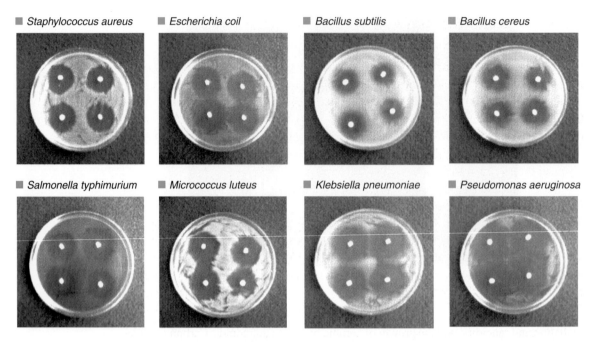

Figure 1.3 Evaluation of antimicrobial plastic films against various bacteria (reproduced courtesy of Micro Science Tech Co., South Korea).

Modified atmosphere packaging

MAP is traditionally used to preserve the freshness of fresh produce, meats and fish by controlling their biochemical metabolism – for example, respiration. Nitrogen flushing, vacuum packaging and carbon dioxide injection have been used commercially for many years. However, current research and development has introduced new modified atmosphere technologies such as inert gas (e.g. argon) flushing for fruits and vegetables, carbon monoxide injection for red meats, and high oxygen flushing for red meats. For a MAP system to work effectively, optimal packaging material with proper gas permeability properties must be selected. The use of MAP systems is attractive to the food industry because there is a fast-growing market for minimally processed fruits and vegetables, non-frozen chilled meats, ready-to-eat meals and semi-processed bulk foods.

MAP dramatically extends the shelf life of packaged food products, and in some cases food does not require any further treatments or any special care during distribution. However, in most cases extending shelf life and maintaining quality requires a multiple hurdle technology system – for example, introducing temperature control as well as MAP is generally essential to maintain the quality of packaged foods. Hurdle technology is therefore important for MAP applications, since the modified atmosphere provides an unnatural gas environment that can create serious microbial problems such as the growth of anaerobic bacteria and the production of microbial toxins. Therefore, an included temperature control system is very important for quality preservation and microbial control.

Edible films and coatings

The use of edible films and coatings is an application of active food packaging, since the edibility and biodegradability of the films are extra functions that are not present in conventional packaging systems (Han, 2002). Edible films and coatings are useful materials produced mainly from edible biopolymers and food-grade additives. Most biopolymers are naturally existing polymers, including proteins, polysaccharides (carbohydrates and gums), and lipids (Gennadios *et al.*, 1997). Plasticizers and other additives are included with the film-forming biopolymers in order to modify film physical properties or to create extra functionalities.

Edible films and coatings enhance the quality of food products by protecting them from physical, chemical, and biological deterioration (Kester and Fennema, 1986). The application of edible films and coatings is an easy way to improve the physical strength of the food products, reduce particle clustering, and enhance the visual and tactile features of food product surfaces (Cuq *et al.*, 1995). They can also protect food products from oxidation, moisture absorption/desorption, microbial growth, and other chemical reactions (Kester and Fennema, 1986). The most common functions of edible films and coatings are that they are barriers against oils, gas or vapours, and that they are carriers of active substances such as antioxidants, antimicrobials, colors and flavors (Guilbert and Gontard, 1995; Krochta and De Mulder-Johnston, 1997). Thus edible films and coatings enhance the quality of food products, which results in an extended shelf life and improved safety.

New food-processing technologies

Besides the traditional thermal treatments for food preservation, many other new thermal and non-thermal processing technologies have been developed recently. These include irradiation, high-pressure processes, pulsed electric fields, UV treatments, antimicrobial packaging etc. Some of these processes have been commercially approved by regulatory agencies for food packaging purposes. These new technologies generally require new packaging materials and new design parameters in order for optimum processing efficiency to occur – for example, packages that undergo an irradiation process are required to possess chemical resistance against high energy to prevent polymer degradation, those that undergo UV treatments require UV light transmittable packaging materials, and retortable pouches should resist pressure changes and maintain seal strength. Since these new technologies each possess unique characteristics, packaging materials should be selected with these characteristics in mind.

These new packaging materials and/or systems not only need to work technically; they should also be examined scientifically to ensure their safety and lack of toxicity, and be approved by regulatory agencies. In some cases, countries may require new regulations and legislation for the use of these new processing and packaging technologies. The globalization of the food industry enforces international standards and compliance with multiple regulations. New technologies should also be examined for their effect on product quality and public health, and the results of these tests should be disclosed to the public, government agencies, processors, and consumer groups. However, some criteria (such as threshold levels, allowable limits, and generally acceptable levels) are decided politically, as are rulings on how to practice and review the policy. Scientific intervention is limited; however, it is important that scientific research results and suggestions be sought and respected during political decision-making.

Future trends in food packaging

A continuing trend in food packaging technology is the study and development of new materials that possess very high barrier properties. High-barrier materials can reduce the total amount of packaging materials required, since they are made of a thin or lightweight materials with high-barrier properties. The use of high-barrier packaging materials reduces the costs in material handling, distribution/transportation and waste reduction.

Convenience is also a "hot" trend in food packaging development. Convenience at the manufacturing, distribution, transportation, sales, marketing, consumption and waste disposal levels is very important and competitive. Convenience parameters may be related to productivity, processibility, warehousing, traceability, display qualities, tamper-resistance, easy opening, and cooking preparation.

A third important trend is safety, which is related to public health and to security against bioterrorism. It is particularly important because of the increase in the consumption of ready-to-eat products, minimally processed foods and pre-cut fruits and vegetables. Food-borne illnesses and malicious alteration of foods must be eliminated from the food chain.

Another significant issue in food packaging is that it should be natural and environmentally friendly. The substitution of artificial chemical ingredients in foods and in packaging materials with natural ingredients is always attractive to consumers. Many ingredients have been substituted with natural components – for example, chemical antioxidants such as BHA, BHT and TBHQ have been replaced with tocopherol and ascorbic acid mixtures for food products. This trend will also continue in food packaging system design areas. To design environmentally friendly packaging systems that are more natural requires, for example, the partial replacement of synthetic packaging materials with biodegradable or edible materials, a consequent decrease in the use of total amount of materials, and an increase in the amount of recyclable and reusable (refillable) materials.

Food science and packaging technologies are linked to engineering developments and consumer studies. Consumers tend continuously to want new materials with new functions. New food packaging systems are therefore related to the development of food-processing technology, lifestyle changes, and political decision-making processes, as well as scientific confirmation.

References

Cuq, B., Gontard, N. and Guilbert, S. (1995). Edible films and coatings as active layers. In: *Active Food Packaging* (M. Rooney, ed.), pp. 111–142. Blackie Academic & Professional, Glasgow, UK.

Debeaufort, F., Quezada-Gallo, J. A. and Voilley, A. (1998). Edible films and coatings: tomorrow's packaging: a review. *Cri. Rev. Food Sci.* **38(4),** 299–313.

Gennadios, A., Hanna, M. A. and Kurth, L. B. (1997). Application of edible coatings on meats, poultry and seafoods: a review. *Lebensm. Wiss. u Technol.* **30(4),** 337–350.

Guilbert, S. and Gontard, N. (1995). Edible and biodegradable food packaging. In: *Foods and Packaging Materials – Chemical Interactions* (P. Ackermann, M. Jägerstad and T. Ohlsson, eds), pp. 159–168. The Royal Society of Chemistry, Cambridge, UK.

Han, J. H. (2000). Antimicrobial food packaging. *Food Technol.* **54(3),** 56–65.

Han, J. H. (2002). Protein-based edible films and coatings carrying antimicrobial agents. In: *Protein-based Films and Coatings* (A. Gennadios, ed.), pp. 485–499. CRC Press, Boca Raton, FL.

Han, J. H. (2003a). Antimicrobial packaging materials and films. In: *Novel Food Packaging Techniques* (R. Ahvenainen, ed.), pp. 50–70. Woodhead Publishing Ltd, Cambridge.

Han, J. H. (2003b). Design of antimicrobial packaging systems. *Int. Rev. Food Sci. Technol.* **11,** 106–109.

Kelsey, R. J. (1985). *Packaging in Today's Society*, 3rd edn. Technomic Publishing Co., Lancaster, PA.

Kester, J. J. and Fennema, O. R. (1986). Edible films and coatings: a review. *Food Technol.* **48(12),** 47–59.

Krochta, J. M. (1997). Edible protein films and coatings. In: *Food Proteins and Their Applications* (S. Damodaran and A. Paraf, eds), pp. 529–549. Marcel Dekker, New York, NY.

Krochta, J. M. and De Mulder-Johnston, C. (1997). Edible and biodegradable polymer films: challenges and opportunities. *Food Technol.* **51(2),** 61–74.

Nestle, M. (2003). *Safe Food: Bacteria, Biotechnology, and Bioterrorism*, p. 1. University of California Press, Berkeley, CA.

Rodrigues, E. T. and Han, J. H. (2003). Intelligent packaging. In: *Encyclopedia of Agricultural and Food Engineering* (D. R. Heldman, ed.), pp. 528–535. Marcel Dekker, New York, NY.

Stilwell, E. J., Canty, R. C., Kopf, P. W., Montrone, A. M. and Arthur D. Little, Inc. (1991). *Packaging for the Environment, A Partnership for Progress*. AMACOM, American Management Association. New York, NY.

Testin, R. F. and Vergano, P. J. (1990). *Food Packaging, Food Protection and the Environment. A Workshop Report*. IFT Food Packaging Division. The Institute of Food Technologists, Chicago, IL.

Tucker, B. J. O. (2003). A view on the trends affecting the current development of food processing and packaging machinery. *Int. Rev. Food Sci. Technol.* **11,** 102–103.

Yokoyama, Y. (1985). Materials in packaging. In: *Package Design in Japan*, Vol. 1 (S. Hashimoto, ed.), pp. 113–115. Rikuyo-sha Publishing, Tokyo, Japan.

2 Mass transfer of gas and solute through packaging materials

Jung H. Han and Martin G. Scanlon

Introduction

There are many applications in the area of food packaging that use mass transfer phenomena. Examples include selecting a packaging material to predict and extend product shelf life, and to control the in-package atmosphere for protection and preservation of food products. Permeation, absorption and diffusion are typical mass transfer phenomena occurring in food packaging systems (Figure 2.1). Permeation is the ability of permeants to penetrate and pass right through an entire material in response to a difference in partial pressure – the gas and water vapor transmission rates of packaging materials give a good indication of permeation. This property of the packaging material may also be referred to as the "permeance" (ASTM, 1993; Segall and Scanlon, 1996). To convert the permeance (which is evidently dependent on the thickness of the film) into an intensive property, the permeance is multiplied by the film thickness to derive the permeability (P) of the film. Thus the mass transfer coefficient for permeation is permeability (P). The mass transfer of a solute from a solution through a (polymeric) material is also a useful way to determine mass transfer coefficients experimentally, because it requires simple permeation apparatus (i.e. a permeation cell) consisting of the high and low concentration solution chambers divided by the test film material. The concentration increase of the substance in the low concentration chamber is measured to determine the permeance. Durrheim *et al.* (1980) showed the

Innovations in Food Packaging
ISBN: 0-12-311632-5

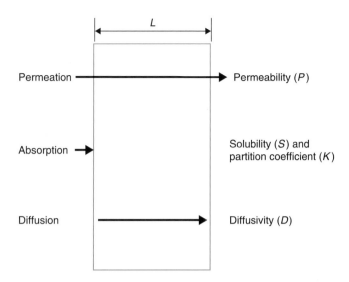

Figure 2.1 Mass transfer phenomena and their characteristic coefficients.

relationship between the chain length of various alkanols and their permeability through mouse skin and human epidermal tissue. It was found that the permeability increased exponentially with increasing length of the carbon chain, up to eight carbons.

Diffusion is the movement of a diffusant in a medium caused by a concentration difference acting as a driving force. Diffusivity (D) is a measure of how well the compound diffuses in the medium. Schwartzberg and Chao (1982) reported the diffusivities of different kinds of food components in food systems. Rico-Pena and Torres (1991) measured the diffusivity of potassium sorbate through a methylcellulose-palmitic acid film under various conditions of water activity and pH. The diffusivity of potassium sorbate through the edible film increased with decreasing pH and with increasing water activity.

Absorption and its counterpart desorption measure the affinity of a given substance for two media with which it comes into contact. Flavor scalping of d-limonene, an orange flavor component, into the sealing layer of a flexible packaging material for orange juice is a good example of absorption in packaging. The d-limonene has a much higher affinity for the plastic layer than for the juice, in which it should preferably reside. The affinity of a substance for a material can be expressed using the solubility (S) or partition (K) coefficient. A microbial stability model for intermediate moisture foods has been suggested based on the concentration distribution of sorbate in edible polysaccharide films (Torres et al., 1985; Torres, 1987). The intermediate moisture foods were also coated with a zein protein-containing sorbate. The microbial stability factor was the sorbate concentration during storage, and the higher concentration of the residual sorbate was assumed to maintain the higher microbial stability of the intermediate moisture food.

The permeability, solubility, and diffusivity have characteristic values for a migrating component through a particular medium. These parameters are therefore essential in simulating the mass transfer profile. This chapter reviews mass transfer phenomena and defines the diffusivity, permeability, solubility, and partition coefficient of transferring molecules. Gases and solutes are considered separately.

General theory

The mass transfer rates of molecules through a package material or through a membrane are often described as irreversible processes (Miller, 1960). A generalized thermodynamic driving force is required to elicit movement of the molecules, which, for the movement of gases and solutes, is the gradient in the chemical potential of the migrating species ($\partial M_i/\partial x$). For most packaging and membrane applications the area through which transfer occurs is large compared to the thickness, so that one-dimensional flow (or flux) is considered. The linear coefficient linking the flux (per unit cross-section) to the driving force (Miller, 1960) can be considered as a resistance of the package or membrane material to the passage of the given species:

$$\text{Flux} = (\text{Material resistance})^{-1} \, (\text{Driving force}) \qquad (2.1)$$

With the appropriate substitutions and assumptions, the gradient in chemical potential is related to the concentration gradient of the migrating species. Figure 2.2 shows an example of mass transfer through a packaging material. The permeation of a molecule is its movement from the region where its concentration is C_1 to the region where the concentration is lower at C_0. The absorption phenomenon explains the transfer from concentration C_1 to C_{s1}; the diffusion phenomenon expresses the movement of molecules from concentration C_{s1} to C_{s0}. Desorption explains the change in concentration

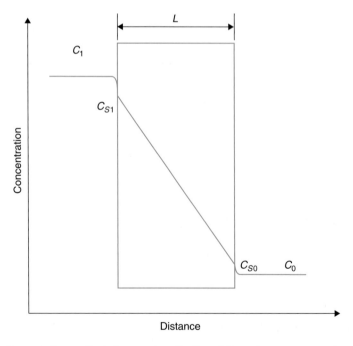

Figure 2.2 Concentration profile during mass transfer. C_1 and C_0 are the gas concentrations in the environment and inside the package respectively, while C_{s1} and C_{s0} are the concentrations of gas at the outer surface of the packaging material and inside of the packaging material respectively.

between C_{s0} and C_0. To evaluate how fast either solutes or gases move in response to the thermodynamic driving force, the factors making up the material resistance are examined separately.

Diffusivity

Events occurring within the material are examined first where diffusion is the dominant factor. Diffusion obeys Fick's law (Crank, 1975), and Fick's first law can be expressed as:

$$J_d = -D\frac{\partial C}{\partial x} \tag{2.2}$$

where J_d, D, C and x are the flux per unit cross-section, the diffusivity, the concentration of the diffusant, and the distance across which the diffusant has to travel, respectively. Fick's second law can be used to analyze unsteady state diffusion with time t:

$$\frac{\partial C}{\partial t} = -D\frac{\partial^2 C}{\partial x^2} \tag{2.3}$$

Crank (1975) introduced various solutions for equation (2.3) with different geometries of the material (e.g. infinite slabs, spheres etc.) and various initial boundary conditions. Analytical solutions of equation (2.3) for the case of heat transfer problems were also presented by Carslaw and Jaeger (1959). After integrating equation (2.2) for the case where the concentrations C_{s1} and C_{s0} remain constant (and provided that D is a constant), the flux of the diffusant in the steady state is given by equation (2.4):

$$Flux,\ J_d = \frac{Q}{A \cdot t} = D\frac{(C_{s1} - C_{s0})}{L} \tag{2.4}$$

where Q is the amount of diffused moving substance (mol or kg), A is the cross-sectional diffusing area, and L is the thickness of the package or membrane. The diffusivity D has units of $m^2 s^{-1}$ and the flux has units of $mol\ m^{-2}\ s^{-1}$ or $kg\ m^{-2}\ s^{-1}$:

$$D = \frac{J_d \cdot L}{\Delta C} = \frac{Q \cdot L}{A \cdot t \cdot \Delta C} = \frac{[kg][m]}{[m^2][s][kg \cdot m^{-3}]}\ or\ \frac{[m^2]}{[s]} = [m^2 \cdot s^{-1}] \tag{2.5}$$

Solubility/partitioning

Henry's law and solubility

Before gas can diffuse through the packaging material from C_{s1} to C_{s0} it must first dissolve into the material. The sorption of a gas component into a packaging material generally has a linear relationship to the partial pressure of the gas as shown in Henry's law under conditions where the gas concentration is lower than its saturation concentration or maximum solubility:

$$p = \sigma X_s \tag{2.6}$$

where p and X_s are the partial pressure of the gas in the atmosphere and the molar fraction of gas in the packaging material respectively, and σ is the Henry's law constant in Pa (Moore, 1972). If the permeable gas molecule has an affinity to the packaging material matrix, or is immobilized in the microvoids of the matrix polymer at a relatively low pressure, the sorption behavior follows a logarithmic non-linear relationship, which is expressed as a Langmuir-type sorption (Robertson, 1993). Equation (2.7) shows the linear relationship between the concentration at the surface of the packaging material and the partial pressure of the gas for the example shown in Figure 2.2:

$$C_s = H^{-1} p_1 \tag{2.7}$$

where p_1 is the partial pressure of the gas on the high concentration (C_1) side. This relationship is compatible with equation (2.6) when dealing with dilute solutions, as in permeation situations, so that $C_{s1} \propto X_s$. The constant H^{-1} can be expressed in the form of equation (2.8):

$$H^{-1} = \frac{C_{s1}}{p_1} = \frac{n}{V_{material} \cdot p_1} = \frac{n}{V_{STP}} \frac{V_{STP}}{V_{material} \cdot p_1} = \frac{S}{V_{STP}} = \frac{[mol][m^3_{gas}]}{[m^3_{gas}][m^3_{material}][Pa]} \tag{2.8}$$

where n is the number of moles of gas that dissolve in the packaging material, and $V_{material}$ is the volume of the packaging material into which the gas dissolves. V_{STP} is the volume occupied by 1 mole of the gas under standard temperature and pressure (STP) conditions (0°C and 1 atm). This has a value of $22\,414 \times 10^{-6} m^3 \, mol^{-1}$ for an ideal gas, and thus can be taken as a constant. The solubility S of the gas in equation (2.8) can be expressed as in equation (2.9):

$$S = \frac{n \cdot V_{STP}}{V_{material} \cdot p_1} \tag{2.9}$$

where S is the solubility in $mol \, m^3_{gas} \, mol^{-1} \, m^{-3}_{material} \, Pa^{-1}$ which is equivalent to Pa^{-1}, which shows the volume ratio of the absorbed component in a material under standard gas conditions (Geankoplis, 1993). The solubility is a constant which is independent of the absorbed concentration at a given temperature (Robertson, 1993). From equations (2.8) and (2.9), the concentrations of gas at the edges of the packaging material can be expressed as:

$$C_{s1} = \frac{S \cdot p_1}{V_{STP}} \text{ and } C_{s0} = \frac{S \cdot p_0}{V_{STP}} \tag{2.10}$$

Convective mass transfer

The driving force for the mass transfer of solute molecules in a medium or material is solely the concentration difference between two positions. The concentration difference at the boundary film layer between the solution and the material is therefore pertinent for describing the driving force for partition of the solute between the solution and the packaging material. In this case, the convective mass transfer coefficient h_m may need to be considered, since this coefficient is dependent on solution agitation and the absorption affinity of the solute for the solution and for the material. Figure 2.3 represents the mass transfer of a solute across a membrane or a packaging material. The convective mass transfer coefficient, h_m, with dimensions of velocity (ms^{-1}), is the parameter which expresses the resistance to transfer of mass at the surface of the material. The parameter is also referred to as the surface mass transfer coefficient. The flux on the solution side of the boundary film layer is given by:

$$J_h = h_{m1} \Delta C = h_{m1} (C_1 - C_1')$$

$$(2.11)$$

As shown in Figure 2.3, there may be more than just resistance to mass transfer on the solution side. On the package or membrane side at the interface, a further resistance to mass transfer may exist:

$$J_h = h_{ms} \Delta C = h_{ms} (C_{s1}' - C_{s1})$$

$$(2.12)$$

The absorption affinity of the migrating solute across the interface itself can be expressed using the partition coefficient K, which is the concentration ratio for the solute between

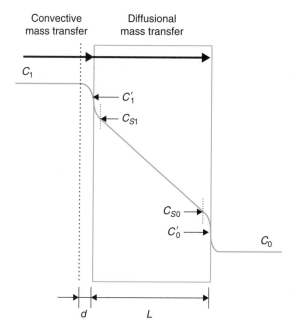

Figure 2.3 Mass transfer of a solute through a membrane.

the two different phases (solution and packaging material). At the interface, the partition coefficient of the solute between the solution and the material is:

$$K = \frac{C'_{s1}}{C'_1} \text{ and } C'_{s1} = K C'_1 \tag{2.13}$$

Similar expressions prevail for desorption. After the solute molecule has been absorbed into the material layer, the next transfer phenomenon that follows is diffusion, which can be represented by Fick's laws (equations (2.2) and (2.3)).

Overall mass transfer of gases and solutes

Gaseous diffusivity and permeability

Since the driving force for gas penetration through a packaging material is the difference in gas concentrations or partial pressures between the two sides of the packaging material, the gas flux J_g of both permeation and diffusion can use the partial pressure term instead of the concentration gradient. In the mass transfer situation of Figure 2.2, the concentration can be substituted for the partial pressure p and the solubility S from equations (2.4) and (2.10):

$$Flux_{gas} \, J_g = D \cdot \frac{C_{s1} - C_{s0}}{L} = D \cdot \frac{S \cdot (p_1 - p_0)}{V_{STP} \cdot L} = P \cdot \frac{p_1 - p_0}{V_{STP} \cdot L} \tag{2.14}$$

where J_g is the flux, and p_1 and p_0 are the partial pressures of the gas on the left- and right-hand sides of Figure 2.2 respectively (C_1 and C_0 sides). The partial pressure p_1 would generally be a constant at ambient conditions. However, p_0 is dependent on the diffusion of the gas inside the food material or on its reactivity with food components. Fast diffusion would displace the absorbed gas into the inside of the food within the package and generate a larger Δp across the package material. Alternatively, in a modified atmosphere package, where the left-hand side represents the inside of the package, C_1 may vary according to the respiratory activities of the food in the package.

From equation (2.14), the permeability, the diffusivity, and the solubility of a gas have the relationship shown in equation (2.15):

$$P = DS \tag{2.15}$$

Therefore, the SI units of P, D and S are $m^2 s^{-1} Pa^{-1}$, $m^2 s^{-1}$ and Pa^{-1}, respectively. Alternatively, equation (2.14) can be rearranged using the flux definition in equation (2.4) to retain all relevant dimensions:

$$P = \frac{Q_{gas} \cdot V_{STP} \cdot L}{A_{material} \cdot t \cdot (p_1 - p_0)} \tag{2.16}$$

so that P has units of $m^3_{gas} \, m_{thickness} \, m^{-2}_{area} \, s^{-1} \, Pa^{-1}$, compatible with units found in many packaging texts for permeability: cubic centimeters (or ml) of gas (under *STP* or other defined conditions) multiplied by package thickness in ml (= 0.001 inch) per 100 square inches per 24 hours per atmosphere (ASTM, 1993).

Solute mass transfer coefficient and overall permeability

For solutes, the overall mass transfer coefficient U includes the resistances to mass transfer at the surfaces and solute diffusivity through the membrane. The total resistance (R_t) is the sum of all the resistances to mass transfer:

$$R_t = \frac{1}{U} = R_{surface} + R_{diffusion} = \frac{1}{h_{ma}} + \frac{L}{D} + \frac{1}{h_{md}} \qquad (2.17)$$

where h_{ma} and h_{md} are the convective mass transfer coefficients associated with absorption and desorption respectively. In the case of a multilayer film, the diffusional resistance will be the sum of the resistances of each layer. Assuming a film of n layers without any interfacial resistance, the diffusional resistance is:

$$R_{diffusion} = \frac{L_T}{D_T} = \frac{L_1}{D_1} + \frac{L_2}{D_2} + \cdots + \frac{L_n}{D_n} \qquad (2.18)$$

where L_T and D_T are overall thickness and overall diffusivity respectively, and L_n and D_n are the thickness and the diffusivity of the nth layer. Vojdani and Torres (1989a) used equation (2.18) in estimating the overall diffusivity of potassium sorbate through methylcellulose and hydroxypropyl methylcellulose multilayer films. In their synopsis of permeability in multilayer materials, Cooksey et al. (1999) recommended using a similar equation for calculating overall permeability from permeability data of individual layers.

The resistance to absorption of the solute into the package, and its desorption on the low concentration side (Figure 2.3), depends on the properties of the solvents, package, and solutes, and it also depends on the degree of agitation of the solvents on both sides of the package material. In the case where resistances are small, such as when agitation ensues on the solution side, and diffusion in the interface region is identical to that through the membrane material as a whole, then $C'_1 \approx C_1$ and $C'_{s1} \approx C_{s1}$, so that only the partition coefficient affects the rate of mass transfer across the interface.

The flux of the overall mass transfer, J_T, is the product of the overall mass transfer coefficient, U, and the concentration difference driving mass transfer, which is:

$$J_T = U\Delta C = U(C_1 - C_0) \qquad (2.19)$$

Equation (2.19) shows the overall mass transfer from the high concentration (C_1) side to the low concentration (C_0) side through a packaging material. In such a mass transfer phenomenon, the total flux J_T is the same as the fluxes at the interfaces, J_h, and the flux of the diffusional mass transfer, J_d. Therefore, the total flux J_T and the overall mass transfer coefficient U are analogous to the flux of permeation and the permeation coefficient (the permeability P), respectively, in the transfer of gas through a package given above, but in this case the driving force is a concentration difference rather than a difference in partial pressure.

$$J_T = J_{ha} = J_d = J_{hd} = U(C_1 - C_0) = h_{ma}\,(C_1 - C_{s1})$$

$$= \frac{D \cdot (C_{s1} - C_{s0})}{L} = h_{md}\,(C_{s0} - C_0) \qquad (2.20)$$

From equations (2.13) and (2.20), where surface resistance is dominated by the partition coefficient, the diffusion flux J_d can be written by substituting C_{s1} with $K\,C_1$ and assuming that the partition coefficient is identical for desorption and absorption. The convective mass transfer and the surface mass transfer coefficient h_m are not as difficult to work with experimentally when the dimensionless partition coefficient K is introduced, as shown in equation (2.21):

$$J_T = \frac{D \cdot (C_{s1} - C_{s0})}{L} = \frac{D \cdot K \cdot (C_1 - C_0)}{L} \qquad (2.21)$$

Washitake *et al.* (1980) measured the permeability of betamethasone and salicylic acid through eggshell membrane using equation (2.21). Smith and Haigh (1992) and Diez-Sales *et al.* (1991) designed diffusion cells for drug penetrations and validated their permeation phenomena with equation (2.21). Durrheim *et al.* (1980) introduced the partition coefficient to determine the permeability of a drug through mouse skin and human epidermal tissue. They converted the flux of equation (2.21) to a finite volume model. The permeability was determined by equation (2.22).

$$P = \frac{J}{A \cdot \Delta C} = \frac{V \cdot \dfrac{dC}{dt}}{A \cdot \Delta C} \qquad (2.22)$$

where V is the finite volume of a drug-accepting chamber, A is the area through which mass transfer takes place, and the flux J in this case was pertinent for their cell – i.e. it was not expressed on a per unit area basis.

Torres and his group (Torres *et al.*, 1985; Torres, 1987) used the partition coefficient K and equation (2.21) to estimate the transfer rates of a preservative into a maize zein coating applied to intermediate moisture foods. However, since the late 1980s the partition coefficient has been ignored and the definition of diffusivity D in $m^2\,s^{-1}$ has been used as a permeability constant P for sorbate in polysaccharide edible films (Vojdani and Torres, 1989a, 1989b, 1990; Rico-Pena and Torres, 1991). Therefore, the considerations of equation (2.21) are based on unhindered equilibration between the concentration in the film and the concentration in the bulk solution (Torres *et al.*, 1985). Keshary and Chien (1984) also used equation (2.21), with the assumption that the partition coefficient $K = 1$, for a nitroglycerine patch.

Summary

The definitions of permeation, diffusion and absorption have been reviewed with separate considerations for gas and solute penetration. Permeation is the mass transfer phenomenon that occurs when a molecule passes through a material or membrane from an area of high concentration to an area of low concentration. Diffusion is the movement

Table 2.1 Summary of mass transfer through packaging material

	Gas transmission		Solute permeation	
	Flux J	**Mass transfer coefficient**	**Flux J**	**Mass transfer coefficient**
Permeation	$P\dfrac{\Delta p}{V_{STP}\cdot L}$	Permeability P $(m^2\,s^{-1}\,Pa^{-1})$ $\left[\dfrac{m^3_{gas}\cdot m}{m^2\cdot s\cdot Pa}\right]$	$P\Delta C$	Permeability P $\left[\dfrac{m}{s}\right]$
Diffusion	$D\dfrac{\Delta C}{L}$	Diffusivity D $\left[\dfrac{m^2}{s}\right]$	$D\dfrac{\Delta C}{L}$	Diffusivity D $\left[\dfrac{m^2}{s}\right]$
Absorption	$S\dfrac{D\cdot \Delta p}{V_{STP}\cdot L}$	Solubility S (Pa^{-1}) $\left[\dfrac{m^3_{gas}}{m^3_{material}\cdot Pa}\right]$	$h_m\Delta C$	Surface mass transfer coefficient h_m $\left[\dfrac{m}{s}\right]$

of molecules within a material caused by a concentration difference. Absorption is the surface sorption of the molecules from the surroundings to the material. The mass transfer coefficients of permeation, diffusion, and absorption are the permeability (P), the diffusivity (D), and the solubility (S) in the case of a gas, or the partition coefficient (K) for a solute, respectively. Table 2.1 summarizes the definitions and the units of the coefficients.

The units and definition of the diffusivity of a molecule are identical regardless of whether the diffusant is a gas or a solute. The diffusivity defines the transfer rate of an amount of the diffusant across a known distance in the material. In both gas and solute diffusion, the diffusion has a driving force of the difference in concentration of the molecule within the material.

The permeation of gas and solute molecules has the same physical phenomenon of penetration as a permeant through a material. However, gas and solute permeation usually have their flux defined differently. Henry's law is applied to relate the surface concentration of a gas component with the partial pressure in the atmosphere in which the packaging material is in contact. On the other hand, solute permeants do not follow Henry's law. The permeability of a solute is directly related to the overall mass transfer coefficient, which includes the surface mass transfer coefficient (h_m) and the diffusional resistance (L/D). Often, the surface mass transfer coefficient is dictated by the partition of the solute between solvent and membrane. The partition coefficient (K) relates the concentration of solute at the surface (C_s) with the concentration of the solute in the solution.

Most permeable substances that affect the quality of food products are gases such as oxygen, carbon dioxide, noble gases, nitrogen and water vapor. These gases affect rancidity, ripening, and hydration/dehydration of a food product, and generally determine the length of a product's shelf life. Therefore, the oxygen transmission rate OTR ($OP\,\Delta p/L$, where OP is oxygen permeability) and water vapor transmission rate WVTR ($WVP\,\Delta p/L$, where WVP is water vapor permeability) are commonly used for quantifying the performance of packaging materials in industry. Because OTR and WVTR are common practical examples for the passage of a substance (oxygen and water vapor) through a material (the package), solute permeation has not been considered as extensively as gas permeation. However, in the case of drug delivery and active packaging systems solute permeation is theoretically important in order to explain active ingredient transfer, and its experimental determination is reasonably straightforward. This warrants studying all the factors that affect how fast solutes will permeate through given materials.

References

ASTM (1993). D 1434: Standard test method for determining gas permeability characteristics of plastic film and sheeting. In: *Annual Book of ASTM Standards*, Vol. 15.09, pp. 274–283. ASTM, Philadelphia, PA.

Carslaw, H. S. and Jaeger, J. C. (1959). *Conduction of Heat in Solids*. Oxford University Press, London, UK.

Cooksey, K., Marsh, K. S. and Doar, L. H. (1999). Predicting permeability and transmission rate for multilayer materials. *Food Technol.* **53(9)**, 60–63.

Crank, J. (1975). *The Mathematics of Diffusion*. Oxford University Press, London, UK.

Diez-Sales, O., Copovi, A., Casabo, V. G. and Herraez, M. (1991). A modelistic approach showing the importance of the stagnant aqueous layers in in vito diffusion studies, and in vitro–in vivo correlations. *Intl J. Pharmaceutics* **77**, 1–11.

Durrheim, H., Flynn, G. L., Higuchi, W. I. and Behl, C. R. (1980). Permeation of hairless mouse skin I: Experimental methods and comparison with human epidermal permeation by alkanols. *J. Pharm. Sci.* **69(7)**, 781–786.

Geankoplis, C. J. (1993). *Transport Processes and Unit Operations*, 3rd edn., pp. 408–413. Allyn and Bacon, Newton, MA.

Keshary, P. R. and Chien, Y. W. (1984). Mechanisms of transdermal controlled nitroglycerin administration. (1): Development of a finite-dosing skin permeation system. *Drug Develop. Ind. Pharmacy* **10(6)**, 883–913.

Miller, D. G. (1960). Thermodynamics of irreversible processes. *Chemical Reviews* **60**, 15–37.

Moore, W. J. (1972). *Physical Chemistry*, 5th edn. Longman Group Limited, London, UK.

Rico-Pena, D. C. and Torres, J. A. (1991). Sorbic acid and potassium sorbate permeability of an edible methylcellulose-palmitic acid film: Water activity and pH effects. *J. Food Sci.* **56(2)**, 497–499.

Robertson, G. L. (1993). Permeability of thermoplastic polymer. In: *Food Packaging, Principles and Practice*, pp. 73–110. Marcel Dekker, New York, NY.

Schwartzberg, H. G. and Chao, R. Y. (1982). Solute diffusivities in leaching processes. *Food Technol.* **36(2)**, 73–86.

Segall, K. I. and Scanlon, M. G. (1996). Design and analysis of a modified-atmosphere package for minimally processed romaine lettuce. *J. Am. Soc. Hort. Sci.* **121**, 722–729.

Smith, E. W. and Haigh, J. M. (1992). In vitro diffusion cell design and validation. II. Temperature, agitation and membrane effects on betamethasone 17-valerate permeation. *Acta Pharm. Nordica* **4(3)**, 171–178.

Torres, J. A. (1987). Microbial stabilization of intermediate moisture food surfaces. In: *Water Activity* (Proceedings of the Tenth Basic Symposium by the Institute of Food Technologists and the International Union of Food Science and Technology), pp. 329–368. IUFoST.

Torres, J. A., Motoki, M. and Karel, M. (1985). Microbial stabilization of intermediate moisture food surfaces. I. Control of surface preservative concentration. *J. Food Process. Preserv.* **9,** 75–92.

Vojdani, F. and Torres, J. A. (1989a). Potassium sorbate permeability of methylcellulose and hydroxypropyl methylcellulose multilayer films. *J. Food Process. Preserv.* **13,** 417–430.

Vojdani, F. and Torres, J. A. (1989b). Potassium sorbate permeability of polysaccharide films: chitosan, methylcellulose and hydroxypropyl methylcellulose. *J. Food Process. Eng.* **12,** 33–48.

Vojdani, F. and Torres, J. A. (1990). Potassium sorbate permeability of methylcellulose and hydroxypropyl methylcellulose coatings: effect of fatty acid. *J. Food Sci.* **55(3),** 841–846.

Washitake, M., Takashima, Y., Tanaka, S., Anmo, T. and Tanaka, I. (1980). Drug permeation through egg shell membranes. *Chem. Pharm. Bull.* **28(10),** 2855–2861.

3

Quality of packaged foods

Rakesh K. Singh and Nepal Singh

Introduction

The quality of packaged foods is a combination of attributes that determine their value as human food. These quality factors include visual appearance (freshness, color, defects, and decay), texture (crispness, turgidity, juiciness, toughness, and tissue integrity), flavor (taste and smell), nutritive value (vitamins A and C, minerals, and dietary fiber), and safety (absence of chemical residues and microbial contamination) (Robertson, 1993). "Keeping quality" is used more commonly than "shelf life" because of consumer demand for product freshness. The deterioration of foods that occurs progressively during storage may result from physical or chemical changes in the food itself, or from the activity of micro-organisms growing in or on the product. Eventually, the cumulative effect of the changes reaches a point at which the consumer rejects the product. Rejection is based on the sensory expectations and perceptions of consumers. Shelf life depends on a multiplicity of variables and their changes, including the product, the environmental conditions, and the packaging. Depending on the product and its intended application, shelf life may be dictated by microbiology, enzymology, and/or physical effects (Brody, 2003). The combined knowledge and experience of processors and those involved in the storage, distribution and retailing of foods enable estimates to be made of the likely shelf life of the product under specific storage conditions. In practice, however, the influencing variables that accelerate or retard shelf life are temperature, pH, water content, water activity, relative humidity, radiation, gas concentration, redox potential, the presence of metal ions, and pressure.

Innovations in Food Packaging
ISBN: 0-12-311632-5

Table 3.1 Undesirable changes that can occur in foods	
Attribute	**Undesirable change**
Texture	a. Loss of solubility b. Loss of water holding capacity c. Toughening d. Softening
Flavor	Development of: e. Rancidity (hydrolytic or oxidative) f. Cooked or caramel flavor g. Other off-flavors
Color	h. Darkening i. Bleaching j. Development of other off-colors
Appearance	k. Increase in particle size l. Decrease in particle size m. Non-uniformity of particle size
Nutritive value	Loss or degradation of: n. Vitamins o. Minerals p. Proteins q. Lipids

Reprinted from Robertson (1993), p. 254, courtesy of Marcel Dekker Inc.

There is no simple, generally accepted definition of shelf life in the food technology literature. In 1978 the National Food Processors Association in the USA recommended, as a working professional definition for internal industry, the following (Hotchner, 1981):

> A product is within its shelf life when it is neither misbranded nor adulterated; when the product quality is generally accepted for its purported use by a consumer; and so long as the container retains its integrity with respect to leakage and protection of the contents.

The Institute of Food Technologists in the USA (Anon, 1974) has defined shelf life as the period between manufacture and retail purchase of a food product, during which time the product is in a state of satisfactory quality in terms of nutritional value, taste, texture and appearance. An alternative definition is that shelf life is the duration of that period between packing of a product and its use for which the quality of the product remains acceptable to the product user. Packaging can affect the rate and magnitude of many of the quality changes shown in Table 3.1. As examples, development of oxidative rancidity can often be minimized if the packaging is an effective oxygen barrier, and the particle size of many food powders can increase if the package is a poor moisture barrier. Thus packaging is commonly used to maintain the quality and improve the shelf life of foodstuffs.

Kinetics

Rates of deteriorative reactions

Quality attributes that define shelf life gradually deteriorate over time. The rates of these deteriorative chemical, biochemical and microbiological reactions depend on both intrinsic and extrinsic factors. The general equation describing quality loss may be written as:

$$-dc/dt = f(I, E) \qquad (3.1)$$

where c is the index of deterioration, t is the time, I is intrinsic variables, and E is extrinsic variables. A negative sign is used if the concentration of c decreases with time. Since the quality of foods and the rate of quality changes during processing and storage depend on intrinsic factors, it is possible in many cases to correlate quality losses with the loss of a particular component. Hence, the quality loss can also be represented as being proportional to the power of the concentration of the reactant or product:

$$dA/dt = -k(A)^n \qquad (3.2)$$

where n is the order of reaction, dA/dt is the change in concentration of A with respect to time (rate of reaction), and k is the rate constant. The negative sign indicates that at a constant temperature, the magnitude of the quality parameter decreases with time. In some cases the opposite is true, as in case of formation of chemicals rather than its disappearance, and the rate should be designated as an activation or formation rate rather than an inactivation or destruction rate. It is important to know the concentration of the reactant or the product so as to determine the acceptability of the final product from the packaging point of view.

Zero-order reactions

In zero-order reactions, such as non-enzymatic browning in certain situations, which follow a linear rate independent of the concentration of reactants, the rate of loss of A is constant with time and independent of the concentration of A:

$$A_e/A_0 = -kt_s \qquad (3.3)$$

or

$$A_e/A_0 = kt_s \qquad (3.4)$$

where A_0 is the initial value (time zero), A_e is the value of attribute A at the end of its shelf life, t_s is the length of shelf life, and k is the rate constant. For a zero-order reaction, the concentration of A remaining with respect to time yields a straight line as shown in Figure 3.1. The slope of this straight line represents the rate constant k having the units of (concentration) (time^{-1}). When $n = 0$, the reaction is said to be pseudo-zero-order with respect to A. These types of reactions include non-enzymatic browning, lipid oxidation, enzymic degradation, etc.

First-order reactions

First-order reactions, $n = 1$, such as microbiological growth, follow a non-linear rate dependent on reactant concentration. In this case:

$$A_e = A_0 e^{-kt} \qquad (3.5)$$

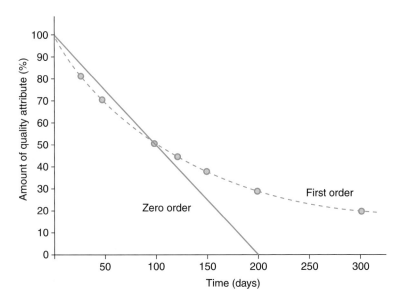

Figure 3.1   Change in quality versus time showing the effect of order of the reaction on the extent of change.

where A_o is the initial value (time zero), A_e is the value at time t, and k is the rate constant. A plot of first-order data as the concentration of A versus time gives a curved line, as shown in Figure 3.1.

The time-dependence kinetic parameter can also be expressed as the D value (decimal reduction time). The D value and the first-order reaction rate constant are related by the equation:

$$k = 2.303/D \qquad\qquad (3.6)$$

In addition to describing the change in a property as a function of time at a fixed temperature it is important to describe the effect of temperature on the property, because the rate of quality loss is temperature dependent. The Arrhenius relationship is often used to describe the temperature dependence of deterioration rate as:

$$k = k_o\, e^{-Ea/RT} \qquad\qquad (3.7)$$

where k is the rate constant for the deteriorative reaction, k_o is the rate constant, independent of temperature, E_a is activation energy in kJ/mol, R is the gas constant (8.314 J/mol K) and T is an absolute temperature ($K = 273 + {}°\text{C}$). Essentially, the rate of product deterioration as a function of temperature generally increases exponentially with linear increase in temperature. From the Arrhenius equation, the activation energy of a reaction or shelf life can be obtained. The Arrhenius equation is also used to predict the shelf life at any unknown temperature. The shelf-life data and the activation energy are the key design parameters for time–temperature indicators (TTI). TTIs are more suitable predictors of product's shelf life.

Shelf life

Shelf-life prediction

The shelf life of novel foods that are not shelf stable may vary from batch to batch. Accurate prediction of shelf life is, therefore, particularly important to ensure product quality. Different approaches to shelf-life estimation have been used. Traditionally the development of packaged foods has been largely an empirical procedure, involving extensive laboratory evaluations of food formulations, packaging materials and various package geometries. In an attempt to speed up the development process and optimize food/package combinations, various mathematical models are able to predict the shelf life of packed foods, and help in designing packaging systems to achieve the desired results.

Mathematical modeling is the representation of real-life phenomena in terms of a set of numerical expressions. The solutions of these mathematical systems are supposed to stimulate the natural behavior of a food/package load. Whiting and Buchanan (1994) proposed a three-level model classification scheme comprised of primary, secondary, and tertiary models. Primary-level models describe the change in microbial numbers over time, and secondary-level models indicate how the features of primary models change with respect to one or more environmental factors – such as pH, temperature and water activity. Tertiary-level models are personal computer software packages that use the pertinent information from primary- and secondary-level models to generate desired graphs, predictions and comparisons.

Predictive microbiology has been used to forecast the growth of spoilage micro-organisms in order to determine the shelf life of a food product. Predictive models for the growth of micro-organisms include temperature, pH and water activity as the main growth-determining factors. However, other factors can significantly influence the growth characteristics of the modeled micro-organisms, such as nitrite content, organic acids and atmosphere (Devlieghere et al., 1999). The effects of initial product contamination, product characteristics, and storage conditions can be included in predictive models. However, specific spoilage organisms and spoilage reactions depend upon intrinsic and extrinsic parameters. Recently, more attention has been paid to the atmosphere as a fourth important growth-determining factor (Farber et al., 1996; Sutherland et al., 1996, 1997; Dalgaard et al., 1997; Fernandez et al., 1997). In these studies the effect of storage temperature was described by the relative rate of spoilage models, and the responses of numerous sensory, microbial, chemical, biochemical, and physical measurements were correlated with remaining product shelf life. Temperature models may predict the slope of the regression lines relating sensory or microbial measurements and remaining shelf life. However, in general the regression models are only applicable to products stored in given conditions (Olley and Ratkowsky, 1973; Gibson, 1985; Bremner et al., 1987). Devlieghere et al. (1999) compared two types of predictive models – an extended Ratkowsky model and a response surface model – for the effect of temperature, concentration of dissolved CO_2, and water activity on the lag phase and the maximum specific growth rate of *Lactobacillus sake* for gas-packed cooked meat product. They reported that both developed models proved to be useful in the prediction of the microbial shelf life of packed meat products.

More recently, multivariate statistical methods have been used to relate the sensory attributes of products (Girard and Nakai, 1994). Risbo (2003) observed that for a packaged raisin–cereal mixture, shelf-life calculations based on the kinetic theory show that the optimal product composition and product permeability for overall product quality can be identified. In food packaging, measurements and predictions of the shelf life of a product independent of the packaging structure are made first and then integrated with the measurements and predictions made regarding the effect of the package. Among the more relevant package properties are water vapor transfer, gas transfer, odor transfer, flavor scalping, and other product/package interactions.

To perform predictive modeling, the food model must be integrated with the package model. The package model must fit the food model – the package should be analyzed for its ability to function as a barrier against those variables that have been identified as critical for the food's shelf life. Distribution parameters should be quantified to determine their variability within the distribution environment. As a first assumption, suppose that the main characteristics are the package barrier properties (such as water vapor, oxygen or carbon-dioxide barrier) and that the package material is unaffected by the packaging operations and distribution. In the predictive model, assume that the internal package environment (water vapor, carbon dioxide, or oxygen) changes as a permeant enters or leaves the package. Transmission of materials will always occur from the high-concentration to the lower-concentration side of the material. Standardize the transmission rate for the package area, gauge, time of measurement, and permeation properties for the package material/structure, the difference of partial pressure of permeant across the package, and the environmental conditions – in particular, temperature. Permeation is a function of dissolution into and diffusion across the package material due to partial pressure P. The general models are based on a standard differential equation which describes mass transfer across a permeable membrane, with the effect on the product being the net gain or loss of the permeant by the product (e.g. loss of carbon dioxide in a carbonated beverage, or loss of oxygen attributable to reaction with a food component). The predictive model for the package takes the general form:

$$\mathrm{d}W/\mathrm{d}t = (k/l)\,A\,\mathrm{d}P \qquad (3.8)$$

where $\mathrm{d}W$ is the change in the weight of the critical food component, t is time, k is the permeation coefficient of the package material, l is the package material thickness, A is surface area of the package, and $\mathrm{d}P$ is the difference in partial pressures of the permeant inside and outside of the package structure. Partial pressure requires knowledge of the permeant: for oxygen external to the package it is 0.209; for carbon dioxide outside the package it is 0.0003; for water vapor on the exterior it is six times the relative of the distribution environment; for flavors and volatile components it is essentially zero, etc. The elementary predictive model is for a monolayer material structure of uniform thickness. When thickness changes, differential equations for the thickness are incorporated into the model. When two or more materials are present, as in a lamination, the net permeation is described by the sum of reciprocals of the individual permeations:

$$1/P = 1/P_1 + 1/P_2 + ... + 1/P_n \qquad (3.9)$$

plus the reciprocal of permeation across the boundary of different materials.

Shelf-life modeling and prediction have been limited by a serious paucity of key variables, such as using only water vapor gain as a measure with no consumer acceptability input. The models do not replace the need for actual storage studies. Storage is still necessary to verify product changes, find unusual effects, and, in some cases, meet compliance criteria with government regulations. The use of these predictive models in the food industry has been warranted by the need for rapid and cost-effective provision of useful information for making decisions regarding product development, prediction of safety and shelf life of a product, and identifying critical control points in quality control tests. The use of models serves to reduce the time and effort necessary to identify optimal packaging protection, and eliminate storage studies on materials with insufficient barrier properties to be reasonable candidates for the product. The ability of the computer models to incorporate all of the known relevant variables permits the food scientist, in concert with the packaging technologist, to predict accurately and precisely the results of a package material and/or structural design change on the shelf life of most product contents.

Storage conditions

The shelf life of a packaged food product is determined by numerous interactions between parameters related to the product itself and/or associated with storage conditions. The quality changes in packaged products could be chemical, physical, enzymatic or microbiological, and are mainly due to mass transfer between foods and their environment. Several studies have shown that barrier efficiency of packaging materials depends not only on chemical composition of packaging material and penetrant, but also on storage conditions (Giacin, 1995). Storage conditions (i.e. temperature, humidity, atmospheric gas composition, and light) affect the appearance, aroma, flavor, texture, and acceptability of packaged food products. Hence, appropriate packaging of shelf-stable foods is essential, particularly during handling, transportation/distribution, extended storage, and marketing.

Temperature

Storage temperature is probably the most important factor in maintaining the quality and extending the shelf life of packaged foods. In most cases, an increase in storage temperature degrades the quality and acceptability of packaged foods. To determine the deleterious effects of temperature, it is often necessary to know how long a food is exposed to that temperature. Biological reactions tend to increase by a factor of two to three for each 10°C increase in temperature (Beaudry et al., 1992; Exama et al., 1993). Changes in environmental temperature create a specific problem in packaged foods design because the respiration rate is more influenced by temperature changes than by the O_2 and CO_2 permeability of the permeable films used to obtain equilibrium modified atmosphere (EMAs). Owing to this fact, it is difficult to obtain an optimum atmosphere inside a package when the surrounding temperature is not constant. A film that produces a favorable atmosphere at the temperature for which the package was designed may cause excessive accumulation of CO_2 and/or depletion of O_2 at higher temperatures caused by an increased respiration rate. This anoxic situation could lead to metabolic disorders such as fermentation, with production of off-flavors (Kader et al., 1989; Silva et al., 1999). In the case of a lower storage temperature, decreased respiratory activity will lead to an accumulation of O_2 above the optimal 3%, resulting in a less effective EMA package. The influence of temperature on respiration rate was first quantified with the Q_{10} value, which is the respiration rate increase for a 10°C rise in temperature:

$$Q_{10} = (Rr_2/Rr_1)^{10/(T2-T1)} \tag{3.10}$$

where Rr_2 is the respiration rate at temperature T_2, and Rr_1 is the respiration rate at temperature T_1. For various products, Q_{10} values may range from 1 to 4 depending on the temperature range (Kader, 1987).

The Arrhenius equation is also used to quantify the effect of temperature on respiration rate. The simultaneous use of this equation to describe the influence of temperature on film permeability simplifies the mathematical modeling of MAP systems (Exama et al., 1993; Susana et al., 2002):

$$Rr = \delta \times \exp(-E/RT) \tag{3.11}$$

The above equation may be rewritten with a reference temperature (Nelson, 1983; Van Boekel, 1996):

$$Rr = \delta_{ref} \times \exp[-E/R \, (1/T - 1/T_{ref})] \tag{3.12}$$

Temperature variations during transportation and storage affect the produce respiration rate (Exama et al., 1993). Kang et al. (2000) observed that temperature fluctuations between 5°C and 15°C did not induce a harmful atmosphere inside the polyolefin package, though high temperatures accelerated the weight loss, stipe elongation, and darkening of the mushrooms. Temperature also has a significant effect on the sensory quality of packaged food products. Enzymatic browning and shriveling, being the most limiting sensory disorders of fresh cut produce, are inhibited by low-temperature storage (Willocx, 1995; Garcia-Gimeno et al., 1998). The barrier efficiency of packages, too, depends on temperature. Labuza and Contreras-Medellin (1981) observed that water permeability decreases when temperature decreases, but may increase again at very low temperatures. Morillon et al. (2000) reported that corrected permeabilities of cellophane and polyethylene films decrease with temperature due to a preponderant effect of temperature on water diffusion. Jacobsson et al. (2003) reported that change of temperature during storage induced changes in smell, flavor, taste, and texture of broccoli, depending on the packaging material used. However, the appearance of the broccoli was found to be more dependent on the atmosphere than on the temperature of storage.

Equilibrium relative humidity

Equilibrium relative humidity (ERH) is critical to the quality and safety of packaged foods. Moisture loss or gain from one region or food component to another region will continuously occur in order to reach thermodynamic equilibrium with the surrounding food components and the environment.

The concept of water activity (a_w) is an important property of food safety. Water activity is widely used to predict the stability of food with respect to the potential for growth of micro-organisms, and also some of the physical, chemical, and enzymic changes that lead to deterioration. In packaged foods, the water-vapor permeability of the packaging material is a decisive factor in controlling changes in moisture content and thus the water activity (a_w) of packaged foods. Water activity is defined as the ratio of partial pressure of water vapor in the product (p) to that in presence of pure water (po). A difference in the water activities of food components, food domains, and the external environment outside the package introduces a driving force for water transport (Labuza and Hyman, 1998). Water transport ceases when differences in water activity have leveled out – i.e. water activities converge to a common equilibrium value. For some systems, the equilibrium water activity is acceptable from the point of view of textural (Katz, 1981; Bourne, 1987), chemical (Leung, 1987), and microbiological stability (Leneovich, 1987), and the shelf life is not limited by water transport. For systems where this equilibrium water activity is undesirable to one or more of the components, the product shelf life is determined by the dynamics of the water transport process.

By measuring the water activity of foodstuffs, it is possible to predict which micro-organisms will be potential sources of spoilage and infection. Non-enzymatic browning reactions and spontaneous autocatalytic lipid oxidation reactions are strongly influenced by water activity. Troller and Christian (1978) observed that at very low water activity levels, foods containing unsaturated fats and exposed to atmospheric oxygen are highly susceptible to the development of oxidative rancidity. The high oxidative rancidity occurs at a_w levels below the monolayer level, and as a_w increases, both the rate and the extent of auto-oxidation increase until a_w in the range of 0.3–0.5 is reached. Above this point the rate of oxidation increases until a steady state is reached, normally at a_w levels in excess of 0.75.

Water activity can also play a significant role in determining the activity of vitamins in packaged foods. The rate of degradation of vitamins A, B_1, B_2 and C increases as a_w increases over the range 0.24–0.65. The rate of ascorbic acid and riboflavin degradation increases with an increase in a_w. Browning reactions generally increase with an increasing a_w at low moisture content, reach a maximum at a_w of 0.4–0.8, and decrease with a further increase in water activity.

Gas atmosphere

The sensory properties of packaged foods depend on different physiological and biochemical pathways that are induced by atmospheric conditions. Lipid oxidation and non-enzymatic browning are the major chemical reactions that lead to deterioration in the sensory quality of packaged foods. Responses to atmospheric modifications are found to vary dramatically among packaged foods, and include both unwanted and beneficial physiological changes. Desirable responses include a reduction in respiration, oxidative tissue damage or discoloration, rate of chlorophyll degradation, and ethylene sensitivity, with the concomitant reduction in the rate of ripening. Undesirable responses include the induction of fermentation, development of disagreeable flavors, a reduction in aroma biosynthesis, induction of tissue injury, and an alteration in the makeup of microbial fauna.

Atmospheric oxygen generally has a detrimental effect on the nutritive quality of foods, and it is therefore desirable to maintain many types of foods at a low oxygen concentration – or at least to prevent a continuous supply of oxygen into the package. However, in the presence of sufficient oxygen, the sensory quality is less affected (Watkins, 2000). Oxygen penetrating the packages affects the rate of carotenoid degradation, and thus the color and ascorbic acid degradation (Nagy, 1980; Gvozdenovic et al., 2000). The severity of off-flavor production will depend on the time of exposure to conditions below the minimum required oxygen concentration and/or above the maximum tolerated carbon dioxide concentration (Beaudry, 1999, 2000; Watkins, 2000).

The diffusion of gases through flexible packaging material depends on the physicochemical structure of the barrier. The gas permeabilities of the common thermoplastic packaging material are presented in Table 3.2.

Table 3.2 Permeability coefficients* for various polymers and permeants

Polymer	PN_2 (30°C)	PO_2 (30°C)	PCO_2 (30°C)	PSO_2 (25°C)	PH_2O (25°C)	$\dfrac{PO_2}{PN_2}$	$\dfrac{PCO_2}{PO_2}$	$\dfrac{PCO_2}{PN_2}$	Nature of polymer
Low-density polymer	19	55	352	200	800	2.9	6.4	18.5	Some crystallinity
High-density polyethylene	2.7	10.6	35	57	130	3.9	3.3	13.0	Crystalline
Polypropylene	–	23	92	7	680	–	4.0	–	Crystalline
Unplasticized poly(vinyl chloride)	0.4	1.2	10	1.16	1560	3.0	8.3	25.0	Slightly crystalline
Cellulase acetate	2.8	7.8	68	–	75 000	2.8	8.7	24.3	Glassy, amorphous
Polystyrene	2.9	11	88	220	12 000	3.8	8.0	30.3	Glassy
Nylon 6	0.1	0.38	1.6	22[a]	7000	3.8	4.2	16.0	Crystalline
Poly(ethylene terephthlate)	0.05	0.22	1.53	2	1000	4.4	7.0	30.6	Crystalline
Poly(vinylidene chloride)	0.0094	0.053	0.29	–	14	5.6	5.5	30.9	Crystalline
Mean						3.8	6.2	23.6	

Reprinted from Robertson (1993), p. 81, by courtesy of Marcel Dekker Inc.
* $P \times 10^{11}$ ml (STP) cm cm^{-2} s^{-1} (cm Hg^{-1}).
[a] Nylon II.

Light

Exposure of packaged foods to both ultraviolet radiation and visible light causes oxidative deterioration of lipids, vitamins, proteins, and colorants in foods, leading to the formation of off-flavors, nutrient losses, and discoloration. Factors influencing the deteriorative effect of light include the intensity of light, duration of light exposure, and light transmittance of the packaging material. The interplay between product, packaging, light, and oxygen in Figure 3.2 shows factors of importance for the light absorption by the product, and the content of oxygen, respectively.

Availability of oxygen is a prerequisite for light-induced oxidation to occur. Hence, vacuum packaging prevents light-induced oxidation by reducing the oxygen in the headspace, thereby minimizing the rate of oxidation processes (Marsh *et al.*, 1994; Hong *et al.*, 1995a). However, the critical concentration (in the package and dissolved in the product) has not yet been determined, and is likely to vary between products and to depend upon packaging and storage conditions. Reducing the oxygen content of the headspace – for example, by vacuum packaging or packaging in modified atmospheres with high carbon dioxide/nitrogen and low residual oxygen, and/or using

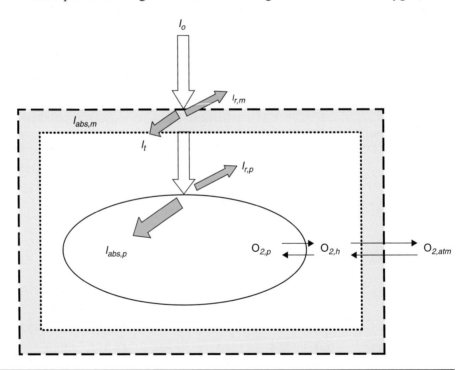

Factors influencing light absorbed ($I_{abs,p}$)	Factors influencing
Intensity of light (I_o)	Initial gas composition ($O_{2,p}+O_{2,h}$)
Spectrum of light source	Product to headspace volume ratio
Absorption ($I_{abs,m}$) and reflection ($I_{r,m}$) of packaging material	Product respiration
Product reflection ($I_{r,p}$)	O_2 transpiration rate of packaging material
Product absorption ($I_{abs,p}$)	
Contents of specific endogenous absorbers	

Figure 3.2 Schematic overview of product/packaging interactions of light with oxygen (reprinted from Mortensen *et al.*, 2004, with permission of Elsevier Ltd).

oxygen impermeable packaging – may actually reduce light-induced oxidation by limiting the oxygen pool available for dissolution, and thereby the reaction in the aqueous phase. Light protection offered by packaging materials depends on numerous factors, including:

1. The inherent absorption characteristics of the material
2. The thickness of the material
3. The material processing conditions
4. The coloration of the material.

These may all be combined to optimize the photo-oxidative protection of specific foods.

Different packaging material categories offer varying degrees of protection against light-induced changes due to differences in reflectance, transmittance, and oxygen permeability determined by molecular composition of the material. Generally, metals offer the best protection, followed by paper/paperboard, various plastics, and finally glass, through which up to approximately 90% of light is transmitted (Bosset *et al.*, 1994; Lennersten and Lingert, 1998). Unbleached paper provides a better light barrier than bleached paper, especially at short wavelengths, due to the removal of and alterations in the light-absorbing pigments (the lignins) during the bleaching process (Mortensen *et al.*, 2003a). Hence, very different light transmission characteristics are notable within different categories – a fact that should be utilized for optimization of packaging with respect to photo-oxidation. For instance, a secondary packaging consisting of cardboard may protect against photo-oxidation as well as providing mechanical stability and ease of handling for the retailers.

Nelson and Cathcart (1984) examined the effect of the wall thickness of polyethylene milk bottles on light transmission characteristics, and found that increasing the wall thickness led to lower light transmission rates. Moreover, material processing (orientation of the polymer, crystallinity, and incorporation of additives) may have an impact on the light transmission characteristics. Incorporation of titanium oxide into plastic materials increases light scattering, thus reducing light transmittance – especially light at wavelengths shorter than 400 nm (Nelson and Cathcart, 1984; Lennersten and Lingert, 1998). Compounds such as carbon black, chalk and talc may also be applied to reduce light transmittance. Cavitation, which may be used in, for example, the production of polypropylene, is another approach to increase the light barrier. Lennersten and Lingert (1998) concluded that cavitied films reflect more light than non-cavitied films, which results in reduced light transmittance.

The catalytic effects of light are most pronounced for light in lower wavelengths of the visible/ultraviolet spectrum (Bekbolet, 1990; Lennersten and Lingert, 1998). Increasing light intensity, i.e. the photon flux, accelerates light-induced oxidation (Deger and Ashoor, 1987; Hong *et al.*, 1995b; Alves *et al.*, 2002). Packaging materials absorb most of the energy-rich ultraviolet light, which is generally not as harmful to the packaged dairy product as is light in the blue–violet region (400–500 nm) of the spectrum. Hansen (1996) noted that monochromatic light at 405 and 448 nm was more detrimental to the examined dairy spread model than was monochromatic light at 460 nm. Evidently, prolonged exposure time increases the light-induced damage (Kristensen *et al.*, 2000; Alves *et al.*, 2002; Mortensen *et al.*, 2002a, 2002b, 2003a, 2003b, 2003c;

Wold *et al.*, 2002; Juric *et al.*, 2003; Kim *et al.*, 2003). These examples show that light can play an important role in deterioration of nutrients. Depending on the light transmission characteristics of the packaging materials, suitable packaging material can offer protection of quality packaged foods by absorption or reflection of all or part of the incident light. Lennersten and Lingnert (2000) observed that the effect on lipid oxidation in low-fat mayonnaise was most pronounced with ultraviolet radiation, although short-wavelength visible light also had a significant effect. Polymer materials offered some protection against lipid oxidation by filtering out the ultraviolet radiation to varying degrees. PEN and PET/PEN copolymer offered better protection than PET. None of the materials offered sufficient protection against color changes.

The total amount of light absorbed by a packaged food can be calculated using the following formula:

$$I_{abs} = I_o I_{fp} \frac{1 - I_{rf}}{1 - I_{rf} I_{rp}} \tag{3.13}$$

where I_{abs} = intensity of light absorbed by the food, I_o = intensity of incident light, I_{fp} = fractional transmission by the packaging material, I_{rp} = fraction reflected by the packaging material, and I_{rf} = fraction reflected by the food.

The fraction of the incident light transmitted by any given material can be considered to follow the Beer–Lambert law:

$$I = I_o e^{-kx} \tag{3.14}$$

where I = intensity of light transmitted by the packaging material, k = a characteristic constant (absorbance) for the packaging material, and x = thickness of the packaging material.

The absorbance k varies not only with the nature of the packaging material but also with the wavelength. Thus the amount of light transmitted through a given package will be dependent on the incident light and the properties of packaging material.

Aseptic packaging

Aseptic packaging is an alternative to conventional canning in the production of shelf-stable packaged food products. The word "aseptic", derived from the Greek word *septicos*, means the absence or exclusion of putrefactive micro-organisms. Aseptic packaging technology is fundamentally different from that of traditional food-processing systems. Canning processes commercially sterilize filled and sealed containers, while in aseptic packaging the presterilized product is filled into sterilized containers that are hermetically sealed in a commercially sterile environment. This form of heat treatment enables the commercial sterilization of the product, and results in minimal loss of product quality and nutrition. A typical flow diagram is shown in Figure 3.3.

In aseptic packaging, raw or unprocessed product is heated, sterilized by holding at a high temperature for a predetermined amount of time, cooled, and then delivered to

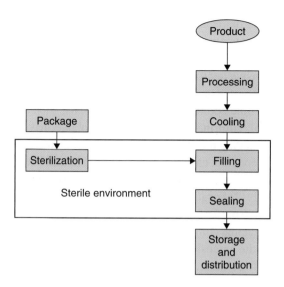

Figure 3.3 Schematic flow diagram of aseptic packaging system.

a packaging unit for packaging. Aseptic packaging offers advantages to the consumer as well as distribution channels (for example, lower distribution and storage costs, an extended shelf life, relief of pressure on chilled cabinets, cost effectiveness, and freedom from additives).

Sterilization of packaging material is a critical step in the aseptic packaging system, and the method used should:

1. Have rapid microbicidal activity
2. Be compatible with the surfaces treated, especially packaging material and equipment
3. Leave minimal residue – i.e. residue should be easily removable from surfaces
4. Present no health hazard to the consumer
5. Have no adverse effects on product quality in cases of unavoidable residue or erroneous high concentration
6. Present no health hazard to operations personnel around the packaging equipment
7. Be compatible with the environment
8. Be non-corrosive to the surface treated
9. Be reliable and economical.

Various methods for the sterilization of packaging materials are currently used in aseptic packaging systems. Hydrogen peroxide sterilization followed by hot air appears to have the most potential for use as in-line sterilization of packaging materials. Sterilization of packaging materials followed by ultraviolet irradiation has also been accepted for industrial application. Ultraviolet irradiation promotes the breakdown of peroxides into hydroxyl radicals, and the overall lethal effect is greater than the sum of the effects of peroxide and irradiation alone.

Ethylene oxide has been used to pre-sterilize paperboard cartons (Hedrick, 1973) and plastic packaging materials (Alguire, 1973). However, no commercial system currently uses ethylene oxide as a sterilizing agent in the aseptic packaging of low-acid foods. An important consideration in bulk sterilization of preformed packaging materials is ethylene oxide's ability to permeate and contact all the surfaces that need to be sterilized. Extensive aeration of the sterilized material is necessary to remove residual ethylene oxide because of its toxic nature, and the length of time necessary precludes its use in in-line applications. Other chemicals, such as peracetic acid, beta propiolactone, alcohol, chlorine and its oxide, and ozone, have potential for use in sterilizing aseptic packaging materials. However, peracetic acid produces off-flavor in food if any residual deposit is enclosed in the container, and beta propiolactone lacks the penetrating power of ethylene oxide – although it is considerably more active against micro-organisms. Only 2–5 mg/l of beta propiolactone is required for sterilization, compared with 400–800 mg/l of ethylene oxide. Ethanol is effective only against vegetative cells, and not against conidia or bacterial spores; therefore its use in packaging of foods is limited to the extension of the shelf life of packaged foods which are normally stored under refrigeration.

Dry heat, saturated steam, and superheated steam can be effective sterilants, but the degree of heat damages many packaging materials and so they have limited application. Sterilization of metal cans and lids by saturated steam under pressure was used as early as 1920 in the USA (Reuter, 1988), and this method is today employed to sterilize thermostable plastic cups. Saturated steam under pressure is used to sterilize moulded polystyrene cups and foil lids at 165°C and 6 bars immediately after deep drawing. At the same time, the external surface of the cup is cooled to limit the heating effect to the internal surface of the cups. Moist heat could cause blistering or delamination of paper-based packaging materials and impair the heat-sealing characteristics of plastics. Saturated steam is a preferred treatment method for sterilizing metal–food contact surfaces downstream from the hold tube, including sterile hold tank, filters, and the aseptic packaging zone. Superheated steam was the method used in the 1950s for sterilization of tin-plate and aluminum cans and lids in the Martin–Dole aseptic canning process. This process continuously sterilizes tin-plate and aluminum cans by passing them through 220–256°C superheated steam at normal pressure for 45 seconds (Larousse and Brown, 1997). Hot air is preferred as a sterilant over superheated steam to sterilize cardboard laminates (i.e. aluminum and paper) at 145°C for 3 minutes (Reuter, 1988). However, the method has only been found suitable for products that have a pH < 4.6. An aseptic packaging system utilizing the heat co-extrusion process for sterilization is the form–fill–seal packaging system, which relies on temperatures reached by thermoplastic resin during the co-extrusion process used to produce multi-layer packaging material to produce a sterile product contact surface. The high temperature (180°C–230°C for up to 3 minutes) produced by extrusion produces sterile surfaces. Reports as to the efficacy of this heating method vary considerably (Reuter, 1988). Aseptic filling into extruded containers appears suitable only for acidic products with a pH less than 4.6. For products with a pH greater than 4.6, extruded containers should be post-sterilized with hydrogen peroxide or with mixtures of peracetic acid.

Ultraviolet radiation at a wavelength of 253.7 nm is an effective germicide against yeast, moulds, bacteria, viruses and algae. Maunder (1977) suggested the use of high-intensity ultraviolet light for aseptic packaging material sterilization. However, dust particles present on the surfaces reduce the effectiveness of ultraviolet irradiation for sterilization of aseptic packaging materials. Doyen (1973) described an aseptic packaging machine that used both an alcohol bath and high-intensity ultraviolet radiation for the sterilization of flexible pouches. Infrared rays ($0.8\text{--}15 \times 10^{-6}$ waveband) have been used to treat the interior of aluminum lids with a plastic coating on the exterior surface, but a temperature rise in the packaging material due to infrared application results in softening of the plastics. Gamma irradiation (25 kGy) has been used to sterilize plastic bags for the bulk packaging of acid foods using an aseptic bag in box system (Nelson, 1984). Ionizing rays are not acceptable, as they have harmful effects on personnel.

The technology of utilizing short pulses of light is attractive for sterilizing packaging materials and processing equipment in aseptic packaging. The spectrum of light used for sterilization purposes includes wavelengths from the ultraviolet to the near-infrared region. The packaging material is exposed to at least one pulse of light having an energy density in the range of about $0.01\text{--}50\,\mathrm{Jcm^{-2}}$ for 1 μs to 0.2 s at the surface (Anonymous, 1994). It is desirable for certain aseptic processes that packaging material be treated with pulses having a relatively high ultraviolet content to minimize the total fluence ($\mathrm{Jcm^{-2}}$) necessary to achieve the desired reduction in microbial population (Dunn *et al.*, 1991). Light pulses have not been adequately studied, so far as their effect on food material is concerned.

The versatility of aseptic technology has given rise to the use of a variety of plastic and polyolefin materials for packaging, and examples commonly used in aseptic packaging are presented in Table 3.3. There are several basic requirements that these packaging materials must meet for successful application:

1. The packaging material must be acceptable for use in contact with the intended product, and must comply with applicable material migration requirements

Table 3.3 Functional attributes of some aseptic packaging materials

Material	Barrier property			Seal quality and adhesion	Durability		
	Oxygen	Moisture	Light		Stiffness	Tear	Puncture
Linear-low density polyethylene		x		x		x	x
Low-density polyethylene		x		x			
Polypropylene		x		x		x	
Polystyrene					x		
Polyvinylidine chloride	x	x					
Ethylenevinyl alcohol	x						
Nylon						x	x
Ethyleneacrylic acid				x			x
Paperboard			x		x		
Aluminum foil	x	x	x				
Metallized film	x		x				

Source: *Encyclopedia of Packaging Technology* (1997).

2. Physical integrity of the package is necessary to assure contamination of the product and maintenance of sterility
3. The package material must be able to be sterilized and be compatible with the method of sterilization used (heat, chemical, or radiation)
4. The package must provide the barrier protection necessary to maintain product quality until it is used.

The structure and composition of aseptic packaging is more complex, and varies depending on product application, package size, and package type. Factors such as strength and integrity, package shape, stiffness, durability, and barrier properties determine the choice and/or combination of materials required. In most applications aseptic packages incorporate more than one material in the structure, and these are assembled by lamination or co-extrusion processes.

Conclusions

Food packaging is an important element in the development of a successful consumer product. An effective package must prevent spoilage, and have oxygen, temperature, light, and water-vapor barrier properties adequate to retard quality deterioration of packaged foods. The packaging must withstand processing conditions and distribution abuse. Further research is still necessary to achieve microbiologically-safe packaged foods, and maintain their nutritional value and original sensory quality. The shelf life attained by packaged foods must be enhanced to allow distribution and marketing. Studies on the influence of different packaging on the shelf life of processed foods need to be carried out. In addition, modeling of the package atmosphere composition, respiration rates, and internal atmospheres in the package throughout storage are of major importance in designing appropriate packaging and predicting the shelf life of packaged foods of different types.

References

Alguire, D. E. (1973). Ethylene dioxide gas sterilization of packaging materials. *Food Technol.* **27**, 64.

Alves, R. M. V., Van Dender, A. G. F., Jaime, S. B. M. and Moreno, I. (2002). Stability of "Requeijão cremoso" in different packages exposed to light. Poster, Congrilait, 26th International Dairy Conference, Paris, France, 24–27 September.

Anon (1974). Shelf life of foods. *J. Food Sci.* **39**, 861.

Anonymous (1994). *The Pure Bright Process*. Pure Pulse Proprietary Information.

Beaudry, R. (1999). Effect of O_2 and CO_2 partial pressure on selected phenomena affecting fruit and vegetable quality. *Postharvest Biol. Technol.* **15**, 293–303.

Beaudry, R. (2000). Responses of horticultural commodities to low oxygen: limits to the expended use of modified atmosphere packaging. *Hort. Technol.* **10**, 491–500.

Beaudry, R. M., Cameron, A. C., Shiraji, A. and Dostal-Lange, D. L. (1992). Modified atmosphere packaging of blueberry fruit: effect of temperature on package oxygen and carbon dioxide. *J. Am. Soc. Hort. Sci.* **117,** 436–441.

Bekbölet, M. (1990). Light effects on food. *J. Food Prot.* **53,** 430–440.

Bosset, J. O., Gallmann, P. U. and Sieber, R. (1994). Influence of light transmittance of packaging materials on the shelf life of milk and dairy products – a review. In: *Food Packaging and Preservation* (M. Mathlouthi, ed.), pp. 222–268. Blackie Academic and Professional, Glasgow, UK.

Bourne, M. C. (1987). *Water Activity: Theory and Application to Food*, Basic Symposion Series, pp. 75–99. Marcel-Dekker, New York, NY.

Bremner, A. H., Olley, J. and Vail, A. M. A. (1987). Estimating time–temperature effects by a rapid systematic sensory method. In: *Seafood Quality Determination* (D. E. Kramer and J. Liston, eds), Proceedings of an International Symposium (University of Alaska Sea Grant College Program, Anchorage, Alaska, USA), pp. 413–435. Elsevier Science Publishers B.V., Amsterdam, The Netherlands.

Brody, A. L. (2003). Predicting packaged food shelf life. *Food Tech.* **57,** 100–102.

Dalgaard, P., Mejlholm, O. and Huss, H. H. (1997). Application of an iterative approach for the development of a microbial model predicting the shelf-life of packed fish. *Intl J. Food Microbiol.* **38,** 169–179.

Deger, D. and Ashoor, S. H. (1987). Light-induced changes in taste, appearance, odor, and riboflavin content of cheese. *J. Dairy Sci.* **70,** 1371–1376.

Devlieghere, F., Van Belle, B. and Debevere, J. (1999). Shelf life of modified atmosphere packed cooked meat products: a predictive model. *Intl J. Food Microbiol.* **46,** 57–70.

Doyen, L. (1973). Aseptic system sterilizes pouches with alcohol and UV. *Food Technol.* **27,** 49.

Dunn, J. E., Clark, R. W., Asmus, J. F. *et al*. (1991). Methods and Apparatus for preservation of foodstuffs, US Patent no. 5,034,235.

Exama, A., Arul, J., Lencki, R. W., Lee, L. Z. and Toupin, C. (1993). Suitability of plastic films for modified atmosphere packaging of fruits and vegetables. *J. Food Sci.* **58,** 1365–1370.

Farber, J. M., Cai, Y. and Ross, W. H. (1996). Predictive modeling of the growth of *Listeria monocytogenes* in CO_2 environments. *Intl J. Food Microbiol.* **32,** 133–144.

Fernandez, P. S., George, S. M., Sills, C. C. and Peck, M. W. (1997). Predictive model of the effect of CO_2, pH, temperature and NaCl on the growth of *Listeria monocytogenes*. *Intl J. Food Microbiol.* **32,** 133–144.

Garcia-Gimeno, R., Castillejo-Rodriquez, A., Barco-Alcala, E. and Zurera-Cosano, G. (1998). Determination of packaged green asparagus shelf-life. *Food Microbiol.* **15,** 191–198.

Giacin, J. R. (1995). Factors affecting permeation, sorption and migration processes in package-product systems. In: *Foods and Packaging Materials. Chemical Interactions* (P. Ackerman, M. Jagerstad and Y. Ohlsson, eds), p. 231. The Royal Society of Chemistry, Cambridge, UK.

Gibson, D. M. (1985). Predicting the shelf life of packed fish from conductance measurements. *J. Appl. Bacteriol.* **58,** 465–470.

Girard, B. and Nakai, S. (1994). Grade classification of canned pink salmon with static headspace volatile patterns. *J. Food Sci.* **59,** 507–512.

Gvozdenovi, J., Popov-Ralji, J. and Curakovi, M. (2000). Investigation of characteristic colour stability of powdered orange. *Food Chem.* **70,** 291–301.

Hansen, E. (1996). Light-induced oxidative changes in dairy spreads. Project report. Copenhagen, Department of Dairy and Food Science, The Royal Veterinary and Agricultural University.

Hedric, T. L. (1973). Aseptic packaging in paperboard containers. *Food Technol.* **27,** 64.

Hong, C. M., Wendorff, W. L. and Bradley Jr, R. L. (1995a). Factors affecting light-induced pink discoloration of annatto-colored cheese. *J. Food Sci.* **60,** 94–97.

Hong, C. M., Wendorff, W. L. and Bradley Jr, R. L. (1995b). Effects of packaging and lighting on pink discoloration and lipid oxidation of annatto-colored cheeses. *J. Dairy Sci.* **78,** 1896–1902.

Hotchner, S. J. (1981). Container shelf life methodology. In: *Shelf Life: A Key to Sharpening Your Competitive Edge Proceedings*, p. 29. Food Processors Institute, Washington, DC.

Jacobsson, A., Nielsen, T., Sjoholm, I. and Wendin, K. (2003). Influence of packaging material and storage condition on the sensory quality of broccoli. *Food Quality Preference* **15(4),** 301–304.

Juric, M., Bertelsen, G., Mortensen, G. and Petersen, M. A. (2003). Light-induced colour and aroma changes in sliced modified atmosphere packaged semi-hard cheeses. *Intl Dairy J.* **13,** 239–249.

Kader, A. A. (1987). Respiration and gas exchange of vegetables. In: *Post Harvest Physiology of Vegetables* (J. Weichmann, ed.), pp. 25–43. Marcel Dekker, New York, NY.

Kader, A. A., Zagory, D. and Kerbel, E. L. (1989). Modified atmosphere packaging of fruits and vegetables. *CRC Crit. Rev. Food Sci. Nutr.* **28,** 1–30.

Kang, J., Park, W. and Lee, D. (2000). Quality of enoki mushrooms as affected by packaging conditions. *J. Sci. Food Agric.* **81,** 109–114.

Katz, E. E. and Labuza, T. P. (1981). Effect of water activity on the sensory crispness and mechanical deformation of snack food products. *J. Food Sci.* **46,** 403–409.

Kim, G.-Y., Lee, J.-H. and Min, D. B. (2003). Study on light-induced volatile compounds in goat's milk cheese. *J. Agric. Food Chem.* **51,** 1405–1409.

Kristensen, D., Orlien, V., Mortensen G., Brockhoff, P. and Skibsted, L. H. (2000). Light induced oxidation in sliced Havarti cheese packaged in modified atmosphere. *Intl Dairy J.* **10,** 95–103.

Labuza, T. P. and Contreras-Medellin, R. (1981). Prediction of moisture requirements for foods. *Cereal Foods World* **26,** 335–343.

Labuza, T. P. and Hyman, C. R. (1998). Moisture migration and control in multi-domain foods. *Trends in Food Sci. Technol.* **9,** 47–55.

Larousse, J. and Brown, B. E. (1997). *Food Canning Technology*. John Wiley & Sons, New York, NY.

Leneovich, L. M. (1987). In: *Water Activity: Theory and Application to Food*, Basic Symposion Series 6, pp. 119–136. Marcel Dekker, New York, NY.

Lennersten, M. and Lingert, H. (1998). Influence of different packaging materials on lipid oxidation in potato crisps exposed to fluorescent light. *Lebensm. Wiss. u. Technol.* **31,** 162–168.

Lennersten, M. and Lingert, H. (2000). Influence of wavelength and packaging material on lipid oxidation and color changes in low-fat mayonnaise. *Lebensm. Wiss. u. Technol.* **33,** 253–260.

Leung, H. K. (1987). In: *Water Activity: Theory and Application to Food*, Basic Symposion Series 6, pp. 27–54. Marcel Dekker, New York, NY.

Marsh, R., Kajda, P. and Ryley, J. (1994). The effect of light on the vitamin B_2 and the vitamin A content in cheese. *Nahrung* **5,** 527–532.

Maunder, D. T. (1977). Possible use of ultraviolet sterilization of containers for aseptic packaging. *Food Technol.* **31,** 36–37.

Michael, S. M. (1997). Aseptic packaging. In: *The Wiley Encyclopedia of Packaging Technology*, 2nd edn. (A. Brody and K. S. Marsh, eds), pp. 41–45. John Wiley and Sons, New York, NY.

Morillon, V., Debeaufort, F., Blond, G. and Voilley, A. (2000). Temperature influence on moisture transfer through synthetic films. *J. Membrane Sci.* **168,** 223–231.

Mortensen, G., Sørensen, J. and Stapelfeldt, H. (2002a). Light-induced oxidation in semi-hard cheeses. Evaluation of methods used to determine levels of oxidation. *J. Agric. Food Chem.* **50,** 4364–4370.

Mortensen, G., Sørensen, J. and Stapelfeldt, H. (2002b). Effect of light and oxygen transmission characteristics of packaging materials on photooxidative quality changes in Havarti cheeses. *Packag. Technol. Sci.* **15,** 121–127.

Mortensen, G., Sørensen, J. and Stapelfeldt, H. (2003a). Effect of modified atmosphere packaging and storage conditions on photooxidation of sliced Havarti cheese. *Eur. Food Res. Technol.* **216,** 57–62.

Mortensen, G., Sørensen, J. and Stapelfeldt, H. (2003b). Response surface models for prediction of photooxidative quality changes in Havarti cheese. *Eur. Food Res. Technol.* **216,** 93–98.

Mortensen, G., Sørensen, J., Danielsen, B. and Stapelfeldt, H. (2003c). Effect of specific wavelengths on light-induced quality changes in Havarti cheese. *J. Dairy Res.* **70,** 413–421. (accepted for publication).

Mortensen, G., Bertelsen, G., Mortensen, B. K. and Stapelfeldt, H. (2004). Light-induced changes in packaged cheeses – a review. *Intl Dairy J.* **14,** 85–102.

Nagy, S. (1980). Vitamin C contents of citrus fruit and their products: a review. *J. Agric. Food Chem.* **28,** 8–18.

Nelson, P. R. (1983). Stability prediction using the Arrhenius model. *Computer Programs Biomed.* **16,** 55.

Nelson, P. E. (1984). Outlook for aseptic bag-in-box packaging of products for remanufacture. *Food Technol.* **38,** 72–73.

Nelson, K. H. and Cathcart, W. M. (1984). Transmission of light through pigmented polyethylene milk bottles. *J. Food Prot.* **47,** 346–348.

Olley, J. and Ratkowsky, D. A. (1973). Temperature function integration and its importance in the storage and distribution of flesh foods above the freezing point. *Food Technol.* **25,** 66–73.

Reuter, H. (1988). *Aseptic Packaging of Food*. Technomic, Lancaster, PA.

Risbo, J. (2003). The dynamics of moisture migration on packaged multi-component food systems I: shelf life predictions for a cereal-raisin system. *J. Food Eng.* **58,** 239–246.

Robertson, G. (1993). *Food Packaging*. Marcel Dekker, New York, NY.

Silva, F., Chau, K. and Sargent, S. (1999). Modified atmosphere packaging for mixed loads of horticultural commodities exposed to two postharvest temperatures. *Postharvest Biol. Technol.* **17**, 1–9.

Susana, C. F., Fernanda, A. R. O. and Jeffrey, K. B. (2002). Modelling respiration rate of fresh fruits and vegetables for modified atmosphere packages: a review. *J. Food Eng.* **52**, 99–119.

Sutherland, J. P., Aherne, A. and Beaumont, A. C. (1996). Preparation and validation of a growth model for *Bacillus cereus*: the effect of temperature, pH, sodium chloride and carbon dioxide. *Intl J. Food Microbiol.* **30**, 359–372.

Sutherland, J. P., Bayliss, A. J., Braxton, D. S. and Beaumont, A.C. (1997). Predictive modelling of *Escherichia coli* O157:H7: inclusion of carbon dioxide as a fourth factor in a pre-existing model. *Intl J. Food Microbiol.* **37**, 113–120.

Troller, J. A. and Christian, J. H. B. (1978). In: *Water Activity and Food*, pp. 13–47. Academic Press, New York, NY.

Van Boekel, M. A. J. S. (1996). Statistical aspects of kinetic modelling for food science problems. *J. Food Sci.* **61**, 477.

Watkins, C. (2000). Responses of horticultural commodities to high carbon dioxide as related to modified atmosphere packaging. *Hort. Technol.* **10**, 501–506.

Whiting, R. C. and Buchanan, R. L. (1994). Microbial modeling. *Food Technol.* **48**, 113–120.

Willocx, F. (1995). Evolution of microbial and visual quality of minimally processed foods; a case study on the product life cycle of cut endive. PhD Thesis, Katholic University of Leuven, Faculty of Agriculture and Applied Sciences, Leuven, Belgium, pp. 228.

Wold, J. P., Jørgensen, K. and Lundby, F. (2002). Non-destructive measurement of light-induced oxidation in dairy products by fluorescence spectroscopy and imaging. *J. Dairy Sci.* **85**, 1693–1704.

Surface chemistry of food, packaging and biopolymer materials

4

Jung H. Han, Yachuan Zhang, and Roberto Buffo

Introduction

Surface properties of food packaging polymers, such as wettability, sealability, printability, dye uptake, resistance to glazing, and adhesion to food surfaces or other polymers, are of central importance to food packaging designers and engineers with respect to product shelf-life, appearance, and quality control. The most commonly used food packaging polymers are low-density polyethylene (LPDE), high-density polyethylene (HDPE), polypropylene (pp), polytetrafluoroethylene (PTFE), and nylon. Surface properties of these polymers have been extensively studied (Rucka *et al.*, 1990; Michalski *et al.*, 1998; Ozdemir *et al.*, 1999).

In recent years, environmental concerns have increased the interest in preparing biodegradable packaging materials (Sánchez *et al.*, 1998). Proteins and polysaccharides are the biopolymers of prime interest, since they can be used effectively to make edible and biodegradable films to replace short shelf-life plastics (Guilbert *et al.*, 1996). These environmentally friendly films must meet a number of specific functional requirements: moisture barrier, solute and/or gas barrier, color and appearance, mechanical and rheological features, and non-toxicity (Guilbert, 2000). The type of film-forming materials and additives will decisively determine whether a biodegradable film can achieve these functional properties. Studies of biopolymer films have focused

Innovations in Food Packaging
ISBN: 0-12-311632-5

on mechanical properties, water vapor transfer rates, and gas exchange control. Relatively little attention has been paid to surface properties, which may be essential to the behavior of the film in relation to water and organic solvents. In addition, they contribute substantially to the structure of the film (Bialopiotrowicz, 2003).

Biopolymer films primarily composed of polysaccharides (e.g. gums, cellulose and derivatives, starch and derivatives) or proteins (e.g., gelatin, zein, gluten) have suitable overall mechanical and optical properties (Guilbert *et al.*, 1996). With regard to gas permeability, biopolymer films present impressive gas barrier properties in dry conditions, especially against oxygen (O_2). For instance, oxygen permeability of wheat gluten films has been reported to be 800 times lower than that of polyamide-6, a well-known high O_2-barrier polymer (Guilbert, 2000). On the other hand, because of their hydrophilic nature biopolymer films are indeed wettable by water. Although they provide efficient barriers against oils and lipids, their moisture barrier properties are poor. As water activity (a_w) increases, film moisture content increases as well due to water absorption, following non-linear absorption isotherms. This induces a decrease in water vapor barrier properties (Guilbert *et al.*, 1997).

Surface properties of biopolymers provide a supplementary understanding of film behavior, leading to an enhanced design of packaging materials for specific applications (McGuire and Yang, 1991). This chapter discusses the physicochemical principles of surface phenomena, and provides an overview of the research regarding surface properties of biopolymers used for the manufacturing of biodegradable films.

Principles of contact angle and surface energy

Liquid and solid surface thermodynamics are based on the fundamental concept of surface tension, defined as half the free energy change due to cohesion of the material in a vacuum. Surface tension depends on a number of relatively independent forces, such as dispersion, dipolar, induction, hydrogen-bonding, and metallic interactions (Fowkes, 1963). According to the traditional theory, the surface tension of colloid-size particles is determined by an apolar component, Lifshitz–van der Waals (LW) forces; a polar component, Lewis acid–base (AB) forces; and an electrostatic (EL) component. The apolar component depends on the unevenness of the electron cloud surrounding the molecules, the polar component is related to the formation of coordinate covalent bonds between Lewis acids (electron pair acceptors) and Lewis bases (electron pair donors), and the electrostatic component is described through the zeta-potential – a measurement of electrical surface charge (van Oss, 1994; Meiners *et al.*, 1995; Rijnaarts *et al.*, 1995; Besseling, 1997).

Contact angle

Wetting properties are essential surface features of packaging materials. The wettability of a solid surface can be determined relatively simply by measuring the so-called contact angle (Kiery and Olson, 2000). This concept, as applied to a liquid drop on a material surface, is the result of the equilibrium between three surface energies: the

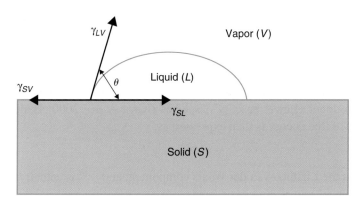

Figure 4.1   The wetting of a solid surface by a liquid drop.

liquid–vapor interfacial energy (γ_{LV}), the solid–vapor interfacial energy (γ_{SV}), and the solid–liquid interfacial energy (γ_{SL}) (Figure 4.1) (Han and Krochta, 1999). The wetting of a solid surface is governed by the Young equation, which establishes the equilibrium relationship between the contact angle θ and the three aforementioned interfacial tensions (Ornebro *et al.*, 2000):

$$\gamma_{LV} \times \cos\theta = \gamma_{SV} - \gamma_{SL} \tag{4.1}$$

According to this equation, a droplet with high surface tension resting on a low-energy solid tends to adopt a spherical shape due to the establishment of a high contact angle with the surface. Conversely, when the solid surface energy exceeds the liquid surface tension, the droplet tends to adopt a flatter shape due to the establishment of a low contact angle with the surface. Thus, wetting means that the contact angle between a liquid and a solid is so nearly zero that the liquid spreads over the solid easily. Similarly, non-wetting means that the angle is greater than 90°, and therefore the liquid will ball up and run off the surface easily (Adamson and Gast, 1997).

Surface energy and spreading coefficient

Another valuable thermodynamic parameter to describe wetting and other surface properties is the surface free energy. The condition for wetting or spreading of a liquid on a solid surface is decided by the spreading coefficient, $S_{L/S}$:

$$S_{L/S} = \gamma_{SV} - \gamma_{LV} - \gamma_{SL} \tag{4.2}$$

This equation means that a liquid will spread out on the surface of a solid if $S_{L/S} > 0$, but will form a droplet with a finite contact angle if $S_{L/S} < 0$. In other words, wetting depends on the surface tension of the liquid and on the free energy of the interfaces (Ornebro *et al.*, 2000). For good adherence to take place, the interfacial energy between the two materials must be small. As it applies to food packaging, the liquid usually denoted is water. Thus, to prevent wetting from occurring, biopolymer packaging materials should be designed to have a high interfacial tension with respect to water (Lawton, 1995).

As mentioned previously, there are two separate and additive attractive forces operating across the solid–liquid interface; the dispersion Lifshitz–van der Waals forces, and the Lewis acid–base forces. These are notated as γ^{LW} and γ^{AB}, respectively. The γ^{LW} forces are mainly due to dipoles induced between adjacent molecules, whereas the γ^{AB} forces are generated by polar interactions between electron-acceptor (Lewis acid: γ^+) and electron-donor (Lewis base: γ^-) species (Kaya et $al.$, 2000; Bialopiotrowicz, 2003). Surface free energy is then expressed as:

$$\gamma = \gamma^{LW} + \gamma^{AB} \tag{4.3}$$

where γ^{LW} is the Lifshitz–van der Waals component, and γ^{AB} is actually a geometric mean between γ^+ and γ^-:

$$\gamma^{AB} = 2(\gamma^+ \times \gamma^-)^{1/2} \tag{4.4}$$

Taking into account equations (4.3) and (4.4), van Oss (1994) derived the following equation describing the work of adhesion of a liquid to a solid:

$$W_{SL} = \gamma_L \times (1 + \cos\theta) = \gamma_S + \gamma_L - \gamma_{SL}$$
$$= 2(\gamma_S^{LW} \times \gamma_L^{LW})^{1/2} + 2(\gamma_S^+ \times \gamma_L^-)^{1/2} + 2(\gamma_S^- \times \gamma_L^+)^{1/2} \tag{4.5}$$

In this equation, W_{SL} represents the work of adhesion of a liquid to a solid (that is, the work necessary to separate a unit area of the interface SL into two liquid–vapor and solid–vapor interfaces); γ_S^{LW} and γ_L^{LW} are the Lifshitz–van der Waals components of solid S and liquid L surface free energies, respectively; γ_S^+ and γ_S^- are the electron-acceptor and electron-donor parameters of the Lewis acid–base component of the surface free energy of a solid; and γ_L^+ and γ_L^- are the electron-acceptor and electron-donor parameters of the Lewis acid–base component of the surface free energy of a liquid. According to Lawton (1995), for good adherence to take place, the interfacial energy between the two materials must be small. This model has been useful to compute bacterial adhesion to surfaces involving acid–base and electrostatic interactions (Michalski et $al.$, 1998). The computation of the above parameters for the unknown solid through equation (4.5) can be made by a three-liquid method, namely two polar and one apolar liquid substances of known surface tension parameters. Thus, the polar components can be evaluated from the settings of the equation using the two polar liquids, whereas γ_S^{LW} is evaluated from the apolar liquid for which the polar components are zero (Kaya et $al.$, 2000; Bialopiotrowicz, 2003).

Zisman equation and critical surface tension

Zisman and coworkers found that $cos\,\theta$ is usually a monotonic function of γ_L for a homologous series of liquids:

$$\cos\theta = a - b\gamma_L = 1 - \beta(\gamma_L - \gamma_C) \tag{4.6}$$

where γ_C is the critical surface tension and is β a constant (Zisman, 1973). The critical surface tension can be obtained from the Fox–Zisman plot ($\cos\theta$ vs. γ_L) by extrapolation to $\cos\theta = 1$ with various liquids of different γ_L values. According to McGuire and Kirtley (1988), γ_C is solely a function of the surface properties of a solid, and related to its true surface energy. Thus γ_C becomes a characteristic property, and has provided a useful means of summarizing wetting behavior and allowing predictions of interpolative nature. Liquids with surface tension $< \gamma_C$ will spread on the surface, completely wetting the solid. Nevertheless, it is important to remember that γ_C is not γ_S; the latter is most probably larger than the former, especially when considering polar materials such as starch (Toussaint and Luner, 1993; Lawton, 1995).

McGuire's theory and equation

In 1972, Fowkes emphasized that work of adhesion (W_{SL}) may also be divided into dispersive energy (d) and polar energy (p) components:

$$W_{SL} = W_{SL}^d + W_{SL}^p \qquad (4.7)$$

With respect to solid–liquid contact, a further identification of W_{SL} was generated as:

$$W_{SL} = W_{SL}^d + W_{SL}^p = 2(\gamma_L^d \times \gamma_S^d)^{1/2} + 2(\gamma_L^p \times \gamma_S^p)^{1/2} \qquad (4.8)$$

From equation (4.5),

$$W_{SL}^p = \gamma_L \times (1 + \cos\theta) - 2(\gamma_L^d \times \gamma_S^d)^{1/2} \qquad (4.9)$$

Contact angle data recorded for a series of diagnostic liquids with known values of γ_L^d on a surface of known γ_S^d should therefore lead to the evaluation of W_{SL}^p. Since γ_L^p of each diagnostic liquid is known, plots of W_{SL}^p vs γ_L^p can be constructed for any material targeted for food contact (McGuire and Kirtley, 1988). Experimentally derived plots led to the finding that the relationship between W_{SL}^p and γ_L^p was linear for all the testing materials, which strongly suggested that the relationship was independent from the diagnostic liquids. McGuire and Kirtley (1989) arranged this fact into mathematical terms as:

$$W_{SL}^p = k\gamma_L^p + b \qquad (4.10)$$

where the slope k (dimensionless) and intercept b (mJ/m^2) are specific to each solid surface.

The value of the polar contribution to the work of adhesion W_{SL}^p depends upon the polar character of both the solid and the liquid that are in contact, and provides either a measure of solid surface hydrophilicity (Yang et al., 1991) or a relative index of surface hydrophobicity/hydrophilicity (McGuire and Sproull, 1990).

Neumann's equation

The solid surface tension, γ_{SV}, can be calculated by combining the Young equation:

$$\gamma_{LV} \cos\theta = \gamma_{SV} - \gamma_{SL} \qquad (4.11)$$

and the Neumann's equation of state for interfacial tensions:

$$\gamma_{SL} = \gamma_{LV} + \gamma_{SV} - 2(\gamma_{LV} \times \gamma_{SV})^{1/2} \times e^{-\beta(\gamma_{LV} - \gamma_{SV})^2} \qquad (4.12)$$

where γ_{LV} is the liquid surface tension, γ_{SL} the solid liquid interfacial tension, and β the Neumann's constant ($\beta = 0.000115\,\text{m}^2/\text{mJ}$). By eliminating γ_{SL} from these equations, we get:

$$\cos\theta = -1 + 2(\gamma_{SV}/\gamma_{LV})^{1/2} \times e^{-\beta(\gamma_{LV} - \gamma_{SV})^2} \tag{4.13}$$

Therefore, solid surface tension γ_{SV} can be computed by equation (4.13) based on the experimentally measured contact angle and the known liquid surface tension (Matuna and Balatinecz, 1998).

Harmonic mean method

Another way to estimate surface energy of solids is derived from liquid–solid contact angle measurements, as follows:

$$(1 + \cos\theta)/\gamma_L = 4\gamma_S^{LW} \times \gamma_L^{LW}/(\gamma_L^{LW} + \gamma_S^{LW}) + 4\gamma_S^{P} \times \gamma_L^{P}/(\gamma_L^{P} + \gamma_S^{P}) \tag{4.14}$$

In this equation, there are two unknowns: γ_S^{LW} and γ_S^{P}. When the contact angle is measured for two different liquids on the solid given that the liquid's γ_L^{LW} and γ_L^{P} are known, the following relationships are produced:

$$(1 + \cos\theta_1)/\gamma_{L1} = 4\gamma_S^{LW} \times \gamma_{L1}^{LW}/(\gamma_{L1}^{LW} + \gamma_S^{LW}) + 4\gamma_S^{P} \times \gamma_{L1}^{P}/(\gamma_{L1}^{P} + \gamma_S^{P}) \tag{4.15}$$

$$(1 + \cos\theta_2)/\gamma_{L2} = 4\gamma_S^{LW} \times \gamma_{L2}^{LW}/(\gamma_{L2}^{LW} + \gamma_S^{LW}) + 4\gamma_S^{P} \times \gamma_{L2}^{P}/(\gamma_{L2}^{P} + \gamma_S^{P}) \tag{4.16}$$

where γ_{L1}^{LW} and γ_{L2}^{LW} are the surface energy for the two probe liquids. In fact, these two equations, known as the harmonic mean method, are able simultaneously to obtain solid surface energy and its polar and apolar components (Lawton, 1995). There are two general criteria for selecting probe liquids for the harmonic mean method: the probe liquids should have very different polarities (water and methylene iodide are the two liquids most often used), and the liquids should be inert with respect to the solid (i.e. the liquid should not react, swell or dissolve the solid being tested) (Lawton, 1995).

Germain's method

An alternative procedure to evaluate liquid-to-solid adhesion phenomena is the Germain's method, based on the following equation:

$$W_{SL} = W_{SL}^{d} + W_{SL}^{P} = 2(\gamma_L^{d} \times \gamma_S^{d})^{1/2} + 2(\gamma_L^{P} \times \gamma_S^{P})^{1/2} + 2(\gamma_L^{h} \times \gamma_S^{h})^{1/2} \tag{4.17}$$

where γ^{h} is the surface tension component due to hydrogen bonds, and γ^{P} the polar Lifshitz–van der Waals component. This model has been used to study the adhesion of ink to polymer surfaces (Michalski et al., 1998).

Techniques for measuring the contact angle

Measuring the contact angle is the most common method to estimate surface hydrophobicity (McGuire and Kirtley, 1988; Jones et al., 1999; Boonaert and

Rouxhet, 2000). There are two basic techniques for measuring contact angles of non-porous solids: goniometry and tensiometry.

Goniometry

Analysis of the shape of a drop of test liquid placed on a solid is the basis for goniometry, also referred to as the sessile drop technique. Clean, dry samples of polymer films are mounted on a glass slide using double-sided sticky tape. Samples should have no surface irregularities (such as seams, lettering or visible scratches). A 5-μl drop of distilled water is placed on the sample using a glass syringe. The tangent to the point of contact at the solid–liquid–vapor interface is then measured using a contact angle goniometer to give the advancing contact angle θ (Jones et al., 1999). Basic elements of a goniometer are a light source, sample stage, and lens, and an image capture device. The contact angle can also be roughly estimated using the following formula:

$$\theta(°) = 2arctg\ (2\ H/W) \tag{4.18}$$

where H is the height of the drop and W is the width of the solid–liquid interface. To use this method, water drops (5-μl) are deposited on the solid surface of interest and photographs are taken within 5 seconds. Negatives are then projected on screen to allow estimation of the droplet dimensions (Sinde and Carballo, 2000).

Goniometry can be used in many situations where substrates can provide regular curvature. A very small quantity of liquid is needed. Sessile drop techniques and associated equipment are currently being used for quality control at food packaging manufacturing plants (e.g. polymeric coatings applied to paper). Tests can be done very easily if solid substrates have a relatively flat portion for testing and can fit on the stage of the instrument (McGuire and Yang, 1991). However, there are some limitations. Measurements have been performed under different experimental conditions and with different procedures in different laboratories, which makes comparison of results very difficult – if not impossible (McGuire and Yang, 1991). It has been found that contact angles decrease with decreasing drop diameter for certain solid–liquid systems (Good and Koo, 1979), and increases with increasing drop diameter in a range below a limiting value identified as the critical drop volume (CDV). CDV varies among materials, and needs to be known to allow measurement standardization (McGuire and Yang, 1991). In addition, contact angles of most test liquids with hydrophilic surfaces tend to be rather small, which increases the possibility of error in experimental measurements (Noda, 1993).

Recently, more advanced equipment has been developed to test contact angles. A digital microscope is connected to a personal computer. Polymer films are glued onto a well-leveled platform, and the microscope is positioned horizontally for capturing the side-view image (Figure 4.2). At the lens position of 10, the side-view image is acquired and converted into a binomial edge-enhancing picture using conventional photo-editing software. Contact angles of the digital images are measured by hand if the sample size is small, or using line-slope calculation software if sample sizes are large (Han and Krochta, 1999).

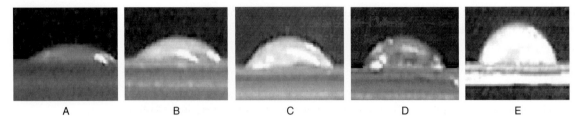

Figure 4.2 Different contact angles of various liquids on polypropylene film surface ($30\,mJ/m^2$ of surface energy). A: DMSO ($44.0\,mJ/m^2$ of surface tension); B: ethylene glycol ($48\,mJ/m^2$); C: formamide ($58.2\,mJ/m^2$); D: glycerol ($63.4\,mJ/m^2$); and E: water ($72.8\,mJ/m^2$).

Tensiometry

The tensiometric method for the determination of contact angles measures the forces present when a sample solid is brought into contact with a test liquid. If the forces of interaction, geometry of the solid, and surface tension of the liquid are known, the contact angle may be calculated. The sample of the solid to be tested is hung on the balance and tared. The liquid, of known surface tension, is then raised to make contact with the solid. When the solid is in contact with the liquid, the change in forces is detected and the height is registered as zero depth of immersion. As the solid is pushed into the liquid, the forces on the balance are recorded. An alternative and probably now more widely used procedure is to raise the liquid level gradually until it just touches the hanging solid suspended from the balance. The change of weight is then noted. A general equation is:

$$\gamma_{LV}\cos\theta = \Delta W/p \qquad (4.19)$$

where ΔW is the change in weight (that is, the force exerted by the solid when it is brought into contact with the liquid), p the geometrical perimeter of the solid, and γ_{LV} is the liquid surface tension. When the solid is advanced into the liquid, this contact angle is the advancing contact angle; otherwise, it is the receding contact angle. At any point on the immersion graph, all points along the perimeter of the solid at that depth contribute to the recorded force measurement. Thus, the force used to calculate contact angles at any given depth of immersion is already an averaged value. This technique is also very useful in studying hysteresis. Variations of contact angles, both advancing and receding, for the entire length of the sample tested, are visualized on the same graph (Adamson and Gast, 1997).

Applied research

Some commonly used food packaging synthetic polymers show rather low surface energies (Table 4.1). Improving these synthetic polymers surface energies is better for package bonding, printing, and coating with other films. Good adhesion in the seal areas is also highly desirable at polymer–polymer or polymer–metal interfaces in food

Table 4.1	Surface energies for some polymers (mJ/m^2)					
PE	PET	PVP	TPX	PMMA	PVCH	PS
32.0 ± 1.6	38.0 ± 2	50.0 ± 2.0	21.5 ± 0.1	40.0 ± 0.2	29.0 ± 1.0	30.0 ± 1.0

PVCH, poly (vinylcyclohexane); PE, polyethylene; PET, poly (ethylene terephthalate); TPX, poly (4-methyl-1-pentene); PS, polystyrene; PVP, poly (2-vinylpyridine); PMMA, poly (methyl methacrylate). From Tirrell (1996).

packages, particularly those composed of laminated materials. Without adequate adhesion, food may become contaminated with food-borne organisms or extraneous materials (Ozdemir et al., 1999).

Some technologies that have been developed to improve the surface energy of food packaging polymers include flame, corona, and plasma treatments, the latter being considered the most effective. Plasma treatments can drastically increase bond strengths (adhesion) of commonly used packaging polymers. For example, the bond strength of LDPE and PP increased 20-fold and 7-fold, respectively, compared to untreated films (Ozdemir et al., 1999). Plasma treatments have also been reported as effective in improving adhesion between a polymer and a metal substrate in the absence of an adhesive (Sapieha et al., 1993; O'Kell et al., 1995). Figure 4.3 shows the peel strength of air- and nitrogen-plasma-treated aluminum/PE/aluminum laminates prepared by direct melting and pressing of PE films onto the aluminum substrates without any adhesives. Peel strength values of air-plasma-treated PE samples were considerably higher than those of nitrogen-plasma-treated counterparts. A 70- to 75-fold increase in peel strength improvement was achieved at the PE/aluminum interface of air-plasma-treated samples, compared to only 3- to 6-fold increases following exposure to nitrogen treatment. The improvement in peel strength was due to the incorporation of O$_2$-containing functional groups into the PE surface during the plasma treatment. In addition, the peel strength improvement between polymer–polymer and/or polymer–metal interfaces without the use of adhesives following exposure to a plasma treatment can significantly limit the use of volatile organic solvents. Eliminating these volatile substances from adhesive formations will reduce damage to the environment and limit health risks from these hazardous solvents (Ozdemir et al., 1999).

Surface properties of conventional plastic films can be modified by biopolymer coating. Whey protein has been coated onto plastic films to increase the surface energy of the plastic films to be hydrophilic (Hong et al., 2004).

With respect to the adhesion of food materials to synthetic package surfaces, most studies in the literature are empirical, without taking into account the existing theoretical adhesion models (Michalski et al., 1997). Oil raises most concerns, because adhesion of fat or oil on packages increases recycling costs and enhances interactions that may alter the food product, leading to a poor product appearance. Michalski et al. (1998) studied the phenomenon using virgin olive oil, refined first-draft sunflower oil, soybean oil, and pure white vaseline oil to test the correlation between model-predicted and experimental adhesion. Low-density polyethylene (LDPE), polyethylene terephthalate

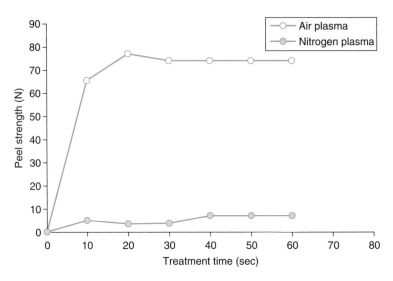

Figure 4.3 Peel strength of air-plasma- and nitrogen-plasma-treated polyethylene/aluminum laminates (modified from O'Kell *et al.*, 1995).

(PET), stainless steel AISI 304, and glass were chosen to be solid package surfaces. Results showed that experimental adhesion correlated well with the McGuire's and Germain's methods. These two approaches seemed to be efficient in predicting bulk adhesion in oils, particularly the latter one.

Unlike research on synthetic films, most research on biopolymer films has focused on water barrier properties, because low water barrier ability is the biggest shortcoming of biopolymer films and seriously limits their application as packaging materials. Biopolymer films are expected to be moderate to excellent O_2 barriers due to their tightly packed, ordered hydrogen-bonded network structure (Fang and Hanna, 2000; Díaz-Sobac and Beristain, 2001), but because of their inherent hydrophilicity, their water vapor permeability (WVP) is high. Thus, lipids or other hydrophobic substances such as resins, waxes, fatty acids or even some non-soluble proteins need to be added to retard moisture transfer, essentially related to the low mobility of fatty acid chains (Fairley *et al.*, 1995; Guilbert *et al.*, 1996; Callegarin *et al.*, 1997; Sherwin *et al.*, 1998). Sebti *et al.* (2002) used hydroxy-propyl-methyl cellulose (HPMC) combined with stearic acid to make packaging films. They found that the incorporation of stearic acid increased the contact angle from 49° to 82°, and decreased WVP by 40%.

Viroben *et al.* (2000) tested the hydrophilicity of films made from pea-protein isolates, adding 1,3-propanediol (PRD) and 1,2-propanediol (PRG) as plasticizers. The results are presented in Table 4.2, which also includes previously obtained data by the same researchers (Guégen *et al.*, 1998) for comparison purposes. The contact angle was much higher for PRG than for PRD, and even higher than those for other plasticizers. This was primarily attributed to the low surface energy of PRG (38 mJ/m^2) compared to that of PRD (49 mJ/m^2). However, PRG-added films also showed a much higher WVP. In order to make pea-protein films more hydrophobic, monoglycerides were incorporated into the film-forming dispersion, but the resulting contact angles

Table 4.2   Effect of plasticizer type on barrier properties of pea-protein films

Plasticizer type	No. of C atoms	Film plasticizer content (%)	Contact angle of water (°)	Water vapor permeability ($\times 10^{10}$g/m$^2\cdot$s$\cdot$Pa)
Ethylene glycol*	2	9.2	21	19
1,2-propanediol (PRG)	3	23.8	67	37
1,3-propanediol (PRD)	3	25.0	32	47
Glycerol*	3	33.9	40	29
Diethylene glycol*	4	35.4	13	21
Triethylene glycol*	6	38.3	21	28
Tetraethylene glycol*	8	40.4	40	28

*Data from Guéguen *et al.* (1998) included for comparison with PRG and PRD (from Viroben *et al.*, 2000).

Table 4.3   Effect of added monoglycerides on contact angle of pea-protein films

Fatty acid	Contact angle (°)
Oleic	21
Linoleic	19
Ricinoleic	18
Myristic	29
Heptanoic	15
Undecenoic	19

From Viroben *et al.* (2000).

were rather low and did not improve hydrophobicity (Table 4.3). The only advantage of using these monoglycerides was that they allowed for the preparation of very soft films, which did not dry after exposure to air at ambient temperature for as long as 30 days, significantly improving their aging behavior.

Starch has received considerable attention because of its totally biodegradable, inexpensive nature. Lawton (1995) tested the hydrophilicity of films and ribbons made from three kinds of corn starch: high amylose, waxy, and normal. Eight different probe liquids with known surface energy γ_L^{LW} and γ_L^p were used to test contact angles and surface energy. It was found that contact angles decreased as surface energy decreased. Critical surface energy ranged from 38 to 43 mJ/m^2, and there were no differences related to the type of starch or processing conditions. The surface energy of starch films and ribbons ranged from 35.7 to 41.2 mJ/m^2 – that is, less than the corresponding γ^C but similar to polyethylene. The author considered that because of the interaction between the polar probe liquid and starch, not enough time was available for the polar probe liquid to get into equilibrium with the starch films or ribbons when testing the contact angle. This led to quick measurements of the contact angle, and large values resulted – comparable to those of non-polar solids. To predict which coating will adhere to starch, the author emphasized that unbiased values for surface energy of starch and its polar and apolar components need to be obtained through more reliable methods than contact angle measurements.

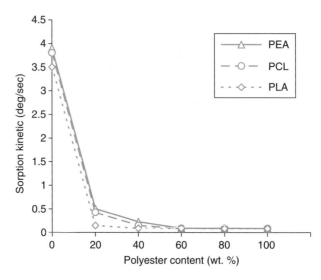

Figure 4.4 Effect of polyester content (PEA, PLA, and PCL, % w/w) on the hydrophilic character of plasticized wheat starch films (from Martin *et al.*, 2001; reprinted with permission of Wiley-VCH).

One strategy to overcome starch hydrophilic weakness is to associate it with a moisture-resistant polymer. Martin *et al.* (2001) reported water contact angle measurements performed on starch–polyester blends as well as on pure polyesters and plasticized starch (Figure 4.4). Starch films, plasticized with glycerol (glycerol/starch ratio = 0.54), were blended with polyesteramide (PEA), polylactic acid (PLA), and polycaprolactone (PCL). All blends showed a rapid decrease of the hydrophilic character as the amount of polyester in the mix increased from 0 to 10%, and kept decreasing until the value of pure polyester was reached. These researchers also performed water immersion tests of the starch-polyester multilayers to check their ability to resist water penetration. They showed good moisture resistance by standing the water for days without delamination of product swell.

Another strategy to overcome the hydrophilicity of biopolymer films is to incorporate lipids into the film matrix. Compared to laminated films, emulsified biopolymer films are less efficient water vapor barriers, but require only a single step in processing, exhibit good mechanical and adhesive properties, and can be made at room temperature (Kamper and Fennema, 1984; Quezada-Gallo *et al.*, 2000). Peroval *et al.* (2002) used arabinoxylans with four different kinds of lipids (palmitic acid, C_{16}; stearic acid, C_{18}; hydrogenated palm oil, OK35; and triolein) to make edible emulsified films. The experimental results are presented in Table 4.4.

Arabynoxylan-only films were characterized by contact angles of about 70°, and did not differ significantly from those with addition of C_{16} or C_{18}. The best hydrophobic surface, showing contact angles >90°, was obtained by adding OK35. These values were similar to those regularly observed with LPDE. Nevertheless, angles of arabinoxylan–OK35 films decrease 10 times faster versus time (data not shown), suggesting a rapid change in surface properties. This study showed that higher water

Table 4.4 Properties of arabinoxylan (AX)-based films and plastic packaging materials

Film	Thickness[a] (μm)	Contact angle[a,b] (°)	WVTR[a] ($\times 10^{-3}$g/m$^2\cdot$s)	WVP[a,c] ($\times 10^{10}$g/m$^2\cdot$s$\cdot$Pa)
AX-C$_{18}$	81.6 ± 5.4b	68.6 ± 3.6b	2.91 ± 0.12d	1.19 ± 0.05c
AX-triolein	66.6 ± 6.7c	39.0 ± 8.5c	3.52 ± 0.15c	1.18 ± 0.05c
AX-OK35	89.8 ± 9.1a	94.4 ± 2.1a	2.71 ± 0.17d	1.24 ± 0.06c
Cellophane	20.0	_d	6.17 ± 0.63a	0.69 ± 0.07d
LDPE	25.0	100.7 ± 11.4a	0.13 ± 0.01e	0.019 ± 0.001e
Arabinoxylan	90.8 ± 6.6a	70.8 ± 5.1b	3.92 ± 0.14b	1.77 ± 0.06a
AX-C$_{16}$	89.9 ± 9.8a	64.0 ± 4.6b	3.37 ± 0.09c	1.52 ± 0.04b

[a] Values in a column followed by the same letter are not significantly different ($p < 0.05$)
[b] Mean of a minimum of 15 measurements of the contact angle of water
[c] Mean of at least three replicates ± standard deviation
[d] Contact angle not measured because of the highly hydrophilic nature of cellophane.
From Peroval *et al.* (2002).

vapor transmission rates and permeability values do not necessarily correspond to lower contact angle values. Ultimately, the addition of fatty acids did not result in important changes on the film surface hydrophobicity.

Future trends

Biopolymer films have been regarded as potential replacements for synthetic films in food packaging applications in respond to a strong marketing trend towards more environmentally friendly materials. However, hydrophilicity is a central limitation that needs to be overcome in order to allow such replacement. To date, most methods of increasing the water barrier ability of biopolymer films depend on the addition of waxes, fatty acids, and lipids. Plasma treatments on synthetic film surfaces have achieved a significant improvement of their surface energy, and thus it should be worth trying them on the surface of biopolymer films to reduce water permeability.

A greater interest in understanding the surface properties of biopolymer films is expected within the next few years. This may be the key to solve the fundamental issue of excessive hydrophilicity, which would allow full-scale commercial utilization of biodegradable films as food packaging materials.

References

Adamson, A. W. and Gast, A. P. (1997). *Physical Chemistry of Surfaces*. John Wiley & Sons, New York, NY.

Besseling, N. A. M. (1997). *Langmuir* **13**, 2113–2122.

Bialopiotrowicz, T. (2003). *Food Hydrocolloids* **17**, 141–147.

Boonaert, C. J. P. and Rouxhet, P. G. (2000). *Appl. Environ. Microbiol.* **6**, 2548–2554.

Callegarin, F., Gallo, J. A. Q., Debeaufort, F. and Voilley, A. (1997). *J. Am. Oil Chem. Soc.* **74,** 1183–1192.

Díaz-Sobac, R. and Beristain, C. I. (2001). *J. Food Proc. Pres.* **25,** 25–35.

Fairley, P., German, J. B. and Krochta, J. M. (1995). *IFT Annual Meeting/Book of Abstracts 39.* Institute of Food Technologists, Chicago, IL.

Fang, Q. and Hanna, M. A. (2000). *Trans. ASAE* **43,** 89–94.

Fowkes, F. M. (1963). Additivity of intermolecular forces at interfaces. I. Determination of the contribution to surface and interfacial tensions of dispersion forces in various liquids. *J. Phys. Chem.* **67,** 2538–2541.

Good, R. J. and Koo, M. N. (1979). *J. Colloid Interface Sci.* **71,** 283–292.

Guéguen, J., Viroben, G., Noireaux, P. and Subirade, M. (1998). *Ind. Crops Prod.* **7,** 149–157.

Guilbert, S. (2000). *Bull. Intl Dairy Fed.* **346,** 10–16.

Guilbert, S., Gontard, N. and Gorris, L.G. M. (1996). *Lebensm. Wiss. u. Technol.* **29,** 10–17.

Guilbert, S., Cuq, B. and Gontard, N. (1997). *Food Add. Cont.* **14,** 741–751.

Han, J. H. and Krochta, J. M. (1999). *Trans. ASAE.* **42,** 1375–1382.

Hong, S .I., Han, J. H. and Krochta, J. M. (2004). Optical and surface properties of whey protein isolate coatings on plastic films as influenced by substrate, protein concentration and plasticizer type. *J. Appl. Polym. Sci.* **92(1),** 335–348.

Jones, C. R., Adams, M. R., Zhdan, P. A. and Chamberlain, A. H. L. (1999). *J. Appl. Microbiol.* **86,** 917–927.

Kamper, S. L. and Fennema, O. (1984). *J. Food Sci.* **49,** 1482–1485.

Kaya, A., Lloyd, T. B. and Fang, H. Y. (2000). *Geotech. Testing J.* **23,** 464–471.

Kiery, L. J. and Olson, N. F. (2000). *Food Microbiol.* **17,** 277–291.

Lawton, J. W. (1995). *Starch* **47,** 62–67.

Martin, O., Schwach, E., Averous, L. and Couturier, Y. (2001). *Starch* **53,** 372–380.

Matuna, L. M. and Balatinecz, J. J. (1998). *Polymer Eng. Sci.* **38,** 765–773.

McGuire, J. and Kirtley, S. A. (1988). *J. Food Eng.* **8,** 273–286.

McGuire, J. and Kirtley, S. A. (1989). *J. Food Sci.* **54,** 224–226.

McGuire, J. and Sproull, R. D. (1990). *J. Food Sci.* **55,** 1199–1200.

McGuire, J. and Yang, J. (1991). *J. Food Prot.* **54,** 232–235.

Meiners, J. M., van der Mei, H. C. and Bussher, H. J. (1995). *J. Colloid Interface Sci.* **176,** 329–341.

Michalski, M. C., Desobry, S. and Hardy, J. (1997). *Crit. Rev. Food Sci. Nutr.* **37,** 591–619.

Michalski, M. C., Desobry, S., Pons, M. N. and Hardy, J. (1998). *J. Am. Oil Chem. Soc.* **75,** 447–454.

Noda, I. (1993). In: *Contact Angle, Wettability and Adhesion* (K. L. Mittal, ed.), pp. 373–381. The Sherwin-Williams Co., Cleveland, OH.

O'Kell, S., Henshaw, T., Farrow, G., Aindow, M. and Jones, C. (1995). *Surface Interface Anal.* **23,** 319–327.

Ornebro, J., Nylandert, T. and Eliasson, A. C. (2000). *J. Cereal Sci.* **31,** 195–221.

Ozdemir, M., Yurteri, C. U. and Sadikoglu, H. (1999). *Food Technol.* **53(4),** 54–58.

Peroval, C., Debeaufort, F., Despre, D. and Voilley, A. (2002). *J. Agric. Food Chem.* **50,** 3977–3983.

Quezada-Gallo, J. A., Debeaufort, F., Callegarin, F. and Voilley, A. (2000). *J. Membr. Sci.* **180,** 37–46.

Rijnaarts, H. H. M., Norde, W., Bouwer, E. J., Lyklema, J. and Zehnder, A. J. B. (1995). *Colloids Surf. B: Biointerface* **4,** 5–22.

Rucka, M., Turkiewicz, B. and Zuk, J. S. (1990). *J. Am. Oil Chem. Soc.* **67,** 887–889.

Sánchez, A. C., Popineau, Y., Mangavel, C., Larre, C. and Guéguen, J. (1998). *J. Agric. Food Chem.* **46,** 4539–4544.

Sapieha, S., Cerny, J., Klemberg-Sapieha, J. E. and Martinu, L. (1993). *J. Adhesion* **42,** 91–102.

Sebti, I., Ham-Pichavant, F. and Coma, V. (2002*). J. Agric. Food Chem.* **50,** 4290–4294.

Sherwin, C. P., Smith, D. E. and Fulcher, R. G. (1998). *J. Agric. Food Chem.* **46,** 4534–4538.

Sinde, E. and Carballo, J. (2000). *Food Microbiol.* **17,** 439–447.

Tirrel, M. (1996). *Langmuir* **12,** 4548–4551.

Toussaint, A. F. and Luner, P. (1993). In: *Contact Angle, Wettability and Adhesion* (K. L. Mittal, ed.), pp. 383–396. The Sherwin-Williams Co., Cleveland, OH.

van Oss, C. J. (1994). *Interfacial Force in Aqueous Media.* Marcel Dekker, New York, NY.

Viroben, G., Barbot, J., Mouloungui, Z. and Guéguen, J. (2000). *J. Agric. Food Chem.* **48,** 1064–1069.

Yang, J., McGuire, J. and Kolbe, E. (1991). *J. Food Prot.* **54,** 879–884.

Zisman, W. A. (1973). In: *Twenty Years of Colloid and Surface Chemistry* (K. J. Mysels, C. M. Samour and J. H. Hollister, eds), p. 109. American Chemical Society, New York, NY.

Active packaging research and development

Introduction to active food packaging technologies

5

Michael L. Rooney

Introduction

The field of active packaging has been the subject of substantial research and development for only the last two decades. It remains largely at that stage of development at the time of writing. The term "active packaging" was first applied by Labuza (1987), and may be defined as packaging which performs some desired function other than merely providing a barrier to the external environment (Hotchkiss, 1994; Rooney, 1995a). Active packaging should not be confused with "intelligent packaging", which informs or communicates with the consumer regarding the present properties of the food, or records aspects of its history.

There are also more expansive definitions in use to elaborate the effects achieved by packaging, but there is the potential for blurring the distinctions between activity and intelligence. A more helpful approach is to use the term "Active and Intelligent Packaging", as used in the literature that resulted from the Actipack Project funded by the European Commission (de Jong, 2003).

The scope of active packaging research and development was consolidated by Rooney (1995a), and its development has been described and evaluated (Gontard, 2000; Brody *et al.*, 2001). Several reviews have discussed the field from differing points of view (Labuza and Breene, 1989; Rooney, 2000; Meroni, 2001).

Innovations in Food Packaging
ISBN: 0-12-311632-5

Traditionally, packaging is desired to assist in the maintenance of the quality of the food at the level achieved at the final stage of its processing. In practice, the quality decreases in packaged storage because, in part, the combination of packaging process and packaging material availability does not exactly match the specific requirements of each food or beverage. Packaging has normally been expected to be inert towards the packaged product, but the potential for packaging to contribute to the quality retention and the convenience of packaged goods was not utilized.

The processes by which food quality is lost often involve interaction with substances taken up from their environment. This may mean a loss or gain of water, ethylene or oxygen, and contamination with micro-organisms. There are also some substances that build up in the packaged food on storage, including the containment and cooking odors resulting from the oxidation of fats and oils. Some of these compounds are normally lost when foods are cooked shortly before serving. The utility of foods can be enhanced significantly if the package contributes to the processes of heating or cooling. These effects are summarized in Tables 5.1 and 5.2.

The principles upon which packaging acts are not limited to any one scientific discipline. Chemical reactions have been used to remove atmospheric gases such as oxygen, carbon dioxide and ethylene. Water is merely absorbed by substances with

Table 5.1 Mechanisms of food quality loss

Quality attribute	Result of presence	Packaging activity
Mold	Microbial spoilage	Antimicrobial surface
		Antimicrobial release
		Oxygen scavenging
Oxidation	Rancidity	Oxygen scavenging
		Odor absorption
		Antioxidant release
	Color change	Oxygen scavenging
	Nutrient loss	Oxygen scavenging
Food chemistry	Odor/flavor formation	Absorption
Water movement	Texture change	Dessication
	Microbial spoilage	Humidity buffering
		Absorption of condensate
Senescence (produce)	Premature ripening	Ethylene scavenging
		1-MCP release
		Atmosphere modification

Table 5.2 Convenience attributes

Effect	Opportunity	Activity
Heating	Convenience meals	Microwave susceptors
	Vending machines	Self-heating cans and cups
	Military rations	
Cooling	Vending machines	Self-cooling cans and cups
	Summer outdoor events	
Gas generation	Gasification of beer	Widgets

high affinity, such as silica gel, dehydrated lime or polyol humectants. Volatile organic compounds are largely adsorbed by porous solids such as zeolites. In some cases the strong adsorption of ethylene onto inorganic solids has been used to retain the vapor in equilibrium with the surrounding environment. Self-heating packages normally involve exothermic or endothermic chemical reactions in a second compartment of the package.

Active packaging is normally designed to address one property or requirement of the food or beverage. The property normally chosen is that most critical as the first limiter of quality or shelf life. To this extent, active packaging is provided to fine-tune the properties of the packaging to meet the requirements of the food. This is not different from the normal aim of the packaging technologist to match the requirements of the food with the properties of the packaging.

Drivers for choice of active packaging

The decision to consider active packaging for a food or beverage is commonly based on factors typically involved in any package selection. These considerations include economic advantage, process engineering limitations, convenience in use, environmental impacts, and secondary effects resulting from some other change in the processing or packaging. The latter effects may result from new product introduction due to lifestyle changes or the availability of technologies that remove a limitation formerly experienced. Some of the constraints on optimizing the processing, packaging and distribution of foods and beverages are shown in Figure 5.1.

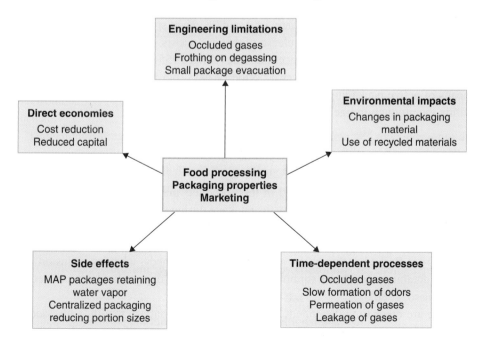

Figure 5.1 Current constraints on food processing, packaging and marketing.

Economic advantage

The optimal passive packaging solution for a particular product sometimes results in an initial quality or shelf life which is at a level below that considered desirable. The packaging may be coupled with a packaging process that introduces costs in terms of line-speed limitation or use of additional processes. Packages of oxygen-sensitive foods can require evacuation followed by inert-gas flushing and evacuation a second time. Introduction of an oxygen scavenger provides the opportunity for removal of two or more of these steps. The general case has been considered by Rooney (1995b).

Removal of the bitter principle, limonin, from orange juice can be achieved by means of a batch adsorption step during processing, but use of an active packaging material may achieve an acceptable effect (Chandler *et al.*, 1968; Chandler and Johnson, 1973). Naringin is a bitter principle in grapefruit juice, and it has been shown that large concentrations of this compound can be removed from the juice by contact with an active packaging material (Soares and Hotchkiss, 1998). Such an approach allows the potential for the time taken to distribute the product to be used to advantage. This may result in avoidance of capital for equipment to remove the naringin in-line.

Process engineering limitations

Oxygen dissolved in beverages can be removed by vacuum treatments or nitrogen flushing. These processes do not always fit well with existing processing equipment due to frothing, so removal of oxygen by means of active packaging is an attractive option. Similar considerations apply to the flushing of small sachets containing low-density powders which are readily made airborne, thus interfering with sealing of the package. An additional characteristic of powders, when spray dried, is the occlusion of air within food particles. The release of this gas occurs slowly, and is not achieved by means of evacuation on the processing line (King, 1955). The gas may be desorbed by equilibration with nitrogen over a period of days, but this is readily achieved by means of active packaging rather than by employing nitrogen-flushed holding tanks within the production line.

Time-dependent processes

The slow release of gases occluded in food products may also be viewed as a time-dependent process best addressed by the package during storage and distribution. Two forms of active packaging have been used to address the particular case of release of carbon dioxide from roasted coffee beans. The incorporation of one-way valves into flexible, gas-flushed coffee packs applies a physical remedy, whereas inclusion of a sachet containing absorbents for both carbon dioxide and oxygen has been utilized for cans (Russo, 1986).

There are other time-dependent chemical processes that are readily addressed by the use of active packaging. These include the scavenging of traces of aldehydes formed during oxidation of fats and oils in foods, and which give the product a rancid odor long before any nutritional damage has been done (Brodie and Visioli, 1994). Other odorous compounds are also formed in foods during storage, and the range of commercial active packaging approaches has been reviewed (Rooney, 1995a; Brody *et al.*, 2001). It is important to avoid odor removal if this could remove indicators of microbial growth in foods.

Besides processes occurring in foods during storage there are is also a time-dependent impact of the permeability of some packaging materials to atmospheric gases. Oxygen and water vapor are of most concern, and substantial research has been directed to development of plastics with an enhanced barrier to oxygen (Rooney, 1995a). The greatest development resources have been applied to reducing the oxygen transmission rate of PET bottles, especially for use with beer and fruit juices. The related problem of oxygen leaking under beer closures has been addressed by oxygen-scavenging closure liners (Teumac, 1995).

Secondary effects

Although marketing considerations are a recognized driver leading to technical changes in the packaging of foods, changes in processing methods commonly lead to changed requirements for the packaging. The beer industry has sought to introduce PET bottles in place of some or all glass bottles for several years (Anonymous, 2000). The flavor of beer is seriously degraded by atmospheric oxidation of some components (Teumac, 1995). PET has an oxygen permeability that is six times too high to provide the required shelf life, so its barrier needs to be increased. The most economic PET-based approach at present appears to be to include an oxygen scavenger in the middle of three layers in a multilayer bottle structure. An added advantage of such a structure is the scavenging of oxygen dissolved in the PET and which is capable of diffusing into the beer. Bottles containing an active barrier can also, to some extent, scavenge some of the oxygen dissolved in the beer. Thus the introduction of the PET bottle has resulted in oxygen removal from the beer – something not achievable by use of glass.

Possibly a more important side effect is found when fresh produce is sealed in poly-olefin bags, either as carton liners or as retail packs, both of which help generate modified atmospheres. The seal allows the permeability of the plastic to regulate the concentrations of oxygen and carbon dioxide involved in the respiration of the pro-duce (Kader *et al.*, 1989; Varoquaux, 2000). Concurrently, the free movement of tran-spired water vapor is substantially reduced compared with that found in an unsealed bag. Consequently the relative humidity rises and there may be condensation of some of this water due to cycling in the cold chain. There is therefore a need for this con-densate to be kept away from the produce, for example by means of super-absorbent pads or, possibly, by inclusion of a humidity buffer in the bag to raise the dew point (Louis and de Leiris, 1991).

Centralized processing of foods can result in changes in the surface-to-volume ratio of foods, as evidenced by cut salads, diced fruit and single cuts of steak. When each of these is packaged there are challenges for the package engineer that are different from those posed by the unprocessed product. Removal of fruit skin introduces more concerns about microbial growth, and so there are opportunities for edible coatings containing approved antimicrobials to be used to advantage. Similar considerations may apply to sliced meat.

Environmental impacts

The application of the "three Rs" (reduce, recycle or reuse) to food package selection introduces the opportunity to question whether the food's requirements are being met by passive packaging alone. The desire to replace glass packaging with PET is driven in part by a requirement to reduce weight during distribution and to reduce the incidence of broken glass. Although PET cannot match all of the desirable properties of glass, some of its properties relevant to specific packaging uses can be upgraded by active packaging. The barrier to oxygen is being raised by several companies using oxygen scavengers within the bottle walls (Anonymous, 2000). Such scavengers bring the added benefit of removing oxygen dissolved in the PET and which is capable of diffusion into the packaged beverage over time.

The introduction of recycled PET has the potential to introduce odors into lightly flavored liquids, and there is potential for development of taint absorbers for use in such packaging.

Enhanced convenience

Consumption of convenience foods, both within and outside the home, has resulted in development of a range of easy-to-use packages, such as "clam-shells", pizza boxes, and lidded disposable cups. Temperature retention of such foods and beverages has proved not to be easy, since the insulation characteristics of such conventional passive packaging appears to be at its economic limit. The perceived need for active packaging in temperature control is found in the event market, for military use, and for vending machine sales. Self-heating cups and cans, and self-cooling cans, are examples of developments designed to overcome this limitation.

The use of microwave susceptors in packaging for pastries and crisp convenience products seems to be limited to little more than microwave popcorn. This form of active packaging appeared to have substantial potential when introduced.

Forms of active packaging

The choice of the form to be taken by the active packaging is made based on three broad considerations. Most important is the requirement of the food, followed by the packaging format, and the requirements of the active agent.

The demands of the food can be visualized by considering the potential application of gas exchange in a retail pack of ground beef. Removal of oxygen or addition of carbon dioxide may be the chosen method of reducing spoilage by aerobic micro-organisms.

This application appears too demanding, as air pockets remain within the product. However, products like shredded cheese or fresh pasta, which are packaged in low-oxygen/high-carbon dioxide atmospheres, would be more amenable to successful atmospheric modification involving package activity. In this case the removal of oxygen could be achieved either by including oxygen-absorbing sachets in the package, or by chemically active plastic packaging.

The relative rates of food degradation and gas exchange between the food and its environment determine whether or not turbulent mixing is necessary. If turbulence is needed it can be provided only with evacuation or gas flushing, as occurs during pack filling. In other cases, the lower energy exchange generated by active packaging occurs only in response to a concentration gradient. This approach is particularly suitable where the kinetics of packaging activity are favorable when compared with those of mold growth, oxidation, and accumulation of odors.

The decision as to whether the source of activity should be localized or spread throughout the packaging material may be limited by the form of the package as well as the requirement of the food/beverage. Commercial active packaging for gas atmosphere modification is available as a variety of inserts, and in some cases is incorporated into the package structure.

Localized effects

Inserts such as sachets, cards, and self-adhesive labels are used to achieve a range of atmospheric-modification effects. The range available commercially in the early days of active packaging have been tabulated and discussed by Abe (1990). Since that time the nature of the inserts has changed in terms of both their content and materials of construction. This has been a significant advance, particularly where finely powdered ingredients have been replaced with the equivalent substances incorporated into plastic strips.

Rigid packs that are sealed with a lid or similar closure offer the opportunity for incorporation of activity into either the body of the pack or the closure. A variety of closure liners have been developed for the market, commencing with oxygen-scavenging closure liners for use on beer bottles (Teumac, 1995). The use of closure liners as the carrier allows both the container and the basic closure to be unchanged despite the introduction of the active function.

Whole package activity

Increasingly, research and development attention has moved from localized effects to achievement of whole-package activity. This is driven by the needs of beverages in particular, and to gain specific benefits with some foods. Beverages sensitive to oxygen are largely packaged in glass. The recent trend towards replacement of glass with polyester in bottles and jars has necessitated an increase in the barrier of the latter to oxygen (and carbon dioxide in the case of beer). The oxygen-barrier enhancement may be achieved by oxygen scavenging within the bottle wall (Brody *et al.*, 2001).

Both beverages and foods may benefit from active packaging which is antimicrobial. Unless the antimicrobial agent is volatile, it is necessary for the packaged product to contact the package surface. The latter may be active because of its antimicrobial surface chemistry, or due to migration of an agent onto the contact layer of the food or beverage (Han, 2000).

Migration of an active agent from a package onto a food can occur by diffusion from one phase to the other. In the case of solid or semi-solid foods, intimate contact may require evacuation (or package collapse due to gas absorption, as occurs with cheeses). If the active agent is required to act on the surface of the food, migration into the latter will impose some conditions of relative diffusivities from the packaging and into the food. This will also apply to pastes and other high-viscosity liquids. In the case of beverages, some agitation of the package contents may be required to achieve the intended effect sufficiently rapidly. The reverse requirements apply when a migrating species is removed from the food/beverage, although diffusion to the package surface may well be the faster process.

Meroni (2001) has proposed that the functionality of packaging be categorized in order that all involved can understand the level of packaging design required. She defines a *basic level* of active packaging which has little or no intelligence added. An *intermediate level* may, for instance, include sensors that might determine the heating time the package with a susceptor might require in a microwave oven. The *superior level* of complexity might involve more combinations of effects, such as releasing color and flavor into the product. Developments in nomenclature of this kind will probably be necessary since such concepts are finding their way into definitions of active packaging, such as in Brody *et al.* (2001).

Edible coatings

When an edible coating contributes to the packaging of a food, the coating performs first as a food component. However, because it is normally made from one or more food constituents it may need protection against microbial activity. Hence if a mobile antimicrobial agent is incorporated therein, the coating can serve several additional functions. These include providing some self-preservation, helping to reduce the microbial load on the food surface, and providing an outer surface with antimicrobial properties. The latter is potentially important when used with finger-foods such as hors d'oeuvres.

The requirements of foods that involve edible coatings can be very complex, so the active contribution provided by the additive may not be a major consideration in the formulation of the coating.

History of active packaging

The recognition of active packaging as a generic approach is a relatively new occurrence, as evidenced by the earliest reviews bringing together the concepts, even if using different descriptors – such as "smart" (Sacharow, 1988; Labuza and Breene,

1989; Anonymous, 2000). The field has been developing largely as a series of niche markets owing to the current approach of the package converting and resin industries of viewing it in terms of a series of market opportunities. The user industries, typified by the food industry, have presented these opportunities in isolation, and this may continue for some years. The approach of considering a range of packaging options (both passive and active) as a whole is not yet common practice.

Many of the developments have been logical consequences of earlier commercial products or of non-commercial research publications. There are, however, some concepts that appear to have established new lines of investigation or commercial development. Any choice of this type is necessarily subjective, but some of these are shown in Table 5.3, and the discussion that follows indicates their significance.

Active packaging for processed foods and beverages

Oxygen scavenging

The earliest approaches to removal of oxygen from canned milk powder involved the use of oxidizable metal powders (Tallgren, 1938). A system with some control over the commencement of oxygen uptake was introduced by Kuhn et al. (1970), who used palladium catalyst attached to the inside of the can lids to catalyze oxidation of hydrogen gas mixed with the nitrogen flush in the canning of milk powder. The process was incorporated into foil laminate pouches by the American National Can Company (Warmbier and Wolf, 1976). This work represented the beginning of the research activity that led to the introduction of flexible oxygen-scavenging packs in 1996 by BP Amoco Chemicals and the Cryovac Division of Sealed Air Corporation. Since that time, refinement of the chemistry used has continued at a steady pace.

Concurrent research designed to meet the needs of the beer industry for trapping oxygen diffusing under closures resulted in a range of polymeric closure-liner modifications. These involved several chemistries, but those used most successfully incorporate sulfites, iron powder and ascorbic acid (Teumac, 1995).

The approach of including oxidizable compounds in porous sachets was concurrently conducted in Japan, resulting in the introduction of Ageless® by Mitsubishi Gas Chemical Co. in 1977 (Smith et al., 1995). Recent developments of these porous sachets have led to a variety of self-adhesive labels and cards for insertion into food packages (Sakakibara, 2000).

Table 5.3 Some seminal technology adaptations in active packaging

Technology	Significance	References
Pd catalysis of hydrogen oxidation	Triggering of oxygen scavenging	King (1955)
Ethylene oxidation by permanganate	Reactive removal of ethylene in produce packs	Scott et al. (1970)
Singlet oxygen reactions in plastics	Light as a trigger in oxygen scavenging	Rooney and Holland (1979)
Side-chain crystallizable polymers	Gas permeability adjustment with temperature in cold chain	Stewart et al. (1994)

Parallel developments for the polyester industry have resulted in a range of approaches to oxygen scavenging by polymeric and low-molecular-weight compounds within polyester bottle walls (Anonymous, 2000). The first approach was the oxidation of MXD-6 nylon by the permeating oxygen in the presence of a transition metal catalyst (Cochran *et al.*, 1991). These developments include multilayer systems as well as blends with PET in monolayer bottles.

Carbon dioxide scavenging or release

Carbon dioxide serves as an antimicrobial gas in modified atmosphere packaging, but can be undesirable when present in excess – as in natural cheeses, coffee or kim-chi. Approaches to its removal have continued to involve the incorporation of lime into sachets, or of one-way valves, as in retail coffee packs (Gaglio, 1986; Abe, 1990). In contrast, release of the gas has been achieved by use of sodium bicarbonate and food acids in sachets. When the combined antimicrobial effects of oxygen scavenging and carbon dioxide release are required, both effects can be achieved where ferrous carbonate is used as the active ingredient.

Removal of odors and flavors

The development of unpleasant flavors as a consequence of food processing can be the result of thermal degradation of components, such as proteins, or of reactions such as the Maillard reaction. Oxidation of fats and oils is also accelerated at processing temperatures. Besides these reactions there can be a slow generation of unpleasant flavors when fruit components are disturbed from their structural components in the fruit. The bitter principle, limonin, builds up in orange juice after pasteurization, and renders juice from some cultivars undrinkable. Chandler and Johnson (1979) showed that substantial quantities of limonin could be removed by acetylated paper, following earlier work involving cellulose acetate gel beads (Chandler *et al.*, 1968).

The concept of odor removal using chemical affinity was further developed by Brodie and Visioli (1994), who used the reaction of aldehydes with amino polymers. The approach of focusing on more specific reactions has been taken a step further by Soares and Hotchkiss (1998), who showed that the naringin content of grapefruit juice could be reduced to acceptable levels by using naringinase immobilized within cellulose triacetate film. This increasing use of specific effects highlights the aim of active packaging to achieve a specific effect without necessarily impacting on other properties of packaging.

Active packaging for produce

Modified-atmosphere packaging of produce has been practiced to varying degrees for several decades, following the successful controlled atmosphere storage of pome fruits. Although some contribution to modification of the package atmosphere by choice of the permeability of the package is regularly achieved, research towards better atmospheres has been conducted for at least two decades. The use of hydration of patches over holes in the packaging was the initial approach, and this was followed by the development of Intellipack® by the Landec Corporation (Stewart *et al.*, 1994). The latter material has a sharp change in gas permeability at specific selected temperatures, allowing compensation for temperature changes during distribution.

Besides the respirational gases, the movement of water transpired by produce in a lined carton or a retail pouch is also a cause of quality loss, or at least build-up of fog on the plastic. Anti-fog coatings were introduced and are still used. Proposals for the removal of water vapor by a form of humidity buffering initially involved porous bags of sodium chloride (Shirazi and Cameron, 1992). This was followed by the introduction of thin pouches containing humectants such as glycols and carbohydrates (Labuza and Breene, 1989). At present, humidity is still largely uncontrolled in wholesale packs for produce. This topic provides a challenge, particularly for desiccant manufacturers, to minimize losses due to build-up of condensate on fruits during distribution.

The aging of produce and flowers can be delayed by the use of MAP, but accelerated by exposure to ethylene. This plant hormone is generated endogenously, especially when a particular item has been injured or is at an advanced stage of ripening. The ethylene generated by one fruit can also trigger the senescence of many others, and so packages require ethylene-scavenging capability rather than just a high permeability which reduces build-up of the gas generated within the pack but dissipates it elsewhere. Early developments involved incorporation of inorganic adsorbents into plastic liner bags, but subsequent research involved sachets of porous solids containing potassium permanganate. Recent research by Chamara *et al.* (2000) demonstrated that the shelf life of Kolikuttu bananas could be extended from 24 days to 54 days at 14°C and 94% RH by incorporation of a potassium permanganate scavenger into polyethylene bags of the fruit. Ripening after storage in this way was found not to differ from that found in the absence of the scavenger. Research from the same laboratories showed that a similar effect could be achieved with Pollock avocados (Chamara *et al.*, 2002). The latter are used commercially, but recent advances have included a metal complex and cyclic compounds that react with ethylene. The latter are tetrazenes, which react very rapidly and quantitatively, with a concurrent change in color (Holland, 1992).

The condensation of water in produce packs noted above has also been found to inhibit the ethylene-adsorbing capacity of a natural zeolite, molecular sieve 5A and activated carbon (Yamashita, 2000). The condensation of water was found to lead to the release of ethylene already adsorbed. The alternative approach found useful was to utilize palladized carbon, a common catalyst used in organic chemistry. This allowed storage of broccoli for a week at 10°C, whereas the product was unsaleable after this time when no ethylene scavenger was used.

More recently, 1-methylcyclopropene (1-MCP) vapor has been found to inhibit the hormonal effect of ethylene on the senescence of produce. Inserts that release 1-MCP into packages have become available commercially, and have some regulatory approval for produce application in the USA.

Active packaging for fresh meat

Packaged unprocessed meats are supplied in several forms, including fresh retail cuts, MAP cuts, and chilled vacuum-packed primals. These packages often exhibit ambient, increased and decreased levels of oxygen respectively. The MAP and vacuum-packaged cuts feature raised carbon dioxide levels. The major chemical route to quality loss involves the formation of brown metmyoglobin at oxygen levels of between around 0.1% and 2.0%. Inclusion of oxygen-scavenging sachets which also have the capacity to release some carbon dioxide enables further enhancement of the maintenance of ideal packaging conditions. Earlier forms of active packaging involved the inclusion of a bicarbonate and an organic acid in pads which absorb weep from cut meat. More recent oxygen scavenging has involved the addition of water to sachets to trigger more rapid oxygen scavenging, because the metmyoglobin formation is rapid and irreversible in retail packs. Oxygen-scavenging plastics have the potential to contribute substantially to removal of oxygen originating from occlusion and permeation of the film material.

The quality of fresh meat is also limited by the growth of slime-forming bacteria, and recent research aimed at providing packaging films that release organic acids offers potential for reducing this effect. Films that release lactic acid are particularly attractive, as this acid is normally present in the meat and can be effective when applied at the cut surface.

Impact on packaging materials and processes

Creation of activity in packaging materials frequently involves the introduction of additional components into otherwise inactive materials, typically plastics. Besides the intended effect, there may be unintended consequences that must be addressed in order to bring the package to market. These additional considerations impact on material properties and the manner in which they must be used either in order to achieve their desired effect or to be processable in the expected manner.

Material properties

Active components may form homogeneous mixtures with the existing plastic, or may occur in a separate phase. The range of options available has been described in detail (Rooney, 2000). Some active polymers may be used as polyblends whereas low-molecular weight additives may dissolve in the base polymer. Components that must be used as powders may interfere with the optical properties of the plastic as well as the propensity for tearing. Some hard particles can lead to reduced sharpness of slitter blades, and possibly increase the wear in extruder barrels.

Chemical effects may impact on material properties where the activity involves substantial chemical change to the active polymer. Such chemical changes may lead to changed compatibility, or degradation (or cross-linking) of the host polymer if the activity involves free-radical reaction mechanisms.

When silicone oxygen-scavenging films based on singlet oxygen reactions were being developed in the author's laboratory, exudation of an insoluble oxidation product was observed within a few hours when an otherwise soluble substrate was oxidized. Such an occurrence is particularly noticeable in silicones that are often poor solvents for organic compounds. In some forms of active packaging, sustained migration of substances in or out of a packaging plastic is necessary to achieve the desired effect. Such systems include, for instance, the release of antimicrobials, flavors or anti-oxidants, and uptake of water vapor, oxygen, ethylene or taints (Floros *et al.*, 1997). The polymer required to support these effects may therefore differ from that used in a plastic with an otherwise similar passive role. The latter roles include that of a heat-seal or barrier layer in a pouch, or a closure liner in bottles. Restrictions imposed by the diffusion of selected species in polymers normally result in the active layer being as close as practicable to the food.

Premature activity

Some active components are unaffected by the presence of air in the distribution environment. Such components include ethylene absorbers, odor absorbers and the like. However, when the air contacts a packaging material containing a compound reactive with oxygen or water vapor, there is the opportunity for premature commencement of activity, resulting in a lower effectiveness when the package is used. For this reason, the concept of activation (or triggering) has been the subject of much patenting. Oxygen-scavenging plastics are now commonly triggered by exposure to ultraviolet light (Rooney and Holland, 1979; Speer *et al.*, 1993; Ching *et al.*, 1994). Active systems consisting of sachets containing desiccants or reduced iron powder are normally kept free from air in barrier laminate pouches, as they become active when exposed to the relevant air components.

Process adaptation

Active packaging may require some level of adaptation at the stages of plastic extrusion, package fabrication, or when the pack is filled and sealed. If the chemistry causing activity is inhibited by the presence of antioxidants, it may be necessary to control their addition more during extrusion. Some additives would be unstable at extrusion temperatures (200–370°C), and so it is likely that they will be applied to the packaging as internal coatings. It has been reported recently that a flavor additive manufacturer in the US, ScentSational, has developed encapsulated food aromas in a form that does not degrade under normal plastics extrusion conditions (Ver-Bruggen, 2003a).

It is more likely that adaptation will be required in the package converting or shortly before filling. This occurs with the insertion of sachets or the application of self-adhesive labels to lidding film for thermoformed trays. Triggering of oxygen-scavenging systems by means of brief exposure to UV light may occur at the film stage, with empty packs as with PET bottles, or after package filling. Various irradiation systems have been described in patents (Rooney, 1993; Speer *et al.*, 1993; Ching *et al.*, 1994).

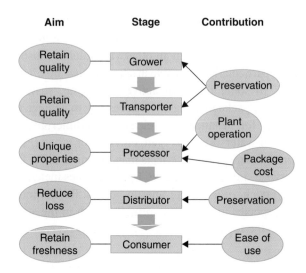

Figure 5.2   The distribution chain: targets and opportunities for active packaging.

Active packaging and the distribution chain

The direct effects of active packaging systems discussed above can be used to differing extents by the participants in the food and beverage distribution chain. The wide
range of technologies, either available or in the process of development, will deliver
benefits that depend upon both the aim of the participant and the drivers that participant
sees in the particular business. This distribution of desired outputs and the prospects
for active packaging inputs are shown diagrammatically in Figure 5.2.

The food chain may be followed from the grower to the consumer. The grower and
distributor need primarily to retain harvested quality until the product reaches (for
instance) the processor. Optimum quality retention may be achieved by selection of
packaging with some activity, such as humidity buffering or ethylene scavenging.
Processors need to differentiate their products from those made by others from the
same raw materials, and so will be expecting contribution from simplification of plant
operations – such as by eliminating juice debittering operations and replacing this
with active packaging. Cost savings on packaging operations, such as reduced
vacuum use, and packaging material cost would be other inputs.

The potential for benefits to be generated along the distribution chain can be visualized using the topics in Figure 5.2, or from other considerations. Space limitations prevent showing the possible benefits, such as being able to market a difficult
fish-oil product because of the oxygen scavenging possible with redesigned
packaging. Another benefit not shown is the possibility of better conformity with
regulations, as with removal of aluminum foil from aseptic brick packs and
replacement with a barrier plastic with an enhanced barrier due to oxygen
scavenging.

Regulatory environment

Integration of active and passive technologies will be assisted when the actions of regulators have been clearly understood by potential petitioners. The European Commission took a step in this direction by funding the Actipack Project (de Jong, 2003). This project involved the coordinated work of six research institutes and three commercial enterprises in evaluating existing active technologies and classifying them in terms of their regulatory features. The migration of ingredients from some active packaging materials was investigated, and results for two iron-based oxygen absorbers have been reported (Lopez-Cervantes *et al.*, 2003). The outcome of the research project was a series of proposals forming a new Framework Directive to replace 89/109/EEC (Ver-Bruggen, 2003b). These amendments, if implemented, will provide clear mechanisms for introduction of further commercial developments in the field. Consideration of these proposed amendments is not expected before late 2004 (Ver-Bruggen, 2003a).

Several packaging technologies based on active concepts have been approved by the USFDA without change to the normal Premarket Notification process. Brody *et al.* (2001) have noted the benefit of the earlier FDA approval of sandwiching post-consumer PET recyclate between virgin PET in their later approval of some oxygen-scavenging compositions. Some zeolites containing silver ions have been approved for use in plastics packaging for food in Japan, and, by the FDA, in the USA. The requirements of the EPA must be considered in the USA, and the position of antimicrobial compositions may be less clear there at present. Additives to packaging plastics may have implications for the environment when the package is returned to the earth via landfill or incineration, so regulations of the EPAs are a consideration in formulating packaging.

References

Abe, A. (1990). *Proceedings of the International Conference on Modified Atmosphere Packaging, Part 1*. Stratford-on-Avon, UK.

Anonymous (2000). *Verpackungs Rundschau* **51(2)**, 6–7.

Brodie, V. and Visioli, D. (1994). Aldehyde scavenging compositions and methods relating thereto. US Patent 5,284,892.

Brody, A. L., Strupinsky, E. R. and Kline, L. R. (2001). *Active Packaging for Food Applications*. Technomic, Lancaster, PA.

Chamara, D., Illeperuma, K., Galapatty, P. T. and Sarananda, K. H. (2000). *J. Hort. Sci. Biotechnol.* **75**, 92–96.

Chamara, D., Illeperuma, K. and Nikapitiya, C. (2002). *Fruits* **57**, 287–295.

Chandler, B. V. and Johnson, R. L. (1979). *J. Sci. Food Agric.* **30**, 825–832.

Chandler, B. V., Kefford, J. F. and Ziemelis, G. (1968). *J. Sci. Food Agric.* **19**, 83–86.

Ching, T. Y., Katsumoto, K., Current, S. and Theard, L. (1994). Ethylenic oxygen scavenging compositions and processes for making same by transesterification in a reactive extruder. International Patent Application PCT/US94/07854.

Cochran, M. A., Folland, R., Nicholas, J. W. and Robinson, M. E. (1991). *Packaging*. US Patent No.5,021,515.

de Jong, A. (2003). *Food Packag. Bull.* **12(3&4)**, 2–5.

Floros, J. D., Dock, L. L. and Han, J. H. (1997). *Food Cosmetics and Drug Packaging* **January,** 10–17.

Gaglio, L. (1986). In: *Wiley Encyclopedia of Packaging Technology* (M. Bakker, ed.), pp. 691–692. John Wiley & Sons, New York, NY.

Gontard, N. (ed.) (2000). *Les emballages actifs*. Editions TEC & DOC, Paris, France.

Han, J. H. (2000). *Food Technol.* **54(3),** 54–65.

Holland, R. V. (1992). Absorbent material and uses thereof. Australian Patent application No.PJ6333.

Hotchkiss, J. H. (1994). In: *Abstracts of the 27th Annual Convention*. Australian Institute of Food Science and Technology, North Sydney, Australia.

Kader, A., Zagory, D. and Kerbel, E. L. (1989). *Crit. Rev. Food. Sci. Nutr.* **28,** 1–30.

King, J. (1955). *Food Manuf.* **30,** 441.

Kuhn, P. E., Weinke, K. F. and Zimmerman, P. L. (1970). *Proceedings of the 9th Milk Concentrates Conference, Pennsylvania State University, 15 September*.

Labuza, T. P. (1987). *Proceedings of the Icelandic Conference on Nutritional Impact of Food Processing,* Reykjavik, Iceland.

Labuza, T. P. and Breene, W. M. (1989). *J. Food Proc. Preserv.* **13,** 1–69.

Lopez-Cervantes, J., Sanchez-Machado, D. I., Pastorelli, S., Rijk, R., Paseiro Louis, P. J. and de Leiris, J-P. (1991). *Active Packaging*. Etude technologique, International Packaging Club, Paris, France.

Meroni, A. (2001). *Food Packag. Bull.* **10(4&5),** 2–7.

Rooney, M. L. (1993). Oxygen scavengers independent of transition metal catalysts. International Patent application PCT/AU93/00598.

Rooney, M. L. (1995a). In: *Active Food Packaging* (M. L. Rooney, ed.), pp. 143–172. Blackie, Glasgow, UK.

Rooney, M. L. (1995b). In: *Active Food Packaging* (M. L. Rooney, ed.), p. 76. Blackie, Glasgow, UK.

Rooney, M. L. (2000). In: *Materials and Development of Plastics Packaging for the Consumer Market* (G. A. Giles and D. R. Bain, eds), pp. 105–129. Sheffield Academic Press, Sheffield, UK.

Rooney, M. L. and Holland, R. V. (1979). *Chem. Ind.* 1979, 900–901.

Russo, J. R. (1986). *Packaging (Chicago)* **31(8),** 26–32, 34.

Sacharow, S. (1988). *Prepared Foods,* **May,** 121–122.

Sakakibara, Y. (2000). *Proceedings of the International Conference on Active Intelligent Packaging*. Campden & Chorleywood Food Research Association, Chipping Campden, UK.

Shirazi, A. and Cameron, A. C. (1992). *Hort. Sci.* **13,** 565–569.

Smith, J. P., Hoshino, J. and Abe, Y. (1995). In: *Active Food Packaging* (M. L. Rooney, ed.), pp. 143–172. Blackie, Glasgow, UK.

Soares, N. F. F. and Hotchkiss, J. H. (1998). *J. Food Sci.* **63,** 61–65.

Speer, D. V., Roberts, W. P. and Morgan, C. R. (1993). Methods and compositions for oxygen scavenging. US Patent 5211875.

Stewart, R. F., Mohr, J. M., Budd, E. A., Lok, X. P. and Anul, J. (1994). In: *Polymeric Delivery Systems – Properties and Applications* (M. A. El-Nokaly, D. M. Pratt and B. A. Charpentie, eds). ACS Symposium Series No. 520, American Chemical Society, Washington, DC.

Tallgren, H. (1938). Keeping food in closed containers with water carrier and oxidizable agents such as zinc dust, Fe powder, Mn dust etc. British Patent 496935.

Teumac, F. N. (1995). In: *Active Food Packaging* (M. L. Rooney, ed.), pp. 193–202. Blackie, Glasgow, UK.

Varoquaux, P. (2000). In: *Les emballages actifs* (N. Gontard, ed.), pp. 89–108. Editions TEC & DOC, Paris, France.

Ver-Bruggen, S. (2003a). *Active & Intelligent Packaging News* **18,** 6–7.

Ver-Bruggen, S. (2003b) *Active & Intelligent Pack News* **18,** 1.

Warmbier, H. C. and Wolf, M. J. (1976*). Modern Packaging* **49,** 38.

Yamashita, I. (2000). In *Les emballages actifs* (N. Gontard, ed.), pp. 219–227. Editions TEC & DOC, Paris, France.

6 Antimicrobial packaging systems

Jung H. Han

Introduction

The quality of foods has been defined as their degree of excellence, and includes factors of taste, appearance and nutritional content. Quality is the composite of characteristics that have significance and make for acceptability. However, acceptability can be highly subjective. Based on deterioration factors and determination procedures, quality may include various aspects such as sensory quality, microbial quality, and toxicological quality. These aspects are not separated from one another – for example, microbial contamination damages sensory quality and safety. Microbial food quality relates to all three groups of factors, since the growth of bacteria generates undesirable odors and life-threatening toxins, changes the color, taste and texture of food, and also reduces the shelf life of the product. Microbial growth on packaged foods significantly decreases the safety of food and the security of public health. In our society, fast foods, convenience foods and fresh foods are essential elements. Many food industries produce minimally processed foods, precut fruits/vegetables, and ready-to-eat foods for the purpose of maximized convenience and freshness. In these environments of food manufacturing on a mass-production scale, food safety is a top priority and issue for producers, the food industry, governments, and public consumers. Improper treatment and accidental cross-contamination of foods can cause major problems of recall and serious food-borne illness. Furthermore, the safety of food and related public health issues can be jeopardized by malicious tampering, extortion for benefits, trials to obtain public attentions for any reasons, and terrorism (Kelly *et al.*, 2003). Since

Innovations in Food Packaging
ISBN: 0-12-311632-5

the 2001 tragedy of the World Trade Center the consideration of food and water safety has assumed major importance, and the security of food chains against bioterrorism is regarded as a significant aspect of public safety (Nestle, 2003). Food safety has become a significant subject of trends in food packaging, logistics, trade, and consumer studies.

Food safety

Spoilage of food products

Most foods are perishable. Spoilage of foods is caused both biologically and chemically. In addition to the chemical degradation of food ingredients through oxidation processes, most spoilage processes are due to biological reasons such as auto-degradation of tissues by enzymes, viral contamination, protozoa and parasite contamination, microbial contamination, and loss by rodents and insects. The growth of micro-organisms is the major problem of food spoilage leading to degraded quality, shortened shelf life, and changes in natural microflora that could induce pathogenic problems. Microbial spoilage of food products is caused by many bacteria, yeast and molds; however, their sensitivities to the spoilage are dependent on nutrients, pH, water activity, and presence of oxygen. Therefore, the many different potential micro-organisms that could contaminate food products and the various growth environments present difficult problems in preventing the spoilage.

For the food industries, the prevention of food spoilage is a very important issue in determining profit. Furthermore, reducing food spoilage can prolong the shelf life of food products and accordingly extend market boundary, resulting in increased profit.

Food-borne illness

Food-borne illness is caused by contamination of food products by micro-organisms or microbial toxins. Many food-processing technologies have been developed to prevent the contamination and inactivation of pathogens. Traditionally, thermal processes such as blanching, pasteurization, and commercial sterilization have been used for the elimination of pathogens from food products. Currently, various new non-thermal technologies are being studied to assess their effectiveness and mechanisms. Such new processes include irradiation, pulsed electric fields, high-pressure processes, and the use of new antimicrobial agents. However, these new technologies cannot completely prevent the contamination and/or growth of pathogens. Some of these methods still require regulatory permission for their commercial use. It is also very hard to control pathogens because of the wide variety of microbial physiology, pathogenic mechanisms, and passage of contamination; the complexity of food composition; their sensitivity to antimicrobial agents; the mass-production nature of food processing; and difficulties in early detection.

There are many prevention systems that are used by the food industry, such as hazard analysis and critical control points (HACCP), sanitation standard operating procedures (SSOP), good manufacturing practice (GMP), and various inspections. The good practice of these quality systems is much more important and effective in eliminating pathogen problems from food system than is the use of reliable new technologies.

Malicious tampering and bioterrorism

One of primary functions of food packaging is to protect food from unintentional contamination and undesirable chemical reactions, as well as providing physical protection. However, recently, the need has arisen for packaging systems to be capable of the protection of the food from intentional contamination – i.e. malicious tampering (Kelly *et al.*, 2003). To secure food safety from tampering and bioterrorism, many systematic protocols are required – such as new food and drug regulation, and new security acts (e.g., Bioterrorism Acts and Homeland Security guideline for imported goods). Enhanced traceability is also required so that any potentially tampered goods can be removed from the food chain. Security against tampering and terrorism can be improved by systematic preparation, practical regulation, inspection, emergency response training, and other systematic protocols. However, issues such as traceability, food security, and safety enhancement involve technical development, such as a database for tracking products, intelligent packaging for monitoring foreign matters in products, robotic automation for warehousing, and new security seals on distribution packages.

The protective function of food packaging has become more significant with respect to safety enhancement. Among all of the potential technologies, practical approaches to enhance the safety of food products include the use of advanced tamper-evident or tamper-resistant packaging, and intelligent packaging that indicates any tampering and contamination and has a high-barrier design (Rodrigues and Han, 2003). The use of tamper-evident packaging and intelligent packaging is highly desirable to minimize the risk of the malicious activities of terrorists endangering public safety through the food product chain. In addition to the visual indication of disintegration and contamination of packages, antimicrobial packaging systems can kill or inhibit the growth of pathogenic micro-organisms that may be injected into packaged foods without any visual evidence of tampering. With the positive contribution to food safety of the new smart design of food packaging, antimicrobial packaging systems can also protect food products more actively.

Tamper-evident packaging and intelligent packaging are considered to be passive monitoring systems. Antimicrobial packaging can reduce the risk of tampering and bioterrorism by eliminating contaminating micro-organisms as well as maintaining the quality of packaged food products by reducing the potential for cross-contamination of food products with spoilage and pathogenic micro-organisms.

Antimicrobial packaging

Antimicrobial packaging is a system that can kill or inhibit the growth of micro-organisms and thus extend the shelf life of perishable products and enhance the safety of packaged products (Han, 2000). Antimicrobial packaging can kill or inhibit target micro-organisms (Han, 2000, 2003a). Among many applications such as oxygen-scavenging packaging and moisture-control packaging, antimicrobial packaging is one of the most promising innovations of active packaging technologies (Floros *et al.*, 1997). It can be constructed by using antimicrobial packaging materials and/or antimicrobial agents inside the package space or inside foods. Most food packaging systems consist of the food products, the headspace atmosphere and the packaging materials. Any one of these three components of food packaging systems could possess an antimicrobial element to increase antimicrobial efficiency.

Antimicrobial packaging research generally started with the development of antimicrobial packaging materials that contain antimicrobial chemicals in their macromolecular structures. However, without the use of alternative packaging materials, common packaging materials can be utilized for antimicrobial packaging systems when there is antimicrobial activity in packaged foods or in the in-package atmosphere. Edible antimicrobial agents can be incorporated into food ingredients, while antimicrobial resources can be interleaved in the in-package headspace in the form of sachets, films, sheets or any in-package supplements, to generate antimicrobial atmospheres.

Besides the use of antimicrobial packaging materials or antimicrobial inserts in the package headspace, gaseous agents have been used to inhibit the growth of micro-organisms. Common gases are carbon dioxide for modified atmosphere packaging, sulfur dioxide for berries, and ethanol vapor for confections. These gases are injected into the package headspace or into palletized cases after shrink-wrapping of a unit load on a pallet. Vacuum, nitrogen-flushing and oxygen-scavenging packaging, which were originally designed for preventing the oxidation of packaged foods, also possess antifungal and antimicrobial properties against aerobic bacteria as a secondary function, since these micro-organisms are restrictively aerobic (Smith *et al.*, 1990; Han, 2003b). However, these technologies, which control the low oxygen concentration to inhibit the growth of aerobic micro-organisms, could cause the onset of anaerobic microbial growth. Controlling anaerobic bacteria in modified atmosphere packaging is a very important issue in maintaining the quality and safety of the products.

Antimicrobial agents

Various antimicrobial agents could be incorporated into conventional food packaging systems and materials to create new antimicrobial packaging systems. Table 6.1 shows potential antimicrobial agents and food-grade preservatives. They can generally be classified into three groups; chemical agents, natural agents, and probiotics.

Chemical antimicrobial agents

For the purpose of food preservation, all packaging ingredients should be food-grade additives. The chemical agents can be mixed with food ingredients, incorporated into packaging additives or inserted into the headspace atmosphere. The antimicrobial agents are in contact with and consumed with the food products in these applications.

Table 6.1 Examples of potential antimicrobial agents for antimicrobial food packaging systems

Classification	Antimicrobial agents
Organic acids	Acetic acid, benzoic acid, lactic acid, citric acid, malic acid, propionic acid, sorbic acid, succinic acid, tartaric acid, mixture of organic acids
Acid salts	Potassium sorbate, sodium benzoate
Acid anhydrides	Sorbic anhydride, benzoic anhydride
Para benzoic acids	Propyl paraben, methyl paraben, ethyl paraben
Alcohol	Ethanol
Bacteriocins	Nisin, pediocin, subtilin, lacticin
Fatty acids	Lauric acid, palmitoleic acid
Fatty acid esters	Glycerol mono-laurate
Chelating agents	EDTA, citrate, lactoferrin
Enzymes	Lysozyme, glucose oxidase, lactoperoxidase
Metals	Silver, copper, zirconium
Antioxidants	BHA, BHT, TBHQ, iron salts
Antibiotic	Natamycin
Fungicides	Benomyl, Imazalil, sulfur dioxide
Sanitizing gas	Ozone, chlorine dioxide, carbon monoxide, carbon dioxide
Sanitizers	Cetyl pyridinium chloride, acidified NaCl, triclosan
Polysaccharide	Chitosan
Phenolics	Catechin, cresol, hydroquinone
Plant volatiles	Allyl isothiocyanate, cinnamaldehyde, eugenol, linalool, terpineol, thymol, carvacrol, pinene
Plant/spice extracts	Grape seed extract, grapefruit seed extract, hop beta acid, Brassica erucic acid oil, rosemary oil, oregano oil, basil oil, other herb/spice extracts and their oils
Probiotics	Lactic acid bacteria

Modified from Han (2000, 2003a, 2003b); Suppakul *et al.* (2003a).

Therefore, the chemical antimicrobial agents should be controlled as food ingredients regardless of where the chemical antimicrobial agents were positioned initially – in the food products, in the packaging materials, or in the package headspace atmosphere. In the case of non-food-grade chemicals, the only way to incorporate the chemical into the food packaging system is through the chemical binding of the antimicrobial agents to packaging material polymers (immobilization). In this case, the migration or residual amount of the non-food-grade chemical in the food products is prohibited by regulation. Therefore, it is necessary to verify that there is no migration of the chemical from packaging materials to foods, and there is no residual free chemical after the immobilization reaction. There will be detailed explanation of the immobilization system later in this chapter.

The most common chemical antimicrobials used by researchers are the various organic acids. Organic acids are widely used as chemical antimicrobial agents because their efficacy is generally well understood and cost effective. Many organic acids, including fatty acids, are naturally existing chemicals and have been used historically. Currently, most of them are produced by chemical synthesis or chemically modified from natural acids. Organic acids have characteristic sensitivities to micro-organisms. For example, sorbic acid and sorbates are very strong antifungal agents, while their antibacterial activities are not effective – they have various antimicrobial mechanisms. Therefore, the correct selection of organic acids is essential to have effective antimicrobial agents. Mixtures of organic acids have a wider antimicrobial spectrum and stronger activity than a single organic acid.

Fungicides are also common antimicrobial agents. Imazalil has been incorporated into the wax coating of oranges and other citrus fruits. Since fungicides are not permitted as a direct food preservative, they cannot be mixed into food ingredients or added to food-contact packaging materials as food-contact substances. Therefore, it is necessary to design antimicrobial food packaging systems when non-food-grade antimicrobial agents, such as fungicides, are used. Food sanitizers and another chemical antimicrobial groups are included in Table 6.1. They are food cleansing agents, food contact substances or food contact surface sanitizers. Residual food sanitizers on foods are permitted, with specific control limits. Thus the use of food sanitizers has many advantages over the use of other non-food-grade antimicrobial agents such as fungicides.

Natural antimicrobial agents

Table 6.1 includes herb extracts, spices, enzymes, and bacteriocins as naturally occurring antimicrobial agents. Looking to the consumers' demand for chemical preservative-free foods, food manufacturers are now using naturally occurring antimicrobials to sterilize and/or extend the shelf life of foods. Herb and spice extracts contain multiple natural compounds, and are known to have a wide antimicrobial spectrum against various micro-organisms. Apart from antimicrobial activity, other advantages they offer include antioxidative activity and their effect as alternative medicines. However, their mode of action and kinetics are generally unknown, and their chemical stability is also of concern. In addition, they create some problems with respect to flavors.

Specificity of enzymes should be considered carefully, since antimicrobial activity is very sensitive to the environments and substrates. For an example, the activity of lysozyme can be significantly affected by temperature and pH. In most cases, lysozyme is not effective against gram-negative bacteria. This is due to the complex cell wall structure of gram-negative bacteria and the specificity of lysozyme for peptidoglycan.

Various bacteriocins, such as nisin, pediocin, lacticin, propionicin etc., can be incorporated into foods and/or food packaging systems to inhibit the growth of spoilage and pathogenic micro-organisms (Daeschul, 1989). The extracted bacteriocins, which are generally small molecular weight peptides, can be utilized in various ways; however, it is very important to characterize their resistance to thermal treatment and pH. In the case of fermented food products, live bacteria which produce bacteriocins can be intentionally added as probiotics in the packaged food system to obtain antimicrobial effectiveness.

Probiotics

Table 6.1 also shows the possible use of probiotics (*Lactobacillus reuteri*) to control *E. coli* O157:H7 (Muthukumarasamy *et al.*, 2003). Various micro-organisms, e.g. lactic acid bacteria, produce bacteriocins and non-peptide growth-inhibiting chemicals such as reuterin. These naturally produced antimicrobials can inhibit the growth of other bacteria. Use of probiotics can therefore effectively control the competitive undesirable micro-organisms. Many traditional fermented food products contain antimicrobial probiotics. There has been much research and development regarding the function of antimicrobial probiotics for the preservation of fermented foods. Currently there is only limited research into the use of probiotics for the purpose of antimicrobial packaging design. With the new technology development for the delivery of live probiotics, the use of probiotics as an antimicrobial source for antimicrobial food packaging will be more popular in future due to its safety and effectiveness.

System design

There are various factors to be considered in designing antimicrobial systems. Antimicrobial systems can be constructed by using antimicrobial packaging materials, antimicrobial inserts (such as sachets) to generate antimicrobial atmosphere conditions inside packages, or antimicrobial edible food ingredients in the formulation of foods. Since antimicrobial packaging systems are designed to control the growth of micro-organisms in packaged foods, the systems essentially consist of packaging materials (or packages), foods, the in-package atmosphere, target micro-organisms, and antimicrobial agents. These five elements are related to one another and to the final system design features.

To study the effectiveness of antimicrobial food packaging systems with respect to the relative effects of these elements, many choices of combinations should be examined using real food systems. The majority of research has been conducted using culture media, which provides the richest nutritional quality and the most favorable environment for microbial growth, and antimicrobial packaging materials. Most micro-organisms in culture media are not stressed compared to micro-organisms under normal conditions in foods. Many antimicrobial systems that have shown strong antimicrobial activity against target micro-organisms in culture media do not demonstrate the same antimicrobial activity when they are actually in the food systems. This is very common, and clearly shows the interactive effects of food ingredients, micro-organisms, and antimicrobial agents. All of these factors have complex systems that cannot be explained by a single chemical mechanism. Therefore, it is strongly recommended that experiments regarding the efficiency examination of an antimicrobial packaging system should be conducted using a real food instead of culture broth or agar media. Table 6.2 lists examples of antimicrobial food packaging systems which have been tried, mostly by university researchers, over the past two decades.

Antimicrobial mechanisms

An antimicrobial agent has specific inhibitory activity and mechanisms against each micro-organism. Therefore, the selection of antimicrobial agents is dependent on their efficacy against a target micro-organism. There is no "magic bullet" antimicrobial agent that will work effectively against all spoilage and pathogenic micro-organisms, because all antimicrobial agents have different activities which affect micro-organisms

Table 6.2 Antimicrobial food packaging systems constructed by researchers

Antimicrobial agents	Packaging materials	Foods	Micro-organisms	Researchers
Organic acids				
Benzoic acids	PE	Tilapia fillets	Total bacteria	Huang *et al.*, 1997
	Ionomer	Culture media	*Pen. spp., Asp. niger*	Weng *et al.*, 1997
Para-benzoate	LDPE	Simulants	Migration test	Dobias *et al.*, 2000
	PE coating	Simulants	Migration test	Chung *et al.*, 2001a
	Styrene-acrylate	Culture media	*S. cerevisiae*	Chung *et al.*, 2001b
Benzoic and sorbic acids	PE-co-Met-acrylate	Culture media	*Asp. niger, Penn. spp.*	Weng *et al.*, 1999
Sorbates	LDPE	Culture media	*S. cerevisiae*	Han and Floros, 1997
	OE, BOPP, PET	Water, cheese	Migration test	Han and Floros, 1998a, 1998b
	LDPE	Cheese	Yeast, mold	Devlieghere *et al.*, 2000a
	MC/palmitic acid	Water	Migration test	Rico-Pena and Torres, 1997
	MC/HPMC/fatty acid	Water	Migration test	Vojdani and Torres, 1990
	MC/chitosan	Culture media		Chen *et al.*, 1996
	Starch/glycerol	Chicken breast		Baron and Sumner, 1993
	WPI	Culture media	*S. cerevisiae, Asp. niger, Penn. rogueforti*	Ozdermir, 1999
	CMC/paper	Cheese		Ghosh *et al.*, 1973, 1977
	PE	Culture media	*S. cerevisiae*, molds	Weng and Chen, 1997; Weng and Hotchkiss, 1993
Sorbate and propionates	PE/foil	Apples	Firmness test	Yakovleva *et al.*, 1999
Acetic, propionic acid	Chitosan	Water	Migration test	Outtara *et al.*, 2000a
Enzymes				
Lysozyme	PVOH	Water	Migration test	Buonocore *et al.*, 2003
Lysozyme, nisin, EDTA	SPI, zein	Culture media	*E. coli, Lactobacillus plantarum*	Padgett *et al.*, 1998
Lysozyme, nisin, propyl paraben, EDTA	WPI	Culture media	*Lis. monocytogenes, Sal. typhimurium, E. coli* O157:H7, *B. thermosphacta, S. aureus*	Rodrigues and Han, 2000; Rodrigues *et al.*, 2000
Immobilized lysozyme	PVOH, nylon, cellulose acetate	Culture media	Lysozyme activity test	Appendini and Hotchkiss, 1996, 1997
Glucose oxidase		Fish		Fields *et al.* 1986
Bacteriocins				
Nisin	PE	Beef	*B. thermosphacta*	Siragusa *et al.*, 1999
	HPMC	Culture media	*Lis. monocytogenes, Staphyl. aureus*	Coma *et al.*, 2001
	HPMC/stearic acid	Culture media	*Lis. monocytogenes, S. aureus*	Sebti *et al.*, 2002
	Corn zein	Shredded cheese	Total aerobes	Cooksey *et al.*, 2000
	Corn zein, wheat gluten	Culture media	*Lactobacillus plantarum*	Dawson *et al.*, 2003
	Ethylene-co-acrylic	Culture media	*Lactobacillus leichmannii*	Leung *et al.*, 2003
Nisin, lacticins	LDPE, poly-amide	Culture media	*M. flavus, Lis. monocytogenes*	An *et al.*, 2000
	LDPE, poly-amide	Oyster, beef	Total aerobes, coli-form	Kim *et al.*, 2002

(Continued)

Table 6.2 *(Continued)*

Antimicrobial agents	Packaging materials	Foods	Micro-organisms	Researchers
Nisin, EDTA	PE, PE-PE oxide	Beef	*B. thermosphacta*	Cutter *et al.*, 2001
Nisin, citrate, EDTA	PVC, nylon, LLDPE	Chicken	*Sal. typhimurium*	Tatrajan and Sheldon, 2000
Nisin, organic acids mixture	Acrylics, PVA-co-PE	Water	Migration test	Choi *et al.*, 2001
Nisin, lauric acid	Zein	Simulants	Migration test	Hoffman *et al.*, 2001
	Soy protein	Turkey bologna	*Lis. monocytogenes*	Dawson *et al.*, 2002
Nisin, pediocin	Cellulose casing	Turkey breast, ham, beef	*Lis. monocytogenes*	Ming *et al.*, 1997
Pediocin	WPI	Culture media	*Lis. innocua*	Quintero-Salazar *et al.*, 2003
Polymers				
Chitosan		Cheese	*Lis. monocytogenes, Lis. innocua*	Coma *et al.*, 2002
	Chitosan/paper	Strawberry	*E. coli*	Yi *et al.*, 1998
Chitosan, herb extracts	LDPE	Culture media	*Lb. plantarum, E. coli, S. cerevisiae, Fusarium oxysporum*	Hong *et al.*, 2000
Chitosan acetate		Culture media	*E. coli, Vibrio vulnificus, Sal. typhimirium, Sal. enteritidis; Shigella sonnei*	Park *et al.*, 2003
UV irradiation, excimer laser	Nylon	Culture media	*Pseudomonas fluorescens, Enterococcus faecalis, S. aureus*	Paik *et al.*, 1998; Paik and Kelly, 1995
Natural extracts				
Grapefruit seed extract	LDPE, nylon	Ground beef	Total aerobes, coli-form bacteria	Ha *et al.*, 2001
	LDPE	Lettuce, soy-sprouts	*E. coli, S. aureus*	Lee *et al.*, 1998
Grapefruit seed extract, lysozyme, nisin	Na-alginate, k-carrageenan	Culture media		Cha *et al.*, 2002
Clove extract	LDPE	Culture media	*L. plantarum, F. oxysporum, E. coli, S. cerevisiae*	Hong *et al.*, 2000
Herb extract, Ag-zirconium	LDPE	Lettuce, cucumber	*E. coli, S. aureus, Leu. mesenteroides, S. cerevisiae, Asp. niger, Asp. oryzae, Pen. chrysogenum*	An *et al.*, 1998
	LDPE	Strawberry	Firmness test	Chung *et al.*, 1998
Cinnam-aldehyde, eugenol, organic acid	Chitosan	Bologna, ham	*Enterobac.*, Lactic acid bacteria, *Lb. sakei, Serratia. spp.*	Outtara *et al.*, 2000a, 2000b
Horseradish oil	Paper in pouch	Ground beef	*E. coli* O157:H7	Nadarajah *et al.*, 2002, 2003
Horseradish extract and *Lactobac. reuteri* (probiotics)	PE/EVOH/PET pouch	Ground beef	*E. coli* O157:H7	Muthukumarasamy *et al.*, 2003
Allyl isothiocyanate	PE film/pad	Chicken, meats, smoked salmon	*E. coli, S. enteritidis, Lis. monocytogenes*	Takeuchi and Yuan, 2002
Green tea extract (catechins)	PVA/starch	Culture media	*E. coli*	Chen *et al.*, 2003
Basil extract	LDPE	Culture media	*E. coli*	Suppakul *et al.*, 2003b
Others				
Benomyl	Ionomer	Culture media		Halek and Garg, 1989
Imazalil	LDPE	Bell pepper		Miller *et al.*, *1984*

Table 6.2 *(Continued)*

Antimicrobial agents	Packaging materials	Foods	Micro-organisms	Researchers
	LDPE	Cheese		Weng and Hotchkiss, 1992
Ageless	Sachet	Bread	Molds	Smith et al., 1989
BHT	HDPE	Breakfast cereal		Hoojjatt et al., 1987
Ethanol	Silicagel sachet	Culture media		Shapero et al., 1978
	Silicon oxide sachet	Bakery		Smith et al., 1987
Hinokithiol	Cyclodextrin sachet	Bakery		Gontard, 1997
Chlorine dioxide	Plastic films		Migration test	Ozen and Floros, 2001
Hexanal, hexenal, hexyl acetate	Modified atmosphere packaging	Sliced apple	E. coli, Sal. enteritidis, Lis. monocytogenes	Lanciotti et al., 2003
Carbon monoxide	Modified atmosphere packaging	Pork chops	Total bacteria, lactic acid bacteria	Krause et al., 2003
Triclosan	Styrene-co-acetate	Culture media	Enterococcus faecalis	Chung et al., 2003
	LDPE	Chicken breast	Lis. monocytogenes, S. aureus, Sal. enteritidis, E. coli O157:H7	Vermeiren et al., 2002
Hexamethylenetetramine	LDPE	Orange juice	Yeast, lactic acid bacteria	Devlieghere et al., 2000b
Silver zeolite, silver nitrate	LDPE	Culture media	S. cerevisiae, E. coli, S. aureus, Sal. typhimurium, Vibrio parahaemolyticus	Ishitani, 1995
Antibiotics	PE	Culture media	Sal. typhimurium, Klebsiella pheumoniae, E. coli, S. aureus	Han and Moon, 2002

MC, methyl cellulose; HPMC, hydroxypropyl methyl cellulose; WPI, whey protein isolate; CMC, carboxyl methyl cellulose; SPI, soy protein isolate; PE, poly ethylene; PVA, polyvinyl acetate; EVOH, ethylene vinyl alcohol; PVOH, polyvinyl alcohol.

differently. This is due to the characteristic antimicrobial mechanisms and the differences in physiology of the micro-organisms. Simple categorization of micro-organisms may be very helpful to select specific antimicrobial agents, which may be categorized by oxygen requirement (aerobes or anaerobes), cell-wall composition (gram-positive and gram-negative), growth-stage (spores or vegetative cells), optimal growth temperature (thermophilic, mesophilic or psychrotrophic), or acid/osmosis resistance. Besides the microbial characteristics, the antimicrobial characteristics of the agent are also important in understanding the efficacy as well as the limits of the activity. For example, some antimicrobial agents inhibit essential metabolic (or reproductive genetic) pathways of micro-organisms, while others alter cell membrane/wall structure. Two major functions of microbial inhibition are microbiocidal and microbiostatic effects.

Microbiocidal

It is expected that antimicrobial packaging systems would kill target spoilage and pathogenic bacteria, since the system could eliminate any micro-organisms from the food/packaging system. Though it is practically very hard to remove all micro-organisms, microbiocidal antimicrobial system may kill the target micro-organisms when the antimicrobial concentration goes above the m.i.c. for a while. With other treatment to encourage the antimicrobial activity of the packaging system, such as refrigeration, the antimicrobial effectiveness will be increased; however, generally refrigeration is not necessary when all of the system factors satisfy the requirements. Refrigeration may be very effective in inhibiting the growth of untargeted (unexpected) micro-organisms. If the initial concentration of microbiocidal antimicrobial systems is lower than the m.i.c. of the target micro-organisms, and the concentration has never been above the m.i.c., the agent may show a microbiostatic instead of a microbiocidal effect. Therefore, it is very important to maintain the antimicrobial concentration above the m.i.c. for certain critical periods, to eliminate the target micro-organisms. If the package has hermetically sealed, the packaged foods may not contain any live micro-organisms even when the concentration decreases to below the m.i.c. due to the migration or loss of the agent after the critical period.

Microbiostatic

Microbiostatic agents can inhibit the growth of micro-organisms above a certain critical concentration (i.e. m.i.c.). However, when the concentration is lower than the critical concentration, or when the agent is removed from the packaging systems through a seal defect, leakage, opening or any other means, the suppressed micro-organisms can grow or their spores germinate. Therefore, it is critical to maintain the concentration of the antimicrobial agent above the m.i.c. during the entire shelf life of the packaged foods. Chemical indicators that show the concentration or microbial growth would be very beneficial in microstatic antimicrobial packaging systems, and this is a concept of intelligent packaging (Rodrigues and Han, 2003).

Functioning modes and volatility

In most packaged solid or semi-solid foods, micro-organisms grow primarily at the surfaces of the foods (Brody et al., 2001). Therefore, antimicrobial activity should be exerted on these surfaces. The antimicrobial activity may exist in packaging materials, in the in-package atmosphere or in the headspace, varying the incorporation methods. The antimicrobial activity should be transferred to the surface of foods to suppress the microbial growth. Therefore, incorporation methods and transferring techniques are critical in designing effective antimicrobial packaging systems. As examples of incorporation methods, antimicrobial agents have been impregnated into packaging materials before final extrusion (Han and Floros, 1997; Nam et al., 2002), dissolved into coating solvents (An et al., 2000), added in edible coating materials (Rodrigues and Han, 2000; Rodrigues et al., 2002), and mixed into sizing/filling materials such as paper and cardboard (Nadarajah et al., 2002). Gaseous antimicrobial agents can also be added to the package atmosphere (Lanciotti et al., 2003; Krause et al., 2003).

The edible coating system has various benefits due to its edibility and biodegradability (Krochta and De Mulder-Johnston, 1997). The edible coating may be either a dry coating or a wet battered coating. Dry coatings can incorporate chemical and natural antimicrobials, and play the role of physical and chemical barriers as well as a microbial barrier (Han, 2001, 2002). Wet coating systems may need another wrapper. However, the wet system can carry many different types of functional agents as well as probiotics and antimicrobials (Gill, 2000). Lactic acid bacteria can be incorporated into the wet coating system and control the competing undesirable bacteria. This new wet coating system may be very beneficial to the fresh produce, meats and poultry industries.

Chemical immobilization covalently binds the agents into the chemical structures of packaging materials when regulations do not permit the migration of agents into foods (Miller *et al.*, 1984; Halek and Garg, 1989; Appendini and Hotchkiss, 1996, 1997). The immobilized antimicrobial agents will inhibit the growth of micro-organisms on the contact surfaces of packaged products.

Non-volatile migration

The mass transfer of non-volatile antimicrobials is dominated by diffusional migration. Non-volatile agents will be positioned initially in the packaging materials or between the package and the surface of the food. If the agent is sprayed onto the surface of food, the initial surface concentration will be very high and starts to decrease due to dissolution and diffusion of the agent towards the center of the food. Therefore, the solubility (or partition coefficient) and diffusion coefficient (diffusivity) of the agent in the food are very important characteristics to maintain the surface concentration above the effective m.i.c. during the expected shelf life. If the agent is incorporated into packaging material initially, it should escape from the packaging material and dissolve into the food before diffusing into the food core. Therefore, the significant characteristic constants of the mass transfer profile are the diffusivity of the agent in the packaging material, the solubility (or partition coefficient) of the agent in the food at the surface, and the diffusivity of the agent in the food. It is important for the food/packaging/antimicrobial agent system to have mass transfer kinetics appropriate to the microbial growth kinetics in order to provide efficient antimicrobial activity. To understand the concentration distribution profile, it is necessary to simulate mass transfer models which have more than two-layer diffusion and interface partitioning (or dissolution). Since the migrating agent is non-volatile, this system requires intact contacting between the packaging materials and the food surface. The food should be a continuous matrix form without significant pores, holes, air gaps or heterogeneous particles, due their interference with the diffusion. One-piece solid, semi-solid (soft solid) foods and liquid products are good examples of the products that can use this non-volatile migrating antimicrobial packaging system.

Practical examples of this system may include cured or fermented meats and sausages battered with antimicrobial agents, natural cheeses sprayed with potassium sorbate before packaging, antimicrobial plastic films for deli products, antimicrobial wax coatings on fruits, and antimicrobial cleansing of fruits/vegetables before packaging. The advantages of this non-volatile migrating system are the simplicity of the system design, which could be installed ahead of the currently existing packaging process without high investment load, and the easy maintenance required to control the effectiveness.

Volatile migration

Many researchers have claimed that it is necessary to have intact contact of the antimicrobial material with the food surface to facilitate the migration of the active agent for maximal effectiveness (Vermeiren *et al.*, 2002; Suppakul *et al.*, 2003a). However, this is not necessary when using volatile antimicrobial agents. To maintain the surface concentration above a certain m.i.c. it is very important to control the headspace gas concentration, since the volatile agent's concentration in the headspace has been equilibrated with the concentration on the food surface and in the packaging materials. Initially the volatile agent is placed in the packaging material, whether it is a film, container, sachet or tray. After packaging the food, the volatile agent is vaporized into the headspace, reaches the surface of the food and is absorbed by the food. The mass transfer of a volatile agent in the packaging system is more dynamically balanced. The release rate of the volatile agent from the packaging system is highly dependent on the volatility, which relates to the chemical interaction between the volatile agent and the packaging materials. There are ways to control the volatility of the agent in the packaging system, including the use of oil, cyclodextrin or microencapsulation. These techniques can control the volatility of the agent and, eventually, the headspace concentration. The absorption rate of headspace volatiles into the food surface is related to the composition of the foods, as the ingredients undergo chemical interactions with the gaseous agents. Since most volatile agents are generally lipophilic, the lipid content of the food is an important factor of the headspace concentration.

The use of volatile antimicrobial agents has many advantages. This system can be used effectively for highly porous foods, powdered, shredded, irregularly shaped, and particulate foods, such as ground beef, shredded cheese, small fruits, mixed vegetables, etc. Because natural herb and spice extracts provide the majority of volatile antimicrobial agents, this system is linked to the nutraceutical research and development area as well as being easily accepted by consumers and governmental regulatory agencies.

Non-migration and absorption

The non-migration system uses non-migratory antimicrobial polymers, where the antimicrobial agent does not migrate out of the polymer due to its covalent attachment to the polymer backbone (Steven and Hotchkiss, 2003). Besides the antimicrobial agents, other bioactive agents (such as enzymes, proteins and other organic compounds) can be attached to the polymer through covalent cross-linkers. Since the agents are not mobile, their activity is limited to the contact surface only. This limitation is more critical in solid or semi-solid foods. However, in liquid foods the disadvantages of this non-migration characteristic may be minimized. This system could be designed for large-size membrane reactors or processing units to convert any pre-existing substrates into valuable compounds using immobilized enzymes. For the future, it is important that this packaging system be evaluated to assess whether it can be used effectively as a unit-operation substituting reaction process during any necessary timed processes such as aging, chilling, tank-holding, etc.

As a food packaging system, this non-migration system has unique advantages in marketing and regulation. Since the active agents are not migrating, this system requires a very small amount of attached agents. It may reduce the overall cost of packaging systems that use very expensive antimicrobial agents. The non-migrating system can include agents that are not permitted as food ingredients or food additives. With the verification of non-migration, the packaging material may contain any food contact substances. From the marketing aspect this system is attractive because the food does not contain any chemical antimicrobial agent throughout its shelf life. However, in contrast to the benefits, this system may have a very limited selection of antimicrobial agents and may also be limited in application to certain types of foods.

Shapes and compositions of systems

Packaging is a system used to contain and protect enclosed products, which consists of a product, a package, and the in-package atmosphere. Antimicrobial agents may be incorporated in the non-food parts of the packaging system, which are the package or the in-package atmosphere. Antimicrobial agents can be incorporated directly in packaging materials in the form of films, over-coating on films, sheets, trays, and containers, or in the in-package space in the form of inserts, sachets or pads. Edible coatings also can contain edible antimicrobial agents, protecting the coated foods from microbial quality degradation (Han, 2001). Figure 6.1 illustrates the possible forms of antimicrobial packaging systems (Han, 2003a).

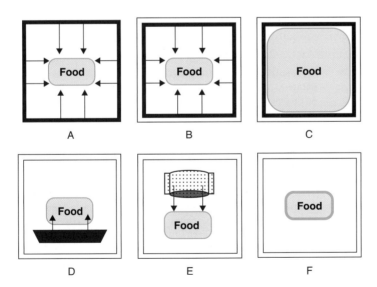

Figure 6.1 Possible ways to construct antimicrobial food packaging systems. A, the use of antimicrobial packaging materials; B, antimicrobial coating on conventional packaging materials; C, immobilization of antimicrobial agents in polymeric packaging materials; D, the use of antimicrobial trays or pads; E, the use of sachet/insert containing volatile antimicrobial agents; F, antimicrobial edible coating on foods (adapted from Han, 2003b).

Commercialization

Some commercial antimicrobial packaging systems are listed in Table 6.3. Most systems consist of silver-containing active agents. Table 6.4 lists potential applications of antimicrobial packaging and food groups. Since foods are complex systems with respect to their composition, there are many factors to be considered in commercializing antimicrobial packaging systems.

Technical factors

Compatibility of process conditions and material characteristics

Film/container casting methods, i.e. extrusion and solvent casting, are important to maintain antimicrobial effectiveness. In the case of extrusion, the critical variables related to the residual antimicrobial activity are extrusion temperature and specific

Table 6.3   Examples of commercial antimicrobial packaging products and manufacturers

Trade name	Active compounds	Manufacturer	References
Piatech	Ag oxide	Daikoku Kasei Co. (Japan)	Brody et al., 2001
Silvi Film	Ag oxide	Nimiko Co. (Japan)	Brody et al., 2001
Okamoto Super Wrap		Okamoto Industries, Inc. (Japan)	Brody et al., 2001
Apacider	Ag zeolite and others	Sangi Co. (Japan)	Brody et al., 2001
Zeomic	Ag zeolite	Shinanen New Ceramics Co. (Japan)	Brody et al., 2001
Bactekiller	Ag zeolite	Kanebo Co. (Japan)	Brody et al., 2001
Zeomix		Mitsubishi Int. Corp. (Japan)	Brody et al., 2001
AgION		AgION Technologies LLC (USA)	Suppakul et al., 2003a; www.agion-tech.com
MicroFree	Ag, copper oxide, zinc silicate	DuPont (USA)	Vermeiren et al., 2002; Brody et al., 2001
Novaron	Ag zirconium phosphate	Milliken Co. (USA)	Vermeiren et al., 2002
Surfacine	Ag-halide	Surfacine Development Co. (USA)	Vermeiren et al., 2002
Ionpure	Ag/glass	Ishizuka Glass Co. (Japan)	Vermeiren et al., 2002
Microban	Triclosan	Microban Products Co. (USA)	Brody et al., 2001
Sanitized, Actigard, Saniprot	Triclosan and others	Sanitized AG/Clariant (Switzerland)	Vermeiren et al., 2002; Suppakul et al., 2003a
Ultra-Fresh	Triclosan and others	Thomson Research Assoc. (Canada)	Vermeiren et al., 2002
WasaOuro	Allyl isothiocyanate	Green Cross Co. (Japan)	Brody et al., 2001
MicroGarde	Clove and others	Rhone-Poulenc (USA)	Brody et al., 2001
Take Guard	Bamboo extract	Takex Co. (Japan)	Brody et al., 2001
Acticap	Ethanol	Freund Industrial Co. (Japan)	Smith et al., 1987
Biocleanact	Antibiotics	Micro Science Tech Co. (Korea)	Han and Moon, 2002; www.biocleanact.com
Microatmosphere	Chlorine dioxide	Southwest Research Institute (USA), Bernard Technologies Inc. (USA)	Brody et al., 2001

mechanical energy input. The extrusion temperature is related to the thermal degradation of the antimicrobial agent, and the specific mechanical energy indicates the severity of the process conditions that also induce the degradation of the agents. Nam *et al.* (2002) showed the severe decrease in the activity of lysozyme in an extruded starch container as the extrusion temperature increased. In the case of the wet casting method (that is, using solvent to cast films and containers such as cellulose films and

Table 6.4 Potential applications of antimicrobial food packaging

Antimicrobials	Meat/poultry	Dairy	Seafood	Produce	Bakery	Beverage	Minimally processed
Organic acids and their salts	Fresh meat, sausage, ham, chicken	Cheese		Fruits, vegetables, jam/jelly		Fruit juice, wine	Precut salad, noodle, pasta, steamed rice, sauce/dressing
Ethanol				Nuts	Bread, cakes, cookies		Noodles, pasta, sandwiches
Bacteriocins	Fresh meat, sausage, ham, chicken	Cheese	Fish, shellfish				Ham/egg sandwiches
Enzymes	Fresh meat, sausage, ham, chicken	Cheese	Fish, shellfish				Ham/egg sand wiches, meatball pasta
Chelating agents	Fresh meat, sausage, ham, chicken	Cheese	Fish, shellfish	Fruits, jam/jelly		Fruit juice	Precut fruits, sauce/dressing
Fungicides				Citrus, berries, nuts			
Sanitizers	Fresh meat, chicken		Fish, shellfish	Fruits, vegetables			Precut salad
Volatile essential oils	Fresh and processed meats, ground beef, chicken nuggets	Shredded cheese	Fish, shellfish, dried fish	Berries, nuts, jam/jelly	Bread, cakes, cookies	Fruit juice	Noodles, pasta, steamed rice, sandwiches, hamburgers, precut salad, sauce/dressing
Spices	Fresh and processed meats, fresh and cooked chicken	Cheese	Fish, shellfish, dried fish			Fruit juice	Noodles, pasta, steamed rice, sandwiches, sauce/dressing
Probiotics	Fresh and processed meats, cured meats	Cheese, yogurt		Fermented vegetables			Deli mix
Oxygen scavengers	Fresh and processed meats, ground beef, dried meats, chicken	Shredded cheese	Dried fish	Nuts, jam/jelly	Bread, cakes, cookies	Fruit juice, wine	Noodles, pasta, steamed rice, sandwiches, hamburgers, sauce/dressing

collagen casing), the solubility and reactivity of the antimicrobial agents and polymers to the solvents are the critical factors. The solubility relates to the homogeneous distribution of the agents in the polymeric materials, while the reactivity relates to the activity loss of the reactive antimicrobial agents.

The chemical properties of the antimicrobial agent, such as solubility, are also important factors. For example, when water-soluble agents are mixed into plastic resins to produce antimicrobial films, the plastic extrusion process may be beset with various problems including crevice hole creation in the films, powder-blooming, the loss of physical integrity, and/or the loss of transparency due to the heterogeneous blending of the hydrophilic agents with the hydrophobic plastics. Therefore, the compatibility of antimicrobial agent and packaging material is an important factor. The pH of the system is also important. Most antimicrobial chemicals alter their activity with different pH. The pH of the packaging system mostly depends on the pH of the packaged foods, and therefore consideration of the food composition along with the chemical nature of the antimicrobial agent is important, as well as consideration of the packaging material reaction with the chemical nature of the agents (Han, 2003b).

Storage and distribution conditions are further significant factors, including the storage temperature and time. This time–temperature integration affects the microbial growth profile, chemical reaction kinetics, and the distribution profile of antimicrobial agents in the food. To prevent microbial growth, storage at the temperature range favorable for microbial growth should be avoided or minimized for the whole period of storage and distribution.

In the case of modified atmosphere packaging with antimicrobial gas, the active gas permeation through the packaging materials may be changed by the temperature and time profile during the whole period of storage and distribution. When the gas composition is altered through active gas permeation, unexpected gas invasion or a seal defect, micro-organisms that are not considered as target micro-organisms may spoil the packaged foods.

Physical properties of packaging materials

The physical and mechanical properties of packaging materials are affected by the incorporated antimicrobial agents. If the antimicrobial agent is compatible with the packaging materials, a significant amount of the agent may be impregnated into the packaging material without any deterioration of its physical and mechanical integrity (Han and Floros, 1997). However, excess antimicrobial agent that is not capable of being blended with packaging materials will decrease physical strength and mechanical integrity (Cooksey, 2000). Polymer morphological studies are very helpful in predicting the possible loss of physical integrity when the antimicrobial agent is added to the packaging material, in the case of polymeric packaging materials. Small-size antimicrobial agents can be blended with polymers, and may be positioned at the amorphous region of the polymeric structure without significant interference with polymer – polymer interactions. If a high level of antimicrobial agent is mixed into the packaging materials, the space provided by the amorphous region will be filled and the mixed agent will start to interfere with the polymer–polymer interactions at the crystalline region. Although there is no damage to the physical integrity after low levels of antimicrobial agent addition, optical properties can be changed – for example, there may be a loss of transparency or a change in the color of the packaging materials (Han and Floros, 1997).

Controlled release technology

The design of an antimicrobial packaging system requires the balanced consideration of controlled release technology and microbial growth kinetics. When the mass transfer rate of an antimicrobial agent is faster than the growth rate of the target micro-organism, loaded antimicrobial agent will be diluted to less than the effective critical concentration (i.e. minimum inhibitory concentration, m.i.c.) before the expected storage period is complete, and the packaging system will lose its antimicrobial activity because the packaged food has almost infinite volume compared to the volume of packaging material and the amount of antimicrobial agent. Consequently, the micro-organism will start to grow following depletion of the antimicrobial agent. On the contrary, when the migration rate is too slow to maintain the concentration above the m.i.c., the micro-organism can grow instantly, before the antimicrobial agent is released. Therefore, the release rate of the antimicrobial agent from the packaging material to the food must be controlled specifically to match the mass transfer rate with the growth kinetics of the target micro-organism. Controversially, in the case of antimicrobial edible coating systems the mass transfer of antimicrobial agents is not desirable, since the migration of the incorporated antimicrobial agents from the coating layer into the food product dilutes the concentration in the coating layer. Compared to the volume of the coating layer, the coated food has almost infinitive volume. Therefore, the migration will deplete the antimicrobial agent in the coating layer, decrease the concentration below the m.i.c., and thus reduce the antimicrobial activity of the coating system. The migration of incorporated antimicrobial agents contributes to antimicrobial effectiveness in the case of packaging systems; on the contrary, no migration is beneficial in the coating system.

The solubility of the antimicrobial agents in the foods is a critical factor of antimicrobial release. If the antimicrobial agent is highly soluble in the food, the migration profile will follow unconstrained free diffusion, while the very low solubility creates the dissolution-dependent monolithic system. For example, when highly soluble potassium sorbate was incorporated in packaging materials (e.g. plastic films or papers) and the antimicrobial packaging materials were used for semi-solid or high-moisture foods, such as paste, yogurt, fruit jelly, soft cheese and sliced ham, the potassium sorbate dissolved in the food immediately after packaging. The potassium sorbate concentration increased very fast on the surface of the foods and the surface concentration decreased slowly as the potassium sorbate diffused into the food. Fast diffusion of the antimicrobial agents in the food decreased the surface concentration rapidly. The maintenance of the surface concentration is highly dependent on the release rate from the packaging materials (diffusivity of packaging materials) and the migration rate of the foods (diffusivity of the foods). Since the flux of the release from the packaging materials decreases as the amount of antimicrobials in the packaging materials reduces

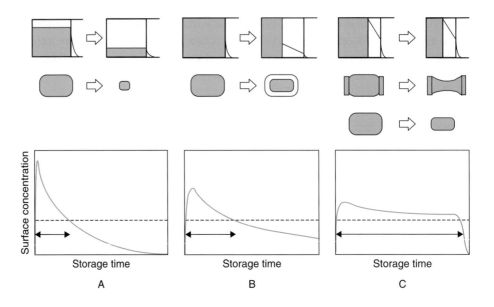

Figure 6.2 Release profiles of antimicrobial agents from various systems. A, unconstrained free diffusion from packaging materials, or fast dissolution from antimicrobial tablets; B, slow diffusion of very low solubility agents from monolithic packaging materials; C, membrane (reservoir) system with constant flux of permeation, slow dissolution from antimicrobial powder/tablets or gaseous agent release from concentrated antimicrobial sachets/tablets with constant volatility in a closed packaging system. Dashed lines and arrows indicate the minimum inhibitory concentration (m.i.c.) of a target micro-organism, and the period of shelf life maintaining the surface concentration over the m.i.c., respectively.

with release time, the period when the surface concentration is maintained above the m.i.c. is carefully estimated, considering its rapid decay profile (Figure 6.2A).

When the solubility of antimicrobial agents in the packaged food is very low, the antimicrobial concentration on the contact food surface is at its maximum solubility. Since the release rate is relatively slower than unconstrained free diffusion, which occurs in high solubility, the period of maintaining the concentration above the m.i.c. in this slow-release system is generally longer than in the free diffusion system. If the antimicrobial agents are impregnated into the polymeric packaging materials, the diffusivity of the agent in the polymeric matrix will control the release rate significantly. This system, shown in Figure 6.2B, may include, for example, sorbic acid anhydride or propyl para benzoic acid in plastic films that wrap high-moisture foods such as fruit jelly or soft cheese. Since the antimicrobial agents are less soluble in water, the release of these agents from hydrophobic plastic will be very slow. It takes more time to reach the maximum peak concentration, and there will be a longer period above the m.i.c. than in the system shown in Figure 6.2A. After the agents initially located on the surface of plastic films have migrated into the food, the release flux will decrease because the agents positioned inside the plastic film should diffuse to the surface of the plastic film. Because of this internal diffusion, the release kinetics do not show a zero-order profile which has a constant release flux. The concentration on the food surface will decrease due to the migration of the agent into the contained food products as well as the reduction of release flux.

Figure 6.2C illustrates the greater period of concentration above the m.i.c. given by membrane systems, which consist of a permeable membrane that controls the release rate. In the case of liquid pharmaceuticals, the release rate will be controlled by the permeability of the liquid agents through the control layer. Until the liquid agent is depleted, the system maintains a zero order of release with a constant permeation flux. This system may include volatile agent concentrates, such as horseradish oil. Horseradish oil contains allyl isothiocyanate, which is a strong volatile flavoring and antimicrobial agent. The volatile agent will split between the oil and the headspace of packages. If there is enough oil in the package, the headspace concentration will be equilibrated and maintained above the m.i.c. until the oil has disappeared. The equilibrated partitioning is an important factor in controlling the release rate as well as the headspace concentration.

Extra advantages

Traditional preservation methods sometimes include antimicrobial packaging concepts – for example, the sausage casings of cured/salted/smoked meats, smoked pottery/oak barrels for fermentation, and brine-filled pickle jars. The basic principle of these traditional preservation methods and antimicrobial packaging is one of hurdle technology applications. The extra antimicrobial function of the packaging system is another hurdle to prevent the degradation of quality and improve the safety of packaged foods, in addition to the conventional protective functions of providing moisture and oxygen barriers as well as physical protection. The microbial hurdle provides the extra function of protection against micro-organisms, which has never been achieved by conventional moisture- and oxygen-barrier packaging materials. Therefore, antimicrobial packaging is one of active packaging and hurdle technology applications. The hurdle technology concept of antimicrobial packaging systems can enhance the efficiency of other sterilization processes, such as aseptic processes, non-thermal processes and the conventional thermal process, where the sterile foods are packaged in the antimicrobial packaging systems.

Since the antimicrobial packaging system can incorporate natural antimicrobial agents such as plant and herb extracts or probiotics, it is considered that natural antimicrobial packaging design development has a connection to nutraceutical research and pharmacognosy. This relationship may be helpful in transferring food packaging knowledge to the area of nutraceutical and pharmaceutical research because of the studies on the barrier properties of materials to volatile active ingredients, and the studies on the clinical effectiveness of natural active agents.

Regulatory, marketing and political factors

The use of antimicrobials should follow the guidelines of regulatory agencies (Meroni, 2000; Brody *et al.*, 2001; Vermeiren *et al.*, 2002; Han, 2003a, 2003b). Antimicrobial agents are additives of packaging material, not food ingredients; however, when the antimicrobial agents migrate into foods, they also require food ingredient approval – as for food contact substances and packaging additives. Therefore, the use of natural

Table 6.5 Factors to be considered for marketing of new antimicrobial packaging systems

Advantages	Disadvantages
Safety enhancement	Changes in culinary culture
Security achievement	Lifestyle changes of consumers
Shelf-life extension	Cost of new materials and systems
Health promoting effect	Regulation conflict
Market attention	Market conflict with conventional packaging
	Political decision-making

antimicrobial agents included in plant extracts or spices is a very promising alternative because of their appeal as natural products, consumers' preference, and no conflict with regulations.

For the commercialization of antimicrobial packaging systems, various marketing factors are involved – for example, logistics, cost, and consumer acceptance (Meroni, 2000). The use of antimicrobial packaging systems should not create any conflict with the current logistic systems of food industry. If the new packaging systems require totally new transportation, distribution and warehousing systems, it is not feasible to commercialize the new systems. The antimicrobial packaging systems should be manageable within current packaging-related logistic systems. Reasonable cost recovery should be promised for the commercialization of new packaging systems. Consumers' acceptance of the use of new antimicrobial packaging systems is critical. This acceptance may be related to the convenience and easiness of the use of a new system, any conflict of the new system with their culture and lifestyles, and other various reasons.

Table 6.5 summarizes the pros and cons of the new antimicrobial packaging systems in terms of marketing.

Since the antimicrobial agent in contact with or migrating into food, the organoleptic property and toxicity of the antimicrobial agent should be satisfied to avoid quality deterioration and to maintain the safety of the packaged foods. The antimicrobial agents may possess a strong taste or flavor, such as a bitter or sour taste, as well as an undesirable aroma that can affect the sensory quality adversely. In the case of antimicrobial edible protein film/coating applications, the allergenicity or chronic disease caused by the edible protein materials, such as peanut protein, soy protein and wheat gluten, should be considered before use (Han, 2001).

The legality of antimicrobial activity in new packaging systems has many critical controversial aspects. For example, research and development departments would not like to claim "antimicrobial activity" on their products and for commercial use, since there is no antimicrobial agent that can eliminate all types of microbial growth. The potential growth of micro-organisms in their new antimicrobial packaging systems could therefore reduce the company's creditability, as well as possible cause serious law suit. However, for marketing purposes there is no point in using a new antimicrobial packaging system, as far as profits are concerned, if the company cannot claim "antimicrobial activity". This example shows that the use of antimicrobial packaging systems possesses a political aspect.

Most foods are perishable, and most medical/sanitary devices are susceptible to contamination. Therefore, the primary goals of an antimicrobial packaging system are:

1. Safety assurance
2. Quality maintenance
3. Shelf-life extension.

This is in reverse order to the primary goals of conventional packaging systems. Nowadays food security is a big issue in the world, and antimicrobial packaging could play a role in food security assurance, comprehensively agreed with industrial sectors, farmers/producers, wholesalers/retailers, governments, and consumer groups. Owing to the political aspects of food safety and security matters, the use of antimicrobial packaging technology is also politically influenced regarding commercialization.

References

An, D. S., Hwang, Y. I., Cho, S. H. and Lee, D. S. (1998). Packaging of fresh curled lettuce and cucumber by using low density polyethylene films impregnated with antimicrobial agents. *J. Korean Soc. Food Sci. Nutr.* **27(4)**, 675–681.

An, D. S., Kim, Y. M., Lee, S. B., Paik, H. D. and Lee, D. S. (2000). Antimicrobial low density polyethylene film coated with bacteriocins in binder medium. *Food Sci. Biotechnol.* **9(1)**, 14–20.

Appendini, P. and Hotchkiss, J. H. (1996). Immobilization of lysozyme on synthetic polymers for the application to food packages. In: *Book of Abstracts*, 1996 IFT Annual Meeting, June 22–26, New Orleans, LA, p. 177. Institute of Food Technologists, Chicago, IL.

Appendini, P. and Hotchkiss, J. H. (1997). Immobilization of lysozyme on food contact polymers as potential antimicrobial films. *Packaging Technol. Sci.* **10(5)**, 271–279.

Baron, J. K. and Sumner, S. S. (1993). Antimicrobial containing edible films as an inhibitory system to control microbial growth on meat products. *J. Food Protect.* **56**, 916.

Brody, A. L., Strupinsky, E. R. and Kline, L. R. (2001). *Active Packaging for Food Applications*, pp. 131–196. Technomic Publishing Co., Lancaster, PA.

Buonocore, G. G., Del Nobile, M. A., Panizza, A., Bove, S., Battaglia, G. and Nicolais, L. (2003). Modeling the lysozyme release kinetics from antimicrobial films intended for food packaging applications. *J. Food Sci.* **68(4)**, 1365–1370.

Cha, D. S., Choi, J. H., Chinnan, M. S. and Park, H. J. (2002). Antimicrobial films based on Na-alginate and kappa-carrageenan. *Lebensm. Wiss. u. Technol.* **30(8)**, 715–719.

Chen, M.-C., Yeh, H.-C. and Chiang, B.-H. (1996). Antimicrobial and physicochemical properties of methylcellulose and chitosan films containing a preservative. *J. Food Process. Preserv.* **20**, 379–390.

Chen, S., Wu, J. G. and Chuo, B. Y. (2003). Antioxidant and antimicrobial activities in the catechins impregnated PVA-starch film. In: *Book of Abstracts*, 2003 IFT Annual Meeting, p. 114. Institute of Food Technologists, Chicago, IL.

Choi, J. O., Park, J. M., Park, H. J. and Lee, D. S. (2001). Migration of preservation from antimicrobial polymer coating into water. *Food Sci. Biotechnol.* **10(3)**, 327–330.

Chung, D., Papadakis, S. E. and Yam, K. L. (2001a). Release of propyl paraben from a polymer coating into water and food simulating solvents for antimicrobial packaging applications. *J. Food Process. Preserv.* **25(1),** 71–87.

Chung, D., Chikindas, M. L. and Yam, K. L. (2001b). Inhibition of *Saccharomyces cerevisiae* by slow release of propyl paraben from a polymer coating. *J. Food Protect.* **64(9),** 1420–1424.

Chung, D., Papadakis, S. E. and Yam, K. L. (2003). Evaluation of a polymer coating containing triclosan as the antimicrobial layer for packaging materials. *Intl J. Food Sci. Technol.* **38(2),** 165–169.

Chung, S. K., Cho, S. H. and Lee, D. S. (1998). Modified atmosphere packaging of fresh strawberries by antimicrobial plastic films. *Korean J. Food Sci. Technol.* **30(5),** 1140–1145.

Coma, V., Sebti, I., Pardon, P., Deschamps, A. and Pichavant, F. H. (2001). Antimicrobial edible packaging based on cellulosic ethers, fatty acids, and nisin incorporation to inhibit *Listeria innocua* and *Staphylococcus aureus*. *J. Food Protect.* **64(4),** 470–475.

Coma, V., Martial-Gros, A., Garreau, S., Copinet, A., Salin, F. and Deschamps, A. (2002). Edible antimicrobial films based on chitosan matrix. *J. Food Sci.* **67(3),** 1162–1169.

Cooksey, D. K., Gremmer, A. and Grower, J. (2000). Characteristics of nisin-containing corn zein pouches for reduction of microbial growth in refrigerated shredded cheddar cheese. In: *Book of Abstracts*, 2000 Annual Meeting, p. 188. Institute of Food Technologists, Chicago, IL.

Cutter, C. N., Willett, J. L. and Siragusa, G. R. (2001). Improved antimicrobial activity of nisin-incorporated polymer films by formulation change and addition of food grade chelator. *Lett. Appl. Microbiol.* **33(4),** 325–328.

Daeschul, M. A. (1989). Antimicrobial substances from lactic acid bacteria for use as food preservatives. *Food Technol.* **43(1),** 164–167.

Dawson, P. L., Carl, G.D., Acton, J.C. and Han, I.Y. (2002). Effect of lauric acid and nisin-impregnated soy-based films on the growth of Listeria monocytogenes on turkey bologna. *Poultry Sci.* **81(5),** 721–726.

Dawson, P. L., Hirt, D. E., Rieck, J. R., Acton, J. C. and Sotthibandhu, A. (2003). Nisin release from films is affected by both protein type and film-forming method. *Food Res. Intl* **36,** 959–968.

Devlieghere, F., Vermeiren, L., Bockstal, A. and Debevere, J. (2000a). Study on antimicrobial activity of a food packaging material containing potassium sorbate. *Acta Alimentaria* **29(2),** 137–146.

Devlieghere, F., Vermeiren, L., Jacobs, M. and Debevere, J. (2000b). The effectiveness of hexamethylenetetramine-incorporated plastic for the active packaging of foods. *Packag. Technol. Sci.* **13(3),** 117–121.

Dobias, J., Chudackova, K., Voldrich, M. and Marek, M. (2000). Properties of polyethylene films with incorporated benzoic anhydride and/or ethyl and propyl esters of 4-hydroxybenzoic acid and their suitability for food packaging. *Food Add. Contam.* **17(12),** 1047–1053.

Fields, C., Pivarnick, L. F., Barnett, S. M. and Rand, A. (1986). Utilization of glucose oxidase for extending the shelflife of fish. *J. Food Sci.* **51(1),** 66–70.

Floros, J. D., Dock, L. L. and Han, J. H. (1997). Active packaging technologies and applications. *Food Cosm. Drug Packag.* **20(1),** 10–17.

Ghosh, K. G., Srivatsava, A. N., Nirmala, N. and Sharma, T. R. (1973). Development and application of fungistatic wrappers in food preservation. Part I. Wrappers obtained by impregnation method. *J. Food Sci. Technol,* **10(4),** 105–110.

Ghosh, K. G., Srivatsava, A. N., Nirmala, N. and Sharma, T. R. (1977). Development and application of fungistatic wrappers in food preservation. Part II. Wrappers made by coating process. *J. Food Sci. Technol.* **14(6),** 261–264.

Gill, A. O. (2000). Application of Lysozyme and Nisin to Control Bacterial Growth on Cured Meat Products. MSc dissertation, The University of Manitoba, Winnipeg, Canada.

Gontard, N. (1997). Active packaging. In: *Proceedings of Workshop sobre Biopolimeros, 22–24 April, Universidade de Sao Paulo* (P. J. do A. Sobral and G. Chuzel, eds), pp. 23–27. Pirassununga, FZEA, Brazil.

Ha, J. U., Kim, Y. M. and Lee, D. S. (2001). Multilayered antimicrobial polyethylene films applied to the packaging of ground beef. *Packag. Technol. Sci.* **14(2),** 55–62.

Halek, G. W. and Garg, A. (1989). Fungal inhibition by a fungicide coupled to an ionomeric film. *J. Food Safety* **9,** 215–222.

Han, J. H. (2000). Antimicrobial food packaging. *Food Technol.* **54(3),** 56–65.

Han, J. H. (2001). Design edible and biodegradable films/coatings containing active ingredients. In: *Active Biopolymer Films and Coatings for Food and Biotechnological Uses,* Proceedings of Pre-congress Short Course of IUFoST, 21–22 April, Seoul, Korea (H. J. Park, R. F. Testin, M. S. Chinnan and J. W. Park, eds), pp. 187–198. IUFoST.

Han, J. H. (2002). Protein-based edible films and coatings carrying antimicrobial agents. In: *Protein-based Films and Coatings* (A. Gennadios, ed.), pp. 485–499. CRC Press, Boca Raton, FL.

Han, J. H. (2003a). Antimicrobial food packaging. In: Novel Food Packaging Techniques (R. Ahvenainen, ed.), pp. 50–70. Woodhead Publishing Ltd., Cambridge, UK.

Han, J. H. (2003b). Design of antimicrobial packaging systems. *Int. Rev. Food Sci. Technol.* **11,** 106–109.

Han, J. H. and Floros, J. D. (1997). Casting antimicrobial packaging films and measuring their physical properties and antimicrobial activity. *J. Plastic Film Sheeting* **13,** 287–298.

Han, J. H. and Floros, J. D. (1998a). Potassium sorbate diffusivity in American processed and Mozzarella cheeses. *J. Food Sci.* **63(3),** 435–437.

Han, J. H. and Floros, J. D. (1998b). Simulating diffusion model and determining diffusivity of potassium sorbate through plastics to develop antimicrobial packaging film. *J. Food Process. Preserv.* **22(2),** 107–122.

Han, J. H. and Moon, W.-S. (2002). Plastic packaging materials containing chemical antimicrobial agents. In: *Active Food Packaging* (J. H. Han, ed.), pp. 11–14. SCI Publication and Communication Services, Winnipeg, Canada.

Hoffman, K. L., Han, I. Y. and Dawson, P. L. (2001). Antimicrobial effects of corn zein films impregnated with nisin, lauric acid, and EDTA. *J. Food Protect.* **64(6),** 885–889.

Hong, S. I., Park, J. D. and Kim, D. M. (2000). Antimicrobial and physical properties of food packaging films incorporated with some natural compounds. *Food Sci. Biotechnol.* **9(1),** 38–42.

Hoojjatt, P., Honte, B., Hernandez, R., Giacin, J. and Miltz, J. (1987). Mass transfer of BHT from HDPE film and its influence on product stability. *J. Packag. Technol.* **1(3),** 78.

Huang, L. J., Huang, C. H. and Weng, Y.-M. (1997). Using antimicrobial polyethylene films and minimal microwave heating to control the microbial growth of tilapia fillets during cold storage. *Food Sci. Taiwan* **24(2),** 263–268.

Ishitani, T. (1995). Active packaging for food quality preservation in Japan. In: *Food and Food Packaging Materials – Chemical Interactions* (P. Ackermann, M. Jagerstad and T. Ohlsson, eds), pp. 177–188. The Royal Society of Chemistry, Cambridge, UK.

Kelly, M., Steele, R., Scully, A. and Rooney, M. (2003). Enhancing food security through packaging. *Intl Rev. Food Sci. Technol.* **11,** 115–117.

Kim, Y. M., Lee, N. K., Paik, H. D. and Lee, D. S. (2002). Migration of bacteriocin from bacteriocin-coated film and its antimicrobial activity. *Food. Sci. Biotechnol.* **9(5),** 325–329.

Kim, Y. M., Paik, H. D. and Lee, D. S. (2002). Shelf-life characteristics of fresh oysters and ground beef as affected by bacteriocins-coated plastic packaging film. *J. Sci. Food Agric.* **82(9),** 998–1002.

Krause, T. R., Sebranek, J .G., Rust, R. E. and Honeyman, M. S. (2003). Use of carbon monoxide packaging for improving the shelf life of pork. *J. Food Sci.* **68(8),** 2596–2603.

Krochta, J. M. and De Mulder-Johnston, C. (1997). Edible and biodegradable polymer films: challenges and opportunities. *Food Technol.* **51(2),** 61–74.

Lanciotti, R., Belletti, N., Patrignani, F., Gianotti, A., Gardini, F. and Guerzoni, M. E. (2003). Application of hexanal, (E)-2-hexenal, and hexyl acetate to improve the safety of fresh-sliced apples. *J. Agric Food Chem.* **51(10),** 2958–2963.

Lee, D. S., Hwang, Y. I. and Cho, S. H. (1998). Developing antimicrobial packaging film for curled lettuce and soybean sprouts. *Food Sci. Biotechnol.* **7(2),** 117–121.

Leung, P. P., Yousef, A. E. and Shellhammer, T. H. (2003). Antimicrobial properties of nisin-coated polymeric films as influenced by film type and coating conditions. *J. Food Safety* **23(1),** 1–12.

Meroni, A. (2000). Active packaging as an opportunity to create packaging design that reflects the communicational, functional and logistical requirements of food products. *Packag. Technology Sci.* **13,** 243–248.

Miller, W. R., Spalding, D. H., Risse, L. A. and Chew, V. (1984). The effects of an imazalil-impregnated film with chlorine and imazalil to control decay of bell peppers. *Proc. Fla State Hort. Soc.* **97,** 108–111.

Ming, X., Weber, G. H., Ayres, J. W. and Sandine, W. E. (1997). Bacteriocins applied to food packaging materials to inhibit *Listeria monocytogenes* on meats. *J. Food Sci.* **62(2),** 413–415.

Muthukumarasamy, P., Han, J. H. and Holley, R. A. (2003). Bactericidal effects of *Lactobacillus reuteri* and allyl isothiocyanate on *Escherichia coli* O157:H7 on refrigerated ground beef. *J. Food Protect.* **66(11),** 2038–2044.

Nadarajah, D., Han, J. H. and Holley, R. A. (2002). Use of allyl isothiocyanate to reduce *Escherichia coli* O157:H7 in packaged ground beef patties. In: *Book of Abstracts,* IFT Annual Meeting, p. 249. Institute of Food Technologists, Chicago, IL.

Nadarajah, D., Han, J. H. and Holley, R. A. (2003). Survival of *Escherichia coli* O157:H7 in ground beef patties containing nondeheated mustard flour. In: *Book of Abstracts*, IFT Annual Meeting, p. 53. Institute of Food Technologists, Chicago, IL.

Nam, S., Han, J. H., Scanlon, M. G. and Izydorczyk, M. S. (2002). Use of extruded pea starch containing lysozyme as an antimicrobial and biodegradable packaging. In: *Book of Abstracts*, IFT Annual Meeting, p. 248. Institute of Food Technologists, Chicago, IL.

Nestle, M. (2003). *Safe Food: Bacteria, Biotechnology, and Bioterrorism*, p. 1. University of California Press, Berkeley, CA.

Ouattara, B., Simard, R. E., Piette, G., Begin, A. and Holley, R. A. (2000a). Diffusion of acetic and propionic acids from chitosan-based antimicrobial packaging films. *J. Food Sci.* **65(5),** 768–773.

Ouattara, B., Simard, R. E., Piette, G., Begin, A. and Holley, R. A. (2000b). Inhibition of surface spoilage bacteria in processes meats by application of antimicrobial films prepared with chitosan. *Intl J. Food Microbiol.* **62(1/2),** 139–148.

Ozdermir, M. (1999). Antimicrobial Releasing Edible Whey Protein Films and Coatings. PhD dissertation, Purdue University, West Lafayette, IN.

Ozen, B. F. and Floros, J. D. (2001). Effects of emerging food processing techniques on the packaging materials. *Trends Food Sci. Technol.* **12(2),** 60–67, 69.

Padgett, T., Han, I. Y. and Dawson, P. L. (1998). Incorporation of food-grade antimicrobial compounds into biodegradable packaging films. *J. Food Protect.* **61(10),** 1330–1335.

Paik, J. S. and Kelley, M. J. (1995). Photoprocessing method of imparting antimicrobial activity to packaging film. In: *Book of Abstracts*, IFT Annual Meeting, p. 93. Institute of Food Technologists, Chicago, IL.

Paik, J. S., Dhanassekharan, M. and Kelly, M. J. (1998). Antimicrobial activity of UV-irradiated nylon film for packaging applications. *Packag. Technol. Sci.* **11(4),** 179–187.

Park, J.-H., Cha, B. and Lee, Y. N. (2003). Antimicrobial activity of chitosan acetate on food-borne enteropathogenic bacteria. *Food Sci. Biotechnol.* **12(1),** 100–103.

Quintero-Salazar, B., Vernon-Carter, J., Guerrero-Legarreta, I. and Ponce-Alquicira, E. (2003). Use of *Pediococcus parvulus* as an alternative to immobilize antimicrobials in biodegradable films. In: *Book of Abstracts*, IFT Annual Meeting, p. 112. Institute of Food Technologists, Chicago, IL.

Rico-Pena, D. C. and Torres, J. A. (1991). Sorbic acid and potassium sorbate permeability of an edible methylcellulose-palmitic acid film: water activity and pH effects. *J. Food Sci.* **56,** 497–499.

Rodrigues, E. T. and Han, J. H. (2000). Antimicrobial whey protein films against spoilage and pathogenic bacteria. In: *Book of Abstracts*, IFT Annual Meeting, p. 191. Institute of Food Technologists, Chicago, IL.

Rodrigues, E. T. and Han, J. H. (2003) Intelligent packaging. In: *Encyclopedia of Agricultural and Food Engineering* (D. R. Heldman, ed.), pp. 528–535. Marcel Dekker, New York, NY.

Rodrigues, E. T., Han, J. H. and Holley, R. A. (2002). Optimized antimicrobial edible whey protein films against spoilage and pathogenic bacteria. In: *Book of Abstracts*, IFT Annual Meeting, p. 252. Institute of Food Technologists, Chicago, IL.

Sebti, I., Pichavant, F. H. and Coma, V. (2002) Edible bioactive fatty acid-cellulosic derivative composites used in food-packaging applications. *J. Agric. Food Chem.* **50(15),** 4290–4294.

Shapero, M., Nelson, D. and Labuza, T. P. (1978). Ethanol inhibition of *Staphylococcus aureus* at limited water activity. *J. Food Sci.* **43,** 1467–1469.

Siragusa, G. R., Cutter, C. N. and Willett, J. L. (1999). Incorporation of bacteriocin in plastic retains activity and inhibits surface growth of bacteria on meat. *Food Microbiol.* **16(3),** 229–235.

Smith, J. P., Ooraikul, B., Koersen, W. J., van de Voort, F. R., Jackson, E. D. and Lawrence, R. A. (1987). Shelf life extension of a bakery product using ethanol vapor. *Food Microbiol.* **4,** 329–337.

Smith, J. P., van de Voort, F. R. and Lambert, A. (1989). Food and its relation to interactive packaging. *Can. Inst. Food Sci. Technol. J.* **22(4),** 327–330.

Smith, J. P., Ramaswamy, H. S. and Simpson, B. K. (1990). Developments in food packaging technology. Part II: Storage aspects. *Trends Food Sci. Technol.* **1990,** 111–118.

Steven, M. D. and Hotchkiss, J. H. (2003). Non-migratory bioactive polymers (NMBP) in food packaging. In: *Novel Food Packaging Techniques* (R. Ahvenainen, ed.), pp. 71–102. Woodhead Publishing Ltd., Cambridge, UK.

Suppakul, P., Miltz, J., Sonneveld, K. and Bigger, S. W. (2003a). Active packaging technologies with an emphasis on antimicrobial packaging and its applications. *J. Food Sci.* **68(2),** 408–420.

Suppakul, P., Miltz, J., Sonneveld, K. and Bigger, S. W. (2003b). Antimicrobial properties of basil and its possible application in food packaging. *J. Agric. Food Chem.* **51(11),** 3197–3207.

Takeuchi, K. and Yuan, J. (2002). Packaging tackles food safety: a look at antimicrobials, Industrial case study (Presentation for 2002 IFT Annual Meeting Symposium). Part of a presentation 'Industrial Case Studies: Meats, Poultry, Produce, Juice, Smoked Salmon' by Holley, R. A., Brody, A. L., Cook, L. K. and Yuan, J. T. C. In: *Book of Abstracts,* IFT Annual Meeting, p. 48. Institute of Food Technologists, Chicago, IL.

Tatrajan, N. and Sheldon, B. W. (2000). Efficacy of nisin-coated polymer films to inactivate *Salmonella typhimurium* on fresh broiler skin. *J. Food Protect.* **63(9),** 1189–1196.

Vermeiren, L., Devlieghere, F. and Debevere, J. (2002). Effectiveness of some recent antimicrobial packaging concepts. *Food Add. Contam.* **19(Suppl.),** 163–171.

Vojdana, F. and Torres, J. A. (1990). Potassium sorbate permeability of methylcellulose and hydroxypropyl methylcellulose coatings: effect of fatty acid. *J. Food Sci.* **55(3),** 841–846.

Weng, Y.-M. and Chen, M.-J. (1997). Sorbic anhydride as antimycotic additive in polyethylene food packaging films. *Lebensm. Wiss u. Technol.* **30,** 485–487.

Weng, Y.-M. and Hotchkiss, J. H. (1992). Inhibition of surface molds on cheese by polyethylene film containing the antimycotic imazalil. *J. Food Protect.* **55(5),** 367–369.

Weng, Y.-M. and Hotchkiss, J. H. (1993). Anhydrides as antimycotic agents added to polyethylene films for food packaging. *Packag. Technol Sci.* **6,** 123–128.

Weng, Y. M., Chen, M. J. and Chen, W. (1997). Benzoyl chloride modified ionomer films as antimicrobial food packaging materials. *Intl J. Food Sci. Technol.* **32(3),** 229–234.

Weng, Y. M., Chen, M. J. and Chen, W. (1999). Antimicrobial food packaging materials from poly(ethylene-co-methacrylic acid). *Lebensm. Wiss. u. Technol.* **32(4),** 191–195.

Yakovleva, L. A., Kolesnikov, B. F., Kondrashov, G. A. and Markelov, A. V. (1999). New generation of bactericidal polymer packaging material. *Khranenie I Pererabotka Ssel'khozsyr'ya.* **6,** 44–45.

Yi, J. H., Kim, I. H., Choe, C. H., Seo, Y. B. and Song, K. B. (1998). Chitosan-coated packaging papers for storage of agricultural products. *Hanguk Nongwhahak Hoechi.* **41(6),** 442–446.

7

Packaging containing natural antimicrobial or antioxidative agents

Dong Sun Lee

Introduction

The most frequent mechanisms of food deterioration are microbial spoilage and oxidation. For protecting foods from these deteriorative changes, modified atmosphere packages or oxygen scavengers have been used applied with high gas-barrier packaging materials (Rooney, 1995; Vermeiren *et al.*, 1999). This technique of oxygen exclusion works against aerobic microbial growth and oxidative quality changes, but it cannot protect foods from spoilage and contamination due to the growth of facultative or obligate anaerobic bacteria. Heat processing of the packaged foods may also help preserve the food products from microbial deterioration. However, thermal processing is sometimes not useful to maintain specific food quality requirements, such as in the case of fresh or minimally processed foods. The concept of active packaging can contribute to the preservation of perishable foods that are sensitive to microbial spoilage or to oxidation (Miltz *et al.*, 1995; Vermeiren *et al.*, 1999). Antimicrobial packaging materials recently emerged as a potential means to assist in the preservation of perishable foods and extend their shelf life. Antioxidants may also be incorporated into packaging films, and will be released to protect the foods from oxidative degradation.

Typically, in antimicrobial and antioxidant packaging systems the antimicrobial or antioxidant substances are incorporated into or coated onto their packaging materials, such as plastic films or papers (Vermeiren *et al.*, 1999; Appendini and Hotchkiss,

2002). Some specialized applications immobilize the agents to the polymers by ionic or covalent bonding. The active agents in the packaging material are designed to be released into the food or designed to function at the surface of the food product. The concept of controlled release of drugs has already been practiced widely in medical and pharmaceutical areas. However, the release of antimicrobial or antioxidant agents in active food packaging is relatively recent, and causes consumer concerns regarding their safety due to their possible migration into foods (Vermeiren *et al.*, 1999; Han, 2003). For this reason, there is a growing consumer preference for natural agents which have been isolated from microbiological, plant, and animal sources (Nicholson, 1998). Active substances of biological origin have a powerful wide-spectrum activity with low toxicity, and are expected to be used for food preservation as a means of active packaging (Han, 2003).

Table 7.1 lists the natural preservatives and antioxidants to be considered for inclusion or adsorption into food packaging materials. For antimicrobial and antioxidative packaging applications, the agents should be incorporated into or coated on the packaging layer, remain stable there for a required time, and finally be released from the packaging surface in a controlled manner. Because most active compounds from biological origins are sensitive to heat, the thermal fabrication of plastics (such as extrusion and injection) offers limited possibilities for the inclusion of these substances directly in the polymer matrix without loss of their activity. Therefore, solvent compounding and solution coating are preferred, and will prevent the loss of these compounds' activity. When selecting the proper solvents and casting conditions, it is important to consider the solubility of the polymers and of the additives. Biopolymers such as proteins and carbohydrates are soluble in water, ethanol, and many other solvents, and thus are good for casting by solvent compounding (Appendini and Hotchkiss, 2002). However, these biopolymer films have limited stability to high humidity conditions, which restricts the range of applications. On the other hand, some agents such as plant extracts and bacteriocins are relatively heat-stable, and can be extruded or pressed with the use of mild heat. In this case, however, some activity loss may occur, and should be accounted for.

Table 7.1   Natural preservatives and antioxidants for active food packaging applications

Type	Agents
Antimicrobials	Plant extracts from grapefruit seed, wasabi (allyl isothiocyanate), hinoki cypress (hinokitiol), bamboo, *Rheum palmatum* and *Coptis chinesis*
	Polypeptides such as nisin and pediocin
	Chitosan
	Enzymes such as lysozyme, glucose oxidase
Antioxidant	Monophenols and phenolic acids (α-tocopherol)
	Organic acids (ascorbic acid)
	Plant extracts (rosemary, sage)
	Maillard reaction products

From Rooney (1995); Ooki (1996); Han (2000).

Antimicrobial packaging

Among the various natural antimicrobial agents, nisin, a polypeptide produced by *Lactococcus lactis*, has been most widely studied for its incorporation into packaging films. Because it is produced by a safe food-grade bacteria, has been consumed by humans for a long period of time without any recognized problems, and can be degraded in the human gastrointestinal tract by proteolytic enzymes, it is ideal for use as a biopreservative in foods. Nisin, a hydrophobic protein, has been granted GRAS (generally recognized as safe) status for its use in cheese products in the USA, and is effective against a wide range of spoilage and hazardous gram-positive bacteria (Hurst, 1983). Many recent studies have reported that nisin-incorporated packaging materials inhibit the growth of gram-positive bacteria, such as *Brochothrix thermosphacta, Lactobacillus helveticus, Listeria monocytogenes, Micrococcus flavus,* and *Pediococcus pentosaceus* (Daeschel *et al.*, 1992; An *et al.*, 2000), and have also extended the shelf life of perishable foods by suppressing the growth of spoilage bacteria (Siragusa *et al.*, 1999; Kim *et al.*, 2002a). Nisin can withstand moderate thermal processing and exposure to acidic environments without major activity loss (Ray, 1992), which confers high merit for the incorporation in food packaging materials. It is stable in a dry form for years, and thus is expected to be safe when incorporated into the polymer structure (Rück and Jager, 1995; An *et al.*, 2000).

Daeschel *et al.* (1992) treated silicon surfaces to absorb nisin so as to retain its antimicrobial activity. *L. monocytogenes* cells exposed to the surfaces failed to grow, and lost their metabolic activity (Bower *et al.*, 1995). Higher amounts of nisin were absorbed onto surfaces with high hydrophobicity, but such nisin possessed less antimicrobial activity than that which was absorbed onto surfaces of lower hydrophobicity. Scannell *et al.* (2000) fabricated cellulose-based packaging inserts absorbed with nisin, which retained its antimicrobial activity and was effective against *L. lactis, Listeria innocua* and *Staphylococcus aureus*. Cutter and Siragusa (1997) immobilized nisin onto edible alginate gels to be ground with beef tissue. Extrusion of low-density polyethylene (LDPE) containing nisin, at moderately low temperatures of 120°C, has been reported to retain the antimicrobial activity of nisin in its plastic structure (Siragusa *et al.*, 1999). The nisin-filled film was seen to inhibit bacterial growth on meat surfaces. However, extrusion or heat-setting at high processing temperatures, above 140°C, resulted in a complete loss of nisin's activity due to its denaturation (Hoffman *et al.*, 1998).

Owing to the possible antimicrobial activity loss during thermal processes, solution coating or solvent casting processes have attempted to incorporate nisin into the packaging matrix. When a preservative is incorporated into or added as a surface layer onto the film structure by coating or solvent casting processes, the total amount of active substance and the corresponding material costs can be reduced. On the other hand, when a preservative is incorporated into the whole film structure by extrusion, a larger amount of the preservative is needed to mix with the entire resin. An LDPE film coated with nisin and a polyamide binder have also suppressed the microbial growth of *M. flavus* and *L. monocytogenes* in phosphate buffer solutions when in contact with the film (An *et al.*, 2000). The plastic films with nisin inhibited the microbial growth of total aerobic and coliform bacteria on shelled oysters and on ground beef at 3° and 10°C, as well as extending the products' shelf life significantly (Kim *et al.*, 2002a). Therefore, it has been proposed that solution coating on the food contact surface can provide more effective antimicrobial activity (An *et al.*, 2000; Kim *et al.*, 2002a). A thinner coating with higher concentrations of nisin possessed higher antimicrobial activity compared to a thicker coating with lower concentrations of nisin (An *et al.*, 2000).

When nisin (0.1% concentration) was incorporated into LDPE by extrusion, it did not migrate into distilled water at room temperature, however, it migrated into saline water that contained a surfactant (Siragusa *et al.*, 1999; Table 7.2). A heating treatment (at boiling temperature) or a mild surfactant treatment caused the migration of nisin from the LDPE film. It was thus suggested that nisin was not chemically bound to the LDPE structure (Siragusa *et al.*, 1999). Kim *et al.* (2000) also showed that the presence of salt in food simulants increased the equilibrated migration level of nisin from a polyamide coating layer. However, the incorporation of sodium chloride, citric acid, and a surfactant into the coating medium did not enhance nisin's migration, nor the film's antimicrobial activity (Kim *et al.*, 2000). It was therefore suggested by Chi-Zhang *et al.* (2004) that controlled releasing nisin-incorporated packaging materials could be combined with direct nisin addition into foods for more effective antimicrobial activity against *L. monocytogenes*.

The type of polymer affects the migration rate and the equilibrium concentration. Kim *et al.* (2002b) reported that a vinyl acetate–ethylene copolymer coating on paperboard caused a higher migration level of nisin compared to an acrylic polymer. The diffusion coefficients of nisin in an acrylic polymer and in a vinyl acetate–ethylene copolymer coating at 10°C were in the range of 1.1×10^{-12} to $4.2 \times 10^{-12}\,\mathrm{m^2\,s^{-1}}$, and 6.8×10^{-12} to $12.2 \times 10^{-12}\,\mathrm{m^2\,s^{-1}}$, respectively. The migration of nisin from polyvinyl alcohol (PVOH) at ambient temperature has a diffusion coefficient ranging from 3.01×10^{-14} to $8.61 \times 10^{-13}\,\mathrm{m^2\,s^{-1}}$, with a partition coefficient of 26–153 as a concentration ratio in the polymer to the simulant (Buonocore *et al.*, 2003). The diffusion coefficient in a swollen polymer matrix decreased with increasing cross-linking

Table 7.2 Activity of migrated nisin from nisin-impregnated LDPE film

Extraction conditions	Film with 0.1% active nisin (7 min. extrusion)	Film with 0.1% active nisin (14 min. extrusion)	Film with 0.05% active nisin (7 min. extrusion)
Water for 315 min. at room temperature	–	–	–
Boiling water (100°C) for 5 min.	+	+	+
Boiling in 0.02N HCl solution for 5 min.	+	+	+
Saline water + 0.5% Tween 20 at room temperature	+	+	+
Saline water + 0.5% Tween 20 at boiling for 5 min.	+	+	+

From Siragusa *et al.* (1999), reproduced with permission.

ratio and partition coefficient. Higher partitioning of nisin in the polymer provides a great benefit to slow-release mechanics (Han, 2003). Cast films of biopolymers demonstrated higher migration profiles of nisin activity compared to the heat-pressed films (Cha *et al.*, 2003).

Although changing the coating materials resulted in significant differences in the suppression of microbial growth in water and in microbial mediums, no antimicrobial activity difference was found in coating materials used in real foods, such as pasteurized milk and orange juice (Kim *et al.*, 2002b; Figure 7.1). The suppression of the microbial growth in real foods does not exhibit the same results as those in food simulants and in microbial mediums (Appendini and Hotchkiss, 2002).

Other bacteriocins, such as pediocin and lacticin, have also been incorporated into packaging films to create antimicrobial packaging systems (Ming *et al.*, 1997; Scannell *et al.*, 2000). Nisin and other bacteriocins can inhibit gram-positive bacteria; however, they are ineffective against molds, yeasts and gram-negative bacteria, with the exclusion of a few strains (Hurst, 1983). There have been some approaches to improving

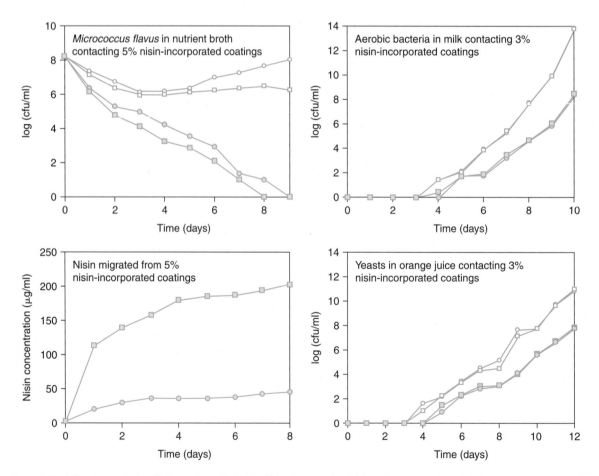

Figure 7.1 Migration and microbial growth profiles in food simulant or microbial medium contacting nisin-coated paper at 10°C. O, acrylic polymer coating only; □, vinyl acetate-ethylene copolymer coating only; ●, acrylic polymer with nisin; ■, vinyl acetate-ethylene copolymer with nisin. From Kim *et al.* (2002b), reprinted with permission of John Wiley & Sons Ltd.

the effectiveness of bacteriocin-incorporated packaging materials by incorporating chelating agents such as EDTA (Nicholson, 1998). A higher hydrophobicity at an acidic pH was reported to exert higher inhibitory activity of nisin against *L. monocytogenes* in the edible films including whey protein, soy protein, egg albumin, and wheat gluten (Ko *et al.*, 2001).

Appendini and Hotchkiss (1997) have immobilized the antimicrobial agent, lysozyme, in PVOH, nylon and cellulose triacetate (CTA) films. CTA films incorporated with lysozyme demonstrated the highest antimicrobial activity against *Micrococcus lysodeikticus*. Higher antimicrobial activities were obtained when larger amounts of lysozyme were incorporated, up to 150–200 mg enzyme/g polymer, while the film thickness did not affect the activity significantly. The antimicrobial activity of lysozyme may be reduced due to its conformational changes when lysozyme is adsorbed onto a solid surface (Bower *et al.*, 1998). Release of lysozyme in PVOH film at ambient temperatures had a diffusion coefficient of 3.83×10^{-15} to $9.98 \times 10^{-13} \, m^2 \, s^{-1}$ with a partition coefficient of 6 to 432 (Buonocore, 2003). Increasing the cross-linking degree decreased the diffusion coefficient and increased the partition coefficient of lysozyme. Glucose oxidase may be immobilized in a polymer matrix to confer antimicrobial activity, due to a catalysis reaction yielding hydrogen peroxide from glucose and oxygen (Appendini and Hotchkiss, 2002).

Chitosan is a biopolymer with good antimicrobial abilities, because it inhibits the growth of a wide variety of fungi, yeasts and bacteria. In addition, it forms a film almost by itself after it has been dissolved in acid solution (Begin and Van Calsteren, 1999). It is a cationic polysaccharide of β-1,4 linkages obtained by the deacetylation of chitin, which is collected from cructaceans or various fungi. Commercial chitosan products usually have a molecular weight ranging from 100 000 to 1 200 000 daltons, but oligomer products with lower molecular weight can be produced by thermal, enzymatic, and/or chemical degradation (Li *et al.*, 1997). Chitosan is readily soluble in various acidic solvents, and has high antimicrobial activity against many pathogenic and spoilage micro-organisms (Tokura *et al.*, 1997; Tsai *et al.*, 2002). However, the antimicrobial activity of chitosan depends on its molecular weight, and the degree of deacetylation and of chemical degradation. Tokura *et al.* (1997) observed growth inhibition of *Escherichia coli* by chitosan of molecular weight 9300 g/mole, but not by that of molecular weight 2200 g/mole. Tsai *et al.* (2002) reported that the antimicrobial activity of chitosan increased with the degree of deacetylation. Both gram-positive and gram-negative bacteria were inhibited by antimicrobial LDPE films incorporated with chitosan above 1.43% in LDPE (Park *et al.*, 2002). Chitosan incorporated in LDPE could be released into the culture broth. Chito-oligosaccharide was immobilized on PVOH by chemical cross-linking, and exhibited reduction in *S. aureus* bacterial counts in culture medium (Cho *et al.*, 2000). Chitosan films can also be used as a delivery system for organic acids to exhibit antimicrobial functions (Ouattara *et al.*, 2000).

Different plant extracts have been used for antimicrobial packaging purposes. Extracts of grapefruit seed, *Rheum palmatum* and *Coptis chinesis* have been included into LDPE films by extrusion (Chung *et al.*, 1998; Lee *et al.*, 1998; Ha *et al.*, 2001). Grapefruit seed extracts in a solution form in glycerol, and in a powder form, have been added in master-batch compounding for film fabrication. Other alcoholic or aqueous plant extracts have been dried by freeze drying or spray drying, and have been blended with polymer pellets to produce films (Chung *et al.*, 1998). The extracts retained partial antimicrobial activity after the thermal extrusion process. Grapefruit seed extract exhibits a wide spectrum of microbial growth inhibition, and has been reported to contain naringin, ascorbic acid, hesperidin, and various organic acids such as citric acid. In addition, in some countries it is permitted for use as a food additive (Sakamoto *et al.*, 1996; Lee *et al.*, 1998). Its water-soluble fraction has anti-microbial activity and contains three reactive components identified by gas chromato-graphy (Cho *et al.*, 1994). Sakamoto *et al.* (1996) identified triclosan and methyl-p-hydroxybenzote to be the active antimicrobial components in grapefruit seed extract. Kim *et al.* (1994) reported the existence of antimicrobial chitinase in the extract, which is stable to heat treatments. The antimicrobial activity of grapefruit seed extract was stable at high temperatures of up to 120°C (Kim and Cho, 1996), and was retained after film fabrication processes (Lee *et al.*, 1998). Films containing grapefruit seed extracts were effective in inhibiting the growth of aerobic bacteria and/or coliforms on fresh produce and on ground beef, when contact packaging was used (Chung *et al.*, 1998; Lee *et al.*, 1998; Ha *et al.*, 2001).

Another plant extract used for commercial antimicrobial packaging is wasabi extract, or Japanese horseradish (Koichiro, 1993). The main active antimicrobial ingredient in wasabi extract is known to be the volatile allyl isothiocyanate (AIT). The AIT gas inhibits many fungi and bacteria. The minimal inhibition concentration of AIT against *Yersinia enterocolitica* and *L. monocytogenes* is in the range of 14–145 ppm at 10–40°C (Kusunoki *et al.*, 1998). The extract has been encapsulated in cyclodextrin to control the volatility of AIT (Figure 7.2). The AIT in the encapsulated powder (dry form) is suggested not to be volatile, but becomes volatile when the AIT–cyclodextrin

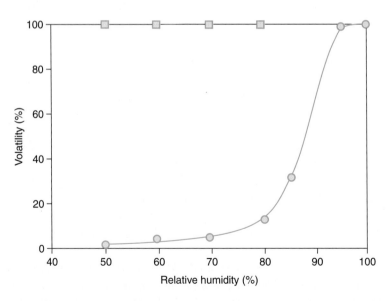

Figure 7.2 Volatility of allyl isothiocyanate encapsulated in different matrixes. ■, zeolite; ●, cyclodextrin. From Koichiro (1993).

complex is exposed to high humidity conditions after the packaging of the food product (Koichiro, 1993). The evaporated AIT then migrates to the food surface, and inhibits the growth of aerobic bacteria such as *Aspergillus niger, Penicillium italicum, S. aureus*, and *E. coli*. Stronger antimicrobial activity of AIT was reported at higher RH, suggesting that moisture has a significant function in AIT's antimicrobial ability (Furuya and Isshiki, 2001). The AIT–cyclodextrin complex powder has been incorporated in packaging materials of drip sheets, polyethylene films, and tablets. These products are commercially available in Japan, and are used for rice lunch boxes, meats, and fresh produce. A design of the drip sheet is shown in Figure 7.3.

Volatile components from essential oils of mustard, cinnamon, garlic and clove have been added to filter papers to inhibit spoilage fungi in breads, such as *Aspergillus flavus, Endomyces fibuliger*, and *Penicillium commune* (Nielsen and Rios, 2000). Nielsen and Rios (2000) have suggested using AIT from mustard as the active component in the packaging of bread.

Packaging materials incorporating a single biopreservative have limitations in ensuring the safety and extending the shelf life of foods, because they can only inhibit specific microbes. For example, nisin-incorporated packaging materials suppress the growth of gram-positive bacteria effectively (Daeschel *et al.*, 1992; Bower *et al.*, 1995; Siragusa *et al.*, 1999; An *et al.*, 2000), but it does not work against gram-negative bacteria such as *E. coli* (Cha *et al.*, 2002). However, a narrow antimicrobial spectrum may, in certain cases, be desirable for packaged foods in which microbial contamination is caused by only a specific type of strain. Nonetheless, generally packaging materials with a wide antimicrobial spectrum are more desirable for universal use and to improve the storage stability of a variety of food products. In an effort to achieve this, Cha *et al.* (2002) incorporated several antimicrobial agents into films made of Na-alginate and κ–carrageenan. The addition of EDTA and grapefruit seed extract to both films gave a strong inhibition against *Micrococcus luteus, L. innocua, Salmonella enteritidis, E. coli*, and *S. aureus*; however, Na-alginate based films with the addition of nisin, lysozyme, and EDTA showed the strongest inhibition against the same bacterial strains. Table 7.3 shows their Na-alginate based film results.

In a similar approach, Lee *et al.* (2003) fabricated a paperboard coated with nisin and chitosan combinations which had been dissolved in a vinyl acetate–ethylene copolymer.

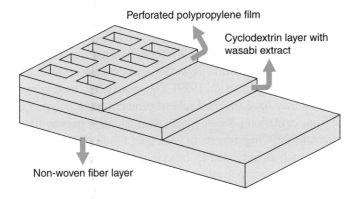

Figure 7.3 Design of drip sheet incorporating wasabi extract. From Koichiro (1993).

Table 7.3 Effects of antimicrobial additives in Na-alginate films on several organisms

Film	Strain				
	Micrococcus luteus ATCC 10240	*Listeria innocua* ATCC 33090	*Salmonella enteritidis* ATCC 14931	*Escherichia coli* ATCC 9637	*Staphylococcus aureus* ATCC 25923
Control without antimicrobial additive	−	−	−	−	−
With nisin	++	−	−	−	−
With lysozyme	+	−	−	−	−
With nisin + EDTA	++	−	−	−	−
With lysozyme + EDTA	+	−	−	−	−
With nisin + lysozyme + EDTA	+++	+++	+++	+++	+++
With EDTA	++	−	−	−	−
With EDTA + grapefruit seed extract	++	++	+	++	++
With grapefruit seed extract	++	−	−	−	−

−, no inhibition; +, weak inhibition (<2 mm in inhibition zone); ++, inhibition zone of 2–5 mm; +++, strong inhibition with diameter >5 mm. From Cha *et al.* (2002), reproduced with permission.

Nisin and chitosan migrated into water and equilibrated at about 8% for nisin and 1% for chitosan. In addition, the release rates and levels were not affected by the presence of other preservatives in the polymer coating. Nisin and chitosan in an acidified vinyl acetate–ethylene copolymer coating layer exhibited strong microbial inhibition against *L. monocytogenes* and *E. coli* O157:H7, and improved the microbial quality of milk and orange juice stored at 10°C. Cha *et al.* (2003) also designed chitosan films containing nisin with a strong antimicrobial activity against *M. luteus*.

Antimicrobial packaging systems containing natural antimicrobial agents can confer specific or broad microbial inhibition, depending on the agents used and their concentrations. Different types of antimicrobial-delivery mechanisms, polymers, and packaging material – food combinations may be applied or used to maximize the effectiveness of the system.

Antioxidative packaging

Antioxidants can be incorporated into or coated onto food packaging materials to control the oxidation of fatty components and pigments, and thus can contribute to the quality preservation of foods (Vermeiren *et al.*, 1999). The antioxidants incorporated into plastic packaging materials may have the dual role of protecting the polymer as well as the packaged food from oxidation (Waite, 2003). Antioxidative packaging can retard the oxidative reactions of fatty ingredients in packaged foods. The incorporation of synthetic antioxidants, such as butylated hydroxytoluene (BHT) and butylated hydroxyanisole (BHA), in high-density polyethylene (HDPE) liners have been shown to protect cereals from oxidation (Miltz *et al.*, 1988; Wessling *et al.*, 2001). However, because of the growing concern about the use of food chemicals, there is greater interest in using

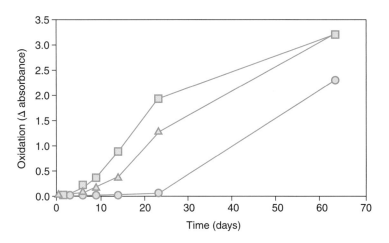

Figure 7.4 Effect of α-tocopherol-incorporated LDPE film on the oxidation of linoleic emulsion at 6°C.
■, no addition; ▲, added in 360 ppm; ●, added in 3400 ppm. From Wessling *et al.* (2000), reprinted with permission of John Wiley & Sons Ltd.

natural antioxidants such as α-tocopherol and ascorbic acid. In particular, α-tocopherol can be easily dissolved into polyolefins and does not break down under polymer processing conditions. A significant concentration of α-tocopherol also usually remains in the final plastic films (Ho *et al.*, 1998; Wessling *et al.*, 1999). Zambetti *et al.* (1995) found that the use of α-tocopherol as an antioxidant in LDPE used in extrusion coating, and in HDPE used in extrusion-blown bottles, improved the sensory performance of the bottled water by preventing the oxidation of the polymer substrates. Wessling *et al.* (1999) found that α-tocopherol incorporated into polypropylene (PP) retained more of its activity than when incorporated into LDPE, when both films were in contact with various foods and food-simulating liquids. The α-tocopherol in LDPE decreased at considerably higher rates with high fat-containing foods or with high-alcohol liquids. Therefore, LDPE containing an active antioxidant can offer the potential of being used as an active packaging because it can help to transfer the incorporated α-tocopherol onto the packaged foods, which will help prevent the oxidation of the food products. PP may be used as oxygen-scavenging packaging in the headspace or at the packaged foods' surface.

LDPE films that incorporated α-tocopherol above 360 ppm delayed the oxidation of linoleic acid emulsions at 6°C (Figure 7.4), but did not do so at 20° and 40°C (Wessling *et al.*, 2000). LDPE films containing α-tocopherol above 700 ppm could not prevent the oxidation of oatmeal at 20°C; this is probably due to the poor contact of the packaging film with the oatmeal product (Wessling *et al.*, 2001). For α-tocopherol-incorporated films to be effective, it is necessary that there is close contact between the packaging material and the food product. The α-tocopherol in the film is degraded chemically without volatiles being loss, while BHT diffuses out of the film and is volatilized on the surface, eventually migrating to the dry food (Miltz *et al.*, 1988; Wessling *et al.*, 2001).

Another concern with α-tocopherol being incorporated into LDPE films is the potential changes it causes in the mechanical properties, color, and oxygen permeability of the film (Wessling *et al.*, 2000).

Ascorbic acid can scavenge oxygen in the headspace air of canned or bottled products. Its ability to scavenge oxygen has made it a possible candidate for scavenging oxygen in active food packaging technologies (Smith *et al.*, 1995). Ascorbic acid in combination with other antioxidants can act as an antioxidant, and can retard rancidity (Yanishlieva-Maslarova, 2001). It can also protect against the peroxidation of lipid components by trapping the peroxyl radical in the aqueous phase. Due to this, the antioxidative property of ascorbic acid has been incorporated into and used in the fabrication of oxygen-scavenging plastics (Rooney, 1995).

Perishable foods that are sensitive to both microbial spoilage and oxidative deterioration may be packaged in polymeric materials containing both antimicrobial and antioxidant additives. Lee *et al.* (2004) produced antimicrobial and antioxidant packaging materials that incorporated nisin and/or α-tocopherol. They examined their effectiveness with an emulsion model system and in milk cream. At 10°C, α-tocopherol migrated into an oil-in-water emulsion at a faster rate and reached an equilibrium level at about 6%, while nisin migrated at about 9%. Therefore, the inclusion of both α-tocopherol and nisin in coated paper could provide antimicrobial and antioxidative functions. However, no synergic or interactive effects in the antimicrobial or antioxidative activities were observed by the use of the combination (Lee *et al.*, 2004). One application combining ascorbic acid and silver zeolite has been reported for antioxidative and antimicrobial activity (Rooney, 1995).

Future potential

Antimicrobial and antioxidative packaging systems containing natural active substances may have high potential for commercial food packaging applications. Consumers would prefer to obtain better food safety of fresh produce and minimally processed foods using this type of packaging system. However, it is necessary that new active packaging systems must satisfy food safety regulations, which are different in each country. Greater food safety and quality assurance may be improved by the use of both antimicrobial and antioxidative packaging systems incorporating natural active agents. Some commercial products, such as films coated with wasabi extract, are already on the market and satisfy the regulations and consumer needs of particular countries. Most materials containing natural active agents are more effective when there is direct contact of the packaging materials with the food product. For new packaging systems to be introduced into the market effectively, careful design is required. Natural antimicrobials and antioxidants are usually costly, and therefore further development of package design using the minimum of active agents is desirable for practical applications. Food contact layers with the appropriate concentrations of the active agents may be laminated or coated onto the barrier layer of the package structure. There are some suggestions that before long many countries would adopt the new active packaging concepts into their packaging regulations. Therefore, new applications of antimicrobial and antioxidative packaging are likely to be available in the market sooner or later.

References

An, D. S., Kim, Y. M., Lee, S. B., Paik, H. D. and Lee, D. S. (2000). Antimicrobial low density polyethylene film coated with bacteriocins in binder medium. *Food Sci. Biotechnol.* **9,** 14–20.

Appendini, P. and Hotchkiss, J. H. (1997). Immobilization of lysozyme on food contact polymers as potential antimicrobial films. *Packag. Technol. Sci.* **10,** 271–279.

Appendini, P. and Hotchkiss, J. H. (2002). Review of antimicrobial food packaging. *Innov. Food Sci. Emerg. Technol.* **3,** 113–126.

Begin, A. and Van Calsteren, M. (1999). Antimicrobial films produced from chitosan. *Intl J. Biol. Macromol.* **26,** 63–67.

Bower, C. K., McGure, J. and Daeschel, M. A. (1995). Protein antimicrobial barriers to bacterial adhesion. *Appl. Environ. Microbiol.* **61,** 992–997.

Bower, C. K., Daeschel, M. A. and McGuire, J. (1998). Suppression of *Listeria monocytogenes* colonization following adsorption of nisin onto silica surfaces. *J. Dairy Sci.* **81,** 2771–2778.

Buonocore, G. G., Del Nobile, M. A., Panizza, A., Corbo, M. R. and Nicolais, L. (2003). A general approach to describe the antimicrobial agent release from highly swellable films intended for food packaging applications. *J. Contr. Release* **90,** 97–107.

Cha, D. S., Choi, J. H., Chinnan, M. S. and Park, H. J. (2002). Antimicrobial films based on Na-alginate and κ-carrageenan. *Lebensm. Wiss. u. Technol.* **35,** 715–719.

Cha, D. S., Cooksey, K., Chinnan, M. S. and Park, H. J. (2003). Release of nisin from various heat-pressed and cast films. *Lebensm. Wiss. u. Technol.* **36,** 209–213.

Chi-Zhang, Y., Yam, K. L. and Chikindas, M. L. (2004). Effective control of *Listeria monocytogenes* by combination of nisin formulated and slowly released into a broth system. *Intl J. Food Microbiol.* **90,** 15–22.

Cho, S. H., Chung, J. H. and Ryu, C. H. (1994). Inhibitory effects of natural antimicrobial agent on postharvest decay in fruits and vegetables under natural low temperatures. *J. Korean Soc. Food Nutr.* **23,** 315–321.

Cho, Y. W., Han, S. S. and Ko, S. W. (2000). PVA containing chito-oligosaccharide side chain. *Polymer* **41,** 2033–2039.

Chung, S. K., Cho, S. H. and Lee, D. S. (1998). Modified atmosphere packaging of fresh strawberries by antimicrobial plastic films. *Korean J. Food Sci. Technol.* **30,** 1140–1145.

Cutter, C. N. and Siragusa, G. R. (1997). Growth of *Brochothrix thermosphacta* in ground beef following treatments with nisin in calcium alginate gels. *Food Microbiol.* **14,** 425–430.

Daeschel, M. A., Mcguir, J. and Al-Makhalafi, H. (1992). Antimicrobial activity of nisin adsorbed to hydrophilic and hydrophobic and hydrophobic silicon surfaces. *J. Food Prot.* **55,** 731–755.

Furuya, K. and Isshiki, K. (2001). Effect of humidity on allyl isothiocyanate antimicrobial activity. *J. Japanese Soc. Food Sci. Technol.* (in Japanese) **48,** 738–743.

Ha, J. U., Kim, Y. M. and Lee, D. S. (2001). Multilayered antimicrobial polyethylene films applied to the packaging of ground beef. *Packag. Technol. Sci.* **14,** 55–62.

Han, J. H. (2000). Antimicrobial food packaging. *Food Technol.* **54(3),** 56–65.

Han, J. H. (2003). Antimicrobial food packaging. In: *Novel Food Packaging Technologies* (R. Ahvenainen, ed.), pp. 50–70. Woodhead Publishing, Cambridge, UK.

Ho, Y. C., Young, S. S. and Yam, K. L. (1998). Vitamin E based stabilizer components in HDPE polymer. *J. Vinyl Add. Technol.* **4,** 139–150.

Hoffman, K. L., Dawson, P. L., Acton, J. C., Han, I. Y. and Ogale, A. A. (1998). Film formation effects on nisin activity in corn zein and polyethylene films. *Activities Report R & D Associates* **50(1),** 238–244.

Hurst, A. (1983). Nisin and other inhibitory substances from lactic acid bacteria. In: *Antimicrobials in Foods* (A. L. Branen and P. M. Davidson, eds), pp. 327–351. Marcel Dekker, New York, NY.

Kim, Y. R. and Cho, S. H. (1996). Antimicrobial activities and effect of grapefruit seed extract on the physiological function of microorganisms. *Korean J. Post-Harvest Sci. Technol. Agric. Prod.* **3,** 187–193.

Kim, W. Y., Cheong, N. E., Jae, D. Y. *et al.* (1994). Characterization of an antimicrobial chitinase purified from the grapefruit seed extract. *Korean J. Plant Pathol.* **10,** 277–283.

Kim, Y. M., Lee, N. K., Paik, H. D. and Lee, D. S. (2000). Migration of bacteriocin from bacteriocin-coated film and its antimicrobial activity. *Food Sci. Biotechnol.* **9,** 325–329.

Kim, Y. M., Paik, H. D. and Lee, D. S. (2002a). Shelf life characteristics of fresh oysters and ground beef as affected by bacteriocin-coated plastic packaging film. *J. Sci. Food Agric.* **82,** 998–1002.

Kim, Y. M., An, D. S., Park, H. J., Park, J. M. and Lee, D. S. (2002b). Properties of nisin-incorporated polymer coating as antimicrobial packaging materials. *Packag. Technol. Sci.* **15,** 247–254.

Ko, S., Janes, M. E., Hettiarachchy, N. S. and Johnson, M. G. (2001). Physical and chemical properties of edible films containing nisin and their action against *Listeria monocytogenes*. *J. Food Sci.* **66,** 1006–1011.

Koichiro, Y. (1993). Keeping freshness by antimicrobial packaging incorporating wasabi extract. *Food Sci.* (in Japanese) **35(11),** 102–107.

Kusunoki, H., Shin, K., Kokubo, Y., Kaneko, S., Sekiyama, Y. and Uemura, T. (1998). Antimicrobial activity of allyl isothiocyanate against low temperature growth bacteria. *Japan. J. Food Microbiol.* (in Japanese) **15,** 107–112.

Lee, D. S., Hwang, Y. I. and Cho, S. H. (1998). Developing antimicrobial packaging film for curled lettuce and soybean sprouts. *Food Sci. Biotechnol.* **7,** 117–121.

Lee, C. H., An, D. S., Park, H. J. and Lee, D. S. (2003). Wide-spectrum antimicrobial packaging materials incorporating nisin and chitosan in the coating. *Packag. Technol. Sci.* **16,** 99–106.

Lee, C. H., An, D. S., Lee, S. C., Park, H. J. and Lee, D. S. (2004). A coating for use as an antimicrobial and antioxidative packaging material incorporating nisin and α-tocopherol. *J. Food Eng.* **62,** 323–329.

Li, Q., Dunn, E. T., Gransmaison, E. W. and Goosen, M. F. A. (1997). Applications and properties of chitosan. In: *Applications of Chitin and Chitosan* (M. F. A. Goosen, ed.), pp. 1–29. Technomic Publishing, Lancaster, PA.

Miltz, J., Passy, N. and Manneheim, C.H. (1995). Trends and applications of active packaging systems. In: *Foods and Packaging Materials – Chemical Interactions* (P. Ackermann, M. Jagerstad and T. Ohlsson, eds), pp. 201–210. The Royal Society of Chemistry, Cambridge, UK.

Miltz, J., Hoojjat, P., Han, J. K., Giacin, J. R., Harte, B. R. and Gray, I. J. (1988). Loss of antioxidants from high-density polyethylene: its effect on oatmeal cereal oxidation. In: *Food and Packaging Interactions* (J. H. Hotchkiss, ed.), pp. 83–93. American Chemical Society, Washington DC, USA.

Ming, X., Weber, G. H., Ayres, J. W. and Sandine, W. E. (1997). Bacteriocins applied to food packaging materials to inhibit *Listeria monocytogenes* on meats. *J. Food Sci.* **62,** 413–415.

Nicholson, M. D. (1998). The role of natural antimicrobials in food/packaging biopreservation. *J. Plast. Film Sheet* **14,** 235–241.

Nielsen, P. V. and Rios, R. (2000). Inhibition of fungal growth on bread by volatile components from spices and herbs, and the possible application in active packaging, with special emphasis on mustard essential oil. *Intl J. Food Microbiol.* **60,** 219–229.

Ooki, K. (1996). Antimicrobial films for food packaging. *Japan. Food Sci.* (in Japanese) **35(10),** 57–67.

Ouattara, B., Simard, R. E., Piette, G., Begin, A. and Holley, R. A. (2000). Diffusion of acetic acid and propionic acid from chitosan based antimicrobial packaging films. *J. Food Sci.* **65,** 768–773.

Park, S. I., Marsh, K. S., Dawson, P. L., Acton, J. C. and Han, I. (2002). Antimicrobial activity of chitosan incorporated polyethylene films. Paper No. 100B–23, 2002 IFT Annual Meeting Book of Abstracts, p. 250. Institute of Food Technologists, Anaheim, CA.

Ray, B. (1992). Nisin of *Lactococcus lactis* ssp. *lactis* as a food biopreservative. In: *Food Biopreservatives of Microbial Origin* (B. Ray and M. Daeschel, eds), pp. 207–264. CRC Press, Boca Raton, FL.

Rooney, M. L. (1995). Active packaging in polymer films. In: *Active Food Packaging* (M. L. Rooney, ed.), pp. 74–110. Blackie Academic & Professional, Glasgow, UK.

Rück, E. and Jager, M. (1995). *Antimicrobial Food Additives*, 2nd edn., pp. 208–213. Springer, Berlin, Germany.

Sakamoto, S., Sato, K., Maitani, T. and Yamada, T. (1996). Analysis of components in natural food additive 'grapefruit seed extract' by HPLC and LC/MS. *Bull. Natl (Japan) Inst. Health Sci.* **114,** 38–42.

Scannell, A. G. M., Hill, C., Ross, R. P., Marx, S., Hartmeier, W. and Arendt, E. K. (2000). Development of bioactive food packaging materials using immobilised bacteriocins Lacticin 3147 and Nisaplin. *Intl J. Food Microbiol.* **60,** 241–249.

Siragusa, G. R., Cutter, C. N. and Willett, J. L. (1999). Incorporation of bacteriocin in plastic retains activity and inhibits surface growth of bacteria on meat. *Food Microbiol.* **16,** 229–235.

Smith, J. P., Hoshino, J. and Abe, Y. (1995). Interactive packaging involving sachet technology. In: *Active Food Packaging* (M. L. Rooney, ed.), pp. 143–173. Blackie Academic & Professional, Glasgow, UK.

Tokura, S., Ueno, K., Miyazaki, S. and Nishi, N. (1997). Molecular weight dependent antimicrobial activity by chitosan. *Macromol. Symp.* **120,** 1–9.

Tsai, G. J., Su, W. H., Chen, H. C. and Pan, C. L. (2002). Antimicrobial activity of shrimp chitin and chitosan from different treatments and applications of fish preservation. *Fisheries Sci.* **68,** 170–177.

Vermeiren, L., Devlieghere, F., Van Beest, M., de Kruijf, N. and Debevere, J. (1999). Developments in the active packaging of foods. *Trends Food Sci. Technol.* **10,** 77–86.

Waite, N. (2003). *Active Packaging*, pp. 75–84. Pira International, Leatherhead, UK.

Wessling, C., Nielsen, T., Leufven, A. and Jagerstad, M. (1999). Retention of α-tocopherol in low-density polyethylene (LDPE) and polypropylene (PP) in contact with food-stuffs and food-simulating liquids. *J. Sci. Food Agric.* **79,** 1635–1641.

Wessling, C., Nielsen, T. and Andres, L. (2000). The influence of α-tocopherol concentration on the stability of linoleic acid and the properties of low-density polyethylene. *Packag. Technol. Sci.* **13,** 19–28.

Wessling, C., Nielsen, T. and Giacin, J. R. (2001). Antioxidant ability of BHT- and α-tocopherol-impregnated LDPE film in packaging of oatmeal. *J. Sci. Food Agric.* **81,** 194–201.

Yanishlieva-Maslarova, N. V. (2001). Inhibiting oxidation. In: *Antioxidants in Food: Practical Applications* (J. Pokorny, N. Yanishlieva and M. Gordon, eds), pp. 23–70. Woodhead Publishing, Cambridge, UK.

Zambetti, P. F., Baker, S. L. and Kelley, D. C. (1995). Alpha tocopherol as an antioxidant for extrusion coating polymers. *Tappi J.* **78(4),** 167–171.

Oxygen-scavenging packaging

8

Michael L. Rooney

Introduction

Traditionally, oxygen-sensitive foods and beverages have been packaged in such a way as to minimize their exposure to oxygen. This oxygen may be in the package at the time of sealing, or may enter the pack by permeation or leakage over the storage life. The first source of oxygen has frequently been addressed by the use of evacuation and inert-gas flushing, while the ingress has been minimized by optimal sealing and the use of high-barrier packaging materials. The approaches used to solve both oxygen problems are now recognized as providing only partial solutions, leaving options for improvements in food quality, production and distribution economics, reduction in environmental impacts, and increased convenience of use.

The impact of oxygen on food quality, and ultimately shelf life, is dependent not only upon the quantity of oxygen available for chemical oxidation or support of growth of organisms but also upon the rate of the reactions which consume the oxygen. This, in turn, will be influenced by the solubility of the gas in the medium provided by the food or beverage. The oxidation of fat in, for example, potato crisps has been shown to be highly dependent upon water activity, with a minimum rate at a_w 0.3–0.4, and the increase in reaction rate above this value is interpreted in terms of the formation of an aqueous multilayer on the food with consequent dissolution of oxygen and enhancement of the oxidation (Taoukis *et al.*, 1988). Much of the trade literature on oxygen-scavenging packaging presents a focus on the quantity of oxygen in a package without consideration of the widely different rates at which food quality can be degraded.

Innovations in Food Packaging
ISBN: 0-12-311632-5

The quantity of oxygen which must be taken up by a food to limit its shelf life to one year has been estimated for a range of foods (Koros, 1990). The quantities lie in a range from a few ppm to a few hundred ppm based on the weight of the food. Removal of oxygen to these small amounts by conventional means is not generally achievable if the food or beverage has components that react rapidly. Beer is one of the most studied beverages, and it has been found that an uptake of around 1 ppm (or a little more) results in the beer reaching its shelf life. A conventional bottle closed with a crown seal allows an uptake or around 750 ppb of oxygen over 3 months and 2000 ppb over 8 months (Teumac, 1995).

Reviews

The various aspects of oxygen-scavenging packaging systems have been reviewed as part of general reviews of active packaging, but most of the reviews of the subject itself are found in the proceedings of conferences (Kline, 1999; Brody et al., 2001). This field is commercially the most developed in active packaging, but most of the papers at conferences still consist of presentations by developers or vendors of systems. There is still a lack of detailed investigation of the comparative performance of the systems on offer to the food industry. Brody et al. (2001) have described most of the commercial plastics-based systems and have given examples of their potential use. Smith et al. (1995) have dealt in depth with the applications of oxygen-absorber sachets, but there have been substantial developments in the area of adhesive oxygen absorber "labels" since that time.

History

The development of oxygen-scavenging systems has followed two lines, depending upon whether the oxidizable substance was designed to be a part of the package or to be inserted into it with the food. The insert approach includes self-adhesive labels, adhesive devices or free sachets included with the food. Modification of packaging materials to confer oxygen-scavenging capability includes monolayer and multilayer materials, and reactive closure liners for bottles and jars.

Package inserts

Oxidation of metals

Brody et al. (2001) refer to early attempts to scavenge oxygen using ferrous sulfate in the 1920s. An early patent was authored by Tallgren (1938), who proposed the use of iron, zinc or manganese to remove oxygen from canned foods. This approach was developed further, but research focused on the use of iron powder in sachets of porous material. Since the oxidation of iron is not inherently rapid, the approach of adding accelerators or adsorbents became popular. Addition of sodium carbonate was patented by

Buchner (1968), and the use of alkali metal halides was patented by the Mitsubishi Gas Chemical Company in 1977, leading to the range of Ageless™ sachets, cards, labels, and closure liners. The development of these concepts into commercial products was accompanied by a progressive movement of the innovation from Europe to Japan, in which the bulk of the current market is found. Dainelli (2003) reported that around 12 billion oxygen absorber units were sold in Japan in 1999, versus 300 000 in the EU.

Iron-based compositions in the form of self-adhesive labels are manufactured by Multisorb Technologies under the trade names FreshMax™ and FreshCard™ in the USA. These adjuncts are applied to the inner wall or lidding film of packages as part of the filling operation. This allows placement of the absorber in a pre-selected position, such as behind the print on the lidding film in packs of smoked, sliced meats in the UK. Thus the esthetic property of the system is enhanced while retaining the functionality. An added benefit is that the absorber is prevented from moving to a position where its access to the headspace would be inhibited by intimate contact with a piece of food.

Other oxidation reactions

Patents for the oxidation of sulfites began appearing as early as 1965, when Bloch proposed the use of bisulfite salts in cloth or paper bags for insertion into cans. This patent was assigned to the US government. He saw the need for prevention of oxidation of dehydrated products in particular, and noted the capability of a variety of sulfite forms to be useful in the presence of an acid such as lactic acid. The reactions required the presence of water and this could be added, even though some might be available as water of crystallization in one or more of the salts. It is clear that the use of oxidase enzymes was already being discussed, and Bloch noted the reactivity of bisulfites at refrigerator temperatures when enzymes were inefficient. It is interesting to note that he recognized the need for intimate mixing of the bisulfite with an activator such as an iron salt, particularly in the presence of a porous support such as carbon or silica gel. Another useful feature was his recommendation of the inclusion of a carbonate salt to release carbon dioxide to maintain pressure in flexible packs as the oxygen is removed. Most of these considerations are emphasized in the more recent literature of oxygen absorbers. Perhaps most important was his recognition that the relative rates of bisulfite oxidation and food degradation might be different.

The reaction of sodium metabisulfite in the presence of lime was patented by Yoshikawa *et al.* (1977). These workers showed that the uptake of oxygen could be triggered by the presence of water in the food. This concept was concurrently being applied in the development of iron-based systems, and marks the key concept in providing a useful commercial oxygen absorber system that could be stored and handled before use.

Oxygen absorbers based on sulfur compounds have been replaced by iron-based compositions, although the former are still used in combination with ascorbic acid in oxygen-scavenging closure liners.

The use of oxidase enzymes for the removal of oxygen from food packages has been discussed by Brody and Budny (1995), and Brody *et al.* (2001). The earliest known application of an oxygen absorber in a sachet was reported to be that of Scott

(1958). This system involved the oxidation of glucose catalyzed by glucose oxidase in the presence of water. The hydrogen peroxide formed was removed by the action of catalase. The concept was commercialized by Scott's company, Fermco Laboratories, using the trade name "Oxyban". A commercial enzymic oxygen scavenger in a sachet is marketed internationally by Bioka, a Finnish company.

The oxidation of hydrogen to form water may appear to be a very attractive mechanism for oxygen scavenging. This process has been used by microbiologists for deoxygenating anaerobic jars for cultivation of anaerobes for many years. The process was proposed for the packaging of milk powder in cans by King (1955), and subsequently by Abbott et al. (1961). The process involved flushing cans of milk powder with a mixture of hydrogen (7%) in nitrogen. The hydrogen reacts with oxygen on the surface of palladium-coated steel attached by means of adhesive tape to the inside of the lid. The particular benefit of this process was the removal of oxygen released from pores in the spray-dried powder over several days following closure.

A variety of other systems, such as catechols, glycols and boron compounds, has been the subject of patents or has been released for commercial sale. Both the Toppan Printing Company and the Mitsubishi Gas Chemical Company have developed ascorbic acid-based sachet systems for use where fast oxygen absorption is required in the presence of carbon dioxide. Generally these compositions are non-metallic, and do not place restrictions on the use of metal detectors on packaging lines.

Packaging materials as oxygen scavengers

Although the performance of oxygen-absorbing sachets was quite satisfactory for a wide range of food storage conditions, a number of limitations to their use in practice were recognized. The esthetics of inserts, coupled with a concern about possible ingestion or rupture, as well as their unsuitability for use with beverages, drove researchers to seek package-based solutions. The approach of using the packaging material as the medium for the oxygen-scavenging chemistry was developed independently in several laboratories and countries. Not surprisingly, the reactions were initially the same as used in the sachet technologies, but eventually it was recognized that the restrictions applying to package inserts need not apply to the package. This has allowed a multiplicity of oxygen access problems, arising from quite disparate packaging factors, to be addressed, thus permitting targeting of problems at their source rather than waiting for the oxygen to enter the package to be absorbed by an insert, such as a sachet. Some of the chemistries and reaction media used in packaging-material-based systems are summarized in Table 8.1.

The use of palladium on alumina as a catalyst is common in organic chemistry. Its use in a sandwich of the powder in a laminate containing a polyvinyl alcohol barrier layer and the polyolefin heat-seal layer offered a number of advantages as a headspace scavenger (Kuhn et al., 1970). The packages of intermediate moisture foods or milk powder were flushed with a mixture of hydrogen, 5%, in nitrogen and sealed. The hydrogen and nitrogen both diffuse through the sealant to the powdered palladized alumina. The water formed on the catalyst surface is absorbed by the alumina. The packaging is expensive to fabricate, and the use of hydrogen mixtures is not popular in the industry. None of this

Table 8.1 Oxygen scavengers with different chemistries

Substrate	Medium	Structure	Application	Commercial
Hydrogen	Pd/alumina	Sandwich	Foil laminates	Briefly
Singlet oxygen acceptors	Plastics	Homogeneous plastic	Plastics packaging	No
Sulfites	Salt blend	Sandwich	Laminates	No
Sulfites	Solution	Sandwich	Bag-in-box	No
Sulfites/ascorbate	Plastics	Compound	Closure liner	Yes
Aromatic nylon	Polymer/blend	Mono/multilayer	PET bottles	Yes
Iron powder	Plastics	Compound	Plastics packaging	Yes
Reducible compounds	Plastics	Blend/polymer	Plastics packaging	No
Polymer-bound olefins	Plastics	Coextrusions	Flexibles	Yes
Polydiene block copolymers	Plastics	Blend/multilayer	PET bottles	Yes

detracts from the inherent high efficiency of the process, which demonstrated quite early that high permeability of the sealant layer is beneficial in oxygen scavenging.

Farrell and Tsai (1985) introduced the concept of dealing with the enhancement of oxygen permeability of plastic packages during and after retorting. They noted that the quantity of water permeating the package was enough to hydrate the sulfite or another hygroscopic salt mixed with it. They took advantage of this fact to provide a concentrated aqueous solution of the oxidizable sulfite between two layers of the packaging material to achieve a high rate of oxygen removal at retort temperature.

The problem of enhanced permeability of EVOH barrier layers to oxygen during retorting has been approached by Tsai and Wachtel (1990) by attempting to keep the EVOH dry, or by Bissot (1990) by including microscopic inorganic platelets in the polymer to introduce a longer diffusion path. It appears that a combination of the oxygen-scavenging packaging coupled with the use of a desiccant might reduce the access of oxygen to the packaged food during the post-retorting period when the EVOH slowly loses water, thereby increasing its barrier.

Scholle (1977) advanced the concept of using an aqueous solution of the sulfite trapped between the layers of a multiwall package, such as bag-in-box. This concept was directed at the rapid removal of headspace and dissolved oxygen from liquid foods and beverages such as wine. This is a particular problem associated with bag-in-box, where it is difficult to avoid air headspaces due to both filling, and that already between the webs used to make up the bag.

Cook (1969) described the use of an all-organic system involving a plastic bilayer separated by a solution of common antioxidants in a minimum quantity of a high-boiling organic solvent. Antioxidants are normally designed to trap free radicals originating from the first step of autoxidation, so the claimed reduction in oxygen permeability is a surprising discovery. This system was designed to reduce the oxygen permeability of the bilayer primarily for meat packaging, with one example being a sandwich between two layers of PVDC.

Plastics with blended sulfites have also been used to remove oxygen diffusing under bottle closures. Teumac (1995) has summarized the commercial developments from 1989, and describes the compositions as including up to 7% sodium sulfite and up to 4% sodium ascorbate. In some patents the use of isoascorbic acid has been proposed.

In the sulfite-free compositions, the quantity of ascorbate determines the scavenging capacity while the amount of catalyst determines the rate. A parallel development in Japan by Toyo Seikan Ltd involved the use of iron in the closure liner.

The success of iron in package inserts gave momentum to research into the potential use of iron in plastic-based compositions. In the early 1990s, several companies launched products based on compounds of iron with polyolefin polymers. These were launched in Japan (Toyo Seikan's Oxyguard) and the USA (Amoco Chemicals' Amosorb 1000 and 2000). The latter became Shelfplus when bought by Ciba Specialty Chemicals. These resins are extrudable under normal conditions of temperature, and are used in oxygen-barrier laminations to packages, particularly for foods subjected to conditions of high temperature and humidity and especially in thermoformed trays. Toyo Seikan produced pouches for blood-product bags with one transparent side and the white-pigmented Oxyguard side. These developments were accompanied by several similar proposed compositions, which did not reach commercialization.

Incorporation of iron into resin strips placed in sachets has been utilized by the Mitsubishi Gas Chemical Company as an alternative to filling the sachets loosely with reduced iron powder. The sachets contain polyolefin strips, which are micro-porous, and the pores allow the oxygen and water vapor increased access to the iron particles compared with the same composition in a continuous film strip. This innovation addresses the need for sachets that do not interfere with microwave reheating of the food, such as semi-aseptic rice. Another welcome benefit is removal of the danger of accidental release of iron powder if the sachet is ruptured when the package is opened, for example with a knife. Brody *et al.* (2001) describe a patent by Kawatiki *et al.* (1992) involving a seemingly related concept involving microvoids generated by stretching an iron-loaded strip of plastic.

Homogeneous plastic structures

The progress from package insert to reactive polymers, via various blends of solids (or inclusion of trapped liquid scavenger solutions), also included homogeneous solutions of reagents in polymers. This provided the opportunity for increased clarity of flexible and rigid packaging, and reduced interference with the inherent properties of the polymers used. This also had the potential to reduce limitations on the component polymers, which might be desired in particular packaging structures.

There was the first multilayer plastic structure in which antioxidants were claimed to function as oxygen "getters". According to Brody *et al.* (2001), the process involved dispersing very minor portions of conventional antioxidants in or between layers in a multilayer. The process appears not to have advanced from that point.

The first plastic to incorporate dissolved reagents with known oxidation chemistry involved the light-energized excitation of oxygen diffusing into the plastic (Rooney and Holland, 1979). The substrate for oxidation does not react with ground-state oxygen, so the oxygen to be scavenged had to be excited to the singlet state. This was achieved by including a photosensitizing dye and exposing the scavenger film to visible light. The process occurs only while the scavenger film is exposed to the light, as shown in Figure 8.1.

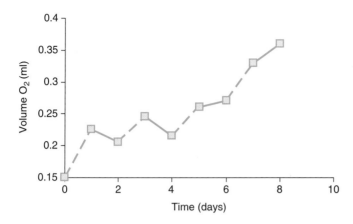

Figure 8.1 Effect of illumination (solid line) and darkness (dashed line) on the cumulative volume of oxygen permeating a scavenger film laminate.

The laminate consisting of polyethylene/nylon 6/cellulose acetate contained a singlet oxygen acceptor, 1,3-diphenylisobenzofuran, 8×10^{-2} M, and methylene blue dye, 10^2 M. The laminate separated the two compartments of a permeability cell, air was placed on the nylon side, and nitrogen with a small oxygen residue was placed on the scavenger side. In darkness the laminate displayed its normal permeability characteristics, but when illuminated it not only prevented permeation but also scavenged oxygen that had permeated during the previous dark period. This process continued until the concentration of scavenger was reduced to a point at which scavenging was now incomplete and some oxygen permeation occurred. This research demonstrated, for the first time, activation by the use of light as well as the finite oxygen-scavenging capacity of film-based scavengers – a topic that has subsequently become commercially important.

Plastics compositions with a higher scavenging capacity for oxygen were subsequently investigated by the same research team (Rooney and Holland, 1979), and the factors affecting the rate of oxygen scavenging by solutions of singlet oxygen acceptors in polymers were elucidated. During the 1980s there appears to have been no reported research other than this academic work focusing on the polymer as a reaction medium for diffusing oxygen molecules (Rooney *et al.*, 1981). It was shown that ascorbic acid could perform as a singlet oxygen acceptor, and that the photochemistry imposed limits on the scavenging rate (Rooney *et al.*, 1982). The use of the one polymer as both the reagent and the reaction medium was investigated using natural rubber (Rooney, 1982). This work was extended to other rubbers which display different inherent reactivities with singlet oxygen while having similar values of oxygen permeability. The process was applied in a poly (furyloxirane) designed to have an even higher reactivity towards singlet oxygen (Maloba *et al.*, 1994).

A more practical approach to creating a polymeric total barrier to oxygen permeation based on the transition-metal-catalyzed oxidation of aromatic nylons was developed by the Carnaud Metal Box Company under the trade name Oxbar™ (Cochran *et al.*, 1991). The key advantage initially seen for such a process was the ability to blend the polymer plus catalyst with PET in the manufacture of bottles for wine and beer.

Few polymers are compatible with PET, so this approach constituted a breakthrough in the development of PET bottles for oxygen-sensitive beverages in general. It was shown that around 200 ppm of cobalt is necessary in a 7% blend of MXD-6 nylon in PET in order to generate a total barrier to oxygen permeation (Folland, 1990). Several developments by competing companies have resulted in approaches that use this chemistry either in a PET matrix or in a separate MXD-6 nylon layer co-injected with two layers of PET. One such patent involves use of a blend of MXD-6 nylon, polyester, and a cobalt salt in the core layer of bottles (Collette, 1991). Other commercial approaches have been to incorporate the cobalt into a thin (10 μm) layer of MXD-6 sandwiched between PET layers in bottles.

An alternative approach to dealing with the PET compatibility issue was devised in Amoco Chemicals (Cahill and Chen, 1997). This approach involved making a block copolymer of a polybutadiene with PET with the trade name Amosorb™ 3000, sold by BP Amoco. The PET caused the compatibilization while the polybutadiene was the oxidizable polymer. The process is catalyzed by means of a transition metal salt. This polymer lent itself to use in PET bottle manufacture, since the catalyst can be added at a late stage in the injection-molding process and premature oxidation can be minimized. Even though the injection-molded preforms have been heated while containing the catalyst, they have a substantial shelf life due to the low permeability of the thick layers of PET, especially when the Amosorb™ 3000 is in a buried layer. Perhaps an improvement would be to include some form of triggering or activation closer to the time of filling the bottle.

Triggering of oxygen-absorbing sachets was an essential feature present from the time they were initially introduced commercially. In that case the trigger was the water necessary for the rusting of iron by the oxygen. In general, this water came from the food and the sachets were not prematurely activated unless exposed to the air for too long. The concept of triggering an otherwise unreactive plastic system was also demonstrated in the 1970s in the singlet oxygen approach to oxygen scavenging (Rooney and Holland, 1979). Around the same time, Rabek and Ranby (1975) showed that the oxidative degradation of a plastic in sunlight was substantial if a photosensitizer and a transition metal salt were dispersed therein. Against this background, the breakthrough concept of single-dose triggering has been developed and introduced commercially. Speer et al. (1993) were the first to claim that films of unsaturated polymers, such as poly(1,2-butadiene), could very effectively scavenge ground-state oxygen provided they contained a transition metal catalyst and a photosensitizer. The use of pendant C=C double bonds was designed to minimize rupture of the polymer backbone during oxidation. This overcame one of the significant drawbacks of the process of Rooney (1982). Speer's process has been developed with a large number of patents, and has been marketed by the Cryovac division of Sealed Air Corporation under the trade name OS 1000™.

The approach of using autoxidation of unsaturated groups on a polymer as a basis for oxygen scavenging has been taken up by Chevron Chemical Company, with a further development of the Cryovac concept. In this case the oxidizable moity is a cyclohexene side group bound to the backbone, for instance by transesterification (Ching et al., 1994). The novelty of this process lies in its apparent "tasteless" achievement of oxygen removal. The earlier approaches involving autoxidation of non-cyclic side groups are claimed to have imparted some taste to the food.

The chemistries employed in the oxidation of aromatic nylons and of hydrocarbon polymers with various side groups have the common theme of light-triggering to produce enough free radicals to remove the antioxidants remaining after extrusion. The transition metal catalysts then facilitate the rupture of the hydroperoxides formed, and accelerate the ongoing chain reactions.

A radically different process also triggered by light has been developed without the use of photo-initiation. This process involves the light excitation of a photoreducible component, such as a quinone, followed by its photoreduction. This new photo-reduced species is then oxidized by the oxygen that it scavenges. In this case the polymer was inactive until exposure to, for example, ultraviolet light – something that might not be applied until immediately before package filling (Rooney, 1993). The species reactive towards oxygen is therefore not present during thermal processes such as extrusion and blow molding.

There have been other approaches to scavenging oxygen from food packs or to enhancing the barrier of packs such as bottles. These have generally included variants on those mentioned here, and some have been reviewed elsewhere (Rooney, 1995; Brody *et al.*, 2001). The processes that have progressed to the market or which are undergoing commercial development are described in ever increasing detail in the forest of patent applications that have been published. Despite the density of this forest, it is clear that several different food distribution problems can be solved by using these scavengers. These problems have their genesis in the nature of the newer packages as well as the requirements of high-speed filling and the use of distribution temperatures that impose demands upon the packaging. Perhaps the greatest demand has been made by the shift from glass and metal cans to substantially plastic packs.

The requirements of foods that impact on package selection involving oxygen scavenging can basically be subdivided into enhancement of barrier and headspace scavenging (Figure 8.2). Frequently both requirements are present, but the most economical approach to package selection may be made by using oxygen scavenging to address one problem or the other.

Barrier enhancement is commonly needed where the product is packaged in an impermeable package with a closure allowing oxygen entry, or when the product has a very low headspace and the oxygen ingress needs to be prevented while maintaining

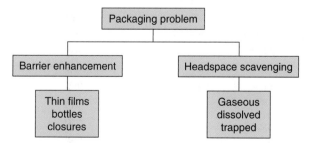

Figure 8.2 Roles for oxygen-scavenging packaging plastics.

a thin barrier layer. The latter limitation may be a requirement of the package for other reasons, such as in aseptic brick packs or with close-fitting thermoformed films used for vacuum packaging. The most recent need for barrier enhancement is found in PET bottles for beer, where beer is packaged with a very low oxygen concentration but is degraded by oxygen permeating the PET.

Headspace scavenging is beneficial when, even though the package may provide a high barrier, the residual oxygen is not readily removed to the necessary level by conventional means. This is typically the case with beverages and products with a porous structure not economically deoxygenated by evacuation or gas flushing. Such products include bakery items, and spray-dried and freeze-dried foods. Beverages that are not readily deoxygenated include those that froth readily yet contain air bubbles due to the presence of fruit pulp.

Application to food and beverage packaging

The interest in, and adoption of, oxygen-scavenging systems has been driven by the wide range of mechanisms by which oxygen can contribute to loss of food quality. These mechanisms include:

- nutrient loss
- discoloration
- microbial spoilage
- rancidity
- organoleptic deterioration
- infestation by insects and vermin.

Pathogen growth is not included in the above list, as measures to prevent their growth should already be in place. Each one of these mechanisms has its own kinetics and level of sensitivity at which the effect on the food becomes unacceptable. Table 8.2 shows a selection of foods that benefit from oxygen scavenging, and indicates the importance of the speed at which quality deterioration occurs. Hence, although a pack of full-cream milk powder which has been nitrogen flushed may have an oxygen content of around 5% after desorption from the pores, the oxidation occurs sufficiently slowly to present

Table 8.2   Oxygen sensitivity of some foods

Food/beverage	Substrate	Rate
Milk powder	Fat	Slow
Cheese	Mold	Slow
Beer	Flavors	Moderate
Wine	Preservatives	Moderate
Juice	Vitamin C	Fast
Fish	Oil	Fast

only a minor problem. Beer, on the other hand, is similarly sensitive to oxidation of oils which cause off-flavor development. However, the damage done by the initial oxygen uptake results in flattening of the flavor and can take some months off the potential shelf life. This is crucial to the introduction of PET packaging for this beverage.

The choice of packaging for preserved fish products is severely limited by the rapid oxidation of the polyunsaturated oils present. Indeed, a variety of fish oil products cannot be packaged in glass jars because the residual oxygen present is sufficient to cause unacceptable levels of both discoloration and rancid odor. Packaging that scavenges oxygen at a rate much greater than the food oxidation calls for rapid scavenging coupled with package design that maximizes the area of scavenger available.

Prevention of mold growth is a major role for oxygen scavenging (Smith *et al.*, 1995). Although mold is one of the major causes of food quality loss, the slow rate of growth in many foods provides an opportunity for oxygen scavenging to reduce oxygen levels to around 0.1%, which inhibits such growth. The capacity of oxygen scavenging to prevent mold growth on processed meat was addressed by Randell *et al.* (1995). They demonstrated that even in the presence of a pinhole in the seal of packages, an oxygen absorber sachet could suppress growth for useful periods. They found that the concurrent degradation of the meat color by the action of the fluorescent display lights on the oxidation reaction was also prevented.

Color loss due to photo-oxidation is not limited to processed meats. It has also been shown that the flavor and color degradation of Havarti cheese is strongly enhanced by fluorescent display lights. Randell *et al.* (1995). These authors showed that by reducing the headspace oxygen in these packs to 0.1%, the formation of pentanol (the indicator of quality loss) could be essentially eliminated, regardless of the illumination conditions tested. The benefit demonstrated by such work is that oxygen scavenging can contribute not just to the package properties but also to the conditions under which a food might be distributed.

The oxidation of vitamin C in orange juice occurs rapidly, and this reaction usually results in substantial deoxygenation of the juice in ambient barrier packaging in aseptic brick packs. The result is the loss of the vitamin, coupled with the consequent generation of browning products several months later. Beverages do not lend themselves to protection by oxygen absorber sachets or labels, so overcoming this problem has been delayed pending the development of plastics with scavenging capability.

The development of these plastics has been described largely in patents, so there are very few peer-reviewed research results published. One such investigation, from the author's laboratory, involved the measurement of dissolved oxygen concentrations, ascorbic acid, and total vitamin C assays, and browning of the juice when stored in an oxygen-scavenging laminate under refrigerated and ambient conditions (Zerdin *et al.*, 2003). The laminate (OS) pouches consisted of EVOH/oxygen-scavenger film/EVA, and the pouches contained orange juice from freshly backed aseptic brick packs. The juice was re-sterilized with dimethyldicarbonate on repacking in the pouches in order to avoid microbial oxygen uptake. The scavenger film was an improved version of one that involved photoreduction of a polymer-bound reducible compound and described previously (Rooney and Horsham, 1998). It was found that the oxygen was scavenged in less than three days at ambient temperature, and that the quantity of ascorbic acid

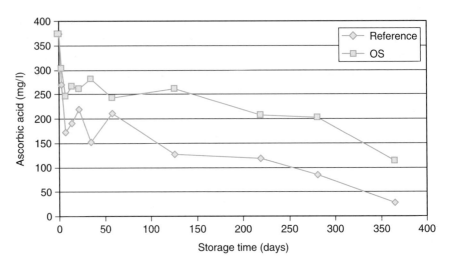

Figure 8.3 Concentration of ascorbic acid in orange juice packed in oxygen scavenging and reference pouches stored at 25°C.

lost during that time was reduced compared with that in the control pouches. The loss of ascorbic acid from the juice packed in the scavenger pouch was consistently less than that in the reference over a period of one year (Figure 8.3). The extent of browning of the juice was reduced by around 33% when the juice was packaged with the scavenger laminate. Results broadly similar to these were reported by Rodgers (2000), whose results were obtained from an independent laboratory and using the metal catalyzed and photoinitiated polymer bearing cyclohexenyl groups.

The results obtained at refrigerator temperature by Zerdin et al. (2003) were similar to those obtained under ambient conditions. The oxygen concentration was similar after three days. The browning was suppressed at low temperature in both the control and test pouches. Thus oxygen scavenging may be significantly more necessary where aseptic packs are distributed under ambient conditions than where short shelf-life juices are distributed chilled.

Besides the direct effects on the food of suppressing oxidation or microbial growth, there are other aspects of food quality that can be influenced by use of oxygen scavengers. Chilled beef is conventionally distributed at the wholesale level as primal cuts weighing several kilograms, vacuum-packaged in shrink bags with low oxygen permeability at 0°C. The drip, or liquid exudates, lost under these circumstances has been minimized, but recent research into the use of non-vacuum packaging with the presence of an oxygen scavenger sachet has revealed a further improvement (Payne et al., 1998). These investigators found that drip loss could be substantially reduced by avoiding the compression effect of evacuation and removing the residual oxygen by using an oxygen absorber sachet (Ageless™ Z50). Results like these are indicative of the potential for oxygen scavenging to contribute to quality and logistic parameters associated with food distribution in the wider sense.

The application of oxygen scavenging has not been limited to plastics packages. The shelf life of cracker biscuits in metal cans, as measured by hexanal formation and peroxide value development, was almost doubled at 25°C (Berenzon and Saguy, 1998). It was found that with the addition of oxygen-absorber sachets, no oxidative odors were observed after 44 weeks at 25°C and 35°C. Results such as these support the view that the method of oxygen scavenging will vary with the food/package combination, and that no one approach is likely to satisfy the varying demands of food and beverage packaging.

Future opportunities

The field of oxygen scavenging using plastics is still largely under development, even though the use of sachets, labels, and closure liners is well established. The introduction of new technologies will depend upon the drivers revealed in the food and packaging industries. The major driver should be the curiosity of food technologists interested in seeking better outcomes when they introduce new or modified products. This in turn may be expected to be a result of the education process. Another driver of importance will be the need to achieve current (or better) quality levels as packaging is changed, especially when newer materials are used. This is being observed already with the introduction of PET bottles and jars in place of rigid metals and glass. Even in the latter packaging there are ample opportunities to enhance the product quality for the consumer by scavenging of the headspace already present therein. The potential impact on canned foods can only be guessed at currently.

Regulation by both food authorities and those caring for the environment will also have major impacts. A variety of scavenging systems have already been approved in Japan, the USA and European countries, among others. The expected amendment of the Food Packaging Directive by the European Commission in late 2004/5 to address active packaging will make the future paths to introduction clearer, even though oxygen removal is not really the main thrust of the expected changes. The potential for reduction in the complexity of multilayer plastics structures and the reduction in rigid packaging use following the use of oxygen scavenging may be expected to have favorable outcomes for the environment, as long as there are not adverse impacts in the manufacturing process.

References

Abbott, J., Waite, R. and Hearne, J. F. (1961). *J. Dairy Res.* **28**, 285–292.

Berenzon, S. and Saguy, I. S. (1998). *Lebensm. Wiss. u. Technol.* **31**, 1–5.

Bissot, T. C. (1990). In: *Barrier Polymers and Structures* (W. J. Koros, ed.), pp. 225–238. ACS Symposium Series No. 423, ACS, Washington, DC.

Brody, A. L. and Budny, J. A. (1995). In: *Active Food Packaging* (M. L. Rooney, ed.), pp. 174–192. Blackie Academic and Professional, Glasgow, UK.

Brody, A. L., Strupinsky, E. R. and Kline, L. R. (2001). *Active Packaging for Food Applications*. Technomic, Lancaster, PA.

Buchner, N. (1968). Oxygen-absorbing inclusions in food packaging. West German Patent 1,267,525.

Cahill, P. J. and Chen, S. Y. (1997). Oxygen scavenging condensation polymers for bottles and packaging articles. PCT Application PCT/US97/16712.

Ching, T. Y., Katsumoto, K., Current, S. and Theard, L. (1994). Ethylenic oxygen scavenging compositions and process for making same by esterification or transesterification in a reactive extruder. PCT Application PCT/US94/07854.

Cochran, M. A., Folland, R., Nicholas, J. W. and Robinson, M. E. R. (1991). Packaging. US Patent 5,021,515.

Collette, W. (1991). Recyclable multilayer plastic preform and container blown therefrom. US Patent 5,077,111.

Cook, J. M. (1969). Flexible film wrapper. US Patent 3,429,717.

Dainelli, D. (2003). *Active Packaging Materials: the CRYOVAC® OS1000 Oxygen Scavenging Film*. Paper presented at "Active and Intelligent Packaging: Ideas for tomorrow and solutions for today" (on CD). TNO, Amsterdam, The Netherlands.

Farrell, C. J. and Tsai, B. C. (1985). Oxygen Scavenger. US Patent 4,536,409.

Folland, R. (1990). In: *Proc. Pack Alimentaire '90, Innovative Expositions Inc., Princeton, NJ*, Session B-2.

Kawatiki, T. T., Kume, K., Nakag, K. and Sugiyama, M. (1992). Oxygen absorbing sheet: molded, stretched low density ethylene copolymer containing iron powder with electrolyte on surface. US Patent 5,089,323.

King, J. (1955). *Food Manuf.* **30**, 441.

Kline, L. (1999). In: *Proc. Oxygen Absorbers: 2000 and Beyond Conference (Chicago IL)*. George O. Schroeder and Assoc., Inc., Appleton, Wisconsin.

Koros, W. J. (1990). In: *Barrier Polymers and Structures* (W. J. Koros, ed.), pp. 1–21. ACS Symposium Series No. 423, ACS, Washington, DC.

Kuhn, P. E., Weinke, K. F. and Zimmerman, P. L. (1970). *Proceedings of 9th Milk Concentrates Conference, Pennsylvania State University, 15 September*.

Maloba, W. F., Rooney, M. L., Wormell, P. and Nguyen, M. (1994). *J. Am. Oil Chem. Soc.* **73**, 181–185.

Payne, S. R., Durham, C. J., Scott, S. M. and Devine, C. E. (1998). *Meat Sci.* **49**, 277–287.

Rabek, J. F. and Ranby, B. (1975). *Photo-oxidation, Photodegradation and Photostabilisation of Polymers*. John Wiley & Sons, London, UK.

Randell, K., Hurme, E., Ahvenainen, R. and Latva-Kala, K. (1995). In: *Food and Packaging Materials – Chemical Interactions* (P. Ackerman, M. Jägerstad and T. Ohlsson, eds), pp. 211–216. Royal Society of Chemistry, Cambridge, UK.

Rodgers, B. D. (2000). Paper presented at "Oxygen Absorbers 2001 and Beyond", a conference organized by George O. Schroeder and Associates, Chicago, 19–20 June.

Rooney, M. L. (1982). *Chem. Ind.* (D. H. Sharp and T. F. West, eds). The Society of Chemical Industry, London, UK. pp. 197–198.

Rooney, M. L. (1993). Oxygen scavengers independent of transition metal catalysts. International Patent application PCT/AU93/00598.

Rooney, M. L. (1995a). In: *Active Food Packaging* (M. L. Rooney, ed.), pp. 143–172. Blackie, Glasgow, UK.

Rooney, M. L. (1995b). *ibid*. In: *Active Food Packaging* (M. L. Rooney, ed.), p. 76. Blackie, Glasgow, UK.

Rooney, M. L. (2000). In: *Materials and Development of Plastics Packaging for the Consumer Market* (G. A. Giles and D. R. Bain, eds), pp. 105–129. Sheffield Academic Press, Sheffield, UK.

Rooney, M. L. and Holland, R. V. (1979). *Chem. Ind.* pp. 900–901. The Society of Chemical Industry, London, UK.

Rooney, M. L. and Horsham, M. (1998). New uses for oxygen scavenging compositions. PCT Application PCT/AU98/00671.

Rooney, M. L., Holland, R. V. and Shorter, A. J. (1981). *J. Sci. Food Agric.* **32,** 265–272.

Rooney, M. L., Holland, R. V. and Shorter, A. J. (1982). *J. Food Sci.* **47,** 291–294, 298.

Scholle, W. R. (1977). Multiple wall packaging material containing sulfite compound. US Patent 4,041,209.

Scott, D. (1958). *Food Technol.* **12,** 7.

Smith, J. P., Hoshino, J. and Abe, Y. (1995). In: *Active Food Packaging* (M. L. Rooney, ed.), pp. 143–172. Blackie, Glasgow, UK.

Speer, D. V., Roberts, W. P. and Morgan, C. R. (1993). Methods and compositions for oxygen scavenging. US Patent 5,211,875.

Tallgren, H. (1938). Keeping food in closed containers with water carrier and oxidizable agents such as zinc dust, Fe powder, Mn dust etc. British Patent 496935.

Taoukis, P. S., El Meskine, A. and Labuza, T. P. (1988). In: *Food and Packaging Interactions* (J. H. Hotchkiss, ed.), pp. 243–261. ACS Symposium Series No. 365, ACS, Washington, DC.

Teumac, F. N. (1995). In: *Active Food Packaging* (M. L. Rooney, ed.), p. 196. Blackie, Glasgow, UK.

Tsai, B. C. and Wachtel, J. A. (1990). In: *Barrier Polymers and Structures* (W. J. Koros, ed.), pp. 192–202. ACS Symposium Series No. 423, ACS, Washington, DC.

Yoshikawa, Y., Ameniya, A., Komatsu, T. and Inoue, T. (1977). Oxygen Scavenger Japan Patent 104:4, 486.

Zerdin, K., Rooney, M. L. and Vermüe, J. (2003). *Food Chem.* **82,** 387–395.

9

Intelligent packaging

Jung H. Han, Colin H. L. Ho and Evangelina T. Rodrigues

Introduction

The primary function of packaging is to secure the food content against physical damage, moisture gain or loss, oxidation, and biological deterioration. The secondary function is to facilitate the distribution of the product (containing the proper information) to the consumer (Side, 2002). Packaging also fosters effective marketing of the food through distribution and sale channels. It is of the utmost importance to optimize the protection of the food product by selecting the right packaging materials, designs, and distribution methods. Other purposes of food packaging are to fulfil changing consumer needs, as well as the customers' desire for quality and convenience. Nowadays, consumers are willing to pay more for better quality foods and for convenient packaged foods with little preparation required. Such foods are "quality convenience foods" that offer the consumer great quality and appearance – better than typical packaged foods (Fito *et al.*, 1997). In recent years, active packaging has evolved due to the growing tendency to develop substances that are incorporated into the package and are active in protecting the foodstuff against contamination.

New packaging systems

Terms such as active, smart, interactive, clever or intelligent are used to describe new packaging methods. These terms lack clear definition, and are interchangeable in some literature. Therefore, it is important to differentiate their meanings. Active packaging

Innovations in Food Packaging
ISBN: 0-12-311632-5

changes the condition of the packed foods with the use of the incorporation of certain additives into the packaging films or inside the packaging headspaces so as to extend the product's shelf life (Day, 1989). In some cases, it also changes the package's permeation properties and/or the concentration of different volatiles and gases in the package headspace during storage. Moreover, this packaging technique actively adds, in small amounts, antimicrobial, antioxidative or other quality improving agents via the packaging materials into the packed food. Active packaging also plays a role in food preservation other than just providing an inert barrier to the external conditions (Rooney, 1995).

Active packaging techniques can be divided into two categories (Ahvenainen, 2003). The first category is absorbers or scavengers, where the systems remove undesired compounds such as oxygen, carbon dioxide, ethylene, and excessive water. The second category is the releasing systems, which actively adds or emits compounds to the packaged foods or into the headspace of the package – such as carbon dioxide, antioxidants and preservatives.

Intelligent packaging

Intelligent packaging refers to a package that can sense environmental changes, and in turn informs the changes to the users (Summers, 1992a). Rodrigues and Han (2003) defined intelligent packaging as having two categories: simple intelligent packaging (as defined by Summers, 1992a), and interactive or responsive intelligent packaging. In the later, the packaging contains sensors that notify consumers that the product is impaired, and they may begin to undo the harmful changes that have occurred in the food product (Karel, 2000; Rodrigues and Han, 2003). Such packaging systems contain devices that are capable of sensing and providing information about the functions and properties of the packaged foods (Day, 2001), and/or contain an external or internal indicator for the active product history and quality determination (Ohlsson and Bengtsson, 2002).

These types of devices can be divided into three groups. The first type is the external indicators. These indicators are attached outside the package, and include time–temperature indicators and physical shock indicators. The second type is the internal indicators, which are placed inside the package. They are either placed in the headspace of the package or attached to the lid – for example, oxygen leak indicators, carbon dioxide, microbial, and pathogen indicators (Ahvenainen, 2003). The third type of devices is the indicators that increase the efficiency of information flow and effective communication between the product and the consumer. These products include special bar codes that store food product information such as use, and consumption date expiration. Product traceability, anti-theft, anti-counterfeiting, and tamperproof devices are also included in this category (Coles *et al.*, 2003).

This chapter will review intelligent packaging technologies and describe different types of devices. Their principal food applications and recent developments will also be highlighted. The intelligent packaging application systems will be discussed in the following categorical order:

1. Those that provide more convenience (quality, distribution and preparation methods)

2. Those that improve product quality and product value (quality indicators, temperature and time–temperature indicators, and gas concentration indicators)
3. Those that change gas permeability properties
4. Those that provide protection against theft, counterfeiting, and tampering (Pault, 1995; Rodrigues and Han, 2003).

For these packaging systems to be practical, they should be easy to use, cost effective, and capable of handling multiple tasks. Color indicating labels, for example, should have color changes that are irreversible, easy to read and easily understood by consumers.

In packaging, "smartness" can have many meanings, and covers a number of functionalities, depending on the product being packaged – food, beverage, pharmaceutical, household products etc. Examples of current and future functions that are considered to have "smartness" would be packages that:

1. Retain the integrity and actively prevent food spoilage (extend shelf life)
2. Enhance product attributes (look, taste, flavour, aroma, etc.)
3. Respond actively to changes in the product or in the package environment
4. Communicate product information, product history or other conditions to the user
5. Assist with opening and indicating seal integrity
6. Confirm product authenticity and act to counter theft.

Indicators are called smart or interactive because they interact with compounds in the food. Microwave heating enhancers, such as susceptors and other temperature regulation methods, are sometimes regarded as intelligent methods as well (Ahvenainen,

Table 9.1 Examples of external and internal indicators and their working principles used in intelligent packaging (adapted from Ohlsson and Bengtsson, 2002)

Technique	Principle/reagents	Information given	Application
Time–temperature indicators (external)	Mechanical, chemical, enzymatic	Storage conditions	Foods stored under chilled and frozen conditions
Oxygen indicators (internal)	Redox dyes, pH dyes, enzymes	Storage conditions Package leak	Foods stored in packages with reduced oxygen concentration
Carbon dioxide indicator (internal)	Chemical	Storage conditions Package leak	Modified or controlled atmosphere food packaging
Microbial growth indicators (internal/external) and freshness indicators	pH dyes, all dyes reacting with certain metabolites	Microbial quality of food (i.e. spoilage)	Perishable foods such as meat, fish and poultry
Pathogen indicators (internal)	Various chemical and immunochemical methods reacting with toxins	Specific pathogenic bacteria such as *Escherichia coli* O157	Perishable foods such as meat, fish and poultry

Table 9.2 Some manufacturers and trade names of commercial smart indicators (adapted from Ohlsson and Bengtsson, 2002)

Manufacturer	Country	Trade name
Time–temperature indicators		
Lifelines Technology Inc.	USA	Fresh-Check
Trigon Smartpak Ltd	UK	Smartpak
3M Packaging Systems Division	USA	MonitorMark
Visual Indicator Tag Systems Ab	Sweden	Vitsab
Oxygen indicators		
Mitsubishi Gas Chemical Co., Ltd	Japan	Ageless Eye
Toppan Printing Co., Ltd	Japan	–
Toagosei Chem. Industry Co., Ltd	Japan	–
Finetec Co., Ltd	Japan	–

2003). Tables 9.1 and 9.2 depict some currently available color indicators, their manufacturers and their trade names.

Intelligent packaging applications and technologies

Intelligent packaging to provide more convenience

Convenience has become a priority for most people due to their busy daily work schedules. Customers are seeking quicker and easier means of accessing better quality foods. In addition to convenience of consumption, the convenience at the distribution level and the product quality level are likely to become more important in the future.

Convenience indicators for quality

The Packaging Steels Development Centre of British Steel Tinplate (South Wales, UK) developed temperature-sensitive thermochromic inks. The ink is printed onto shrink sleeves before they are added to the steel beverage cans ("Smart Cans"; Berragan, 1996). The thermosensitive ink changes from a white to a blue color as the temperature of the can decreases, and the words "READY TO SERVE" appear when the temperature is between 5° and 8°C, thus allowing the consumers to see if the product is ready to be served.

Hite beer, from Korea, utilizes thermochromic inks for beer bottles and cans. At low temperatures the green ink on beer bottles indicates the optimum temperature for consumption (Figure 9.1), and on beer cans it also indicates the level of cold beer inside the can.

Convenience indicators for storage, distribution, and traceability

The need to develop a management integrated information system is crucial for the whole chain, from production to distribution. Information about the location and the time of any

Figure 9.1   A thermochromic indicator label on a bottle of beer. The label has an indicator that looks like a "traffic light signal" with red and green lights. If the temperature decreases to around 7°C the indicator turns on the green light, indicating that the product is ready to be consumed.

mishandling is essential for food recalls, and so that causes, responsibilities, liabilities, and improvements can be traced back, determined, and resolved (Ahvenainen, 2003).

Shoplifting is a concern in the retail sector. A modern EAS (electronic article surveillance) device is paper-thin and is the size of a postage stamp. The EAS is attached to the package and can set off an alarm when the active device is passed through an EAS detection system (Brody, 2001). The common EAS system uses radio frequency technology.

In recent years, radio frequency identification (RFID) has been commercialized to optimize the information flow within the network. RFID is a non-contact, wireless data communication system, where tags are programmed with unique information and attached to the objects for identification and tracking purposes (Jansen and Krabs, 1999). The tag can contain a variety of information, such as location, product name, product code, and expiration date. The tag scanner can also read multiple tags at the same time. Although an EAS system can only trigger an alarm system, RFID can identify the article as a unique product, which enhances the product recall and tracing ability. The RFID system has passive tags, which consists of a chip and an antenna. When a host system sends power to a tag, the tag responds to the reader by revealing the information it contains inside the chip. The reader then sends out electric signals to the antenna within the tag for communication (Brody, 2002a). The data stored in the memory of the tag can be read and transmitted back to the reader, and then sent to the processing unit for signal analysis. Tags can be active or passive; passive tags do not use batteries, whereas active tags have a battery and are used to record information during shipment or to obtain information from a distance.

The Finnish company Rafsec produces RFID transducers, the antenna, and the fixing of the chip to the tag. The chip transducers can operate at temperatures from $-25°$ to 70°C (Ahvenainen, 2003), and can even stand a temperature of 150°C for a short period of time. Because of this, it is possible for them to survive an injection molding operation. A weakness of RFID systems is that their signal transmission interferes with metal, such that signals cannot be read in a metal cage (Ahvenainen, 2003). However, a benefit is that these systems can reduce multiple labeling. When a product is produced, the information offered by the tag includes the product identification, the batch number, and the production date. Although the tag cannot replace the entire printed label on the food product, it does enhance the information that the consumers would normally get. In addition, because of these tags the effectiveness and efficiency of the supply chain can be improved by having fewer labels printed and by reducing the number of barcode printers and label applicators in plants and warehouses.

The growing use of microchip technology has advanced the automation of computer-integrated manufacturing, and makes the whole process less labour intensive. These systems made the use of transducers (chip plus transmitter) – which is the technology used to control microwave – more popular (Linnemann *et al.*, 1999). The improvement in electronic data exchange not only simplifies the packaging process by using "robot" technology, but also increases the ability to trace products during the different stages of distribution (Moe, 1998). Since the emergence of the Bovine Spongiform Encephalitis (BSE) crisis, greater attention has been paid to chain traceability. Traceability is essential in quality management, as it provides the ability to track a product's history from harvest through transport, storage, processing, distribution marketing, and sales (Moe, 1998). Chain traceability is also important in efficient food recall procedures and in minimizing losses, as well as in avoiding unnecessary repetition of measurements in two or more successive steps. Internal traceability, on the other hand, provides a correlation between the product data with raw material quality characteristics. It also improves the process control mechanisms and prevents the mixing of high- and low-quality raw materials (Brody, 2002a). Two examples of this technology are already in place. The first system is the Farm Advisory Services Team (FAST). This system can trace fruits from field farms to customers. Information such as weight, time, and location is recorded for each item and stored in a database. An identification number is then assigned to each fruit (Daniel, 2000). The second system is the Scottish Courage (UK). This technology uses radio frequency data tags to obtain information on beer kegs, such as their content, weight, time, and the location of the beer-can fill process (Daniel, 2000).

With global demand for the use of recycled materials to reduce waste, food packaging industries are adopting new technologies to trace the reusable or returned empties. Automatic identification packages are being developed, and are necessary to ensure an efficient process. The existing barcode systems are not sufficient to store large quantities of data (Schilthuizen, 1999). There is a need to include much more information about the product – for example, distribution tracers, and microwave heating time- and temperature-codes. Symbol Technologies have developed a new generation of barcodes that can accommodate a fingerprint and a photograph (Yam, 2000). The new barcode can be scanned vertically and horizontally, and stores more than two kilobytes of data (over 100 times more than a traditional one).

Improved convenience for preparation and cooking methods

Self-heating and self-cooling of beer and soft drinks currently reflects consumer habits. "Instant Cool" is a technological development where a condenser, an evaporation chamber, and a salt-based drying agent are integrated into packaging systems for cooling purposes (Anonymous, 2002). This technique can be used in cans, bottles, and bags. The system is capable of lowering the temperature of the packaging and its content by 17°C within a few minutes. Crown Cork & Seal is developing a self-chilling beverage can in conjunction with Tempra Technologies. The technology uses the latent heat of water to produce the cooling effect. The water is bound in a gel layer coating that coats a separate container within the beverage can. To activate the system, the consumer needs to twist the base of the can to open a valve, which exposed the water to a desiccant held in a separate evacuated external chamber (Anonymous, 2002). This then initiates evaporation of the water at room temperatures, and thus achieves a cooling effect as the heat is removed from the system.

Another development is "CoolBev". This technique consists of a small add-on device that can be integrated into standard bottles, cans or carton packages; however, it takes up one-third of the container's volume. It is a vinyl bag filled with water, where the fluid inside is pressurized and evaporates when the bag is opened (Anonymous, 2002). The vapor absorbs heat from the product, thus cooling it. The vapor is then precipitated inside the bag and is absorbed by a clay-based drying agent. It is claimed that the product temperature is lowered by 18°C within two to three minutes (Brody, 2002b).

Ontro Cans is making self-heating systems using the heat generated by the reaction of calcium oxide with water to heat coffee, tea, soup, and hot chocolate. Before calcium oxide and water are mixed, they are separated in adjacent chambers within the can. Consumers are required to invert the can to trigger the puck that breaks the internal aluminium foil seal to release the water. The external polypropylene layer can resist boiling water temperatures, and can maintain the elevated temperatures for about 20 minutes (Brody, 2002b).

A microwaveable steamer bag has been developed to provide rapid heating of the food product, by inserting a water pad in the package and using steam generation to heat the product (Johns, 2002). The package is an airtight, stretchable plastic bag, and consists of the food product that is to be heated in a microwave oven, along with a separate pad. This pad, which contains a water absorbent material, is positioned between the upper portion of the food product and the upper portion of the bag (Johns, 2002). The pad serves to shield the food product from direct microwave radiation, and provides a source of steam generated by the microwaves.

The idea of a "smart kitchen" has been introduced, and this consists of a microwave oven with a built-in barcode scanner and a microprocessor for data processing (Yam, 2000). The package serves as an intelligent messenger that carries vital information about the food and the package through a printed barcode that is able to retain large amounts of information. The scanner transfers the scanned signal to the microwave so that it will adjust the magnetron and the turntable accordingly (Yam, 2000). This new invention has the ability to customize different foods with different thermal properties. Also, different packages of various sizes and shapes can be identified and their optimum cooking time calculated automatically. No cooking instructions are necessary. In addition, nutritional data, recent food recalls, and current food allergen alerts can be provided by this system. The information obtained can also be stored for future use. Furthermore, there may be interaction between the consumer and the system via touch screen or voice recognition (Louis, 1999).

Intelligent packaging for improving product quality and product value

Freshness and microbial indicators

Freshness indicators are used to indicate if the product's quality has been impaired due to exposure to unfavorable conditions during storage and transportation (Summers, 1992a). The package is usually equipped with a reversible color-changing device that tells the consumers if the package has undergone deterioration, along with a partial or complete history of the product. Some of the examples described below are internal or external indicators used to monitor elapsed time, time–temperature, humidity, shock abuse, and gas concentration changes (Ahvenainen and Hurme, 1997).

Shockwatch™ indicators, produced by 3M Corp., provide an example of a physical shock indicator. The indicator consists of a closed glass capillary tube. At one end of the tube there is a red liquid, and in the other there is a dispersive material. When the contents of the two ends mix, due to shock and or vibrations, the tube turns red (Summers, 1992b).

COX Technologies' "FreshTag" color-indicating tags consist of a small label attached to the outside of the packaging film. It is used to monitor the freshness of seafood products, and consists of a reagent-containing wick contained within a plastic chip (Millers *et al.*, 1999). As the seafood ages, spoils, and generates volatile amines in the headspace, these are allowed to contact the reagent, causing the wick in the tag to turn bright pink.

Hydrogen sulfide indicators can be used to determine the quality of modified atmosphere packaged poultry products. It is based on detecting the color change of myoglobin caused by the production of hydrogen sulfide. During the aging process of packaged poultry meats, hydrogen sulfide is released from the meats. The indicator correlates with the color of myoglobin, which correlates with the quality deterioration of the poultry product (packaged fresh broiler cuts; Ahvenainen *et al.*, 1997). In addition to hydrogen sulfide indicators, there are also indicators sensitive to microbial metabolites. Cameron and Talasila (1995) investigated the detection of respiration changes in packaged produce by measuring the ethanol in the headspace of the packages, using alcohol oxidase and peroxidase.

Another example of a freshness indicator is a diamine dye-based sensor system. Diacetyl is a volatile metabolite emitted from microbially spoiled meat. Diacetyl can migrate through the permeable meat package to react with the dye and change the color of the indicator (Honeybourne, 1993).

In addition to indicators dependent on microbial metabolites, there are also other types of indicators that are based on other food deterioration factors. DeCicco and Keeven (1995) described an indicator based on the color change of chromogenic substances due to enzymes produced by contaminating bacteria. This kind of indicator is suitable for detecting contamination in liquid health-care products. Kress-Rogers (1993) invented a knife-type freshness probe for meats. The freshness of the meat product is assessed based on the glucose gradient on the surface of that product. During microbial growth the surface glucose is consumed, and therefore, as glucose is being consumed, the probe can detect the level of bacterial contamination and hence the product's freshness (Ahvenainen, 2003).

Toxin Guard, produced by Toxin Alert Inc., is a system built on a polyethylene-based packaging material containing immobilized antibodies which detect the presence of pathogenic bacteria (*Salmonella, Escherichia coli* O157, and *Listeria*). When the bacterial toxin is in contact with the material it will be bound to the immobilized antibodies, which are then identified by a printed characteristic pattern (Bodenhamer, 2000). Another example of microbial indicators is a new barcode made by Food Sentinel Systems. This system is based on immunochemical reactions that occur in the barcode (Goldsmith, 1994), and the barcode will become unreadable when a particular micro-organism is present.

The Lawrence Berkeley National Laboratory has developed a sensing material for the detection of *Escherichia coli* O157 enterotoxin (Quan and Stevens, 1998). The material is composed of cross-polymerized polydiacetylene molecules that can be incorporated into the packaging film. As the toxin binds to the molecules, the color of the film changes permanently from blue to red (Smolander, 2000).

Color indicators on nylon or polyethylene films are newly developed for the packaging of kimchi (fermented Korean cabbage or radish). The films are composed of calcium hydroxide as a carbon dioxide absorbent, bromocresol purple or methyl red as a chemical dye, and a mixture of polyurethane and polyester dissolved in organic solvents as a binding medium (Hong, 2002). This is applied on a nylon film and then laminated with a polyethylene film to form the printed indicator. During the distribution process, the kimchi products undergo natural fermentation. Carbon dioxide, a by-product of fermentation, becomes the marker of kimchi ripeness, since the carbon dioxide concentration is correlated with the pH and the titratable acidity of the product (Hong, 2002). The absorption of carbon dioxide in calcium hydroxide changes the pH of the indicator components and, accordingly, the color of the indicating chemical dye.

The use of pH dyes (e.g. bromothymol blue) as indicators to monitor the formation of carbon dioxide due to microbial growth is one of the most frequent applications in the food packaging industry. The increase in carbon dioxide levels can be used to detect microbial contamination in some products due to pH dyes that can react to the presence of this by-product. Other pH dye reagents include xylenol blue, bromocresol purple, cresol red, phenol red, and alizarin. Besides carbon dioxide, other metabolites (such as SO_2, NH_4, volatile amines, and organic acids) have been used as target monitoring molecules for pH-sensitive indicators (Ahvenainen, 2003).

Other spoilage detection methods include the incorporation of small analytical tools into the package, such as biosensors for the detection of biogenic amines. The increase in diamines in poultry meat can be detected with a putrescine oxidase reactor combined with an amperometric hydrogen peroxide electrode (Ahvenainen, 2003). The system can also be applied in the detection of histamine from rainbow trout meat and biogenic amines from fish muscles.

The "electronic nose" is an analytical tool composed of an array of sensors which respond to volatile compounds by changing their electrical properties (Blixt and Borche, 1999). The samples can then be classified as acceptable or unacceptable, referencing a sensory evaluation or microbiological analysis catalog. The response of the electronic nose has been found to be consistent with microbiological analysis and volatile concentration determination of the product (Gram and Huss, 2000). It has also been proved to be successful in the quality evaluation of fresh Yellowfin tuna and vacuum-packaged beef (Blixt and Borche, 1999).

"Doneness" indicators are convenience-, quality- and temperature-indicating packaging systems. They detect and indicate the state of readiness of heated foods. The "ready" button indicators are commonly placed in poultry products. When a certain temperature has been reached, the material expands and the button pops out, telling the consumer that the poultry product is fully cooked (Ahvenainen, 2003). There are also labels that change color when the desired temperature is reached. The drawback of "doneness" indicators is the difficulty of observing a color change without opening the oven. As a result, other innovative package design utilizes the generation of a whistling sound when the foodstuff has been cooked (Ohlsson and Bengtsson, 2002).

Time–temperature indicators

The temperature variations in a food product can lead to changes in product safety and quality. There are two types of temperature indicators; the simple temperature indicators and the time–temperature integrators (TTIs) (Pault, 1995; Ahvenainen and Hurme, 1997). Temperature indicators show whether products have been heated above or cooled below a reference (critical) temperature, warning consumers about the potential survival of pathogenic micro-organisms and protein denaturation during, for example, freezing or defrosting processes (Pault, 1995). To determine the changes in the shelf life of the food product, the second type of indicator, a TTI, should be employed (Pault, 1995). These indicators display a continuous temperature-dependent response of the food product. The response is made to chemical, enzymatic or microbiological changes that should be visible and irreversible, and is temperature dependent. TTIs provide an overall temperature history of the product during distribution. The changes shown by TTI can easily be measured, and have been proven to give consistent responses under the same temperature conditions and when not under the influence of light, humidity, and contaminants. Ideal TTIs are able to adjust the residual shelf life of the food product by assessing the already possible quality deterioration. However, one drawback is that these integrators measure the surface temperature and not the actual temperatures of the product (Ahvenainen and Hurme, 1997).

The US Army Natick Laboratories have developed a TTI that is based on the color change of an oxidable chemical system controlled by temperature-dependent permeation through a film.

3M Monitor Mark is a diffusion-based indicator label. The action is activated by a blue-dyed fatty acid ester diffusing along a wick (Ahvenainen and Hurme, 1997). When the two tapes are brought together, the visco-elastic material on the one tape migrates into a light reflective receptor on the other tape, at a temperature-dependent rate. This activates a progressive change in the light transmission of the reflective porous matrix and induces a color change. The melting temperature of the colored fatty acid ester determines the range of temperatures the food is to be stored at (de Kruijf et al., 2002).

I Point™ labels, made by I Point AB Technology, are enzymatic time–temperature indicators whose color changes as a result of pH variations due to the enzymatic hydrolysis of lipid substrates (Summers, 1992b). The label is composed of a capsule which contains a lipolytic enzyme and lipid substrates separated in two compartments. The enzyme and the substrates mix when the barrier between the two compartments is broken upon activation, and hydrolysis of the substrates causes acid release. Hence, the pH drops and is recorded by a color change in a pH indicator. The color change can be compared with a reference standard printed on another label on the package.

Lifelines' Freshness Monitor and Fresh-Check™ indicators rely on a solid-state polymerization reaction to give color changes. The indicator consists of a small circle of a polymer surrounded by a printed reference ring. The inside polymer circle darkens if the package has experienced unfavorable temperature exposures (Summers, 1992b), and the intensity of the color is measured and compared to the reference color scale on the label (de Kruijf et al., 2002). The faster the temperature increases, the faster the color changes occur in the polymer. Consumers are advised not to consume or purchase the product, regardless of the "use-by" date. These indicators have been used on fruit cake, lettuce, milk, chilled fresh produce, and orange juice.

Gas concentration indicators

Internal gas-level indicators are placed into the package to monitor the inside atmosphere (Ahvenainen and Hurme, 1997). Most of these indicators induce a color change as a result of gas generated due to enzymatic and chemical reactions (Ahvenainen and Hurme, 1997). Damage to individual packages can be determined by a fast visual check without opening the package (Pault, 1995). In addition, the rapid label check can allow consumers to view the quality of the food inside the package by examining common redox dyes (e.g. methylene blue, which is used as a leak indicator).

Oxygen indicators interact with oxygen penetrating the package through leakages to ensure that oxygen absorbers are functioning properly. Ageless Eye™, made by the Mitsubishi Gas Chemical Corporation, is also the producer of oxygen-scavenging sachets (Ageless™). Ageless Eye sachets contain an oxygen indicator tablet in order to confirm the normal functioning of Ageless absorbers. When oxygen is absent in the headspace ($\geq 0.1\%$), the indicator displays a pink color. When oxygen is present ($\leq 0.5\%$), it turns blue (Ahvenainen and Hurme, 1997). Carbon dioxide indicators are also used in modified atmosphere packages (MAP) in which high carbon dioxide levels are desired. The indicators display the desired concentrations of carbon dioxide inside the package (Ahvenainen and Hurme, 1997). This allows incorrectly packaged product to be immediately repacked, and eliminates the need for destructive, labor-intensive and time-consuming quality control procedures. The disadvantage of these oxygen and carbon dioxide indicators is that the color changes are reversible, which may cause possible false readings. For example, if the food is contaminated by micro-organisms, they will consume the oxygen inside the package and produce carbon dioxide, which will maintain the carbon dioxide levels in the headspace high even though the package has been compromised. Therefore, the food is no longer safe to be consumed as a result of microbial contamination; however, the indicator still displays a "normal" status color, resulting in a false reading (Ahvenainen and Hurme, 1997).

Cryovac Sealed Air Ltd has produced a label containing a visible carbon dioxide indicator. It can be used in MAP to identify machine faults and gas flushing problems. The desired gas mixture composition (oxygen and carbon dioxide) can also be checked by this indicator (de Kruijf et al., 2002).

Moonstone Co. has designed a label containing a gas-sensitive dye, which can be inserted into a package. The dyes produce different colors at different gas concentrations. When carbon dioxide has leaked or diffused out of the MAP, the dye changes from dark blue to a permanent yellow color (Summers, 1992b). The dyes function to indicate any leakage through the seal, as well as carbon dioxide level increases due to microbial growth. Table 9.3 summarizes examples of intelligent indicating systems used to improve quality and product value.

Intelligent packaging to change gas permeability properties

In order to preserve the color and quality of foods for an extended period of time, intelligent polymers with gas permeabilities that vary depending upon temperature changes can be used (Hoofman, 1997). The permeability of breathable films can change radically and reversibly due to relatively small changes in the temperatures when phase transition polymers are used in the packaging materials. As a result, the internal atmosphere of the food product can be regulated by matching the permeability of the film to the respiration of the particular food in the container. The following is an example of a food package composed of polymers with a thermally responsive permeability.

Intellipac polymeric package materials, manufactured by Landec Corp. (Menlo Park, California), are side-chain-crystallizable (SSC) polymers which have the ability to adjust their permeability to gases such as oxygen, carbon dioxide, and water vapor at different temperatures (Stewart, 1993). The melting point of the liquid–crystalline polymers can be altered by changing the length of the side chains. Thus, gas transmission rates can be increased with an increase in temperature and decreased with a decrease in temperature (Hoofman, 1997). In addition, the carbon dioxide-to-oxygen permeability value (β-value) of the package membrane can be uniquely selected by coating the membrane with another substance such that the material can be tailored to the exact requirements of the packaged food content (Brody, 2000). Some products are best stored in high carbon dioxide environments to reduce micro-organisms, whereas others (such as fresh and fresh-cut produce, including broccoli and cauliflower) are suitable for high respiratory-rate conditions (Stewart, 1993). Therefore, it would be tremendously beneficial to use polymeric materials that control the carbon dioxide and oxygen permeabilities individually to generate various characteristic β-values for MAP.

Table 9.3 Examples of freshness and contamination indicator systems for food packages (adapted from Ahvenainen, 2003)

Manufacturer/patent holder/trade name	Compound detected	Principle of the indicator
Freshness indicators		
AVL Medical Instruments	CO_2, NH_4, amines, H_2S	Color change of CO_2, NH_4 and amine-sensitive dyes, formation of color of heavy-metal sulfides (H_2S)
Biodetect Corporation	e.g. Acetic acid, lactic acid, acetaldehyde, ammonia, amines	Visually detectable color change of a pH-dye (e.g., phenol red, cresol red)
VTT Biotechnology	H_2S	Color change of myoglobin
Aromascan	Volatile compounds	Miniaturized electronic component with electrical properties affected by volatile compound associated with spoilage
Visual Spoilage Indicator Company	CO_2	Color change
Cox Technologies (Fresh Tag)	Volatile amines	Color change of a food dye
Sealed Air Ltd. (Tufflex GS)	CO_2	Color change
Time–temperature indicators		
Lifelines Technology Inc. (Fresh-Check)	Time–temperature	Color change due to chemical polymerization
Trigon Smartpak Ltd. (Smartpak)	Time–temperature	
3M Packaging Systems Division (MonitorMark)	Time–temperature	Physical diffusion of a chemical solute causing a color change
I Point AB (I Point)	Time–temperature	A lipid substrate indicator becomes hydrolyzed, causes a pH change and consequently a permanent color change
Oxygen indicators		
Mitsubishi Gas Chemical Co., Ltd. (Ageless Eye)	O_2	Color change
Toppan Printing Co., Ltd. (Fertilizer)	O_2	Color change
Togagosei Chemical Industry Co., Ltd. (Vitalon)	O_2	Color change
Finetec Co., Ltd. (Sanso-Cut)	O_2	Color change
Pathogen indicators		
Food Sentinel System (Sira Technologies)	Various microbial toxins	Barcode detector comprising a toxin printed onto a substrate and indicator color irreversibly bound to the toxin; in the presence of toxin the barcode is illegible
Toxin Guard (Toxin Alert)	Various pathogens	Formation of a colored pattern when a target analyte is first bound to labeled antibody and subsequently to capture antibody
Lawrence Berkeley National Laboratory	*E. coli* 0157 enterotoxin	Color change of polydiacetylene-based polymer

Intelligent packaging to provide protection against theft, counterfeiting, and tampering

Although theft and counterfeiting are not too common in the food industry, they do pose a huge economical burden in other industries. Electronic Article Surveillance (EAS) is used to deter the theft of high-priced goods, and leads the technology in electronic tagging systems. Tampering is another global issue, and therefore more sophisticated anti-tampering devices or packages with responsive technology are necessary to control and minimize these problems.

Andrew Scully, of Food Science Australia of CSIRO (New South Wales, Australia), has invented a technology for flexible packaging that creates a large-diameter color change on the surface of the package after tampering. Even a pinhole can be detected with the highly visible mark, so as to alert customers about potential damage to the contents of the package (Brody, 2002a). Irreversible thermochromic materials can provide a closure which "bruises" during any attempts to tamper, thus alerting the customer before the goods are purchased (Ahvenainen, 2003).

The main purposes in lowering the risks of tampering are first to eliminate tampering, and secondly to locate the already tampered products on the shelf by integrated identification. Aluminum and plastic closures can help customers identify if the package has been opened. The tamper-proof band will split when the package is opened, which provides tamper evidence for consumers. Alcoa CSI of Germany has invented the special cap that has "skirts" which curl under the bottom edge of the closure. The segments of the tamper-evident ring ("skirt") rupture when the container is opened (Jansen and Schelhove, 1999). Furthermore, ITW Envopak has produced the rectangular bolt seal that cannot be rotated. The bright coating on the bolt functions to increase the visibility of bottle seals and prevent counterfeiting (Jansen and Schelhove, 1999).

Holograms, thermochromic inks, tear labels and tapes, micro-tags, and diffraction devices are used to deter counterfeiting. They can be incorporated into films, transfer foils, tear tapes, and labels. Holograms cannot be duplicated by copiers, scanners, or printers. Their visual effects cannot be reconstructed or simulated (Daniel, 2000). Tamper-evidence labels or tapes are invisible before tampering occurs, but change their color permanently and leave behind a "STOP" message when the package is opened (Pault, 1995).

Consumer acceptance and legislative issues

General consumer behavior regarding the use of separate non-edible objects such as sachets and inserts in the headspace of food packages is one of the primary concerns that the intelligent packaging industry faces today. Labels and laminated structures are more easily accepted by consumers, since they are attached to the package (Ohlsson and Bengtsson, 2002). The food industry is currently studying consumer attitudes, prejudices, fears, and expectations about the quality improvement and assurance of color indicators. Manufacturers must educate the public about new packaging technologies in order to increase consumer confidence in the safety of the foods they purchase. Table 9.4 outlines some potential problems and solutions in introducing new smart packages.

Current packaging legislation requires that the migration of substances from food-contact materials is low. The acceptability of smart packaging techniques is considered from a migration point of view. The techniques can be divided into two groups:

1. Group 1, consisting of external indicators fixed on an outer surface of a package, such as time–temperature indicators

Table 9.4 Problems and solutions encountered with introducing new products using intelligent packaging techniques (adapted from Ahvenainen, 2003)

Problem/fear	Solution
Consumer attitude	Consumer research: education and information
Doubts about performance	Storage tests before launching; consumer education and information
Increased packaging costs	Used in high-quality products; marketing tool for increased quality assurance
False sense of security, ignorance of date markings	Consumer education and information
Mishandling and abuse	Active compound incorporated into label or packaging film; consumer education and information
False complaints and returns of packs with color indicators	Color automatically readable at the point of purchase
Difficulty of checking every color indicator at the point of purchase	Barcode labels: intended for quality assurance for retailers only

2. Group 2, consisting of internal indicators intended to be placed in the headspace of a package, such as oxygen and carbon dioxide indicators (Ohlsson and Bengtsson, 2002).

Migration does not occur in Group 1 indicators, since there is no direct contact of the indicator with the food product. Indicators in Group 2 are not intended to come into direct contact with the packaged foods. However, they are placed in the free headspace of a package or fixed to the inner surface of the lid (de Kruijf *et al.*, 2002). In the case of sachets, migration is also a problem, since in practice the system is in direct contact with the food product – especially with moist or fatty foods – and sachet materials are generally very porous.

Conclusions

Recommendations

In the future, intelligent packages can contain more complex invisible messages. Some types of intelligent tags can replace the current barcode system, such as electronic labeling consisting of a chip with ink technology in a printed circuit (Ahvenainen, 2003). In addition, built-in battery radio frequency identity tags will be popular in food supplies. It is also expected that in the future all necessary indicators (time–temperature, leak, freshness etc.) as well as product handling information will be contained within the same electronic tag. Warnings regarding potential allergens, the expiry date, and health monitoring tools for diet management (such as calories, fats and sugars) will also be readily available to consumers through easily readable devices placed in shopping carts at supermarkets. For freshness and pathogen identification indicators, more influential metabolite compounds and bacterial quantification will be carried out via advances in biosensors and biotechnologies applied in food packaging systems.

Increasing marketing efforts will not only contribute to entertainment expansion but also contribute to enhance brand-packaging recognition via information features such as sound and image interactivity (Fito *et al.*, 1997). In the future, packages may talk to consumers. Consumer interactions and error prevention will be improved with the introduction and enhancement of refrigerator recordings, auto-cooking features on microwaves, dosage devices for pharmaceutical products, etc. Furthermore, thermochromic materials will be used to produce containers that darken in the sunlight to protect the product and return to a clear, transparent package under normal lighting. This can also enhance product visibility, such as for soft drink containers (Berragan, 1996). Glow-in-the-dark packages using thermosensitive inks will be adopted to monitor serving temperatures (Berragan, 1996). Consumers will no longer have to worry about cooking times, since time and temperature sensors installed in the packages will be able to tell if the meal is cooked either by a color change or by a displayed message (e.g. "cooking is completed") (Anonymous, 1994).

Conclusions

Intelligent packaging is an emerging and existing area of food technology that can provide better food preservation and extra convenience benefits for consumers. The introduction of quality and freshness indicators (temperature indicators, time–temperature integrators, and gas-level controls), the increased convenience of product manufacturing and distribution methods, the smart permeability films, and the theft and counterfeiting evidence systems maximize the safety and quality of food products. Thus, intelligent packaging systems will improve the product quality, enhance the safety and security of foods, and consequently decrease the number of retailer and consumer complaints.

Disclaimer

Commercial products and manufacturers' names listed in this article are not the only ones that have been commercialized and developed in the world. The authors are not related to any of the commercial products/manufacturers that are referenced in this article.

References

Ahvenainen, R. (2003). *Novel Food Packaging Techniques*, pp. 11–12, 88–89, 108–113, 134–135, 536–541. Woodhead Publishing, Cambridge, UK.

Ahvenainen, R. and Hurme, E. (1997). Active and smart packaging for meeting consumer demands for quality and safety. *Food Add. Contam.* **14(6/7),** 753–763.

Ahvenainen, R., Pullinen, T., Hurme, E., Smolander, M. and Siika-aho, M. (1997). Package for Decayable Foodstuffs. PCT International Patent Application WO 98/21120. (VTT Biotechnology and Food Research, Espoo, Finland).

Anonymous (1994). Talking boxes, food temperature sensors and other "smart" packaging is not far off. *Quick Frozen Foods Intl* **36(2)**, 114.

Anonymous (2002). The new role of packaging: active, smart and intelligent. *Italian Food Bev. Technol.* **27**, 52–61.

Berragan, G. (1996). Innovation accelerates. *Soft Drinks Manag. Intl*, **October,** 26, 29.

Blixt, Y. and Borche, E. (1999). Using an electronic nose for determining the spoilage of vacuum-packaged beef. *Intl J. Food Microbiol.* **46**, 123–134.

Bodenhamer, W. T. (2000). Method and Apparatus for Selective Biological Material detection. US Patent 6,051,388. (Toxin Alert, Inc., Canada).

Brody, A. L. (2000). Innovations in active packaging – a United States perspective. In: *International Conference on Active and Intelligent Packaging, Conference Proceedings, 7–8 September*, pp. 1–16. Campden & Chorleywood Food Research Association Group, Chipping Campden, UK.

Brody, A. L. (2001). What's active about intelligent packaging? *Food Technol.* **55(6),** 75–78.

Brody, A. L. (2002a). Active and intelligent packaging: the saga continues. *Food Technol.* **56(12),** 65–66.

Brody, A. L. (2002b). Packages that heat and cool themselves. *Food Technol.* **56(7),** 80–82.

Cameron, A. C. and Talasila, P. C. (1995). Modified atmosphere packaging of fresh fruits and vegetables. In: *IFT Annual Meeting Book of Abstracts*, p. 254. Institute of Food Technologists, Chicago, IL.

Coles, R., McDowell, D. and Kirwan, M. J. (2003). *Food Packaging Technology*, p. 284. Blackwell Publishing, Oxford, UK.

Daniel, C. D. (2000). Intelligent packaging – an overview. In: *International Conference on Active and Intelligent Packaging, Conference Proceedings, 7–8 September*, pp. 1–16. Campden & Chorleywood Food Research Association Group, Chipping Campden, UK.

Day, B. P. F. (1989). Extension of shelf-life of chilled foods. *Eur. Food Drink Rev.* **4**, 47–56.

Day, B. P. F. (2001). Active packaging – a fresh approach. *J. Brand Technol.* **1(1),** 32–41.

DeCicco, B. T. and Keeven, J. K. (1995). Detection System for Microbial Contamination in Health-care Products. US Patent 5,443,987.

Fito, P., Rodriguez, E. O. and Canovas, G. V. (1997). *Food Engineering 2000*, pp. 362–363. International Thomson Publishing, New York, NY.

Goldsmith, R. M. (1994). Detection of Contaminants in Food. US Patent 5,306,466.

Gram, L. and Huss, H. H. (2000). Fresh and processed fish and shellfish. In: *The Microbiological Safety and Quality of Foods* (D. M. Lund, A. C. Baird-Parker and G. W. Gould, eds), pp. 472–506. Aspen Publishers Inc., Gaithersburg, MD.

Honeybourne, C. L. (1993). Food Spoilage Detection Method. PCT International Patent Application WO 93/15403.

Hong, S. I. (2002). Gravure-printed color indicators for monitoring kimchi fermentation as a novel intelligent packaging. *Packag. Technol. Sci.* **15**, 155–160.

Hoofman, A. S. (1997). Intelligent polymers. In: *Controlled Drug Delivery Challenges and Strategies* (K. Park, ed.), pp. 485–497. American Chemical Society, Washington, DC.

Jansen, R. and Krabs, A. (1999). Automatic identification in packaging – radio frequency identification in multiway systems. *Packag. Technol. Sci.* **12**, 229–234.

Jansen, R. and Schelhove, G. (1999). Tamper evidence – a vital issue in packaging development. *Packag. Technol. Sci.* **12,** 255–259.

Johns, J. (2002). Microwavable steamer bags. US Patent 6455084, 2002.

Karel, M. (2000). Tasks of food technology in the 21st century. *Food Technol.* **54(6),** 56–58, 60, 62, 64.

Kress-Rogers, E. (1993). The maker concept: frying oil monitor and meat freshness indicator. In: *Instrumentation and Sensors for the Food Industry* (E. Kress-Rogers, ed.), p. 523. Butterworth Heinemann, Boston, MA.

de Kruijf, N., van Beest, M., Rijk, R. and Sipilainen Malm, T. (2002). Active and intelligent packaging: applications and regulatory aspects. *Food Add. Contam.* **19(Suppl.),** 144–162.

Louis, P. (1999). Review paper-food packaging in the next millennium. *Packag. Technol. Sci.* **12,** 1–7.

Linnemann, A. R., Meerdick, G. and Meulenberg, M. T. G. (1999). Consumer-oriented technology development. *Trends Food Sci. Technol.* **9,** 409–414.

Millers, D. W., Wilkes, J. G. and Conte, E. D. (1999). Food Quality Indicator Device. PCT International Patent Application WO 99/0456.

Moe, T. (1998). Perspective on traceability in food culture. *Trends Food Sci. Technol.* **9,** 211–214.

Ohlsson, T. and Bengtsson, N. (2002). *Minimal Processing Technologies in the Food Industry*, pp. 88–89, 110–111. Woodhead Publishing, Cambridge, UK.

Pault, H. (1995). Brain boxes or simply packed? *Food Processing UK* **64(7),** 23–24, 26.

Quan, C. and Stevens, R. (1998). Protein Coupled Colorimetric Analyte Detectors. PCT International Patent Application WO 95/33991.

Rodrigues, E. T. and Han, J. H. (2003). Intelligent packaging. In: *Encyclopaedia of Agricultural, Food and Biological Engineering* (D. R. Heldman, ed.), pp. 528–535. Marcel Dekker, New York, NY.

Rooney, M. L. (1995). *Active Food Packaging*. Blackie, Glasgow, UK.

Schilthuizen, S. F. (1999). Communication with your packaging: possibilities for intelligent functions and identification functions and identification methods in packaging. *Packag. Technol. Sci.* **12,** 225–228.

Side, C. (2002). *Food Product Development Based on Experience*, pp. 151–152. Iowa State Press, Ames, IA.

Smolander, M. (2000). Freshness indicators for direct quality evaluation of packaged foods. In: *International Conference on Active and Intelligent Packaging, Conference Proceedings, 7–8 September*, pp. 1–16. Campden & Chorleywood Food Research Association Group, Chipping Campden, UK.

Stewart, F. (1993). Food Package Comprised of Polymer with Thermally Responsive Permeability. US Patent 5,254,354.

Summers, L. (1992a). *Intelligent Packaging*, Centre for Exploitation of Science and Technology, London, UK.

Summers, L. (1992b). Intelligent packaging for quality. *Soft Drinks Manag. Intl*, **May,** 32–33, 36.

Yam, K. L. (2000). Intelligent packaging for the future smart kitchen. *Packag. Technol. Sci.* **13(2),** 83–85.

Modified atmosphere packaging of fresh produce, meats, and ready-to-eat products

Introduction to modified atmosphere packaging

John D. Floros and Konstantinos I. Matsos

Introduction

Modified atmosphere packaging (MAP) is an integral part of the food industry. In this chapter, an historical perspective and the principles of the technology are presented. Examples of product applications are given, and the effects of MAP on product shelf life, safety and nutritional value are briefly discussed. Finally, several possible combinations of MAP with other processing methods are explored.

Historical developments

"Controlled atmosphere" and "modified atmosphere" are terms implying the addition or removal of gases from storage rooms, transportation containers or packages in order to manipulate the levels of gases such as oxygen, carbon dioxide, nitrogen, ethylene etc., and achieve an atmospheric composition different to that of normal air around the food (Floros, 1990). Although "controlled" and "modified" are often used interchangeably, they don't have the same meaning. In the case of controlled atmosphere, the gas

Innovations in Food Packaging
ISBN: 0-12-311632-5

composition around the product (usually in storage rooms or transportation containers) is continuously monitored and controlled. Modified atmosphere indicates that the composition of the storage atmosphere is not closely controlled.

Controlled/modified atmosphere is a relatively old process. Based on ancient writings, certain forms of modified atmosphere storage were used in China, Greece, and other early civilizations. According to some such reports, fruits were sealed in clay containers together with fresh leaves and grass. The high respiration rates of leaves and grass, combined with that of the fruit, quickly modified the atmosphere in the container, creating an environment high in carbon dioxide and low in oxygen, which helped retard fruit ripening.

However, it was not until 1820 that the effect of atmosphere on fruit ripening was studied (Floros, 1990). About 100 years later, the effect of carbon dioxide and oxygen concentration on the germination and growth of fruit-rotting fungi at various temperatures was investigated (Brown, 1922). Similarly, in the 1930s a number of research studies were published regarding the effect of carbon dioxide and storage temperature on the inhibition of microbial growth on meat surfaces and the resulting extended product shelf life (Ooraikul and Stiles, 1991).

Despite the mounting scientific evidence for the potential of modified atmospheres in food preservation, the technique evolved very slowly and its commercialization was very limited. A few cases of controlled atmosphere storage of fruits and vegetables existed, and only some degree of atmosphere modification was used in the bulk transportation of beef and beef carcasses (Inns, 1987; Ooraikul and Stiles, 1991). The first significant trials of retail size modified atmosphere packaging took place in the late 1950s, with vacuum-packed meat, fish and coffee (Inns, 1987). The interest in gas preservation techniques increased dramatically in the 1970s and 1980s, and commercial applications of MAP have steadily increased since then.

This success has been the result of several factors. Consumer trends toward more natural, fresh-like and minimally processed foods became popular. Preservatives came under attack and their use became increasingly restricted. As a result, the industry was in need of alternative preservation methods. At the same time, the cost of raw food products, labour, and energy increased, which negatively affected many traditional preservation techniques such as canning and freezing. Moreover, advances in material science allowed the development of many new packaging films with various physical properties. In addition, a wide range of sophisticated packaging equipments was developed to take advantage of the improved machinability and properties of the new films. They allowed inexpensive and easy MAP of foods in small and convenient retail units. Consequently, many MAP studies were conducted (Ooraikul and Stiles, 1991) and a large number of commercial applications were introduced.

At the same time, consumers were looking for convenient food products due to their busy lifestyles. Health concerns resulted in an elevated interest in the quality rather than the quantity of food. The demand for "fresh" and "natural" products without the addition of "dangerous" chemicals increased dramatically. MAP seemed to be the ideal method of preservation of many foods, because it could extend the shelf life of the product significantly, without affecting its "fresh" or fresh-like characteristics. Nowadays, MAP has become an integral part of the food industry, particularly the fresh produce industry, and it is more important than freezing and canning combined.

Principles of MAP

The objectives of MAP are to extend the shelf life of food products and to prevent (or at least retard) any undesirable changes in the wholesomeness, safety, sensory characteristics, and nutritive value of foods. MAP achieves the above objectives based on three principles:

1. It reduces undesirable physiological, chemical/biochemical and physical changes in foods
2. It controls microbial growth
3. Just like any other packaging technique, it prevents product contamination.

The three main gases used in MAP are nitrogen (N_2), oxygen (O_2), and carbon dioxide (CO_2). At sea level, the approximate composition of atmospheric air is 78.1% N_2, 20.9% O_2, and 0.03% CO_2. The role and the importance of each gas in MAP are related to its properties. Nitrogen is an inert and tasteless gas, without any antimicrobial activity. It is not very soluble in water, and it is primarily used to displace oxygen and prevent package collapse. Oxygen inhibits the growth of anaerobic micro-organisms, but promotes the growth of aerobic microbes. Additionally, oxygen is responsible for several undesirable reactions in foods, including oxidation and rancidity of fats and oils; rapid ripening and senescence of fruits and vegetables; staling of bakery products; color changes; and spoilage due to microbial growth. Due to oxygen's negative effects on the preservation of the food quality, it is generally avoided in the MAP of many products. However, its presence in small quantities is at times necessary for some products. For example, a minimum oxygen concentration is required by many fruits and vegetables in order to sustain their basic process of aerobic respiration. In other cases, such as red meats, high oxygen concentration is used to initiate "bloom" and preserve the bright red color of fresh meats. Finally, carbon dioxide is soluble in both water and lipids, and its solubility increases with decreasing temperatures. The dissolution of CO_2 in the product can result in package collapse. Carbon dioxide has a bacteriostatic effect, and it slows down the respiration of many products. All three gases are common and readily available, safe, economical, and are not considered to be chemical additives. However, the optimum level of each gas for each food product must be determined and used in order to maximize the positive and minimize the negative effects.

In some cases, additional gases are used in combination with the above mentioned gases. For example, carbon monoxide (CO) is sometimes added to inhibit microbial growth. However, CO is toxic to humans, and its application is limited. Similarly, sulfur dioxide (SO_2) may be used to prevent oxidative browning and to control the growth of bacteria and molds. Ethanol has also been used to enhance the firmness of tomatoes, improve flavor, and reduce fungal activity, and argon has been used to reduce microbial growth.

MAP techniques

The atmosphere inside a package can be modified by either passive or active means (Floros, 1990). In the first case, the rate of change and the final gas composition in the package depend largely on both the packaged product and the permeability of the packaging material. It is well known that most foods come from living entities and continue to live after harvesting. Fruits and vegetables, for example, consume the oxygen of the surrounding environment and release carbon dioxide via the respiration process. Similarly, the natural microflora of many products also consume oxygen. Besides the biochemical and physiological processes that utilize the available gases, oxidative reactions also take place during the storage period of foods, resulting in a reduced oxygen concentration over time.

If the above phenomena occur in a sealed package, impermeable or permeable to gases, the gas composition inside the package will change. If the container is impermeable to gases (e.g. glass jars, rigid metal cans, barrier plastic films), the rate of gas production and/or consumption will dictate the gas composition at any time. However, if the container is permeable to gases (e.g. flexible non-barrier plastic package), the gas exchange with the environment will also take place through the package (Figure 10.1). As a result, the gas composition inside the package will be further modified and the final gas composition will be different in the two containers.

The main disadvantage of the passive atmosphere modification method is that the desired atmosphere is achieved very slowly (Figure 10.2a). This can sometimes result in uncontrolled levels of oxygen, carbon dioxide or ethylene, with a detrimental effect on the quality of the product. Active modification of the atmosphere can provide a

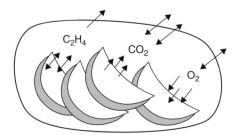

Figure 10.1   Gas exchange between a product and its surrounding atmosphere in a permeable package.

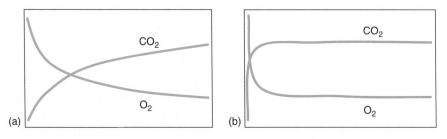

Figure 10.2   Graphical illustration of (a) passive and (b) active atmosphere modification.

solution to this problem. This is usually accomplished by first creating a vacuum and then incorporating the desired gas mixture in the package. Compared to the passive method, active atmosphere modification is practically instantaneous and takes place at the beginning of storage. The atmosphere should then remain practically unchanged, provided that the proper barrier material is used and there is no leakage through pinholes or poor seals (Figure 10.2b).

The process of applying a vacuum can be considered a method of active atmosphere modification (Floros, 1990). It is commonly used in packaging techniques such as canning, or bottling in glass containers. The main purpose of such vacuum application is to reduce the residual oxygen in the headspace of a package, which eventually retards oxidative chemical reactions and aerobic microbial growth. When a vacuum is used with flexible packages, the packaging material collapses around the product and practically eliminates the existence of the headspace. A vacuum is also often applied to storage rooms and transportation containers, a technique called hypobaric or low-pressure storage. In this method, a slight vacuum is maintained in the storage room or container, which reduces the partial pressure of oxygen and continuously removes undesirable gases such as ethylene. As a consequence, oxidative and physiological reactions are retarded.

Finally, relatively recent technological innovations allow for in-package control of a specific gas (oxygen, carbon dioxide or ethylene). Such "active" packaging systems are designed to remove or add certain gases selectively. This is usually achieved by using a substance that can bind (scavenger) or release (emitter) certain molecules as a result of chemical or enzymatic reactions. The "active" substance can be placed in the package (e.g. in a sachet), or in more sophisticated systems it can be incorporated into the packaging material itself. The method is referred to as active packaging. An example of this technique is the addition of small sachets in cans of coffee. The sachets contain a mixture of iron oxide and calcium hydroxide that binds oxygen and enables the control of the package environment without gas-flushing or the application of a vacuum (Floros, 1990).

Advantages and disadvantages of MAP

MAP offers many advantages to consumers and food producers. To the consumer, it offers convenient, high-quality food products with an extended shelf life. It also reduces and sometimes eliminates the need for chemical preservatives, leading to more "natural" and "healthy" products. At the same time, producers also enjoy the benefits of increased shelf life. By using MAP many products can be packaged centrally, and their distribution cost is reduced because fewer deliveries over longer distances become possible. Moreover, because of the extended shelf life, MAP allows transportation of foods to remote destinations and increases product markets.

MAP also has several disadvantages. Usually, each MAP product needs a different gas formulation. This requires the use of specialized and expensive equipment. At the same time, production staff must receive special training. For most products, storage temperature control is required and product safety must be established. Furthermore, MAP causes larger package volumes, which leads to increased transportation and retail display space needs. All the above add a noticeable cost, which must be paid by the consumers. Finally, another disadvantage of MAP is that it loses all its benefits once the consumer opens the package.

Effect of MAP on shelf life

The deterioration mechanisms vary widely among various products. Consequently, the effect of MAP on the shelf life of different products varies and depends upon the product. Some examples of modified atmosphere applications for different products are given below in order to demonstrate the effect of MAP on shelf life. Table 10.1 shows the recommended conditions to maximize the shelf life of various products.

Meats and poultry

The first commercial application of modified atmosphere preservation of meat was developed in the 1930s. Based on scientific evidence, refrigerated beef carcasses were transported by ship from New Zealand and Australia to the UK in a carbon dioxide-enriched environment. However, MAP of retail-size meat packages was not introduced until the 1950s, in the form of vacuum packaging. Much later, in 1981, Marks & Spencer introduced a wide range of fresh meats in the market, packaged in gas-flushed plastic trays (Inns, 1987).

One of the main quality attributes of red meat, and an important indicator of product "freshness", is its color. The bright red color of fresh meat desired by consumers is due to the presence of oxymyoglobin, a myoglobin pigment in its oxygenated form. Over time, however, oxymyoglobin gradually changes to metmyoglobin, which has an undesirable brownish color. Preservation of the desired bright red color can be achieved by packaging the meat in oxygen-enriched atmospheres.

Meat is a highly perishable food, and the major factor that limits its shelf life is microbial growth. It was noted earlier that carbon dioxide slows down the growth rate of microbes. Thus, atmospheres high in carbon dioxide should be used to suppress microbes and extend shelf life. Consequently, MAP of meat in a gas mixture with high amounts of both oxygen and carbon dioxide is ideal for preserving the product's quality and extending its shelf life. The recommended atmosphere compositions for beef and pork are shown in Table 10.1.

In the case of poultry, the presence of oxygen is not necessary. Sometimes, such as in the case of turkey meat, the presence of oxygen may even be harmful because it causes off-flavours (Inns, 1987). The shelf life of poultry products under MAP increases as the concentration of carbon dioxide in the atmosphere increases. This is probably due to the suppression of microbial growth. However, it has been observed that using carbon dioxide in the gas mixture beyond a certain point results in product discoloration and has negative effects on shelf life.

Table 10.1   Recommended MAP conditions of various products

Commodity	Temperature range (°C)	O_2 (%*)	CO_2(%*)	Source
Baked products:				
Bread (sliced)	Ambient	0	100	Smith *et al.*, 1990
Croissants	Ambient	0	100	Smith *et al.*, 1990
English muffins	Ambient	0	60	Smith *et al.*, 1990
Pitta bread	Ambient	0	99	Smith *et al.*, 1990
Pizza (crust)	Ambient	0	90	Smith *et al.*, 1990
Cheeses:				
Hard cheeses	1–4	0	100	Subramanian, 1998
Soft cheeses	1–4	0	20–40	Subramanian, 1998
Fish:				
Lean	0–2	30	40	Garthwaite, 1992
Oily and smoked	0–2	0	60	Garthwaite, 1992
Fruit:				
Apple (whole)	0–5	2–3	1–5	Kader, 1986
Apple (sliced)	0–5	10–12	8–11	Lay, 1997
Avocado	5–13	2–5	3–10	Kader, 1986
Banana	12–15	2–5	2–5	Kader, 1986
Kiwifruit	0–5	2	5	Kader, 1986
Mango	10–15	5	5	Kader, 1986
Pineapple	10–15	5	10	Kader, 1986
Strawberry	0–5	10	15–20	Kader, 1986
Meats:				
Beef	−1–2	60–80	20–40	Blakistone, 1998
Pork	−1–2	30	30	Floros, 1990
Poultry	−1–2	0	25–35	Blakistone, 1998
Processed foods:				
Dried foods	Ambient	0	0–100	Floros, 1990
Low-moisture foods	Ambient	0	0–100	Floros, 1990
Fats and oils	Ambient	0	0	Floros, 1990
Vegetables:				
Asparagus	0–5	20	5–10	Kader, 1986
Broccoli	0–5	1–2	5–10	Kader, 1986
Cabbage	0–5	3–5	5–7	Kader, 1986
Lettuce (head)	0–5	2–5	0	Kader, 1986
Lettuce (shredded)	0–5	1–2	10–12	Price, 1992
Mushrooms	0–5	21	10–15	Kader, 1986
Spinach	0–5	21	10–20	Kader, 1986
Tomatoes (mature)	12–20	3–5	0	Kader, 1986

* Volume or mole percentages; the remainder is nitrogen.

Fish and shellfish

The shelf life of fish products packaged under modified atmospheres can be extended significantly when the atmosphere is high in CO_2 and the temperature is kept below 2°C. This is mainly due to the inhibiting effect of carbon dioxide on the growth of *Pseudomonas* spp. and other psychrotrophic spoilage micro-organisms. In the case of shellfish (crustaceans), MAP provides the additional benefit of reducing the formation of black spots on the shells (Garthwaite, 1992).

The composition of the atmosphere used for MAP of fish depends on the fat content of the product. In general, an atmosphere of 30% O_2, 40% CO_2, and 30% N_2 is used for lean fish. The elevated amount of oxygen contributes to the shelf-life extension of marine fish through the reduction of trimethylamineoxide (TMAO), an osmo-regulator, to trimethylamine (TMA) (Devlieghere *et al.*, 2002). TMA is the major component responsible for the undesired "fishy" odor. In the case of oily and smoked fish, however, the exclusion of O_2 is recommended to avoid oxidative reactions and rancidity development. Gas mixtures consisting of 60% CO_2 and 40% N_2 are often used.

A negative effect of the application of modified atmospheres in fish packaging is the increase of drip inside the package. This is due to the nature of the fish muscle, which absorbs a great amount of carbon dioxide as dissolved CO_2 (Inns, 1987). As a result, the pH of the muscle is reduced, with a subsequent reduction in the water-holding capacity of the muscle proteins. This problem can be easily solved with the addition of an absorbing pad (usually made out of cellulose) beneath the product.

The dissolution of carbon dioxide in the fish muscle lowers the internal pressure of the package, and may even cause package collapse. The incorporation of nitrogen into the gas mixture of the package may solve this problem.

Bakery products

The main factor that limits the shelf life of bakery products is microbiological spoilage, in particular mold growth, which is responsible for significant economic losses (Smith and Simpson, 1996). Because of its bacteriostatic and fungistatic properties, and its inhibitory effect on insect growth, high carbon dioxide MAP can be used with bakery products to eliminate growth of molds and subsequently to extend the product shelf life.

However, the high solubility of carbon dioxide in water and fats may lower the pH of the product and lead to some flavor changes. Another potential problem is that the absorption of carbon dioxide by the product may result in package collapse. The latter problem can be easily solved by using nitrogen as a filler gas. Nitrogen is also used in MAP of certain bakery and snack products with low water activity in order to replace oxygen and prevent rancidity problems (see Table 10.1).

An additional problem in MAP of some bakery products is that aerobic spoilage may take place even when a very small amount of residual oxygen exists (e.g. trapped air in the food, poor gas flushing). This problem has been addressed with the use of novel MAP techniques such as active packaging, and in particular with the use of oxygen absorbents. Oxygen absorbents eliminate all residual oxygen, and extend the mold-free shelf life of certain products by up to 900% as compared to traditional MAP techniques (Smith *et al.*, 1990). More information on oxygen absorbents/scavengers can be found in Chapter 8 of this book.

Fruits and vegetables

The live tissue of fresh fruits and vegetables respires and transpires. The respiration rate varies greatly among different species, and depends heavily on temperature. As a result, MAP of fresh produce requires a different approach as compared to other products. The main goal of modified atmosphere applied to fruits and vegetables is to minimize the respiration rate of the product. This includes suppressing the production of ethylene, a gas responsible for accelerating ripening and deterioration, and hastening the onset of senescence in fruits and vegetables.

Reduced levels of oxygen and increased levels of carbon dioxide in the atmosphere around fresh produce appear to have several positive effects. Among others, they preserve product quality because they slow down respiration, decrease softening rates, and improve chlorophyll and other pigment retention. In addition, as in other products, elevated levels of carbon dioxide reduce the rate of microbial growth and spoilage.

On the other hand, the use of MAP techniques with fresh produce, especially fruits, has a few potential hazards. The complete elimination of oxygen from the package quickly results in anaerobic respiration, the production of ethylene, and, subsequently, a fast and dramatic deterioration of the product quality. This is normally due to the accumulation of acetaldehyde, ethanol, and organic acids, the development of off-flavours, and, finally, the discoloration and the softening of the tissue. However, in some cases a more important hazard is the potential growth of *Clostridium botulinum*, which can grow under anaerobic conditions and may produce its deadly toxin.

Since the mid-1990s, the application of MAP to minimally processed fruits and vegetables has grown dramatically, and it has now become a multi-billion dollar industry in the USA alone. The treatments involved in preparing and handling minimally processed fruits and vegetables result in fast deterioration and quality decline due to the dramatic increase of physiological and biochemical activities in the tissue (Price and Floros, 1993). These negative effects can be offset or significantly reduced by the use of MAP and refrigeration. MAP extends the shelf life, maintains the high quality (particularly the fresh-like characteristics), and improves the convenience of minimally processed fruits and vegetables. As a result, MAP has contributed significantly to the increased consumption of fresh fruits and vegetables by the average North American consumer.

Cheese

The quality of cheese deteriorates primarily because of surface mold growth and/or lipid oxidation. Both problems can easily be addressed with the application of a vacuum and the removal of air. This has been done successfully for many years. However, vacuum packaging of some cheeses has certain disadvantages – for example, the package cannot be opened easily, and the vacuum packaging of soft cheeses or cheeses with a crumbled texture may damage the product. The use of MAP overcomes these problems without sacrificing the benefit of extended shelf life.

Hard cheeses like Cheddar are usually packaged under 100% carbon dioxide conditions. In the case of soft cheeses, a mixture of 20–40% carbon dioxide with 60–80% nitrogen is preferred, to prevent package collapse. Solutions become more complicated for MAP of mold-ripened cheeses or cheeses that produce carbon dioxide. In the first case, for example, the complete elimination of oxygen from the package would ultimately result in the death of useful fungi. On the other hand, high oxygen concentrations in the package would result in uncontrolled mold growth. Due to the large number of cheese types and their wide range in composition and shelf life, the packaging of each type of cheese should be considered separately (Subramaniam, 1998).

Other products

The list of products that utilize MAP techniques is not limited to the above examples. Many other applications exist, including coffee, nuts, snacks, pasta, delicatessen products, yoghurt, milk, and milk powders. In each case, extended shelf life is achieved by the application of MAP principles.

Effect of MAP on micro-organisms – safety issues

The extension of product shelf life by low oxygen atmospheres also results in the control of aerobic bacteria, which are mainly responsible for spoilage. Moreover, high concentrations of carbon dioxide also extend the shelf life of MAP foods. The antimicrobial effect of carbon dioxide occurs at or near a 10% level, and increases with higher concentrations. Using 20% carbon dioxide can control the growth of several aerobes, including *Pseudomonas*, *Acenatobacter* and *Moraxella*, but higher concentrations in the ranges of 20–40% usually give better results. However, very high concentrations of carbon dioxide may stimulate the growth of the anaerobe *Clostridium botulinum*. Although the mechanism of carbon dioxide's antimicrobial activity is not fully understood, it appears that CO_2 extends the lag phase in the microbial growth (Ray, 2001) in many ways: it penetrates the microbial cell wall and alters the cell permeability; it solubilizes inside the cells and produces carbonic acid (H_2CO_3), which reduces the pH of the cell; and finally, it interferes with several enzymatic and biochemical pathways inside the microbial cells, and thus reduces their growth rate. The effect of MAP on each micro-organism and its rate of growth also depend on the food itself, the amount of oxygen dissolved or entrapped in the product, the available nutrients, and the presence of reducing components.

The potential for *Clostridium botulinum* (type E and non-proteolytic type B) growth in refrigerated MAP products is of major concern. Other food-borne pathogens that belong to the non-sporeforming psychrotrophic micro-organism group can also grow under modified atmospheres. This is especially true for facultative anaerobes, including *Listeria monocytogenes* and *Yersinia enterocolitica*. Moreover, if refrigerated products were subject to temperature abuse, mesophilic facultative microbes like certain *Salmonella* strains and *Staphylococcus aureus* can also grow in MAP foods.

The inhibition of spoilage bacteria combined with the possible growth of psychrotrophic pathogens makes the safety problems of MAP products more serious, because the food may be unsafe for human consumption well before it becomes organoleptically unacceptable. In order to improve the safety of MAP products and eliminate such hazards, additional measures are necessary. For example, the continuous and tight control of the temperature is of paramount importance, not only to inhibit the growth of mesophilic pathogens but also to maximize the antimicrobial effect of carbon dioxide and minimize the growth rate of psychrotrophs. Another requirement for improving the safety of MAP products is the use of high-quality raw materials and their subsequent handling, according to strict guidelines for hygiene and sanitation.

Effect of MAP on nutritional quality

As mentioned above, MAP can significantly extend the shelf life of many products. However, the effect of MAP on the nutritional quality of foods during storage is not well understood. Recently, Devlieghere *et al.* (2002) reviewed the subject, but reported that only limited information is available. One of the well-known effects of low oxygen MAP is the retardation of oxidative reactions (Farkas *et al.*, 1997), which in turn helps to maintain the nutritional quality of foods. According to Devlieghere *et al.* (2002), no information is available regarding the effect of high oxygen MAP on the nutritional quality of products such as meat and fish.

Regarding the effect of MAP on the nutritional quality of respiring foods, the antioxidant constituents of fruits and vegetables are of special interest. Lee and Kader (2000) reported that the loss of vitamin C after harvesting can be reduced by storing apples and vegetables in atmospheres with reduced oxygen levels, and carbon dioxide levels of up to 10%. Above this limit, and depending on the commodity, the loss of vitamin C is accelerated. MAP of fruits and vegetables with either increased levels of carbon dioxide or reduced levels of oxygen appears to prevent the loss of provitamin A (Kader *et al.*, 1989), but it also inhibits the biosynthesis of carotenoids. Studies on the retention of carotenoids in fresh-cut products under MAP did not show any loss (Devlieghere *et al.*, 2002). However, it appears that MAP has a negative effect on the preservation of anthocyanins. Finally, the available information on the loss of glucosinolates is very limited, and the effect of MAP is still unclear.

Combination of MAP with other technologies

Refrigeration

Many foods, including fresh produce, meats, and other minimally processed products, are handled and marketed under refrigerated conditions. In these cases, MAP is used synergistically to provide additional benefits and increase positive effects, such as reduced rates of enzymatic, oxidative, and other undesirable chemical reactions; slower respiration rates; delayed ripening; and suppressed microbial growth. The combination of high carbon dioxide atmospheres with refrigeration is of particular importance. This is because the amount of dissolved carbon dioxide and its antimicrobial effectiveness increase as temperature decreases (Gould and Jones, 1989).

Freezing

MAP of frozen products in the form of vacuum packaging is often necessary. Although chemical reactions proceed very slowly under freezing temperatures, they may still cause significant quality changes over time. The oxidation of vitamins and pigments continues to take place, and results in the loss of color, flavor and nutritive value. Another common problem in frozen products, the so-called freezer burn, is due to water lost from the surface of the product to the surrounding environment. All of the above problems can be significantly reduced or completely eliminated by the application of vacuum packaging.

Irradiation

One of the big advantages of MAP is the extended shelf life it provides, while preserving the fresh state of the product. On the other hand, MAP brings some concerns about the microbial safety of products. However, the combination of MAP with irradiation treatments can minimize such safety concerns. Irradiation is a physical treatment that exposes the food to small doses of ionizing radiation and results in the reduction of microbial loads. Of particular importance is the fact that irradiation is not a heat treatment and, just like MAP, it preserves the fresh like characteristics of products. Furthermore, another advantage of irradiation is that it can be applied to pre-packaged MAP products to reduce the risk of post-contamination.

Sometimes the combination of the two treatments leads to complications. The respiration rate of fruits and vegetables increases temporarily during and after the irradiation treatment. This must be taken into account when designing the package. The film permeability must be high enough to avoid the complete consumption of oxygen, which may lead to anaerobic respiration, off-flavour development, and the production of fermentation by-products.

Furthermore, radiation treatments may negatively affect the packaging material itself (Ozen and Floros, 2001). The package and its seal should both be able to withstand the radiation treatment, or the resulting defects may cause recontamination and loss of the modified atmosphere. Research studies show that the negative effects of irradiation are minimal at low doses ($<10\,kGy$), but some plastic materials are susceptible even at such low doses and therefore should not be used with irradiation. The FDA has a list of approved packaging materials that can be used with different irradiation doses.

Hurdle technology

In hurdle technology, an intelligent combination of various preservation techniques is applied to the product. The "hurdles" are chosen carefully in order to maximize the preserving effect while minimizing the impact on product quality. Usually, this is achieved by combining processes with synergistic effects. MAP can act synergistically with a number of technologies, such as refrigeration, freezing, canning, aseptic processing etc., resulting in a better quality product with longer shelf life and increased stability.

Edible coatings

Edible coatings are extensively discussed in Chapter 15, and can be considered as a special form of MAP. Their application on the surface of products limits gas exchange and moisture transportation between the food and the surrounding environment (Kester and Fennema, 1986). This results in the formation of an internal modified atmosphere; however, in many cases this is not enough to provide the desired preserving effects, and therefore an additional package is needed. However, the combination of edible coatings with other packaging materials can lessen the overall need for packaging materials and may reduce waste. Similarly, the combination of edible coatings with traditional MAP can reduce the total packaging cost.

Biological control

Biological control of a certain micro-organism implies the use of antagonistic bacteria, fungi, and/or yeast in order to retard or inhibit its growth. This alternative method recently received significant attention due to the environmental and health concerns that arise from using chemicals to control micro-organisms. MAP alone retards microbial growth, but its combination with biological methods may control the growth of undesirable micro-organisms much better. This requires that the optimum growth conditions of the antagonist are the same or similar to the recommended storage conditions of the product. Further research is necessary to establish viable combinations of MAP with biological control methods, and to test consumer perception and acceptance of such products (Dock *et al.*, 1998).

References

Blakistone, B. A. (1998). Meats and poultry. In: *Principles and Applications of Modified Atmosphere Packaging of Foods* (B. A. Blakistone, ed.), pp. 240–290. Blackie, London, UK.

Brown, W. (1922). On the germination and growth of fungi at various temperatures and in various concentrations of oxygen and carbon dioxide. *Ann. Bot.* **36,** 257–283.

Devlieghere, F., Gil, M. I. and Debevere, J. (2002). Modified atmosphere packaging (MAP). In: *The Nutrition Handbook for Food Processors* (C. J. K. Henry and C. Chapman, eds), pp. 342–370. CRC Press, Boca Raton, FL.

Dock, L. L., Nielsen, P. V. and Floros, J. D. (1998). Biological control of *Botrytis cinerea* growth on apples stored under modified atmospheres. *J. Food Prot.* **61,** 1661–1665.

Farkas, J. K., Floros, J. D., Lineback, D. S. and Watkins, B. A. (1997). Oxidation kinetics of menhaden oil with TBHQ. *J. Food Sci.* **62,** 505–507, 547.

Floros, J. D. (1990). Controlled and modified atmospheres in food-packaging and storage. *Chem. Eng. Prog.* **86,** 25–32.

Garthwaite, G. A. (1992). Chilling and freezing of fish. In: *Fish Processing Technology* (G. M. Hall, ed.), pp. 89–113. Blackie, Glasgow, UK.

Gould, G. W. and Jones, M. V. (1989). Combination and synergistic effects. In: *Mechanisms of Action of Food Preservation Procedures* (G. W. Gould, ed.), pp. 401–421. Elsevier, Barking, UK.

Inns, R. (1987). Modified atmosphere packaging. In: *Modern Processing, Packaging and Distribution Systems for Food*, Vol. 4 (F. A. Paine, ed.), pp. 36–51. Blackie, Glasgow, UK.

Kader, A. A. (1986). Biochemical and physiological basis for effects of controlled and modified atmospheres on fruits and vegetables. *Food Technol.* **40,** 99–104.

Kader, A. A., Zagory, D. and Kerbel, E. L. (1989). Modified atmosphere packaging of fruits and vegetables. *Crit. Rev. Food Sci. Nutr.* **28,** 1–30.

Kester, J. J. and Fennema, O. R. (1986). Edible films and coatings – a review. *Food Technol.* **40,** 47–59.

Lay, S. V. (1997). Maximizing the shelf life of minimally processed apple slices by modified atmosphere and ascorbic acid treatment. MSc thesis, Department of Food Science, Purdue University, West Lafayette, IN.

Lee, S. K. and Kader, A. A. (2000). Preharvest and postharvest factors influencing vitamin C content of horticultural crops. *Postharvest Biol. Technol.* **20,** 207–220.

Ooraikul, B. and Stiles, M. E. (1991). Introduction: review of the development of modified atmosphere packaging. In: *Modified Atmosphere Packaging of Food* (B. Ooraikul and M. E. Stiles, eds), pp. 1–17. Ellis Horwood, Chichester, UK.

Ozen, B. F. and Floros, J. D. (2001). Effects of emerging food processing techniques on the packaging materials. *Trends Food Sci. Technol.* **12,** 60–67.

Price, J. L. (1992). Optimization of oxygen and carbon dioxide levels for controlled/modified atmosphere packaging of shredded lettuce. MSc thesis, Department of Food Science, Purdue University, West Lafayette, IN.

Price, J. L. and Floros, J. D. (1993). Quality decline in minimally processed fruits and vegetables. In: *Food Flavors, Ingredients, and Composition: Proceedings of the 7th International Flavor Conference* (G. Charalambous, ed.), pp. 405–427. Elsevier, Amsterdam, The Netherlands.

Ray, B. (2001). Control by modified atmosphere (or reducing O-R potential). In: *Fundamental Food Microbiology* (B. Ray, ed.), pp. 471–476. CRC Press, Boca Raton, FL.

Smith, J. P., Ramaswamy, H. S. and Simpson, B. K. (1990). Developments in food packaging technology. Part II. Storage aspects. *Trends Food Sci. Technol.* **1,** 111–118.

Smith, J. P. and Simpson, B. K. (1996). Modified atmosphere packaging. In: *Baked Goods Freshness. Technology, Evaluation, and Inhibition of Staling* (R. E. Hebeda and H. F. Zobel, eds), pp. 205–238. Marcel Dekker, New York, NY.

Subramaniam, P. J. (1998). Dairy foods, multi-component products, dried foods and beverages. In: *Principles and Applications of Modified Atmosphere Packaging of Foods* (B. A. Blakistone, ed.), pp. 158–193. Blackie, London, UK.

Internal modified atmospheres of coated fresh fruits and vegetables: understanding relative humidity effects*

Luis Cisneros-Zevallos and John M. Krochta

Introduction

Coatings applied to the surface of fruits and vegetables can be formed from one or more components, and are commonly called "waxes" (Hagenmeier and Shaw, 1992). However, coating materials used are actually various mixtures of lipids, proteins, carbohydrates, plasticizers, surfactants, additives, and solvents like water and alcohols. The different materials used and the properties they possess provide a wide range of characteristics possible for the coating systems.

Although research in this area is extensive, much of the work has been empirical (Banks *et al.*, 1993). Commercial coatings are used mainly to prevent water loss or impart gloss to fruits and vegetables. However, coatings have not been widely

---

*This chapter has been reprinted from the authors' previous article, published in 2002 in the Journal of Food Science **67(8)**, 2792–2797, with copyright permission from the IFT (Institute of Food Technologists).

Innovations in Food Packaging
ISBN: 0-12-311632-5

implemented on a commercial scale for specific atmospheric modification, in part due to the variation in the response of fruits to apparently similar coatings and to uncontrolled factors which influence the coating performance. Different factors that can affect the performance of coating-fruit systems are type of fruit, coating surface coverage, coating thickness and permeability, and temperature.

Banks *et al.* (1993) proposed a mathematical model to explain the interaction between a fruit and a coating system. Through this study they tried to explain the variability of results found in the literature. They described coatings performing either as a film wrap or by blocking pores. The first mechanism is called the loosely adhering coating (*lac*) model. In the *lac* model the coating is assumed to function like a film wrap covering the fruit skin, where coating and fruit resistances act in series (Ben-Yehoshua and Cameron, 1989; Hagenmeier and Shaw, 1992). In this model, fruit resistance is the result of pore and cuticle resistance operating in parallel. The modified atmosphere (MA) generated in the coated fruit will depend on coating permeability and coating thickness. The second mechanism is called the tightly adhering coating (*tac*) model. The *tac* model assumed considers that the coating tightly covers pores and cuticle, so the pore/coating and cuticle/coating resistances are added in parallel. In this model, pore blockage is more important than coating characteristics. Thus, MA will depend on the differing proportions of pores blocked by the coating. Between the *lac* and *tac* models, the real mechanism is still unknown.

One factor not considered throughout the literature is the relative humidity (RH) of storage conditions. RH is known to affect the permeability values of hydrophilic films (McHugh *et al.*, 1993, 1994; McHugh and Krochta, 1994a, 1994b). Usually fruits are stored between 90 and 95% RH; however, values lower than these are common in commercial practice and in research studies.

The transmission of molecules through polymer films, known as permeability, is a basic function of the chemical structure and other factors such as polymer morphology, including density, crystallinity, and polymer orientation. Higher densities, higher degrees of crystallinity, and cross-linking will all decrease the permeability of a polymer film. The type of solvent used in casting films and the drying rate will also influence the permeability coefficient (Pauly, 1989). Plasticizers are often added to increase film flexibility; however, this will also increase film permeability (McHugh *et al.*, 1994; McHugh and Krochta, 1994a; Mate and Krochta, 1996). Water is the most effective plasticizer in hydrophilic films, and the amount of moisture in the films is related to the RH of the environment through their moisture sorption isotherm behavior (Gontard *et al.*, 1996). Thus, permeability of hydrophilic films and films formed as coatings increases at higher RH due to increasing moisture concentration within the film (Mark *et al.*, 1966; Elson *et al.*, 1985; Rico-Pena and Torres, 1990; Hagenmeier and Shaw, 1991; McHugh and Krochta, 1994a). This increase in permeability is also related to a decrease in the glass transition temperature of the film. An amorphous polymer matrix may exist as a viscous glass or a liquid-like rubber state. The change from one state to another will strongly depend on water content or other plasticizers. The glass–rubber transition will affect molecular mobility of the system which can be observed as changes in viscosity, diffusivity or flexibility of the system (Roos and Karel, 1991; Buera and Karel, 1993; McHugh and Krochta, 1994a).

Reports of RH effects on coated fruits are few. Elson *et al.* (1985) reported how RH affected the O_2 and CO_2 permeability characteristics of a hydrophilic coating (Nutri-save) and the resulting coating effect on apples. Meheriuk and Lau (1988) related the loss of normal ripening in coated pear fruits stored at low RHs to a reduction in coating permeation. Most of the work of RH effects has been on film permeabilities. For example, Hagenmeier and Shaw (1991) showed that O_2 permeability of shellac coatings was dependent on RH. Coatings cast from ethanol, compared to water, had lower permeability values and were less dependent on RH. Similar RH dependency of gas permeability for fruit coating waxes was reported (Hagenmeier and Shaw, 1992). It was stated that this RH dependence is in large part due to the polar components used to raise pH and solubilize the polymer. Rico-Pena and Torres (1990) found an exponential dependence of O_2 permeability with RH for an edible methylcellulose–palmitic acid film. Whey protein films have a similar dependence of O_2 and aroma permeability on RH conditions (McHugh and Krochta, 1994a; Miller and Krochta, 1997, 1998; Miller and others, 1998). RH dependence on gas permeabilities for amylose starch films (Mark *et al.*, 1966) and hydroxypropylated starch films (Roth and Mehltretter, 1967) were also reported. More recently, Gontard and others (1996) and Mujica-Paz and Gontard (1997) showed the O_2 and CO_2 permeability dependence on RH for wheat gluten films.

The objective of this study was to use steady-state mathematical models to understand the influence of relative humidity (RH) on the performance of hydrophilic films formed as coatings on fruits, and to explain in part the conflictive results found in the literature.

Theoretical approach

Fruit-coating model system

The fruit system model is analyzed as a Modified Atmosphere (MA) model system. The MA model is described as a function of the internal atmospheres. In this study, we used the *lac* model and the *tac* model for the specific case of complete surface coverage and gas exchange mainly through pores. Gas exchange depends on coating permeability and thickness for achieving internal MA. The assumptions considered for the model are steady-state condition, constant temperature, no effect of CO_2 on respiration rate, and no internal atmosphere composition gradients within the fruit. The model consists of equations that describe fruit respiration rate, diffusion through skin and coating (Banks *et al.*, 1993), and the coating permeability dependence on RH.

Equations of the fruit model system are those proposed by Banks *et al.* (1993) for apple fruits. However, the approach can be extended to other fruits as well:

(a) *Fruit oxygen uptake* (r_{O_2}, $cm^3 kg^{-1} s^{-1}$)

$$r_{O_2} = \frac{V_m \, p_{O_2 i}}{(K_m + p_{O_2 i})} \qquad (11.1)$$

where V_m is the maximum rate of O_2 consumption ($cm^3 kg^{-1} s^{-1}$); K_m is the Michaelis–Menten constant (atm) and $p_{O_2 i}$ is the internal oxygen partial pressure (atm) (Banks *et al.*, 1993). K_m is the equivalent to the substrate concentration that yields half maximal rate of O_2 consumption and indicates the affinity of the enzyme system to the substrate.

(b) *Respiratory CO_2 production (aerobic and anaerobic), (r_{CO_2}, $cm^3\,kg^{-1}\,s^{-1}$)*

$$r_{CO_2} = RQ^\infty V_m \left[\frac{p_{O_2i}}{(K_m + p_{O_2i})} + \frac{10^{-10}}{(p_{O_2i} + a)^b} \right] \tag{11.2}$$

where RQ^∞ is the respiratory quotient when O_2 is unlimited ($= 1$), constant $a = 0.007$, and constant $b = 5$.

(c) *Internal oxygen partial pressure, p_{O_2i}*

At steady state, $r_{O_2} = J_{O_2}$, where J_{O_2} is the diffusive flux of O_2 between internal and external atmospheres per unit weight of fruit. According to Fick's law:

$$J_{O_2} = (p_{O_2e} - p_{O_2i}) \frac{A^{fruit}}{R_{O_2}^{total}\,W} \tag{11.3}$$

where A^{fruit} is the fruit surface area (cm^3), p_{O_2e} is the external oxygen concentration (atm), W is the fruit weight (kg) and $R_{O_2}^{total}$ is the O_2 diffusion resistance between the fruit's internal and external atmospheres (s atm cm^{-1}). Resistance values will include skin and/or coating thickness.

Combining equations (11.1) and (11.3) and solving for p_{O_2i}:

$$p_{O_2i} = \frac{P_{O_2e} - K_m - R_{O_2}^{total}\dfrac{V_m W}{A^{fruit}}}{2} + \frac{\sqrt{\left[P_{O_2e} - K_m - R_{O_2}^{total}\dfrac{V_m W}{A^{fruit}}\right]^2 + 4K_m\,P_{O_2e}}}{2} \tag{11.4}$$

(d) *Internal carbon dioxide partial pressure, obtained from equations (11.2) and (11.3), is:*

$$p_{CO_2i} = \left[\frac{R_{CO_2}^{total}\,W}{A^{fruit}}\right] RQ^\infty V_m \left[\frac{p_{O_2i}}{(K_m + p_{O_2i})} + \frac{10^{-10}}{(p_{O_2i} + a)^b} \right] \tag{11.5}$$

Here, p_{CO_2i} is the internal carbon dioxide partial pressure (atm), $R_{CO_2}^{total}$ is the resistance to CO_2 diffusion between the fruit's internal and external atmospheres (s atm cm^{-1}).

(e) *The lower oxygen limit*

This is defined as the p_{O_2i} under which fermentation is initiated, being product specific. *LOL* is determined when $r_{CO_2}/r_{O_2} = RQ > 1.1$ in this study, where RQ is the respiration quotient. The *LOL* is calculated as 0.012 atm for the fruit model system.

(f) *For the lac model, the coating resistance (R^{coat}) and fruit resistance (R^{fruit}) are added in series.*

Thus:

$$R^{total} = R^{fruit} + R^{coat} \tag{11.6}$$

A similar relationship describes the *tac* model in the case of complete surface coverage and gas exchange mainly through pores. In this latter case, $R^{fruit} = R^{pores}$.

Permeability dependence on RH

The coating system used in this study was plasticized, edible whey protein film (McHugh and Krochta, 1994a) with an exponential O_2 permeability dependence on RH (25°C), defined in this study as:

$$P_{O_2} = j\ 10^{k\ RH} \tag{11.7}$$

where RH is relative humidity (%), P_{O_2} is O_2 film permeability (cc μm/m^2 d kPa) and j and k are constants. For example, values of j for WPI-sorbitol films (WPI/Sorb, ratio 60/40) and WPI-glycerol films (WPI/Gly, ratio 70/30) are 0.026 and 1.208, respectively. Values of k for WPI/Sorb and WPI/Gly films are 0.049 and 0.037, respectively.
  Coating resistance to O_2 ($R^{coat}_{O_2}$, s atm cm^{-1}) is defined as:

$$R^{coat}_{O_2} = H/n\ P_{O_2} \tag{11.8}$$

where H is coating thickness (cm) and n is the conversion factor (n = 1.1727×10^{-11} m^2 d kPa/μm cm s atm) used to obtain P_{O_2} values in cm^2/atm s.
  A general assumption for the coating system is that it equilibrates completely with the RH of the external atmosphere (no moisture concentration gradient within the film) at a constant temperature (25°C). It is also assumed that the surrounding air is well mixed; therefore, the bulk air RH is similar to the air RH at the film/air interface. The film thicknesses used were 0.75, 1.3 and 2.6 μm. The ratio between CO_2 and O_2 film permeability (P_{CO_2}/P_{O_2}), also known as β film value, was assumed to be 2, 4 or 10. It was also assumed that the fruit skin resistance (R^{fruit}) to gas exchange was not affected by changes in RH.
  Assumed values for the different variables in this study are as follows: A = 160 cm^2, W = 160 g, K_m = 0.02 atm, p_{O_2e} = 0.21 atm, $R^{fruit}_{O_2}$ = 5996 atm s/cm, R^{fruit}_{CO2} = 6535 atm s/cm, V_m values used were 5, 20 and 30 cm^3/kg h (from Banks *et al.*, 1993).
  By using this steady-state approach we can study the effects of RH on p_{O_2i}, p_{CO_2i}, r_{O_2}, r_{CO_2} and RQ of coated fruits if we know values for V_m, K_m, A, W, H, P_{O_2} and R^{fruit}.

Results and discussion

RH and coating thickness effects on internal gas pressure of coated fruits

Oxygen permeability of whey protein film coatings increases with RH in an exponential form according to equation (11.7) (Figure 11.1). The same tendency is shown for CO_2 permeability for an assumed β value of 4.

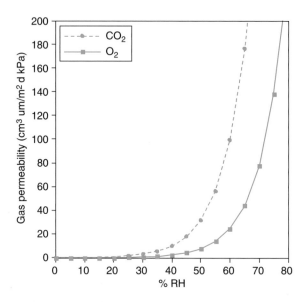

Figure 11.1 Exponential relationship of O_2 and CO_2 permeability with relative humidity (% RH) for a plasticized whey protein film (60% whey protein/40% sorbitol) at 20°C.

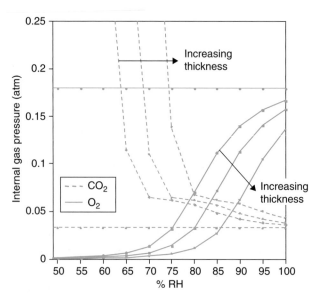

Figure 11.2 Predicted $p_{O_2 i}$ and $p_{CO_2 i}$ of WPI/Sorb coated fruits stored under different RH conditions and for different coating thickness. Film $\beta = 4$ and film coating thicknesses are 0, 0.75, 1.3 and 2.6 μm. Fruit V_m is 20 cm^3/kg h at 20°C.

When a hydrophilic film is formed as a coating on a fruit, the model predicts $p_{O_2 i}$ and $p_{CO_2 i}$ change as the external RH decreases (Figure 11.2). Internal oxygen is reduced and carbon dioxide increasingly accumulated within the fruit as RH decreases. Thus, storing fruits at a given RH (e.g. 65–70% RH) can deplete the $p_{O_2 i}$ inside the fruit and induce a large $p_{CO_2 i}$ increase. This effect will be influenced by coating thickness. For a given RH, the p_{O_2i} within the coated fruit is lowered and p_{CO_2i} is increased with an increase in thickness. As a result of this, fruits coated with thick films may reach large p_{CO_2i} at higher RH compared to fruits coated with thinner films.

Gas exchange will show similar tendency, with a decrease in r_{O_2} and r_{CO_2} when RH decreases. The explanation is that as RH decreases, film O_2 and CO_2 permeance decreases, thereby lowering the p_{O_2i} levels within the fruit. Increasing film thickness will reduce even more the O_2 and CO_2 permeances and lower even more the p_{O_2i} within the fruit. When a lower oxygen limit (*LOL*) is reached, anaerobic respiration will dominate, inducing large r_{CO_2} production and accumulation of large p_{CO_2i} within the fruit at a "critical" RH. Thus, anaerobic conditions can be reached at different RHs depending on coating film thickness.

Although fruit respiration has not been studied as a function of film thickness, investigations have been based on the amount of coating material applied or coating solution concentration (Ben Yehoshua *et al.*, 1970; Banks *et al.*, 1993). Film thickness has been related to coating solution concentration (Park *et al.*, 1994a, 1994b) and will probably be influenced by physicochemical properties of the coating solution. Coating thicknesses have been reported to be in the range of 1–5 μm (Trout *et al.*, 1953; Brusewitz and Singh, 1985; Elson *et al.*, 1985; Hagenmeier and Shaw, 1992) and from 4.5–13 μm (Park *et al.*, 1994b). Our model predictions may explain in part the variability observed in the past when studying coated fruit systems. Changes in RH and variation in coating thickness can interact, inducing large effects in internal MA of coated fruits.

RH and maximum respiration rate effects on internal gas pressure of coated fruits

We considered a fruit model with three different V_m values (low = 5, moderate = 20, and high = 30 cm^3/kg h), which represent fruits with differing V_m values or fruits stored at different temperatures. Coated fruits have a constant film thickness.

In this case the model predicts that coated fruits with higher V_m would yield lower internal p_{O_2i} as RH decreases, compared to coated fruits with lower V_m. The trend is similar to the effect observed when film thickness decreases in coated fruits (Figure 11.2). Similarly, internal p_{CO_2i} will be larger for fruits with higher V_m values as RH decreases. For all fruits there is a RH at which a large increase in p_{CO_2i} will occur. This critical RH will be lower for fruits with lower V_m, and higher for fruits with higher V_m.

Oxygen uptake, r_{O_2}, for coated fruits with differing V_m will also be affected by RH in similar fashion.

We hypothesize that coated fruits with a large V_m will reach the *LOL* at higher RH values compared to fruits with a lower V_m. Even though film permeability declines with decreasing RH, p_{O_2i}, r_{O_2} and r_{CO_2} for fruits with lower V_m remains fairly constant down to lower RH values. This makes it possible to store fruits with low V_m over a wider range of RH conditions. Fruits with a large V_m are severely affected by RH when coated with hydrophilic films. Large decreases in p_{O_2i}, r_{O_2} and r_{CO_2} are observed with small drops in RH. This response can be explained because of the higher demand of O_2

by the fruit and the decreasing film gas permeability with lower RH. Therefore, coated fruits with a large V_m can only be stored over a short range of RH before anaerobic conditions are reached.

Internal CO_2 vs O_2 plots as influenced by RH, thickness, maximum respiration rates and film β values: their importance in coating design

Carbon dioxide vs oxygen plots for internal gas composition of coated fruits can be used in a similar way to the CO_2 vs O_2 plots of external gas composition used for modified atmosphere packaging design (Mannaperuma *et al.*, 1989). The latter plots are used to target external p_{O_2e} and p_{CO_2e} combinations or "windows" for storing fruits under external modified atmospheres. For coated fruits, we use the p_{O_2i} and p_{CO_2i} within the fruits, and the target "windows" would be the combinations of internal gases appropriate for storage. These plots can also give us information on the internal *LOL* values, above which fruits can be stored safely and under which anaerobic conditions are generated.

By changing external RH conditions for a coated fruit, we can generate a p_{CO_2i} vs p_{O_2i} plot. Each point of the curve corresponds to a different RH condition. The position of the curve will be influenced by the fruit's V_m, producing curves closer to the X-axis for those fruits which have lower V_m (Figure 11.3).

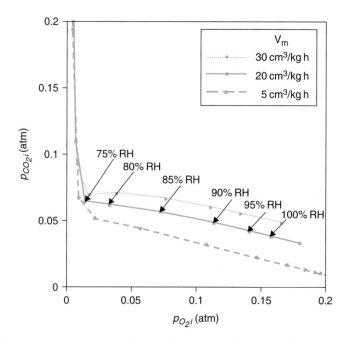

Figure 11.3 Predicted p_{CO_2i} vs p_{O_2i} plots generated by storing WPI/Sorb coated and uncoated fruits under different RH conditions and different V_m values. Fruit V_m values are 5, 20 and 30 cm³/kg h. Film thickness is 1.3 μm and Film β = 4.

The ratio between O_2 and CO_2 permeabilities (P_{CO_2}/P_{O_2}) also has an effect on these plots. Film β values are generally in the range of 3–5 for coating materials (Hagenmeier and Shaw, 1992). Coated fruits with larger film β values will yield curves closer to the X-axis and have smaller slopes, compared to coated fruits with smaller film β values.

More recently, Gontard *et al.* (1996) have shown that β increases with increasing RH depending on film type. A generated curve assuming non-constant β value will have similar trend as those curves generated assuming constant β values (Figure 11.3).

The composition of the film coating will influence the moisture sorption properties of the coating material, and may cause the dependence of film gas permeability and β values on RH (Gontard *et al.*, 1996). The trend of hydrophilic films is to have similar moisture sorption properties and gas permeability dependence on RH. However, gas permeabilities of non-hydrophilic films (e.g. pure lipid films) are not affected by RH since their moisture sorption properties are negligible or independent of RH (McHugh *et al.*, 1993). For emulsified wax films (e.g. commercial waxes) the gas permeability dependence on RH will depend on film composition. For example, an emulsion film may consist of lipid particles dispersed in a hydrophilic matrix, giving a gas permeability dependence on RH between that of a hydrophilic film and a pure lipid film. This gas permeability dependence may also be influenced by lipid particle size, particle arrangement within the matrix, and lipid concentration (Shellhammer and Krochta, 1997).

Curves similar to those in Figure 11.3 can also be constructed for coated fruits stored at only one RH. For a coated fruit with a given V_m and constant film β value, changes in film thickness or film permeability will give points on the same curve. For example, thicker films will give points shifted towards the left, while films with a larger permeability will give points shifted to the right. This indicates the possibility of tailoring films with appropriate β values, thickness and permeability characteristics, allowing us to reach specific combinations of $p_{O_2 i}$ and $p_{CO_2 i}$ or avoid detrimental anaerobic conditions.

These $p_{CO_2 i}$ vs $P_{O_2 i}$ plots can be used as tools to design coatings which can target specific combinations of O_2 and CO_2 for coated fruits (Banks *et al.*, 1997). Coating design for fruits with differing V_m values would be possible by selecting coating materials with appropriate film β values, permeabilities and thicknesses, and by controlling RH conditions.

Possible sources of variation in internal gas composition

One assumption in this model system is that the hydrophilic coating equilibrates completely with the RH of the external atmosphere, with no moisture concentration gradients within the film. Film permeability is a function of moisture concentration, which is related to RH of the surrounding air. Thus, film permeability would be constant throughout the film with our assumption. This assumption may be valid if the fruit skin resistance to water vapor is high and the film resistance to water vapor is low. In this way the moisture concentration within the film would be similar to that of the coating surface facing the atmosphere. However, fruits with low skin resistance to water vapor may generate large gradients of moisture concentration within the coating. For example, fruits with low skin resistance that are exposed to low RH may lose moisture readily, exposing the inner fruit/coating surface to large moisture concentration values and the coating/air interface to smaller moisture concentration values. If this occurs, then the average coating permeability would be in between the permeability values for the conditions at the fruit/coating and the coating/air surfaces, depending on the moisture sorption isotherm of the biopolymer. In the extreme case of fresh cut surfaces of fruits (Baldwin *et al.*, 1995), coatings facing the fruit/coating surface will most likely be exposed to a water activity of ≈ 1 ($\approx 100\%$ RH). This will indeed increase the overall coating permeability and reduce its barrier properties. Thus, we propose that similar coatings applied to different fruit surfaces may have quite different barrier performances.

It was also assumed that there is no boundary layer resistance to water vapor transmission. The existence of a boundary layer would increase the moisture concentration of the coating due to an increase in the surrounding water vapor pressure adjacent to the film/air surface. This increase in film moisture concentration will increase the overall coating gas permeability affecting the internal gas composition. The assumption of a negligible boundary layer will hold if the air velocity of the surrounding atmosphere is high enough to reduce it. In commercial storage conditions as well as in research, this assumption is not always met.

Another source of variation is the RH gradients experienced in commercial storage facilities, as well as in research techniques used for controlling RH. In the latter, it is common to adjust RH of the air inlet in flow-through systems when measuring respiration of fruits in jars. However, large RH gradients may still occur inside the jars, depending on the speed and RH of the air inlet, the amount of fruits in the jar and the fruit water loss rate.

Other sources of variation to consider are temperature fluctuations and film thickness variations. Changes in temperature may affect the fruit V_m, the film gas permeability and the external RH conditions. Film thickness may vary on individual fruits, changing the total gas resistance of the fruit.

In summary, the predicted trends obtained by varying RH conditions support the idea that film coating permeability and coating thickness are important in modified atmospheres of coated fruits, following either a *lac* model or a *tac* model for the specific case of complete pore coverage. However, if incomplete pore blockage is achieved, then film coating permeability and thickness may not be important in modifying the atmosphere within the fruit (Banks *et al.*, 1993). Thus, for the incomplete pore blockage model, RH would not have a major effect in the MA of coated fruits.

Conclusions

It is possible, when using the approach described in this paper, to observe and understand the interaction of storage conditions (e.g. RH) and composition of film coatings applied to a fruit surface. This analysis can explain, in part, the variability of results observed when coatings are applied on a commercial scale and in research. The approach supports the idea that film coating thickness and permeability play an important role in internal gas modification of coated fruits. Therefore, it is necessary to understand the physicochemical properties of the materials used as edible coatings and their moisture sorption behavior. Using this approach will make it possible to predict their performance once applied to a real fruit system and to tailor appropriate coatings to achieve a target gas composition.

References

Baldwin, E., Nisperos-Carriedo, M. and Baker, R. (1995). Edible coatings for lightly processed fruits and vegetables. *Hort. Sci*. **30**, 35–38.

Banks, N., Dadzie, B. and Cleland, D. (1993). Reducing gas exchange of fruits with surface coatings. *Postharvest Biol. Technol*. **3**, 269–284.

Banks, N., Cutting, J. and Nicholson, S. (1997). Approaches to optimizing surface coatings for fruits. *NZ J. Crop Hort. Sci*. **25**, 261–272.

Ben-Yehoshua, S. and Cameron, A. (1989). Exchange determination of water vapor carbon dioxide, oxygen, ethylene and other gases of fruits and vegetables. In: *Gases in Plant and Microbial Cells. Modern Methods of Plant Analysis*, new series 9 (H. F. Linskens and J. F. Jackson, eds), Springer-Verlag, New York. pp. 178–193.

Brusewitz, G. and Singh, P. (1985). Natural and applied wax coatings on oranges. *J. Food Process. Preserv*. **17**, 31–45.

Buera, P. and Karel, M. (1993). Application of the WLF equation to describe the combined effects of moisture and temperature on non-enzymatic browning rates in food systems. *J. Food Process. Preserv*. **17**, 31–45.

Elson, C., Hayes, E. and Lidster, P. (1985). Development of the differentially permeable fruit coating Nutri-Save® for the modified atmosphere storage of fruit. In: *Proceedings of the 4th National Controlled Atmosphere Research Conference (Dept Horticultural Science, North Carolina State University, Raleigh, NC). Horticultural Report* (S. M. Blankenship, ed.), pp. 126, 248–262.

Gontard, N., Thibault, R., Cuq, B. and Guilbert, S. (1996). Influence of relative humidity and film composition on oxygen and carbon dioxide permeability of edible films. *J. Agric. Food Chem*. **44**, 1064–1069.

Hagenmeier, R. and Shaw, P. (1991). Permeability of shellac coatings to gases and water vapor. *J. Agric. Food Chem*. **39**, 825–829.

Hagenmeier, R. and Shaw, P. (1992). Gas permeability of fruit coating waxes. *J. Am. Soc. Hort. Sci*. **117**, 105–109.

Mannapperuma, J., Zagory, D., Singh, P. and Kader, A. (1989). Design of polymeric packages for modified atmosphere storage of fresh produce. Proc. 5th Intl Controlled Atmosphere Research Conference 14–16 June, Wenatchee, WA, Vol. 2.

Mark, A., Roth, W., Mehltretter, C. and Rist, C. (1966). Oxygen permeability of amylomaize starch films. *Food Technol*. **January**, 75–77.

Mate, J. and Krochta, J. M. (1996). Comparison of oxygen and water vapor permeabilities of whey protein isolate and β-lactoglobulin edible films. *J. Agric. Food Chem*. **44**, 3001–3004.

McHugh, T., Aujard, J. and Krochta, J. M. (1994a). Plasticized whey protein edible films: water vapor permeability properties. *J. Food Sci*. **59**, 416–419, 423.

McHugh, T. and Krochta, J. M. (1994b). Sorbitol-vs glycerol-plasticized whey protein edible films: integrated oxygen permeability and tensile property evaluation. *J. Agric. Food Chem.* **42**, 841–845.

McHugh, T. and Krochta, J. M. (1994c). Milk-protein based edible films and coatings. *Food Technol.* **48**, 97–103.

McHugh, T., Avena-Bustillos, R. and Krochta, J. M. (1993). Hydrophilic edible films: modified procedure for water vapor permeability and explanation of thickness effects. *J. Food Sci.* **58**, 899–903.

Meheriuk, M. and Lau, O. (1988). Effect of two polymeric coatings on fruit quality of Bartlett and d'Anjou pears. *J. Am. Soc. Hort. Sci.* **113**, 222–226.

Miller, S. and Krochta, J. M. (1997). Oxygen and aroma barrier properties of edible films: a review. *Trends Food Sci. Technol.* **8**, 228–237.

Miller, S. and Krochta, J. M. (1998). Measuring aroma transport in polymer films. *Trans. ASAE* **41**, 427–433.

Miller, S., Upadhyaya, S. and Krochta, J. M. (1998). Permeability of d-limonine in whey protein films. *J. Food Sci.* **63**, 244–247.

Mujica-Paz, H. and Gontard, N. (1997). Oxygen and carbon dioxide permeability of wheat gluten film: Effect of relative humidity and temperature. *J. Agric. Food Chem.* **45**, 4101–4105.

Park, H. J., Chinnan, M. and Shewfelt, R. (1994a). Edible coating effects on storage life and quality of tomatoes. *J. Food Sci.* **59**, 568–570.

Park, H. J., Bunn, J., Vergano, P. and Testin, R. (1994b). Gas permeation and thickness of the sucrose polyester Semperfresh coatings on apples. *J. Food Proc. Preserv.* **18**, 349–358.

Pauly, S. (1989). Permeability and diffusion data. In: *Polymer Handbook*, 3rd edn. (J. Brandrup and E. H. Immergut, eds), pp. IV: 435–449. Wiley, New York, NY.

Rico-Pena, D. and Torres, A. (1990). Oxygen transmission rate of an edible methyl cellulose-palmitic acid film. *J. Food Proc. Eng.* **13**, 125–133.

Roos, Y. and Karel, M. (1991). Applying state diagrams to food processing and development. *Food Technol.* **45**, 66–71, 107.

Roth, W. and Mehltretter, C. (1967). Some properties of hydroxypropylated amylomaize starch films. *Food Technol.* **21**, 72–74.

Shellhammer, T. and Krochta, J. M. (1997). Whey protein emulsion film performance as affected by lipid type and amount. *J. Food Sci.* **62**, 390–394.

Trout, S., Hall, E. and Sykes, S. (1953). Effects of skin coatings on the behavior of apples in storage. I. Physiological and general investigations. *Aus. J. Agric. Res.* **4**, 57–81.

Modified atmosphere packaging of ready-to-eat foods

12

Kevin C. Spencer

Introduction

Ready-to-eat meals

Ready-to-eat meals (or ready meals) can be broadly defined as complex assemblages of precooked foodstuffs, packaged together and sold through the refrigerated retail chain in order to present the consumer with a rapid meal solution. They may include such diverse items as pasta and sauce combinations, with and without the inclusion of meat; rice-based dinners containing sauce, vegetables and/or meats; etc. Because of the wide range of possible ingredients included, production of ready meals involves tremendous operational challenges. Many different types of packaging need to be employed so as to contain the food effectively and also present an attractive product to the consumer.

The production process for ready meals is generally located in geographically concentrated areas so as to provide logistical efficiencies in their distribution. This concentration occurs because of ready meals' high intrinsic value and cost of preparation, but, as importantly, because of the need to maximize the shelf-life benefits that accrue from shortening the chill transport and storage chain.

Ready meals cost more to be manufactured, but also have higher sales value than their component parts. Due to their complexity and the high level of manipulation involved in their preparation, they are also much more subject to spoilage than their

Innovations in Food Packaging
ISBN: 0-12-311632-5

constituent ingredients alone. Therefore, questions of microbial growth and relatively short shelf life are paramount to the successful introduction and sales of these products, as well as to the safety of the end user.

Modified atmosphere packaging

Modified atmosphere packaging (MAP) is the imposition of a gas atmosphere, typically containing an inert gas such as nitrogen combined with an antimicrobially active gas such as carbon dioxide, upon a packaged food product to extend its shelf life. MAP can significantly extend the shelf life of food products, thus prolonging the distribution chain and diminishing the need for centralized production. MAP provides an added barrier against spoilage, and can therefore improve shelf life and enhance product safety. MAP is inexpensive, easy to apply, and suitable for a wide range of packaging machinery and production venues. However, food manufacturing and transport to the end user are complex processes to which MAP contributes only a partial solution regarding shelf-life problems: a constant chill chain and good hygiene in manufacture remain key to maintaining the efficacy of the MAP process (Brody, 1989a).

Applying MAP to ready meals in fact presents considerable practical difficulties. First, many types of packaging which are optimal for containing and presenting the meals to the consumer are most definitely not optimal for gas-modified packaging. For example, a tray may have several partitions to keep the ingredients separate in a complex dish, necessary both for an appealing, neat appearance, and for preventing degeneration of the product through the mixing of various incompatible food components. Adversely, these partitions are barriers to the flow of gas across and through the product, and make application of MAP inefficient or even impractical on high-speed, high-throughput packaging machines.

One solution to this barrier problem is to apply a vacuum across the tray prior to gas flushing in order to facilitate the movement of gas into the product and tray spaces over the partitions. However, there is a trade-off in that vacuum applications can distort product shapes, draw off desirable flavors and volatiles, boil off moisture, and deform the packaging. In addition, the vacuum cycle also lowers the production-line speed significantly. Another problem unique to ready meals is that many are based upon three-dimensional textural components, such as pasta in many shapes, or a bed of rice. These features retain large amounts of air in their interior spaces, causing a problem that can be compounded by partially enveloping the entrapped air with overlaid sauces and cheeses. In order to remove these air pockets, nitrogen MAP must be delivered at very high pressures and flow rates, and vacuum cycles should be necessarily longer. As discussed below, the development of argon MAP has addressed these limitations, permitting the easy use of MAP to remove oxygen in most packaging situations without sacrificing line speed and without causing packaging deformations or product disruption.

In the next section, the utility of classical nitrogen MAP in preventing the oxidation and deterioration of ready-meal food products will be discussed. The mechanism by which nitrogen MAP is known to function and its benefits/limitations in practice will be outlined. These limitations were the impetus for the development of argon MAP, which has been found to solve many of these problems. After the discussion of nitrogen MAP and consideration of important common elements of basic gas packaging, a detailed case study that describes the development of argon MAP in Europe will be presented.

Classical nitrogen MAP

Packaging foods in atmospheres containing an inert gas and carbon dioxide inhibits both oxidative reactions and microbial spoilage. Oxidation of foods is a largely chemical process driven by oxygen, which can be blocked by the displacement of oxygen by nitrogen. Microbial spoilage is caused by the growth of microbes upon foodstuffs, which can be blocked by the microbiocidal and microbiostatic effects of CO_2.

Inhibition of oxidation of food products by MAP

Oxidation is driven by the concentration of oxygen in direct contact with reactive chemicals within the food product (Richardson and Finley, 1985). Chemical oxidation of foods causes deleterious degradation of the flavor, aroma, and color of foodstuffs. Enzymatic oxidation of foods, catalyzed by lipoxygenase and oxidase, can cause rapid degradation in product quality (Shahidi and Cadwallader, 1997). The reaction mechanisms for both enzymatic oxidation and chemical oxidation (autoxidation and photooxidation) of foods are well understood (Shahidi, 2000).

In order to reduce the concentration of oxygen in contact with foods, the oxygen must be removed by vacuum and/or displaced by an inert gas. The efficiency of this process depends upon the design of the packaging machinery, the package (design and materials), and the physical properties of the MAP gas. Removal and exclusion of oxygen by displacing air with MAP gas inhibits the rate of oxidation (Lingnert, 1992), which directly slows the rate of degradation of valuable flavor and color components (Civille and Dus, 1992; Shahidi, 2000). To block the oxidation entirely, however, the complete removal of entrained and dissolved oxygen is necessary.

Air is composed of 78% nitrogen and 21% oxygen. Therefore, MAP mixtures based on nitrogen are of a similar density to air. Application of nitrogen across airspaces will introduce a great deal of turbulence as the two columns of similar density interact. Therefore, the application of nitrogen MAP in packaging machine operations depends upon applying sufficient force to the incoming MAP gas to both displace the air in the package completely and overcome the turbulence that will draw in more air. The force must be sufficient to introduce a mass of at least eight times the volume of the airspace into the package, which makes the whole process tremendously inefficient. Applying a vacuum cycle can increase the displacement efficiency by lowering the volume of the air mass which is required to be displaced. However, complete removal of the air in the package, especially air entrained within the product itself, requires long vacuum cycles at high vacuum. These long cycles result in concomitant damage to the package and product, and slow line speeds, as described above. There are thus two limitations in nitrogen MAP: the inefficiency of using a large volume of nitrogen gas, and the time requirements of vacuum cycles. These limitations cannot be overcome easily when employing nitrogen as a MAP gas.

Nitrogen : CO_2 : O_2 MAP for ready meals

Despite its limitations, nitrogen has been used extensively as a MAP gas because it is the least expensive and the most readily available inert gas approved for use in food packaging. Classical MAP packaging utilizes combinations of nitrogen with carbon dioxide, and oxygen (Brody, 1989b; Ooraikul and Stiles, 1991). For a wide range of food products, particularly chilled items such as meats and ready meals, CO_2 is added to nitrogen to create MAP atmospheres (Boyd, 1997) designed to inhibit the growth of spoilage micro-organisms, and the biochemical degradation caused by the microbial activities.

Carbon dioxide is an active agent in controlling microbes. Carbon dioxide acts by becoming solubilized in water, where it dissociates to form carbonic acid. The resulting decline in pH inhibits the growth of most micro-organisms. Carbonic acid also disrupts microbial cell membranes and inhibits respiratory enzymes directly. Carbon dioxide is a competitive feedback inhibitor of aerobic respiration, and is especially effective in an oxygen-starved atmosphere. Carbon dioxide thus depresses the respiration in micro-organisms as well as in respiring foods (Stiles, 1991). Unlike nitrogen, then, which is inert, carbon dioxide is highly reactive to biological systems.

Carbon dioxide controls many microbes well at concentrations above 20%, including molds and gram-negative aerobes (e.g., *Pseudomonas* spp.), but has been found to be much less effective in controlling yeasts or lactic acid bacteria (e.g. *Lactobacillus* spp.) (Fierheller, 1991).

Conditions for nitrogen MAP application for ready-to-eat products include the use of impermeable package into which a nitrogen : CO_2 mixture is introduced with the maximum possible efficiency. In order to control microbes well, the highest level of CO_2 that the product can tolerate is added. For respiring products, or where the risk of anaerobic pathogen growth is extreme, a small titer of oxygen (usually 5%) is added, or is alternatively allowed to remain from entrained air.

MAP applications have been developed, most extensively in Europe, for a great range of food products (Haard, 1992), including meats, produce, vegetables and fruits, bakery, and other products (Day, 1992; Betts, 1996). MAP is used not only to confer quality advantages, but also to ensure food safety without added risk (Anonymous, 1995). Packaging of foods in anaerobic MAP environments such as nitrogen : CO_2 presents no greater risk than the anaerobosis that results from temperature abuse of sealed packages (Brody, 1989a).

Nitrogen MAP is a very low-cost, economical process, utilized with varying success for many types of ready-to-eat foods (Brody, 1989b; Coulon and Louis, 1989; Fierheller, 1991; Jenkins and Harrington, 1991). Claims for the shelf-life extension vary among the several hundred product types currently packaged in nitrogen MAP gas, ranging from a 5–100% longer shelf life than in air. However, most of these products are not, strictly speaking, ready-to-eat products. Claims for shelf-life extension of the subset of products that are ready-to-eat are more modest, and average about 25–50% compared to the products packaged in air.

Nitrogen, CO_2, and O_2 are all accorded GRAS status, and may be used freely, alone or in mixtures, in the USA for food preservation. These gases are also all considered safe for use in the European Union, and are given EU numbers. They may be used freely providing the package is labeled as a preservative atmosphere. Argon is also accorded GRAS status, has an EU number, and is approved for use in the EU. As discussed below, argon provides significant advantages over nitrogen in the practical control of oxidation and spoilage micro-organism growth.

Packaging machinery

A large number of gas-packaging machinery types are currently in use around the world, and these may be categorized generally by the basic functional process used to prepare and seal the package. Form and fill machines create usually flexible packages by pulling a two-dimensional film along a forming guide and sealing it into a three-dimensional pouch. These machines can be either horizontal or vertical in format, and can produce flexible sacks, bags, cylinders, and square or rectangular pouches. Tray-wrapping and tray-overwrapping machines are similar in that they form a gas-tight enclosure around or over an input tray. Tray-lidding machines accept incoming preformed rigid trays and seal a gas-tight film over the tray. Thermoforming machines create a semi-rigid package by using a heated mold to form a tray or other pack on line, which is sealed after the introduction of the product by imposing a covering film in the case of a tray or heat-sealing the finished pack. Bulk gas packaging machines use large plastic sacks or other forms into which product is introduced in large quantities and heat-seal them. A brief review of the basic types of gas packaging machines is given in Coulon and Louis (1989). A more comprehensive discussion of food packaging machinery, including illustrations, is given in Jenkins and Harrington (1991: 65–99).

Packaging films and plastics

A wide variety of metallized plastic films and paperboard products are used in MAP packaging. Characteristics of the most commonly used materials are given in Jenkins (1991: 35–63). Films cited therein that are employed for gas packaging include PET, CPET, OPP, PP, HDPE, and occasionally LPDE, PVC, and PVDC composed as monolayers, multilayered, laminated, or metallized filmstock and trays. Among the critical properties of these major packaging resins, the most important characteristic for the purpose of gas packaging is their relative permeability to oxygen, carbon dioxide, and nitrogen or argon. The CO_2 and O_2 gas transmission-rate properties of films are generally well known and can be obtained from the materials specification sheets provided by the manufacturer. The gas transmission rate may also be measured directly by forming packages with included gas of known composition and sampling changes over time with a commercially available oxygen and/or CO_2 meter. However, accurate data on nitrogen permeabilities are less widely available, and are generally calculated by difference measurement – that is, by measuring that which remains after CO_2 and O_2 have been accounted for. Permeation of argon and other gases is less well known, but some very good data exist (Meyer *et al.*, 1988).

Gas delivery systems

MAP gases can be delivered either as pre-mixtures in cylinders, or as bulk gases to be mixed on site. Bulk gases may be delivered as compressed gases in cylinders, or as cryogenic liquids. Nitrogen : CO_2 : O_2 systems may also depend upon membrane-generated gas. A calibrated mixing panel or gas mixer is required to prepare the appropriate MAP mixture for introduction into the gas packaging machine. In addition, valves along with pressure-regulating systems are used to step pressure down as it passes from the source tank, through the mixer, to the point of introduction. Most packaging machines rely upon a solenoid valve to control gas timing and mass delivery at point of pack; however, continuous form and fill machinery may use a simple lance with appropriate pressure regulation.

Argon MAP

Development of argon MAP

In extensive laboratory testing, argon has been found to be superior to the MAP gases used previously. Building upon these results, commercial development of the use of argon gas for modified atmosphere packaging (MAP) has proceeded in Europe for a wide range of food products (Spencer and Humphreys, 2002). Currently, over 200 products, including a wide range of ready-to-eat packaged foods, are offered at retail in the UK and elsewhere in Europe. The commercial development of argon as a MAP gas has required the re-engineering of gas-handling systems, modification of packaging equipment, and building a base of practical knowledge and detailed know-how regarding the practical application of argon in factory environments. Considerable engineering has been necessary to maximize the efficiency of argon use in view of its cost, and to ensure the reliability of the product shelf-life outcome.

As discussed above, deterioration of ready-to-eat foods is caused by enzymatic and oxidative changes over time as well as microbial spoilage. Packaging these products in an inert gas atmosphere displaces air and hence oxygen, and thus, prevents oxidation. Admixing a titer, usually 20–70%, of CO_2 to the inert MAP gas reduces the rate of microbial growth.

Argon is an inert gas which has been used in modified atmosphere packaging and processing of a variety of foodstuffs (Spencer, 1993; Spencer and Rojak, 1993; Spencer and Steiner, 1997). The use of argon improves food quality for a number of reasons. While both nitrogen and argon are inert gases, their physical properties differ in significant ways. Since argon displaces and excludes oxygen more efficiently than nitrogen, it provides better control against oxidation of flavor and color components of foods. In addition, using argon results in the use of less CO_2 to control microbes. As CO_2 is reactive and generally deleterious to product quality, the quality of the product is better with the use of argon rather than nitrogen.

Exploiting differences between argon and nitrogen for use in MAP

Control of oxidation

Argon is a much denser gas than nitrogen, being 1.43 times as dense (1.650 vs 1.153 kg/m^3). As a denser column of matter, argon can thus be made to flow in a laminar fashion, like a liquid, through air space, whereas nitrogen cannot. Since air is close in density to nitrogen (being in fact composed of 78% nitrogen), as discussed above, nitrogen moves through airspaces inefficiently, creating much turbulence and mixing. Without the application of great pressure to the incoming nitrogen mass, random mixing occurs upon contact with air in a packaging tray. Thus, one volume of incoming nitrogen will mix completely with one volume of air to displace one-half of the air (and one-half of the oxygen). Applying 1 liter of nitrogen to a 1-liter tray of air will change the residual oxygen from 20.8% to 10.4%; the second liter added will drop the O_2 to 5.2%, etc. It therefore requires eight volumes of nitrogen to bring the residual O_2 to substantially zero.

As a denser mass of gas, argon can be introduced into the bottom of a tray, and as it fills the tray the argon will lift the air column upwards. If applied with minimal pressure, filling the tray can be accomplished with very little mixing of argon with air. Therefore, if properly introduced, one volume of argon can displace one volume of air completely, resulting in a residual oxygen level of zero. In practice, between one and two volumes of argon are used, depending upon the level of control available to the machine operator through the gas-handling system. Thus, argon is up to eight times more efficient than nitrogen in the filling process. While the cost of argon has been generally four to five times higher than that of nitrogen, because of its greater efficiency of use argon can be profitably employed as a MAP gas.

Beyond its greater efficiency, argon is also much more soluble in both water and oils than is nitrogen, making it much more efficient at displacing entrained and dissolved oxygen from food media. Argon is more than three times as soluble as nitrogen in water, and nearly twice as soluble as nitrogen in oils (Lide, 2001; Spencer and Humphreys, 2002). It is nearly half again as soluble as oxygen in both water and oil. Since argon is more soluble in liquids than is nitrogen, it is possible to use it to displace oxygen from liquid media more rapidly, and this results in a lower final oxygen residual level. Displacement of oxygen depends not only upon the final level of inert gas solubilized, but also upon the rate of dissolution of inert gas into the liquid food product. Using argon MAP therefore allows the use of much lower introduction pressures and total volumes of applied gas, avoiding outgassing and stripping-off of essential aroma and flavor volatiles during sparging or packaging operations.

Besides its physical effects in media, argon is also effective in controlling biological enzymatic oxidations, whereas nitrogen is not. Argon is a competitive inhibitor of respiratory enzymes, including oxidases (Spencer and Humphreys, 2002). Noble gases have been shown to have an effect upon enzyme reaction rates and yields (Spencer *et al.*, 1995, 1998), and this effect depends upon the atomic size of the noble gas. Since this effect can only be manifested if the noble gas interacts with the enzyme at the active site, or alters the conformation of the enzyme through penetration of protein interstices, the inhibitory effect of the gas is proportional to its molar concentration. Additionally, argon has been shown to inhibit oxidases even when some oxygen is present, whereas nitrogen does not. Therefore, argon can control post-harvest respiration and also microbial respiration better than can nitrogen. Antimicrobial results for argon alone have been shown to be particularly effective for yeasts and molds.

Contributing to argon's biological effects are other atomic properties. Argon has a polarizability closer to oxygen than to nitrogen, and a higher ionization potential than nitrogen. Argon has a van der Waals' constant closer to that of oxygen than of nitrogen, and is closer in size to oxygen than is nitrogen. These properties affect liganding of gases to active sites, penetration of gases through membranes, and biological reaction potentials. Argon has been found to behave in biological systems much more similarly to oxygen, which is biologically reactive, than does nitrogen (Spencer *et al.*, 1995, 1998).

Control of microbial growth

Inclusion of CO_2 in MAP mixtures is used to control microbes which contribute to food spoilage and shortened shelf life. While inert gases can be used to defeat oxidative respiration, CO_2 is the microstatic/microbiocidal element of standard MAP mixtures. Its mode of action has been discussed above. However, CO_2 has deleterious effects upon many food products, causing discoloration and other damage. It generates carbonic acid in water solution, leading to acid effects such as color changes and production of off-tastes and aromas. CO_2 is also an oxidant, causing bleaching effects and oxidation of volatiles and flavor components. Despite these significant limitations, CO_2 is included in MAP because it is a very dense and quite inexpensive gas.

Atmospheric air is composed of 20.946% oxygen, 78.084% nitrogen, and 0.934% argon (Lide, 2001). Increasing argon concentration to high levels (70–100%) in MAP subjects foods to a completely novel atmospheric environment. Whereas in non-living systems carbon dioxide is a reactive gas, argon, like nitrogen, is chemically inert. Using argon, however, allows the inclusion of lesser amounts of CO_2 than that required with nitrogen MAP to achieve the desired levels of antimicrobial effects. The same or superior levels of microbial control can therefore be achieved, while lessening the chemical and acidic side effects to food products.

Argon MAP for ready meals

Because of the physical properties discussed above, argon provides an easy solution to the limitations of the packaging of ready meals in nitrogen-based MAP. Because of its higher density, argon flows in a laminar fashion, like a liquid, into package spaces. Therefore, filling trays from the bottom towards the top, argon displaces air with great efficiency. As argon washes into the bottom spaces of a tray in this fashion, it completely displaces all of the entrained air from the complex structural components of ready meals, leaving exceedingly low residual oxygen levels in the package. Since this action of argon depends upon the density of the gas rather than on the pressure of delivery, low pressure of introduction of gas may be used, limiting the product disruption by minimizing the flow of gas. Since very low oxygen residuals are obtained because of the efficient displacement of air by the denser argon, vacuum cycles can be shortened or eliminated altogether, conferring valuable advantages in line speed of production.

Modifications of packaging machinery and tools to accommodate argon

Gas delivery system

First, since argon is heavier than nitrogen, larger-diameter tubing and pressure lines must be used to reduce frictional pressure loss and to assure adequate delivery pressures. The use of argon in nitrogen MAP systems will often result in lower line pressures and delivery of an insufficient mass of argon to the point of packaging. Unless the gas mixer is also recalibrated to accommodate these differences, serious errors will become manifest in the composition of the final MAP mixture. Furthermore, the packaging machine solenoid may not be able to deliver sufficient argon to the package in its normal pulse. The argon systems referred to here use larger-bore supply lines, mixers, solenoids, and lances, and have redundant pressure-regulation systems to ensure adequate mass flow rate.

Second, the problems associated with turbulence in nitrogen MAP can be avoided with the application of argon, but the removal of air (oxygen) becomes the key to successful application. Adding exhaust vents to MAP tooling provides the needed escape routes for the oxygen. When set up this way, no vacuum cycle is needed for removal of air, and line speeds are maximized. Upper speeds depend only upon the fill, forming, and seal functions of the packaging machine.

Films

A wide range of commonly available gas packaging machinery was used in these example experiments, and all were successfully and easily modified as above to accommodate argon. A wide range of plastic films and trays were used, and most of these were found to be satisfactory. The key point was to use highly impermeable films and trays. A very large array of tests was conducted to determine the best economical fit of plastic packaging for various product constraints. Generally, excellent results were obtained with highly impermeable (O_2 transmission rate 0–5 cm^3/day/M^2/atm SPTRH) CPET/PET trays and either high or medium barrier APET/PET, EVA/PET or PP/PET film combinations (O_2 transmission rate 75–100 cm^3/day/M^2/atm SPTRH). The results reported here were produced on a tray-lidding machine using heat-sealable peelable topfilms.

Example experiments

Sample preparations

Ready meals were produced on site in one of several factories, using a wide range of different packaging machine types. The types of ready meals included a wide variety of components, assembled as meat pasta dishes, vegetable pasta dishes, rice-based dishes with meat and vegetable sauces, curry dishes, meat plus side dishes, vegetarian entrees, and dishes with added cheese, potatoes, etc. All meals were precooked and deposited automatically in plastic trays under normal production conditions, and placed in packaging materials as described above. A medium barrier topweb film was heat-sealed onto the trays. For the results reported below, all production was carried out at normal line speed. Gas use measurements confirmed that four to eight times less gas was used with argon than with nitrogen in each comparative production cycle.

Organoleptic assays

Ready meals were assessed in-house at a dedicated facility by panels of volunteers, randomly selected, who were asked to score each product using a hedonic five-point scale (Stone and Sidel, 1993), where 5 was preferred, and 3 was the cut-off for acceptability. In order to indicate complete spoilage (score of 0) and thus to exclude toxic samples from testing, a six-point scale was used in some assessments. Five organoleptic parameters were scored – overall acceptability, appearance, flavor, aroma, and texture. Each of these five parameters was assessed in randomized tests pairing a treated ready meal and a control. Two tests were used; an independent paired comparison discrimination test, and a paired comparison hedonic acceptance test of the product against control. A panel was comprised of enough testers to assure that all results were reproducible, and statistically significant for each assay replicate. Most panels were of 16, 24 or 30 people. At least once for each product type, a pooled result of N = 49 was obtained by establishing seven panels of seven panelists each; these results reported as the pooled average of the panels (the differences between the panels were determined to be non-significant). Statistical significance was set at the 5% probability level (alpha $P < 0.05$). Significance was assessed using binomial, chi-squared, or reported probability tables (Roessler et al., 1978; Stone and Sidel, 1993). Differences were robust across the factors of choice of panel and location, date of replication, and method of test.

In addition to the above tests, the products were also assessed organoleptically at the point-of-pack and post-distribution point by professional factory production personnel, under the guidance of the QA/QC function. Pairwise difference tests were used, and involved the same scales as described above and in sufficient number to retain statistical significance. Full-scale production was also subject to sample trials using in-store packages as well as post-production product. Difference tests were conducted on treatment packages and controls sharing similar dating (verified by production date codes). Sample packages were obtained off the shelf or taken directly from distribution. Organoleptic differences displayed robust statistical discrimination as well as depth.

The samples used in the above tests were prepared, stored, and shipped in accordance with good manufacturing practice and standard protocols for ready meals in normal retail distribution. Temperatures and hygiene protocols were maintained normally, and checked throughout the production, storage, and shipment stages. Treatment samples and controls were subjected to precisely the same specifications, except for the type of MAP used. All samples were subject to precise gas measurement and tested for seal integrity. In a separate set of tests, a wide range of handling, microbial, and temperature-abuse studies were conducted. These tests allowed us to conclude that argon MAP products do not suffer a greater rate of degradation than those under nitrogen MAP, and nor is the rate of hazard greater.

Finally, as production is ongoing, normal QA/QC sampling of all products continues at an appropriate sampling rate, providing an enormous statistical depth to the results described.

Microbial assays

Microbial assays were conducted on all products during production and storage according to good manufacturing practice, as well as in accordance with UK and EU regulatory standards. Standard methods required by regulatory oversight were used, and all assays were repeated first in pentuplicate and then in either triplicate or pentuplicate. In all cases, total viable count (TVC) was assayed, and in most cases yeasts, molds, *Pseudomonas* spp., *Lactobacillus* spp., and *Enterobacter* spp. were counted. In addition, for selected products, total anaerobes, molds, and pathogens were assayed, depending upon the specific commercial concerns for each product. Microbial assays were conducted concomitantly with parallel sensory assessment panels for all products. All microbial tests were performed by external independent laboratories, and were always blind-coded.

Results

Relative utility of argon MAP vs Nitrogen MAP in displacing package oxygen

In order to measure the relative efficacy of nitrogen and argon in removing air from packages in normal production, ready meals were packaged under optimal production conditions as described above. In a standard run, 1000 packages prepared under each of the two MAP atmospheres ($Ar:CO_2 = 75:25$; $N_2:CO_2 = 75:25$) were assayed for oxygen content after packaging, using an oxygen gas analyzer calibrated to $\pm0.2\%$ O_2.

Figure 12.1 shows the results for a test of residual oxygen in ready meals packaged under nitrogen MAP and argon MAP. The results show an average residual O_2 concentration of 0.5% for argon MAP, and 98.5% of all packages having less than 1% O_2

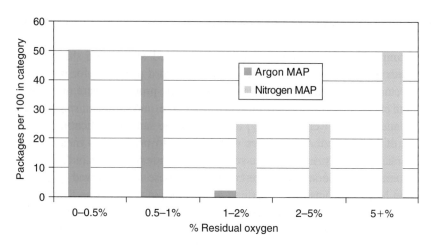

Figure 12.1   Residual oxygen in ready meals packaged under nitrogen MAP and argon MAP.

residual. The results for nitrogen MAP show an average of 5% O_2 residual, with approximately 50% of packages exceeding the average, and no packages having less than 2% residual O_2. Similar results were obtained when this experiment was repeated for each of 130 different ready-meal/packaging combinations. It was in fact possible with argon to achieve any desired level of residual oxygen concentration by slowing the packaging cycle on the production line in order to allow time for the argon to flow through all of the interstitial spaces in the product. Such complete substitution was unnecessary, however, as it provided removal of only the final 0.1–0.3% residual O_2 entrained within the product. It was not ordinarily possible when nitrogen was used to achieve residual levels of the same order as those obtained with argon, except by the use of a combination of impractically fast flow rates and excessively retarded packaging cycles.

The variance of the results with nitrogen was very high. Under normal line speed production, the chance of any given package having an oxygen content more than 25% different from its neighbor was greater than 50%. The range of oxygen residuals obtained was 0–20%, with 95% of the packages having greater than 2%. In addition, the entire distribution was skewed strongly toward higher percentages. Under practical conditions, it was not possible to enforce a specification of 5% residual oxygen. Under argon MAP, on the other hand, the range of oxygen residuals obtained was far smaller, 0–2%, and showed no skewness. The mean target of 0.5% O_2 was obtained in 98.5% of all packages. A specification of <1.0% residual oxygen was set and met for all products.

Effectiveness of argon MAP in controlling microbial growth in ready meals

The effect of argon MAP compared to nitrogen MAP or air control on microbial growth in ready meals was assessed in over 130 types of ready meals by random sampling of ready meals produced under normal operating conditions. Gas composition was checked on all samples.

Figures 12.2a–c show representative results for the effect of argon MAP upon microbes across three different ready-meal types. Figure 12.2a shows the total viable count (TVC) in a pasta carbonara ready-meal product under air, nitrogen MAP ($N_2 : CO_2 = 75 : 25$) or argon MAP ($Ar : CO_2 = 75 : 25$) atmosphere. While nitrogen MAP is effective in inhibiting microbial growth, significant additional repression of microbial growth is seen with argon MAP. As the shelf-life cut-off for this product is a TVC of 10^7, the shelf-life extension using nitrogen MAP was found to be 1 day (12%), while with argon MAP it is 6 days (120%) longer than air control. In another example (Figure 12.2b), a vegetable pasta ready meal, the shelf life of argon MAP was 15 days (88% increase), while an 8-day shelf life was obtained under nitrogen MAP. It was not possible to obtain shelf-life extension for a curried vegetable and rice-based ready meal using nitrogen MAP, as the air entrained within the rice base under the sauce could not be effectively flushed out. The problem was solved using argon, to provide a 13-day shelf life compared to 4 days for the air control (Figure 12.2c).

The effect of argon MAP on TVC is replicated with *Pseudomonas* spp., *Enteromonas* spp., lactic acid bacteria, yeasts, and molds. In a typical product, for *Pseudomonas* spp., argon held CFU below 10^3 throughout an 18-day shelf life, whereas the air control passed that mark on day 4 and the nitrogen MAP passed it on day 6. For *Enteromonas* spp., both air and nitrogen passed 10^3 on day 10. For yeasts, argon MAP provided suppression below the 10^5 cut-off throughout shelf life for all products, whereas the air control failed at 50% of life and nitrogen MAP failed at 67% of life for more than three-quarters of the products. Lactic acid bacteria were very well controlled by argon MAP in about half of all products overall, especially those with acidic sauces. Further assessment is required, although lactic acid bacteria were usually not a determinant of shelf life in the ready-meal products.

Ability of argon MAP to preserve organoleptic characteristics of ready meals

Organoleptic assessment of argon MAP as compared to nitrogen MAP and air control was carried out for all products. Tests were initially carried out at each sampling day post-packaging, but later tests focused on the peak and end of the shelf life. As an example, Figure 12.3a shows the results for the pasta carbonara ready meal at peak of life, where panelists preferred the argon MAP product in direct choice tests for all parameters. Differences were significant and reproducible across multiple panels. Differences found between nitrogen MAP and air control samples were more difficult for panels to discriminate, and discrimination among samples along individual parameters were more variable. For both a vegetable pasta ready meal and a meat ready meal presented at day 6, results (Figure 12.3b and 12.3c) show a significant preference for the argon MAP over the nitrogen MAP product. Figure 12.4 shows the progression of preference discrimination through storage for a complex ready meal with a 14-day (in argon MAP) shelf life. In this example, scores from three panels are averaged, and a score of 10 is the limit for saleability. The positive effect of argon upon all organoleptic parameters increased steadily throughout the shelf life of the product.

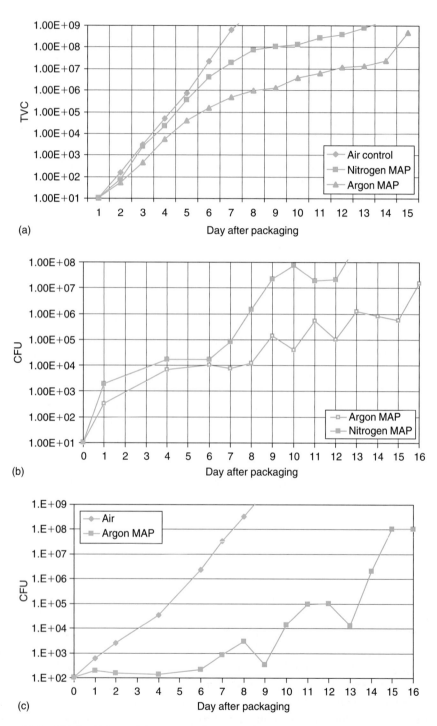

Figure 12.2 (a) Microbial growth in a pasta carbonara ready meal under different MAP atmospheres; (b) Microbial growth in a vegetable ready meal packaged in nitrogen MAP and argon MAP; (c) Microbial growth in a curried vegetable and rice ready meal packaged in argon MAP and air control.

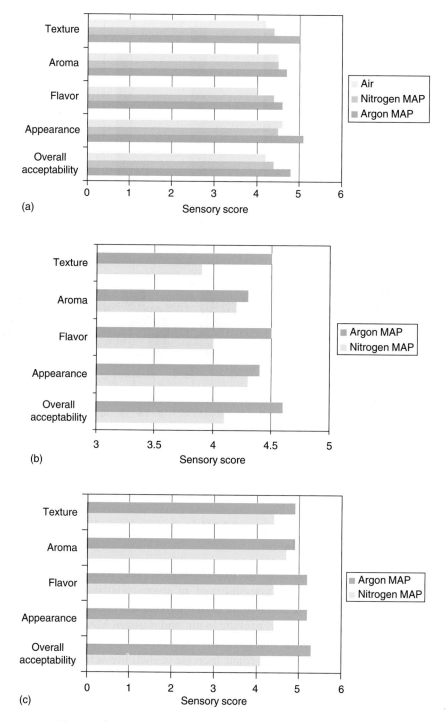

Figure 12.3 (a) Organoleptic scores for a pasta carbonara ready meal packaged under different modified atmospheres; (b) Sensory scores for a vegetable ready meal packaged in nitrogen MAP and argon MAP; (c) Sensory scores for a meat ready meal packaged in nitrogen MAP and argon MAP.

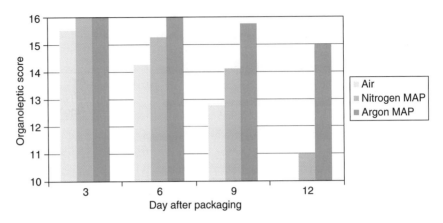

Figure 12.4   Changes in ready meal organoleptic score over time.

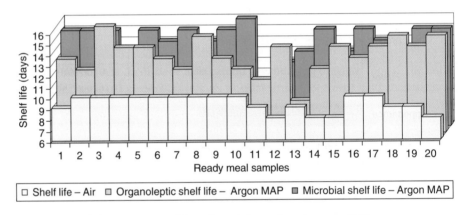

Figure 12.5   Shelf-life improvement of 20 ready meals in practice using argon MAP.

Practical application of argon MAP in extending the shelf life of ready meals

Figure 12.5 shows the average shelf-life improvement of ready meals in practice when using argon MAP. In the initial assessment of an argon installation at one factory, 20 disparate ready meals were assessed for both microbial and organoleptic limitations on shelf life. Current practice was packing in air, and nitrogen MAP had been found not to significantly add to shelf life, as results were excessively variable. Installation of an argon system provided an immediate increase in shelf life as measured either by organoleptic acceptability or by microbial cut-off (as indicated in Figure 12.5). All shelf-life specifications were increased for the 20 ready-meal products. The average specified shelf life of the packaged products before the introduction of argon MAP was 9 days. With argon MAP, the microbial shelf life was increased by an average of an extra 5 days (56%) to an average of 14 days, while the organoleptically-determined shelf life was increased by an average of an extra 4 days (44%) to 13 days. These results were extended to a full range of over 100 products at several manufacturing sites. The resulting extension of the shelf life provided retailers with an extra level of protection against food safety concerns, and allowed for a complete reorganization of production scheduling and distribution logistics. Besides this significant improvement in the cost profile of the production facilities concerned, it was possible to increase profitability significantly through the control of wastage and out-of-stock at the retail store level.

Conclusions

Oxygen catalyzes the oxidative degradation of foods, particularly of flavor, aroma, and color components. Microbial growth in packaged foods, especially under aerobic atmospheres, causes spoilage of the product. Modified atmosphere packaging (MAP) is effective in lowering the residual oxygen levels in ready-meal packages. Nitrogen, usually combined with CO_2, has been used for many years as the MAP gas of choice in controlling oxidation of foods, and extending its organoleptically determined shelf life. Carbon dioxide confers antimicrobial action to the MAP atmosphere, and extends the microbially determined shelf life of foods by inhibiting the growth of micro-organisms.

Owing to the different physical properties of argon, especially its higher density, it is far superior to nitrogen in removing and/or excluding oxygen from packages, and much lower levels of residual oxygen may be obtained in practical application. Since argon, unlike nitrogen, inhibits oxidase activity, argon-based MAP mixtures effectively depress respiration in microbes. Less CO_2 is therefore needed in argon MAP than in nitrogen MAP to control microbial growth, and using argon combined with CO_2 provides a higher level of antimicrobial action than that obtainable with nitrogen MAP. As CO_2 is destructive to food products, the use of less CO_2 means that there will be a better preservation of food quality. Thus, while nitrogen MAP is a cost-effective means of preserving foods, argon MAP is significantly better at controlling oxidation and microbial growth, and results in a significant improvement in the shelf life of a wide range of ready meals. Argon MAP has also been successfully tested, developed, and profitably commercialized for the preservation of ready meals and other products.

References

Anonymous (1995). *The Freshline Guide to Modified Atmosphere Packaging (MAP)*. Air Products plc., Crewe, UK.

Betts, G. D. (1996). *Code of Practice for the Manufacture of Vacuum and Modified Atmosphere Packaged Chilled Foods*, Guideline No. 11. Campden & Chorleywood Food Research Association (CCFRA), Chipping Campden, UK.

Boyd, L. C. (1997). Influence of processing on the flavor of seafoods. In: *Flavor and Lipid Chemistry of Seafoods* (F. Shahidi and K. R. Cadwallader, eds), pp. 9–19. ACS Symposium Series 674, American Chemical Society, Washington, DC.

Brody, A. L. (ed.) (1989a). *Controlled/Modified Atmosphere/Vacuum Packaging of Foods*. Food & Nutrition Press, Inc, Trumbull, CT.

Brody, A. L. (ed.) (1989b). Microbiological safety of modified/controlled atmosphere packaging of foods. In: *Controlled/Modified Atmosphere/Vacuum Packaging of Foods* (A. L. Brody, ed.), pp. 159–174. Food & Nutrition Press, Inc, Trumbull, CT.

Civille, G. V. and Dus, C. A. (1992). Sensory evaluation of lipid oxidation in foods. In: *Lipid Oxidation in Food* (A. J. St Angelo, ed.), pp. 279–291. ACS Symposium Series 500; American Chemical Society, Washington, DC.

Coulon, M. and Louis, P. (1989). Modified atmosphere packaging of precooked foods. In: *Controlled/Modified Atmosphere/Vacuum Packaging of Foods* (A. L. Brody, ed.), pp. 135–148. Food & Nutrition Press, Inc., Trumbull, CT.

Day, B. P. F. (1992). *Guidelines for the Good Manufacturing and Handling of Modified Atmosphere Packed Food Products*, Technical Manual No. 34. Campden & Chorleywood Food Research Association (CCFRA), Chipping Campden, UK.

Fierheller, M. G. (1991). Modified atmosphere packaging of miscellaneous products. In: *Modified Atmosphere Packaging of Food* (B. Ooraikul and M. E. Stiles, eds), pp. 246–257. Ellis Horwood Ltd, Chichester, UK.

Haard, N. F. (1992). Control of chemical composition and food quality attributes of cultured fish. *Food Res. Intl* **25,** 289–307.

Jenkins, W. A. and Harrington, J. P. (1991). *Packaging Foods with Plastics*. Technomic Publishing, Lancaster, PA.

Lide, D. R. (ed.) (2001). *CRC Handbook of Chemistry and Physics*, 82nd edn. CRC Press, Boca Raton, FL.

Lingnert, H. (1992). Influence of food processing on lipid oxidation and flavor stability. In: *Lipid Oxidation in Food* (A. J. St Angelo, ed.), pp. 292–301. ACS Symposium Series 500, American Chemical Society, Washington, DC.

Meyer, B. U., Schroter, W., Zuchner, K. and Hellige, G. (1988). Seperatormembranen für die Blutgas-Massenspektrometrie: Gaspermeabilität im Hinblick auf die Messung von Indikator-gasen Zur Bestimmung der Organdurchblutung. *Biomed. Technik* **33,** 66–72.

Ooraikul, B. and Stiles, M. E. (eds) (1991). *Modified Atmosphere Packaging of Food*. Ellis Horwood Ltd, Chichester, UK.

Richardson, T. and Finley, J. W. (eds) (1985). *Chemical Changes in Food during Processing*. AVI Van Nostrand Reinhold Co., New York, NY.

Roessler, E. B., Pangborn, R. M., Sidel, J. L. and Stone, H. J. (1978). Expanded statistical tables for estimating significance in paired-preference, paired-difference, duo-trio and triangle tests. *Food Sci.* **43,** 940–947.

Shahidi, F. (2000). Lipids in flavor formation. In: *Flavor Chemistry, Industrial and Academic Research* (S. J. Risch and C-.T. Ho, eds), pp. 24–43. ACS Symposium Series 756, American Chemical Society, Washington, DC.

Shahidi, F. and Cadwallader, K. R. (eds) (1997). Flavor and lipid chemistry of seafoods: an overview. In: *Flavor and Lipid Chemistry of Seafoods* (F. Shahidi and K. R. Cadwallader, eds), pp. 1–8. ACS Symposium Series 674, American Chemical Society, Washington, DC.

Spencer, K. C. (1993). Method of improving the aroma and flavor of stored beverages and edible oils. *Intl Patent. Publ. WO 93/19626.*

Spencer, K. C. and Humphreys, D. J. (2002). Argon packaging and processing preserves and enhances flavor, freshness, and shelf life of foods. In: *Freshness and Shelf Life of*

Foods (K. R. Cadwallader and H. Weenen, eds), pp. 270–291. ACS Symposium Vol. 836, American Chemical Society, Washington, DC.

Spencer, K. C. and Rojak, P. A. (1993). Method of preserving foods using noble gases. *Intl Patent Publ. WO 93/19629.*

Spencer, K. C. and Steiner, E. F. (1997). Method of disinfecting fresh vegetables by processing the same with a liquid containing a mixture of argon:carbon dioxide. US Patent 5,693,354.

Spencer, K. C., Schvester, P. and Boisrobert, C. E. (1995). Method for improving enzyme activities with noble gases. US Patent 5,462,861.

Spencer, K. C., Schvester, P. and Boisrobert, C. E. (1998). Method for regulating enzyme activities by noble gases. US Patent 5,707,842.

Stiles, M. E. (1991). Scientific principles of controlled/modified atmosphere packaging. In: *Modified Atmosphere Packaging of Food* (B. Ooraikul and M. E. Stiles, eds), pp. 18–25. Ellis Horwood Ltd, Chichester, UK.

Stone, H. and Sidel, J. L. (1993). *Sensory Evaluation Practices*, 2nd edn., pp. 118–147. Academic Press, New York, NY.

13 Preservative packaging for fresh meats, poultry, and fin fish

Alexander O. Gill and Colin O. Gill

Introduction

The aim of any packaging system for fresh muscle foods is to prevent or delay undesirable changes to the appearance, flavor, odor, and texture. Deterioration in these qualities can result in economic losses due to consumer rejection of the product.

The appearance of fresh muscle foods greatly affects the purchasing decisions of consumers (Cornforth, 1994). Products that appear dull or discolored are generally considered undesirable, and will be rejected when products that look fresh can be obtained. Thus, a preservative packaging that accelerates irreversible discoloration of a muscle food product is unsuitable for use with that product, irrespective of the other preservative effects of the packaging.

During the storage of raw muscle foods at chiller temperatures, various enzyme-mediated and non-enzymic chemical reactions will affect the flavor, odor, and texture of the tissues. The changes produced by some of the enzymic reactions are desirable – for example, the increased tenderness of aged beef (Jeremiah *et al.*, 1993). However, most of the changes caused by enzymic activities are undesirable. Therefore, a preservative packaging should ideally inhibit undesirable enzymic activities, but not interfere with, or inhibit, activities that are beneficial. The non-enzymic reactions that affect the organoleptic qualities of raw meats are invariably undesirable, so these should preferably be slowed or prevented by a preservative packaging.

Innovations in Food Packaging
ISBN: 0-12-311632-5

The rates of chemical reactions, whether or not they are catalyzed by enzymes, and the rates of bacterial growth increase with increasing temperature. However, the increase in rate with increasing temperature will differ for different processes, and may differ for the same process in different products. That being so, a statement of a storage life for a raw muscle tissue product is only meaningful if both the type of deterioration that renders the product unacceptable and the storage temperature are identified. Otherwise, the reported storage life provides no information of general value.

As most changes during storage of raw muscle foods are undesirable, and changes are slowed by decreasing temperature, it follows that the optimum temperature for storing raw muscle products is the minimum that can be indefinitely maintained without the tissue freezing. That temperature is in practice $-1.5 \pm 0.5°C$ (Gill *et al.*, 1988). For proper understanding of product storage stability under different circumstances, it is desirable that consideration be given to the storage life at the optimum temperature, as well as at temperatures that might be encountered during storage, distribution, and display.

In addition to the preservative effects, the commercial function of the packaging must be considered. The cost of a packaging is obviously a matter of major commercial concern. A relatively expensive packaging that confers extended storage stability is unlikely to be adopted when a storage life adequate for a market can be obtained with a cheaper, less effective packaging. Different forms of packaging may be required, depending on whether the packaging will be used for retail display of the product, or only for storage and distribution before further fabrication and/or repackaging. If the conditions during storage and distribution are known and consistent, a less robust packaging system is required than when conditions are uncertain. It is then apparent that the commercially optimal packaging need not be the packaging that confers the greatest storage stability on a product.

There are four categories of preservative packaging that can be used with raw muscle foods (see Table 13.1). These are vacuum packs (VP), high oxygen modified atmosphere packs (high O_2 MAP), low oxygen modified atmosphere packs (low O_2 MAP), and controlled atmosphere packs (CAP).

Table 13.1   Forms of modified atmosphere packaging

Type	Packing gases	Film gas permeability	Atmosphere	Notes
Vacuum pack	Residual air	Low	Anaerobic	Residual O_2 consumed by enzymatic reaction or respiration
High O_2 MAP	O_2, CO_2, N_2	Low	Aerobic	O_2 decreases over time
Low O_2 MAP	CO_2, N_2, residual O_2	Low	Anaerobic	Residual O_2 consumed by enzymatic reaction or respiration
Controlled atmosphere packaging	CO_2, N_2	Impermeable	Anaerobic	Stable atmosphere; oxygen scavenger may be used

Vacuum packs comprise evacuated pouches and vacuum skin packs, in which a film of low gas permeability is closely applied to the surface of the product. Preservative effects are achieved by the development of an anaerobic environment within the pack. Residual oxygen in any remaining atmosphere or dissolved in the product is removed by enzymic reactions of the muscle tissue, or other chemical reactions with tissue components. As the capacity of the muscle tissue for removing oxygen is limited, the amount of oxygen remaining in the pack at the time of closure must be very small if the product is to be effectively preserved. This will not be the case if the shape of the product units gives rise to bridging of the film, or otherwise allows vacuities to be formed within the pack. Items themselves that encompass spaces, such as whole carcasses, cannot be reliably preserved in VP. Bone-in items also present difficulties for VP, as bone ends can puncture packaging films. Though puncture-resistant materials to cover bone ends may be used, they complicate the packaging process, may not be consistently effective, and will certainly increase packaging costs.

High O_2 MAPs contain atmospheres of oxygen and carbon dioxide and, often, nitrogen. Only oxygen and carbon dioxide have preservative effects. However, carbon dioxide is highly soluble in both muscle and fat tissues (Gill, 1988), while oxygen may be respired by tissues and bacteria and, at high initial concentrations, will tend to be lost through the packaging film. Consequently, nitrogen is often included in high O_2 MAP atmospheres as an inert filler, to guard against pack collapse. Because of the interactions between the preservative gases and the product, a large gas-volume to product-weight ratio is required if an atmosphere adequate for preservation is to be maintained. An atmosphere of 3 l/kg of product is probably necessary for extended storage (Holland, 1980). Even so, it is usually found that, after the initial dissolution of carbon dioxide in the product, the carbon dioxide contents of high O_2 MAP atmospheres change little while the oxygen concentration progressively decreases (Nortje and Shaw, 1989).

Low O_2 MAPs contain atmospheres of carbon dioxide and nitrogen usually with some residual atmospheric oxygen remaining in the pack at the time of closure. As nitrogen has no preservative function in low O_2 MAP atmospheres, the initial volume of the pack atmosphere need only be sufficient to allow for dissolution of carbon dioxide in the product without pack collapse and crushing of the product. Low O_2 MAP atmospheres may be supplemented with carbon monoxide at concentrations less than 0.5% for the purpose of imparting a cherry-red color to the meat (Sørheim et al., 1999).

CAPs contain atmospheres that do not change during storage of the product. To achieve this, the films used for such packaging must have very low, preferably immeasurable, gas permeability (Kelly, 1989). CAP atmospheres may contain carbon dioxide and/or nitrogen. If carbon dioxide is used, the initial gas volume must be sufficient to allow for dissolution of carbon dioxide in the product. As an anoxic atmosphere is required for product preservation, but complete removal of oxygen is not practicable, oxygen scavengers may be used to accelerate the removal of residual oxygen from the pack atmosphere (Doherty and Allen, 1998). Otherwise, the residual oxygen is removed by reactions with product components.

The specific attributes of different raw meat tissues determine which of these various packaging systems is best suited for preservation.

Preservation of meat appearance

Red meats

The muscle pigment myoglobin, which is more or less abundant in the muscle tissue of red meats, is involved in the transfer of oxygen from the blood to muscle cells. The pigment therefore reacts rapidly and reversibly with oxygen to form bright red oxymyoglobin when oxygen concentrations are relatively high; but reverts to dull purple myoglobin under anoxic conditions (Livingstone and Brown, 1981). In addition, myoglobin can react with oxygen to form the oxidized pigment metmyoglobin (Faustman and Cassens, 1990). The stable brown metmyoglobin is formed from both myoglobin and oxymyoglobin, but the reaction with myoglobin is far more rapid than with oxymyoglobin (Ledward, 1970). Thus, contrary to what might be expected, the oxidized pigment metmyoglobin is formed rapidly when oxygen concentrations are low, and the pigment is preserved against oxidation by high oxygen concentrations. Muscle with high respiratory activity tends to have low color stability because of low concentrations of oxygen in the tissues near the surface (O'Keefe and Hood, 1982) (Figure 13.1).

Although metmyoglobin is stable, it is slowly reconverted to myoglobin by enzymic reactions collectively termed metmyoglobin reduction activity. Metmyoglobin reduction activity involves the oxidation of reduced coenzymes (Echevarne *et al.*, 1990). When these coenzymes are exhausted in muscle tissue, the reduction of metmyoglobin ceases. Muscle with high metmyoglobin reduction activity tends to have high color stability, because pigment oxidation at the surface is countered while the activity persists (Hood, 1980).

Myoglobin can react with ligands other than oxygen to form stable compounds. Among such reactants is carbon monoxide, which forms bright red carboxymyoglobin with the muscle pigment (Lanier *et al.*, 1978).

Consumers prefer red meats to have the bright red color of oxymyoglobin, which is considered indicative of fresh meat (Young *et al.*, 1988). The dull purple color of unoxygenated myoglobin is generally considered unattractive, while meat with the

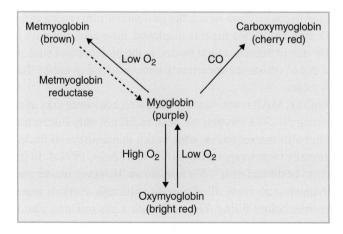

Figure 13.1 Effect of atmospheric gases on myoglobin color.

brown color of metmyoglobin is often considered unacceptable (Renerre, 1990). Ideally, packaging for red meats should preserve a desirable red color for the muscle tissue. If this is not possible, the packaging must prevent the product from being grossly discolored by the formation of metmyoglobin during storage, so that the product will retain the ability to bloom to a bright red color when it is displayed.

Vacuum packaging is widely used with primal cuts and ground red meats. During the relatively short times for storage and national or continental distribution of VP meat (Gill et al., 2002), the muscle tissue will not usually discolor and the meat will retain the ability to bloom when VPs are opened. Color, and the ability to bloom, is often retained as well during the longer times required for surface freighting of meat to overseas markets (Gill, 1989). However, materials used for VP are usually not wholly gas impermeable, so oxygen that diffuses into the pack will react with pack contents. When pockets of exudate form in packs, as is usual during prolonged storage, the incoming oxygen can oxidize myoglobin in the exudate, which may then precipitate onto and discolor meat surfaces (Jeremiah et al., 1992).

After storage for two or three weeks, respiration and metmyoglobin reduction activity have usually been lost by muscle tissue (O'Keefe and Hood, 1980–81). Therefore, after that time the color stability of different muscles is similarly low and comparable with the color stability of ground meat, in which metmyoglobin reduction activity is destroyed by the grinding process (Madhavi and Carpenter, 1993). Thus meat that has been stored in VP for two weeks or more is likely to discolor more rapidly when displayed than is meat of a lesser age.

Exposed spongy bone of bone-in cuts that have been stored in VP is also likely to darken and even blacken rapidly after the meat is exposed to air because of the oxidation of haemoglobin from the bone marrow (Gill, 1990). Haemoglobin is released from red cells that disintegrate as storage is extended. The accumulation of the oxidized pigment at cut bone surfaces gives rise to much darker colors than those that develop at spongy bone surfaces of fresh cut meat.

The use of VP with consumer cuts offered for retail sale has generally been unsuccessful because of the unattractive color of anoxic muscle tissue, although VP primal cuts are apparently saleable at retail. Attempts to remedy this situation by packing meat in vacuum-skin packs, from which the gas-barrier film can be stripped to expose a gas-permeable film before the meat is displayed, have also met with little commercial success. The lack of success might be due to the high costs of such packaging, but it might also be due to consumer uncertainty about packaging unlike that traditionally used with fresh meats.

In contrast, high O_2 MAP is now widely used with consumer cuts of red meats. Gas mixtures containing 50–70% oxygen are used to fill not only lidded trays containing consumer cuts but also master packs, which each contain several packs of consumer cuts in conventionally over-wrapped trays (Gill and Jones, 1994a). In lidded trays the input gas is likely to be diluted with < 5% residual air. However, master packs are usually formed by withdrawing air from filled pouches, through snorkels inserted into each pouch, for a set time before filling the pouch with a gas mixture, also for a set time. The withdrawing of air must cease before the pouch collapses around and crushes the retail trays, but the extent to which the pouches are filled with product and air varies.

Thus more or less large volumes of air are inevitably present in master packs when the preservative gas is added, and so the initial concentrations of both oxygen and carbon dioxide in master pack atmospheres are often well below the concentrations in the input gas. Thus, in practice, master pack atmospheres may do little to preserve red meat color, but will protect overwrapped trays from mechanical damage during their distribution to retail outlets from central cutting facilities.

Provided that pack atmospheres contain concentrations of oxygen well above that present in air, the fraction of myoglobin in the form of oxymyoglobin and the depth of the oxygenated layer will both be greater than in air (Taylor, 1985). The desirable red color of the meat will thus be intensified. In commercial practice, the initial enhancement of meat color, rather than extension of storage life, may be the principal benefit sought from packaging in lidded trays with high oxygen atmospheres.

If the oxygen concentration in the pack atmosphere is maintained, the formation of metmyoglobin at the meat surface can be retarded sufficiently to prolong the acceptable appearance for two to three times longer than in air (Renerre, 1989). Discoloration of the product is the factor that usually limits the storage life of red meats in high O_2 MAP, since bacterial spoilage is delayed by the carbon dioxide in the atmosphere (Gill and Jones, 1994b).

Because metmyoglobin forms rapidly at the surfaces of red meats exposed to low concentrations of oxygen, low O_2 MAP is generally not used with red meats. However, an acceptable color for red meats in low O_2 MAP can be preserved if a small fraction of carbon monoxide, usually $<0.5\%$, is present in the input gas (Sørheim et al., 1999). Carbon monoxide binds rapidly and essentially irreversibly with myoglobin to confer a permanent red color to the product. As carbon monoxide has a strong affinity for, and prevents the proper functioning of the oxygen transporting pigments of both blood and muscle, the gas is highly toxic. Consequently, most regulatory authorities concerned with food safety do not allow carbon monoxide to be used with meat. However, it has been shown that the risks to consumer health from the small amounts of carbon monoxide associated with meat are likely trivial (Sørheim et al., 1977). The use of carbon monoxide with red meats is permitted in Norway; where lidded trays of retail-ready red meats are commonly filled with a gas mixture containing 60% carbon dioxide, 40% nitrogen, and about 0.4% carbon monoxide (Sørheim et al., 1999).

The deterioration of red meat color during storage can be wholly prevented if oxygen is completely excluded from the pack atmosphere. This can be achieved by packaging meat in a gas-impermeable film, such as a laminate composed of two layers of metallized material, and filling the pack with carbon dioxide or nitrogen to obtain a CAP.

A major problem that arises in the use of CAP with red meats is the removal of residual air, as any remaining oxygen will react rapidly with the unoxygenated myoglobin to grossly discolor the product. In practice, very low concentrations of oxygen in the pack atmosphere at the time of pack closure can be achieved by evacuating a pouch through snorkels with the pack beneath a hood which is simultaneously evacuated (Gill, 1990). Partial breaking of the vacuum within the hood before the pouch is filled with gas allows the pack volume to be minimized before gassing, without crushing the contained product or retaining pockets of air between items that are pressed together.

When such packaging procedures are used, the oxygen content of the initial pack atmosphere can be kept to a few hundred parts per million. However, if the volume of the atmosphere is relatively large compared to the product volume, which it must be if a high carbon dioxide atmosphere is used, then the amount of residual oxygen may be sufficient to cause extensive discoloration (Gill and McGinnis, 1995a). Discoloration may be mitigated or prevented by the inclusion of oxygen scavengers in a pack, but for these to be effective they must rapidly strip oxygen from the atmosphere as the muscle tissue itself is a very effective oxygen scavenger (Gill and McGinnis, 1995b).

Although discoloration of the meat immediately after pack closure may be unavoidable, that need not present a problem if the meat is fresh and it is stored for longer than two or three days. In those circumstances, the metmyoglobin formed as oxygen was stripped from the atmosphere is reconverted to myoglobin by the metmyoglobin reduction activity of the muscle tissue. Provided the amount of metmyoglobin formed does not exceed the metmyoglobin reducing capacity of the muscle, no metmyoglobin will be present in the muscle tissue after storage for a few days (Gill and Jones, 1996). Then, when the meat is exposed to air it will bloom to the bright red color of freshly cut meat.

CAP is used for red meat products that must be preserved for periods beyond those attainable with MAP but for which VP is ineffective or inappropriate. Examples of such products are whole lamb carcasses surface-freighted to overseas markets, or retail packs of sheep meats widely distributed in the USA from a few central cutting facilities.

Poultry

The concentrations of myoglobin in poultry muscle are generally low, while rates of oxygen consumption are high (Millar et al., 1994). As a result, poultry muscle exposed to air does not exhibit the bright red color of oxymyoglobin; instead it has the duller tones imparted by metmyoglobin and the unoxygenated pigment. Of course, consumers find that usual color of the meat wholly acceptable. There is therefore no benefit in packing poultry in high O_2 MAP, and that type of packaging is not used with poultry in commercial practice.

Because the appearance of poultry is unaffected by anaerobic conditions, or low concentrations of oxygen in pack atmosphere, all other forms of preservative packaging are usable with poultry for storage or display.

Fish

There are two distinctly colored forms of muscle in fin fish; white muscle and dark muscle (Love, 1988). White muscle, which usually forms the greater part of the muscle present in a fish, contains low concentrations of myoglobin. Dark muscle, which is used for continuous movement, is rich in haem pigments, predominantly myoglobin. In most species of commercially exploited fin fish, the dark muscle is localized in a band running under the skin along the flank of the fish. The proportion and distribution of dark muscle varies between species, and is related to the type of swimming activity of each (Love et al., 1977).

Though changes in the form of myoglobin will occur in fish muscle packaged in VP or low O_2 MAP, this has little effect on the appearance and acceptability to consumers of fish in which there is little dark muscle (Silva and White, 1994; Hong *et al.*, 1996). The exception is fish with high dark muscle content, such as active predators like tuna. In those fish the distinct red color of dark muscle is considered attractive, so preservation of the red color is desirable. The dark muscle in fillets of such fish discolors, as does other red muscle tissue, and discoloration can be retarded or prevented by packaging in VP, CAP, or high O_2 MAP (Oka, 1989). It should be noted that, due to the localization of dark muscle, it may not be visible to the consumer in whole fish and certain fillet cuts. The effects of packaging on the appearance of dark muscle may then have little effect on the acceptability of some forms of fish, even when dark muscle is present.

With fish such as salmon, in which the intense color of the muscle tissue is due to carotenoid pigments assimilated from the fishes' food, any colors imparted by the muscle pigment are immaterial. Carotenoid colors are unaffected by VP; but colors may become lighter during prolonged storage in atmospheres with high carbon dioxide concentrations (Randell *et al.*, 1999).

Consumer acceptance of whole fish is dependent on the appearance of tissues other than muscle. The appearance of such features as belly flaps, gills, and cornea of the eye has been reported to be adversely affected by high CO_2 atmospheres (Chen *et al.*, 1984; Haard, 1992). However, no deterioration in the appearance of whole salmon was observed with 100% CO_2 by Sivertsvik *et al.* (1999). Low concentrations of carbon monoxide can be used to preserve gill color in anoxic atmospheres, but the same regulatory difficulties that limit its use in red meats exist (Rosnes *et al.*, 1998).

To avoid the pack collapse and possible bleaching of muscle which may occur with high concentrations of carbon dioxide, high O_2 MAP is often used with lean fish (Gibson and Davies, 1995). Even so, the effects of VP and low O_2 MAP on the appearance of fish can be small, and the appearance of most fish is not greatly enhanced by the inclusion of O_2 in pack atmosphere (Dhananjaya and Stroud, 1994). There appears to be no use of CAP with fresh fish, perhaps because of the supposed adverse effects of high concentration of carbon dioxide. However, fatty fish are packaged in low O_2 MAP, with input gases containing up to 80% CO_2, apparently without adverse effects on the appearance of product (Sivertsvik *et al.*, 2002).

Preservation or development of desirable eating qualities

Red meats

Both the texture and the flavor of meat may alter during storage. Consumer perceptions of the eating qualities of red meats are largely determined by the tenderness of the muscle tissue (Jeremiah *et al.*, 1993). Various intrinsic and extrinsic factors affect the tenderness of muscle tissue (Dransfield, 1994). Among these is the increase of tenderness with aging. The tenderizing of beef with time in storage does not apparently continue indefinitely, but after peaking starts to decline exponentially with time. Maximum tenderization is attained after two or three weeks (Dransfield, 1994). Preservative packaging appears to have little effect on the rate or extent of tenderizing with age.

Although tenderness in beef may not increase after the first few weeks of storage, lamb stored for prolonged periods in VP may lose texture and develop an undesirable, meal-like consistency. Moreover, in both lamb and beef the breakdown of proteins with the release of peptides and free amino acids continues during extended storage in VP (Rhodes and Lea, 1961). The accumulation of proteolysis products imparts bitter and liver-like flavors to the meat, which many consumers find undesirable. The deterioration of texture and flavor during prolonged storage of lamb does not occur when the meat is in CAP with a carbon dioxide atmosphere (Gill, 1989). The storage life of red meats in high O_2 MAP is too short for any deleterious effects of proteolysis to become evident.

The flavor of meat can also be adversely affected by lipid oxidation, which gives rise to rancid and stale odors and flavors. The susceptibility of lipids to oxidation depends upon the degree of unsaturation of the fatty acid residues in the lipid molecules (Lillard, 1987). Oxidation is accelerated by pro-oxidants such as ferric ions, and is inhibited by antioxidants such as vitamin E (Lillard, 1987). Generally, red meats contain less polyunsaturated fatty acids than poultry or fish, and so develop oxidative rancidity relatively slowly (Allen and Foegeding, 1981).

As lipids are not oxidized in the absence of oxygen, rancidity generally does not develop in red meats in VP or CAP. The increased concentration of oxygen in high O_2 MAP might be expected to accelerate lipid oxidation in red meats, but rates of lipid oxidation were found to be similar in pork stored under oxygen-rich atmospheres or air (Ordonez and Ledward, 1977). However, grinding of red meat can greatly accelerate lipid oxidation, and ground red meats may develop rancid odors and flavors more rapidly in high O_2 MAP than in air (Sánchez-Escalante et al., 2001).

Poultry

The pre-slaughter handling of birds and the method of carcass processing substantially affect the tenderness of poultry meat (Jones and Grey, 1989). Tenderizing proceeds rapidly after the development of rigor, with 80% of the maximum tenderness being attained within a day (Dransfield, 1994). Thus, tenderizing during storage is not a factor of much importance for the eating quality of poultry meats.

The oxidative rancidity is accelerated in the meat of birds fed on diets high in polyunsaturated fatty acids, and is more rapid in dark leg meat than in white breast meat from the same carcass (Lin et al., 1989). However, the development of rancidity in poultry meats will be inhibited by the low levels or lack of oxygen in preservative packaging for poultry. In such packaging, the storage life of poultry meats is usually determined by microbial spoilage.

Fish

Unlike red meats, cooked fish is usually perceived as tender by consumers (Cardello *et al.*, 1982). Consequently, aging does not enhance the acceptability of fish muscle. The primary factor determining the texture of fish is the final pH, which affects the water content of the muscle. Water content increases with increasing pH, and muscle of unusually high pH can have an undesirable "sloppy" texture (Love, 1988). High CO_2 concentration can have the reverse effect, as CO_2 dissolved in the tissue lowers pH, thus reducing the water-holding capacity and consequently increasing water extrusion (Davis, 1998). Dipping fish fillets in a sodium chloride solution can reduce this effect (Pasteroriza *et al.*, 1998).

Spoilage of fish due to changes in flavor and odor during storage is almost invariably a consequence of microbial activity. However, oxidative rancidity may contribute to spoilage, and for this reason high O_2 MAP is not appropriate for use with fatty fish (Sivertsvik *et al.*, 2002). The early development of rancid odors and flavors in fatty fish such as salmon and trout is delayed or prevented by packaging in VP or low O_2 MAP (Randell *et al.*, 1999).

Delay of microbial spoilage

Red meats

Muscle tissue provides a rich medium for the support of bacterial growth, and spoilage bacteria can grow at temperatures below that at which muscle tissue starts to freeze (Lowry and Gill, 1984). Consequently, meat that is stored at chiller temperatures will inevitably be spoiled by the activities of a spoilage microflora, whether or not it has been previously rendered unacceptable to consumers by undesirable changes in appearance, odor, or flavor.

The composition of the bacterial flora at the onset of spoilage is determined by the qualities of the tissue on which the bacteria are growing, the atmospheric composition, and the numbers and composition of the microflora at the time of packaging. A spoilage flora will generally be dominated by those organisms that can grow most rapidly in the environment provided by the tissue and the pack atmosphere (Gill, 1986). If the initial numbers of bacteria on the meat are small, the growth rate advantage of the dominant species will be expressed over a relatively large number of generations before the onset of spoilage. Then, slower-growing organisms will be a trivial fraction of the flora present at spoilage. However, if the initial number of bacteria is high, slower-growing organisms may persist as a substantial part of the flora and contribute to the spoilage process.

The spoilage flora of red meats exposed to air is invariably dominated by species of *Pseudomonas* (Gill and Newton, 1977). These organisms are strictly aerobic and preferentially utilize glucose for growth, although many other substances can be utilized as well. While glucose is available to the pseudomonads, the utilization of other substances is prevented by catabolite repression (Nychas *et al.*, 1988). Red meat muscle

Table 13.2	Final pH of fresh muscle foods
Muscle	**Final pH**
Red meat	
Normal	5.3–5.8
Dark firm dry	6.2–6.8
Pale soft exudative	5.2–5.4
Poultry	
Breast	5.3–5.8
Legs	6.2–6.4
Fish	6.0–6.9

tissues of pH $\leqslant$ 6.0 usually contain small amounts of glucose, and the growth of pseudomonads is unaffected by pH values down to 5.4, which can occur in post-rigor muscle (see Table 13.2). In air, the pseudomonads grow on the surface of meat at the maximum rate for the temperature, with glucose as the primary substrate. When the rate of diffusion of glucose from the underlying tissue can no longer meet the demands of the bacteria, they switch to catabolism of amino acids.

The by-products of glucose catabolism by pseudomonads are inoffensive, so while glucose is available the meat remains unspoiled. However, by-products of amino acid catabolism, such as ammonia, amines, and organic sulfides, impart offensive odors and flavor to meat when they are present in even small quantities. On muscle tissue of normal pH (5.5), the aerobic flora attains numbers about $10^8/cm^2$ before glucose is exhausted. With such high numbers, organoleptically detectable quantities of by-products are formed rapidly when amino acids are attacked. Consequently, in those circumstances, the onset of spoilage is abrupt (Gill, 1981).

In dark, firm, dry (DFD) muscle tissue of high pH (>6.0), little or no glucose may be present. Glucose concentrations at the surfaces of fat tissue are also low (Gill and Newton, 1980), and glucose consumed by bacteria cannot be replenished by glucose diffusing from the underlying tissue. In these circumstances, amino acids are utilized when bacterial numbers are still low, and spoilage becomes apparent when the flora reaches numbers that are sufficient to produce offensive by-products in organoleptically detectable quantities. This usually occurs when the numbers of the aerobic spoilage flora are about $10^6/cm^2$.

The growth of pseudomonads is suppressed when meat is vacuum packaged in a film of low gas permeability to obtain a pack free of vacuities. Then, the oxygen concentration at the meat surface will be too low for any substantial growth of pseudomonads to occur. Under these conditions the organism that can grow most rapidly are lactic acid bacteria, particularly leuconostocs (Newton and Gill, 1978). On muscle tissue of normal pH the growth of other spoilage organisms is inhibited by anaerobic conditions, and the spoilage flora is composed of little other that lactic acid bacteria.

Lactic acid bacteria ferment glucose and a few other minor components of muscle tissue. Growth ceases when these substrates are exhausted, which usually occurs as numbers reach about $10^8/cm^2$ (Dainty *et al.*, 1979). However, the lactic acid bacteria

Table 13.3   Principle spoilage bacteria of fresh muscle foods

Organism	Gram reaction	Oxygen requirement	Spoilage characteristics	Growth characteristics
Brochothix thermosphacta	+	Facultative anaerobe	Diacetyl, acetic isovaleric and isobutyric acids	No anaerobic growth below pH 5.8
Enterobacteriaceae	−	Facultative anaerobe	Sulphides	No anaerobic growth below pH 5.8
Lactic acid bacteria	+	Aerotolerant anaerobe	Lactic acid and ethanol	Ferments a restricted range of substrates
Photobacterium phosphoreum	−	Facultative anaerobe	Sulphides, TMAO reduction	Requires 100 mM Na^+
Pseudomonas	−	Aerobe	Sulphides, amines and ethyl esters	Aerobic growth only
Shewanella putrefaciens	−	Facultative anaerobe	Sulphides, TMAO reduction	No anaerobic growth below pH 6.0

Spoilage characteristics from Whitfield (1998).

generally do not produce grossly offensive by-products. Instead, meat with a lactic flora at maximum numbers will usually develop mild acidic, dairy odors and flavors only some time after the maximum numbers are attained.

If muscle tissue is of a pH > 5.8, or oxygen is available to the microflora at low concentrations, then various facultative anaerobes may grow. When present in substantial numbers, such organisms will usually spoil the meat as the flora approaches its maximum numbers. The facultative anaerobes found in meat spoilage flora include *Brochothrix thermosphacta*, which imparts a strong, stale, "sweaty socks" odor and a distinctive flavor to meat by the production of acetoin from glucose (Grau, 1983); enterobacteria, that decarboxylate amino acids to produce organic amines which give meat strong putrid odors and flavors (Gill and Penney, 1986); and *Shewanella putrefaciens,* which preferentially utilizes the amino acid cysteine with the production of hydrogen sulfide and organic sulfides, which give meat strong "rotten egg" odors and flavors, and which may react with myoglobin to produce green discoloration of the meat (McMeekin and Patterson, 1975; see also Table 13.3). Thus the storage life of high-pH meat in VP is generally substantially less than that of normal pH meat stored under the same conditions.

Pseudomonads can grow at their maximum rate for the prevailing conditions with oxygen concentrations of 1% or less (Clark and Burki, 1972). Thus, the growth of these organisms is unaffected by the concentrations of oxygen in the atmosphere of either high O_2 or low O_2 MAP. Instead, growth of pseudomonads on MAP meats is inhibited by carbon dioxide. The growth rates of pseudomonads decrease with increasing concentrations of carbon dioxide in MAP atmospheres up to about 20% (Gill and Tan, 1980). At that concentration of carbon dioxide, the rate of growth is about half the rate of growth in air. Higher carbon dioxide concentrations have little further effect upon the rate of growth. Therefore, the time before microbial spoilage of meat in MAP is, at most, twice that of meat stored in air.

The slowed growth of pseudomonads usually allows lactic acid bacteria to predominate in the flora of meat in MAP. However, growth of pseudomonads and other strictly aerobic organisms, such as acinetobacteria, is not wholly prevented, and facultatively anaerobic organisms continue to grow. Therefore, the flora of meat in MAP usually includes more or less substantial fractions of some of those organisms, with the meat being spoiled by offensive by-products from pseudomonads, enterobacteria, and/or *B. thermosphacta* (Stiles, 1991).

In CAP with a nitrogen atmosphere, spoilage develops as in VP. However, in an atmosphere of carbon dioxide, growth of facultative anaerobes on high-pH meat is inhibited either severely or wholly at temperatures near $-1.5°C$, the optimum for storage. Under such conditions, a flora of lactic acid bacteria develops irrespective of the meat pH, and the storage life can be considerably longer than that of similar meat in VP stored at the same temperature (Gill and Penney, 1988).

Poultry

The pH of poultry breast muscle is usually 5.5, but leg muscle and skin are invariably of higher pH – 6.2–6.4 and 7 respectively (McMeekin, 1977). Vacuum packaging of whole carcasses is unsatisfactory, because of the vacuity of the body cavity. Vacuum packaging of bone-in portions is also often unsatisfactory because of bridging of the packaging film. Moreover, poultry meat is always relatively heavily contaminated with spoilage bacteria during carcass dressing and breaking processes (Barnes, 1976). In these circumstances, the flora that develops on the meat in VP, high O_2 MAP, or low O_2 MAP is similar. The flora is rich in lactic acid bacteria, but includes large, sometimes dominant, fractions of enterobacteria which cause putrid spoilage of the product after relatively short storage times (Jones *et al.*, 1982; Bohnsack *et al.*, 1988).

However, when poultry is packaged in CAP under CO_2, growth of the enterobacteria is inhibited and a flora dominated by lactic acid bacteria develops (Gill *et al.*, 1990). Slow growth of the enterobacteria does occur after extended storage, and the meat is ultimately spoiled by the activities of those organisms. Nonetheless, the storage life of poultry in CAP under CO_2 is three to four times that of the same product in VP.

Fish

The initial microflora of fish is far more variable than that of red meats and poultry. The flora can differ greatly, in both numbers and composition, depending on the source of the fish. The microflora on fish from colder sea waters generally numbers $< 10^5\,cfu/cm^2$, and is composed mostly of gram-negative species that grow most rapidly in media with sodium chloride at concentrations about 2% (ICMSF, 2000). Such organisms include *Pseudomonas, Psychrobacter, Vibrio, Shewanella,* and *Photobacterium.* Fish from warm sea waters generally carry larger numbers of bacteria, with gram-positive organisms such as *Baccillus, Clostridium, Lactobacillus, Micrococcus* and *B. thermosphacta* forming large, or predominant, fractions of the flora, along with gram-negative organisms of the types found on fish from colder waters. Fish from fresh waters carry similar flora, although *Vibrio* and *Photobacterium* are usually absent while *Aeromonas* is often present (ICMSF, 2000).

The microbial spoilage of fish in air is similar to that of red meats, and it is also due to the activities of gram-negative species, particularly pseudomonads, *S. putrefaciens*, and *Photobacterium phosphoreum*. These organisms degrade amino acids to produce ammonia, amines, organic sulfides, and hydrogen sulfide. However, microbial spoilage occurs faster in fish than in red meat, as the final pH of fish muscle is normally greater than pH 6 and little glucose is available in fish muscle. This situation results in fast growth rates for pH-sensitive spoilage organisms and early exhaustion of available glucose, with initiation of catabolism of amino acids and other nitrogenous compounds. Longer shelf lives are associated with fish from tropical waters, in which the initial flora is dominated by gram-positive organisms, which generally cause less offensive spoilage.

A major factor contributing to the spoilage of fish is the osmoregulatory molecule trimethylamine oxide (TMAO) in fish muscle (Whitfield, 1998). Gram-negative bacteria, such as *S. putrefaciens* and *P. phosphoreum*, use TMAO as an alternative terminal electron acceptor to oxygen during respiration. That reduces the odorless TMAO, $(CH_3)_3NO$, to trimethylamine (TMA), $(CH_3)_3N$, which imparts a strong ammonia odor to the meat. The TMAO reductases of *S. putrefaciens*, and *P. phosphoreum* are constitutively expressed, but expression is repressed in the presence of O_2, resulting in increased TMA production in micro-aerobic and anaerobic conditions (Easter *et al.*, 1983). The activity of TMAO reductase is pH sensitive, and at high CO_2 concentrations may be reduced because of acidification of fish muscle by dissolved CO_2 (Gibson and Davis, 1995). An increase in pH, consequent on the production of ammonia from amino acids, also inhibits TMAO reductase and reduces its contribution to spoilage (Gibson and Davis, 1995).

The elasmobranchs (sharks and rays) use urea rather than TMAO as an osmoregulator, and muscle from these fish may contain up to 2.5% urea. Microbial spoilage then occurs primarily because of the production of ammonia as a result of bacterial urease activity (Jemmi *et al.*, 2000).

Because of the high pH of most fish muscle, growth of the facultatively anaerobic *S. putrefaciens* is not prevented by vacuum packaging, so that organism may dominate the spoilage process in vacuum-packaged fish (Jorgensen and Huss, 1989). However, *P. phosphoreum*, utilizing TMAO to maintain respiratory metabolism, may be largely responsible for the spoilage of fish such as cod in VP (Dalgaard *et al.*, 1993). The rate of growth for *P. phosphoreum* and *S. putrefaciens* is not greatly reduced by anaerobic conditions (Jorgensen *et al.*, 1988). Consequently, the extension of shelf life achieved by vacuum packaging of fish is often small (Davis, 1998).

There has been some investigation of the use of high O_2 MAP with fish, with the aim of reducing TMA production by repressing TMAO reductase expression. Boskou and Debevere (1997) have demonstrated that TMA production by *S. putrefaciens* growing on a fish extract can be inhibited by 10% oxygen. Guldager *et al.* (1998) have also reported inhibition of TMA formation on refrigerated cod fillets under 40% oxygen. However, such systems can do little to delay other forms of spoilage caused by gram-negative organisms.

Although the growth of *S. putrefaciens* is greatly inhibited by high concentrations of carbon dioxide, low O_2 MAP also may give only a modest extension of fish storage life (Dalgaard, 1995). This can occur because *P. phosphoreum* is relatively insensitive to carbon dioxide and may spoil fish in MAP after a storage life little longer than that of the same product stored in air.

There appear to be no reports on the use of CAP with fresh fish, although a large extension of storage life has been reported for smoked fish packaged in CAP under carbon dioxide (Penney *et al.*, 1994). Fish packaged in MAP under 100% carbon dioxide have been reported to develop spoilage flora rich in lactic acid bacteria, whereas with lower concentration of the gas flora were composed mainly of gram-negative species (Lannelongue *et al.*, 1982).

Microbiological safety

Muscle tissue from healthy mammals, poultry, and fish is sterile. Consequently pathogens, like spoilage flora, are introduced to flesh during processing and handling of the carcass. Thus, two groups of pathogens may contaminate muscle foods. Those groups are pathogens present in the environment of the animal and transferred to the carcass from the skin or the intestinal tract during processing, and human-associated pathogens transferred during handling.

Concerns about the microbiological safety of raw meats in preservative packaging are related to the possibility that pathogenic organisms may be less sensitive to preservative atmospheres than are spoilage bacteria. If that should be the case, then pathogens may reach higher numbers than usual, or initiate toxin production, while the meat remains unspoiled as assessed by microbiological or sensory criteria (Hintlian and Hotchkiss, 1986). Also, it has been suggested that a growth-rate advantage conferred on a pathogen by a preservative atmosphere might be enhanced by suppression of the usual competitive inhibition of the pathogen by the developing microflora (Farber, 1991).

Mesophilic pathogens can grow on meats in VP and MAP, but some are inhibited by high concentrations of CO_2 near their minimum growth temperatures (Siliker and Wolfe, 1980; Gill and DeLacy, 1991). Thus, whether or not their growth is advantaged relative to that of the spoilage flora, the safety of raw meats in preservative packaging is dependent on the growth of mesophilic pathogens being certainly prevented by storage of the product at chiller temperatures.

At chiller temperatures, most meat-associated pathogens can not grow, with the exception of the cold tolerant *Aeromonas hydrophila*, *Listeria monocytogenes* and *Yersinia enterocolitica* (Table 13.4). However, their growth on muscle tissue with a pH of under 6 is inhibited or prevented by storage under anaerobic conditions at chiller temperatures (Palumbo, 1988; Grau and Vanderlinde, 1990). The growth of cold-tolerant pathogens is also inhibited by high concentrations of CO_2 (Garcia de Fernando *et al.*, 1995). In CAP under CO_2, growth of these pathogens does not occur on high-pH meat stored at 5°C or lower (Gill and Reichel, 1989). Suppression of competitive inhibition is not a factor of importance on rich substrates, such as meat, as interactions between bacteria do not affect rates of growth until the flora approaches maximum

Table 13.4 Characteristics of major bacterial pathogens associated with fresh muscle foods

Organism	Gram reaction	Oxygen requirement	Minimum temperature (°C)	Associated muscle foods	Notes
Aeromonas hydrophila	−	Facultative anaerobe	0	Red meats, poultry and fish	
Campylobacter	−	Facultative anaerobe	28	Red meats and poultry	
Clostridium botulinum					
Proteolytic A, B, F	+	Obligate anaerobe	10	Red meats and fish	Produces toxin during growth
Non-proteolytic B, E, F	+	Obligate anaerobe	3	Red meats and fish	Produces toxin during growth
Enterobacteriaceae					
E. coli EHEC	−	Facultative anaerobe	7–10	Red meats	Enterohaemorrhagic E. coli
Salmonella	−	Facultative anaerobe	28	Red meats and poultry	
Shigella spp.	−	Facultative anaerobe	10	Red meats, poultry and fish	
Yersinia enterocolitica	−	Facultative anaerobe	−1	Pork	
Listeria monocytogenes	+	Facultative anaerobe	0	Red meats and fish	
Vibrio spp.	−	Facultative anaerobe	5	Fish	

numbers (Gill, 1986). It can be concluded that storage of normal pH red meat and poultry in VP or low O_2 MAP, at chiller temperatures, will not present an increased risk from infectious pathogens. However, on DFD meat and fish, *A. hydrophila*, *L. monocytogenes* and *Y. enterocolitica* may grow slowly to increase the potential risk to consumers, unless high CO_2 CAP is used.

A major concern about the microbiological safety of raw meats in preservative packaging relates to the possibility of growth and toxin production by strictly anaerobic *Clostridium botulinum*. Proteolytic types A and B cannot grow at chiller temperatures and so are controlled, as are other mesophilic pathogens, by appropriate refrigerated storage (Hauschild *et al.*, 1985). However, *C. botulinum* types E and F, and non-proteolytic type B, can grow at temperatures of about 3°C, and so might form toxins in chilled foods (Hauschild, 1989). The risks from cold-tolerant *C. botulinum* are seen to be greatest with fish, which are not only commonly contaminated with *C. botulinum* type E, but also have been implicated in outbreaks of type E botulism (Genegeorgis, 1985).

The production of toxins by *C. botulinum* on raw meat or fish in preservative packaging has been extensively studied using muscle foods inoculated with the various types of the pathogen (Skura, 1991). Although findings have varied greatly, it has been shown with some types of product under some experimental circumstances that toxin formation could precede organoleptic spoilage (Post *et al.*, 1985). However, it has also been shown that anaerobic niches where *C. botulinum* may grow can exist in packages of raw meat whether or not oxygen is present in the pack atmosphere (Eklund, 1982; Lambert *et al.*, 1991). The risks to consumers from *botulinum* toxin production in raw meats are then probably little different for products stored in air or in preservative packaging, with risks being very small when products are properly stored at chiller temperatures.

Summary

Few foods are as rapidly perishable as fresh muscle foods. Although freezing allows long term preservation of meat and fish, frozen products do not command the premium price and consumer acceptance of fresh products. Preservative packaging under atmospheres other than air can dramatically increase the shelf life of muscle foods while preserving sensory characteristics, allowing producers to access markets that are unreachable without these technologies.

The success of preservative packaging for controlling the spoilage of fresh muscle foods depends largely on the three factors of pack atmosphere, temperature of storage, and product pH. The maximum achievable shelf life with any packaging system requires chiller temperatures approaching $-1.5°C$. The reliability of any packaging system for delaying spoilage is only as good as the temperature control in the distribution chain for the product. Temperature is also the critical factor in assuring product safety. The potential threat posed by *C. botulinum* can be eliminated by ensuring the product temperature remains below $3°C$. Similarly, though cold-tolerant pathogens can grow slowly at temperatures below $4°C$, their growth is inhibited by chiller temperatures and high CO_2.

The preservation of red meats and poultry of normal pH (<6) in preservative packaging is accomplished by establishing conditions in which the spoilage flora will be composed primarily of lactic acid bacteria, as opposed to gram-negative bacteria with higher spoilage potentials. This is achieved by the creation of an anaerobic environment and/or slowing the growth of gram-negative organisms with CO_2. Unfortunately, consistent establishment of a spoilage flora dominated by lactic acid bacteria requires a pH of below 6. Consequently, the shelf life for meat of high pH and fish is generally much less than that of normal pH meat, although substantial extensions in shelf life compared to air storage can still be achieved.

Of the three gases commonly used in MAP systems, only CO_2 plays a role in controlling microbial spoilage. Oxygen at high concentrations preserves the color of myoglobin, but when it is present in atmospheres the growth of gram-negative organisms of high spoilage potential will occur. High oxygen concentrations can also be used with fish, to inhibit TMAO reduction, but the gains in storage life are limited since other spoilage activities of bacteria are not similarly inhibited. Nitrogen serves solely to prevent pack collapse. CAP systems containing carbon dioxide from which all oxygen has been removed consistently achieve the longest shelf lives. For chilled meats in preservative packaging, the only limitation on the use of CAP is myoglobin color changes. These may be prevented by the rigorous exclusion of oxygen or the use of low concentrations of carbon monoxide. Concerns about the safe use of carbon monoxide with red meat and fish can apparently be overcome. Should regulations regarding the use of carbon monoxide change, the need for high O_2 MAP could disappear and systems of that type could become redundant.

References

Allen, C. E. and Foegeding, E. A. (1981). Some lipid characteristics and interactions in muscle foods. *Food Technol.* **35(5)**, 253–257.

Barnes, E. M. (1976). Microbiological problems of poultry at refrigerator temperatures – a review. *J. Sci. Food Agric.* **27,** 777–782.

Bohnsack, U., Knippel, G. and Höpke, H.-U. (1988). The influence of a CO_2 atmosphere on the shelf-life of fresh poultry. *Fleischwirtsch. Intl* **68,** 1553–1557.

Boskou, G. and Debevere, J. (1997). Reduction of trimethylamine oxide by *Shewanella* spp. under modified atmospheres in vitro. *Food Microbiol.* **14,** 543–553.

Cardello, A. V., Sawyer, F. M., Maller, O. and Digman, L. (1982). Sensory evaluation of the texture and appearance of 17 species of North Atlantic fish. *J. Food Sci.* **47,** 1818–1823.

Chen, H. M., Meyers, S. P., Hardy, R. W. and Biede, S. L. (1984). Color stability of astaxanthin pigmented rainbow trout under various packaging conditions. *J. Food Sci.* **49,** 1337–1340.

Clark, D. S. and Bruki, T. (1972). Oxygen requirements of strains of *Pseudomonads* and *Achromobacter*. *Can. J. Microbiol.* **18,** 321–326.

Cornforth, D. (1994). Color–its basis and importance. In: *Quality Attributes and their Measurement in Meat, Poultry and Fish Products* (A. M. Pearson and T. R. Dutson, eds), pp. 34–78. Blackie Academic, London, UK.

Dainty, R. M., Shaw, B. G., Harding, C. O. and Michanie, S. (1979). The spoilage of vacuum packaged beef by cold tolerant bacteria. In: *Cold Tolerant Microbes in Spoilage and the Environment* (A. D. Russell and R. Fuller, eds), pp. 83–100. Academic Press, London, UK.

Dalgaard, P. (1995). Modelling of microbial activity and prediction of shelf life for packed fresh fish. *Intl J. Food Microbiol.* **26,** 305–317.

Dalgaard, P., Gram, L. and Huss, H. H. (1993). Spoilage and shelf-life of cod fillets packed in vacuum or modified atmospheres. *Intl J. Food Microbiol.* **19,** 283–294.

Davis, H. K. (1998). Fish and shelfish. In: *Principles and Applications of Modified Atmosphere Packaging of Foods*, 2nd edn. (B. A. Blakistine, ed.), pp. 194–239. Blackie Academic, London, UK.

Dhananjaya, S. and Stroud, G.D. (1994). Chemical and sensory changes in haddock and herring stored under modified atmospheres. *Intl J. Food Sci. Technol.* **29,** 575–583.

Doherty, A. M. and Allen, P. (1998). The effect of oxygen scavengers on the color stability and shelf-life of CO_2 master packaged pork. *J. Muscle Foods* **9,** 351–363.

Dransfield, E. (1994). Tenderness of meat, poultry, and fish. In: *Quality Attributes and their Measurement in Meat, Poultry and Fish Products* (A. M. Pearson and T. R. Dutson, eds), pp. 287–315. Blackie Academic, London, UK.

Easter, M. C., Gibson, D. M and Ward, F. B. (1983). The induction and location of trimethylamine-N-oxide reductase in *Alteromonas* sp. NCIMB 400. *J. Gen. Microbiol.* **129,** 3689–3696.

Echevarne, C., Renerre, M. and Labas, R. (1990). Metmyoglobin reductive activity in bovine muscle. *Meat Sci.* **27,** 161–172.

Eklund, M. W. (1982). Effect of CO_2 atmospheres and vacuum packaging on *Clostridium botulinum* and spoilage organisms of fisheries products. In: *Proceedings of the 1st National Conference on Seafood Packaging and Shipping* (R. E. Martin, ed.), pp. 298–331. National Fisheries Inst, Washington, DC.

Farber, J. M. (1991). Microbiological aspects of modified-atmosphere packaging technology – a review. *J. Food Prot.* **54,** 58–70.

Faustman, C. and Cassens, R. G. (1990). The biochemical basis for discoloration in fresh meat – a review. *J. Muscle Foods* **1,** 217–243.

Garcia de Fernando, G. D., Mano, S. B., Lopez, D. and Ordóñez, J. A. (1995). Effect of modified atmospheres on the growth of pathogenic psychrotrophic microorganisms on proteinaceous foods. *Microbiologia* **11,** 7–22.

Genegeorgis, C. (1985). Microbiological safety of the use of modified atmospheres to extend the storage life of fresh meat and fish. *Intl J. Food Microbiol.* **1,** 237–251.

Gibson, D. M. and Davis, H. K. (1995). Fish and shellfish products in sous vide and modified atmosphere packs. In: *Principles of Modified Atmosphere and Sous Vide Product Packaging* (J. M. Farber and K. L. Dodds, eds), pp. 153–174. Technomic Publishing, Lancaster, PA.

Gill, C. O. (1981). Meat spoilage and evaluation of the potential storage life of fresh meat. *J. Food Prot.* **46,** 444–452.

Gill, C. O. (1986). The control of microbial spoilage in fresh meats. In: *Advances in Meat Research*, Vol. 2 (A. M. Pearson and T. R. Dutson, eds), pp. 49–88. AVI Publishing, Westport, CT.

Gill, C. O. (1988). The solubility of carbon dioxide in meat. *Meat Sci.* **22,** 65–71.

Gill, C. O. (1989). Packaging meat for prolonged chilled storage: the Captech Process. *Br. Food J.* **91,** 11–15.

Gill, C. O. (1990). Meat and modified atmosphere packaging. In: *The Encyclopedia of Food Science and Technology* (Y. H. Hui, ed.), pp. 1678–1683. Wiley, New York, NY.

Gill, C. O. and DeLacy, K. M. (1991). Growth of *Escherichia coli* and *Salmonella typhimurium* on high-pH beef packed under vacuum or carbon dioxide. *Intl J. Food Microbiol.* **13,** 21–30.

Gill, C. O. and Jones, T. (1994a). The display life of retail-packaged beef steaks after their storage in master packs under various atmospheres. *Meat Sci.* **38,** 385–396.

Gill, C. O. and Jones, T. (1994b). The display life of retail-packs of ground beef after their storage in master packs under various atmospheres. *Meat Sci.* **37,** 281–295.

Gill, C. O. and Jones, T. (1996). The display life of retail-packaged pork chops after their storage in master packs under atmospheres of N_2, CO_2 or O_2 + CO_2. *Meat Sci.* **42,** 203–213.

Gill, C. O. and McGinnis, J. C. (1995a). The effects of residual oxygen concentration and temperature on the degradation of the color of beef packaged under oxygen-depleted atmospheres. *Meat Sci.* **39,** 387–394.

Gill, C. O. and McGinnis, J. C. (1995b). The use of oxygen scavengers to prevent the transient discoloration of ground beef packaged under controlled, oxygen-depleted atmospheres. *Meat Sci.* **41,** 19–27.

Gill, C. O. and Newton, K. G. (1977). The development of spoilage flora on meat stored at chill temperatures. *J. Appl. Bacteriol.* **43,** 189–195.

Gill, C. O. and Newton, K.G. (1980). Development of bacterial spoilage at adipose tissue surfaces of fresh meat. *Appl. Environ. Microbiol.* **39,** 1076–1077.

Gill, C. O. and Penney, N. (1986). Packaging conditions for extended storage of chilled, dark, firm, dry beef. *Meat Sci.* **18,** 41–53.

Gill, C. O. and Penney, N. (1988). The effect of the initial gas volume to meat weight ratio on the storage life of chilled beef packaged under carbon dioxide. *Meat Sci.* **22,** 53–63.

Gill, C. O. and Reichel, M. P. (1989). Growth of the cold-tolerant pathogens *Yersinia ente-rocolitica*, *Aeromonas hydrophila* and *Listeria monocytogenes* on high-pH beef packaged under vacuum or carbon dioxide. *Food Microbiol.* **6**, 223–230.

Gill, C. O. and Tan, K. H. (1980). Effect of carbon dioxide on growth of meat spoilage bacteria. *Appl. Environ. Microbiol.* **39**, 317–319.

Gill, C. O., Phillips, D. M. and Harrison, J. C. L. (1988). Product temperature criteria for shipment of chilled meats to distant markets. In: *Refrigeration for Food and People*, pp. 40–47. International Institute of Refrigeration, Paris, France.

Gill, C. O., Harrison, J. C. L. and Penney, N. (1990). The storage life of chicken carcasses packaged under carbon dioxide. *Intl J. Food Microbiol.* **11**, 151–157.

Gill, C. O., Jones, T., Rhan, K. *et al.* (2002). Temperatures and ages of boxed beef pack-aged and distributed in Canada. *Meat Sci.* **60**, 401–410.

Grau, F. H. (1983). Microbial growth on fat and lean surfaces of vacuum packaged beef. *J. Food Sci.* **48**, 326–329.

Grau, F. H. and Vanderlinde, P. B. (1990). Growth of *Listeria monocytogenes* on vacuum packaged beef. *J. Food Prot.* **53**, 739–741.

Guldager, H. S., Bøknæs, N., Østerberg, C., Nielsen, J. and Dalgaard, P. (1998). Thawed cod fillets spoil less rapidly than unfrozen fillets when stored under modified atmos-phere packaging at 2°C. *J. Food Prot.* **61**, 1129–1136.

Haard, N. F. (1992). Technological aspects of extending prime quality of seafood: a review. *J. Aquatic Food Prod. Tech.* **1**, 9–27.

Hauschild, A. H. W. (1989). *Clostridium botulinium*. In: *Foodborne Bacterial Pathogens* (M. P. Doyle, ed.), pp. 111–189. Marcel Dekker, New York, NY.

Hauschild, A. H. W., Poste, L. M. and Hilsheimer, R. (1985). Toxin production by *Clostridium botulinium* and organoleptic changes in vacuum-packaged raw beef. *J. Food Prot.* **48**, 712–716.

Hintlian, C. B. and Hotchkiss, J. H. (1986). The safety of modified atmosphere packaging: a review. *Food Technol.* **40(12)**, 70–76.

Holland, G. C. (1980). Modified atmospheres for fresh meat distribution. In: *Proceedings of the 33rd Meat Industry Research Conference*, pp. 21–39. American Meat Institute Foundation, Arlington, VA.

Hong, L. C., LeBlanc, E. L., Hawrysh, Z. J. and Hardin, R. T. (1996). Quality of Atlantic mackerel (*Scromber scrombus* L.) fillets during modified atmosphere storage. *J. Food Sci.* **61**, 646–651.

Hood, D. E. (1980). Factors affecting the rate of metmyoglobin accumulation in pre-pack-aged beef. *Meat Sci.* **4**, 247–265.

ICMSF (2000). Fish and fish products. In: *Microbial Ecology of Food Commodities*, Inter-national Commission On Microbiological Specifications For Foods, pp. 130–189. Aspen Publishers, Gaithersburg, MD.

Jemmi, T., Schmitt, M. and Rippen, T. E. (2000). Safe handling of seafood. In: *Safe Handling of Foods* (J. M. Farber and E. C. D. Todd, eds), pp. 105–165. Marcel Dekker, New York, NY.

Jeremiah, L. E., Penney, N. and Gill, C. O. (1992). The effects of prolonged storage under vacuum or CO_2 on the flavor and texture profiles of chilled pork. *Food Res. Intl* **25**, 9–19.

Jeremiah, L. E., Tong, A. K. W., Jones, S .D. M. and McDonell, C. (1993). A survey of Canadian consumer perceptions of beef in relation to general perceptions regarding foods. *J. Consumer Studies Home Econ.* **17**, 13–37.

Jones, J. M. and Grey, T. C. (1989). Influence of processing on product quality and yield. In: *Processing of Poultry* (G. C. Mead, ed.), pp. 127–181. Elsevier, London, UK.

Jones, J. M., Mead, G. C., Griffiths, N. M. and Adams, B. W. (1982). Influence of packaging on microbiological, chemical and sensory changes in chill-stored turkey portions. *Br. Poultry Sci.* **23**, 25–40.

Jorgensen, B. R. and Huss, H. H. (1989). Growth and activity of *Shewanella putrefaciens* isolated from spoiling fish. *Intl J. Food Microbiol.* **9**, 51–62.

Jorgensen, B. R., Gibson, D. M. and Huss, H. H. (1988). Microbiological quality and shelf life prediction of chilled fish. *Intl J. Food Microbiol.* **6**, 295–307.

Kelly, R. S. A. (1989). High barrier metallized laminates for food packagings. In: *Plastic Film Technology*, Vol. 1 (K. M. Findlayson, ed.), pp. 146–152. Technomic Publishing, Lancaster, PA.

Lambert, A. D., Smith, J. P. and Dodds, K. L. (1991). Effects of initial O_2 and CO_2 and low-dose irradiation on toxin production by *Clostridium botulinum* in MAP fresh pork. *J. Food Prot.* **54**, 939–944.

Lanier, T. C., Carpenter, J. A., Toledo, R. T. and Reagan, J. O. (1978). Metmyoglobin reduction in beef systems as affected by aerobic, anaerobic and carbon monoxide-containing environments. *J. Food Sci.* **43**, 1788–1792.

Lannelongue, M., Hanna, M. O., Finne, G., Nickelson, R. and Vanderzant, C. (1982). Storage characteristics of fin fish fillets (*Archosargus probatocephalus*) packaged in modified gas atmospheres containing carbon dioxide. *J. Food Prot.* **45**, 440–444.

Ledward, D. A. (1970). Metmyoglobin formation in beef stored in carbon dioxide enriched and oxygen-depleted atmospheres. *J. Food Sci.* **35**, 33–37.

Lillard, D. A. (1987). Oxidative deterioration in meat, poultry, and fish. In: *Warmed-Over Flavor of Meat* (A. J. St Angelo and M. E. Bailey, eds), pp. 41–67. Academic Press, London, UK.

Lin, C. F., Gray, J. I., Asghar, A., Buckley, D. J., Booren, A. M. and Flegal, C. J. (1989). Effects of dietary oils and α-tocopherol supplementation on lipid composition and stability of broiler meat. *J. Food Sci.* **54**, 1457–1460, 1484.

Livingston, D. J. and Brown, W. D. (1981). The chemistry of myoglobin and its reactions. *Food Technol.* **35(5)**, 244–252.

Love, R. M. (1988). *The Food Fishes*, p. 276. Ferrand Press, London, UK.

Love, R. M., Munro, L. J. and Robertson, I. (1977). Adaptation of the dark muscle of cod to swimming activity. *J. Fish Biol.* **11**, 431–436.

Lowry, P. D. and Gill, C. O. (1984). Mould growth on meat at freezing temperatures. *Intl J. Refrig.* **7**, 133–136.

Madhavi, D. L. and Carpenter, C. E. (1993). Aging and processing affect color, metmyoglobin reductase and oxygen consumption of beef muscles. *J. Food Sci.* **58**, 939–947.

McMeekin, T. A. (1977). Spoilage association of chicken leg muscle. *Appl. Environ. Microbiol.* **33**, 1244–1246.

McMeekin, T. A. and Patterson, J. T. (1975). Characterization of hydrogen sulphide producing bacteria isolated from meat and poultry plants. *Appl. Microbiol.* **29**, 165–169.

Millar, S., Willson, R., Moss, B. W. and Ledward, D. A. (1994). Oxymyoglobin formation in meat and poultry. *Meat Sci.* **36,** 397–406.

Newton, K. G. and Gill, C. O. (1978). The development of the anaerobic spoilage flora of meat stored at chiller temperatures. *J. Appl. Bacteriol.* **44,** 91–95.

Nortje, G. L. and Shaw, B. G. (1989). The effect of aging treatment on the microbiology and storage characteristics of beef in modified atmosphere packs containing 25% CO_2 plus 75% O_2. *Meat Sci.* **25,** 43–58.

Nychas, G. J., Dillon, V. M. and Board, R. G. (1988). Glucose, the key substrate in the microbiological changes occurring in meat and certain meat products. *Biotechnol. Appl. Biochem.* **10,** 203–231.

Oka, H. (1989). Packaging for freshness and the prevention of discoloration of fish fillets. *Packag. Technol. Sci.* **2,** 201–213.

O'Keefe, M. and Hood, D. E. (1980–81). Anoxic storage of fresh beef. 2: Color stability and weight loss. *Meat Sci.* **5,** 267–281.

O'Keefe, M. and Hood, D. E. (1982). Biochemical factors influencing metmyoglobin formation in beef from muscles of differing color stability. *Meat Sci.* **7,** 204–228.

Ordonez, J. A. and Ledward, D. A. (1977). Lipid and myoglobin oxidation in pork stored in oxygen and carbon dioxide-enriched atmospheres. *Meat Sci.* **7,** 204–228.

Palumbo, S. A. (1988). The growth of *Aeromonas hydrophila* K144 in ground pork at 5°C. *Intl J. Food Microbiol.* **7,** 41–48.

Pasteroriza, L., Sampedro, G., Herrera, J. J. and Cabo, M. L. (1998). Influence of sodium chloride and modified atmosphere packaging on microbiological, chemical and sensorial properties in ice storage of slices of hake (*Merluccius merluccius*). *Food Chem.* **61,** 23–28.

Penney, N., Bell, R. G. and Cummings, T. L. (1994). Extension of the chilled storage life of smoked blue cod (*Parapercis colias*) by carbon dioxide packaging. *Intl J. Food Sci. Technol.* **29,** 167–178.

Post, L. S., Lee, D. A., Solberg, M., Furgang, D., Specchio, J. and Graham, C. (1985). Development of botulinal toxin and sensory deterioration during storage of vacuum and modified atmosphere packaged fresh fillets. *J. Food Sci.* **50,** 990–996.

Randell, K., Hattula, T., Skytta, E., Silvertsvik, M., Bergslien, H. and Ahvenainen, R. (1999). Quality of filleted salmon in various retail packages. *J. Food Qual.* **22,** 483–497.

Renerre, M. (1989). Retail storage and distribution of meat in modified atmospheres. *Fleischwirtsch. Intl* **1,** 51–53.

Renerre, M. (1990). Factors involved in the discoloration of beef meat. *Intl J. Food Sci. Technol.* **25,** 613–630.

Rhodes, D. N. and Lea, C. M. (1961). Enzymatic changes in lamb's liver during storage 1. *J. Sci. Food Agric.* **12,** 211–227.

Rosnes, J. T., Sivertsvik, M., Skipnes, D., Nordtvedt, T. S., Corneliussen, C. and Jakobsen, Ø. (1998). Transport of superchilled salmon in modified atmosphere. In: *Hygiene, Quality and Safety in the Cold Chain and Air-Conditioning*, pp. 229–236. International Institute for Refrigeration, Paris, France.

Sánchez-Eschalante, A., Dejenane, D., Torrescano, G., Beltran, J. A. and Roncalés, P. (2001). The effects of ascorbic acid, taurine, carrosine and rosemary powder on color and lipid stability of beef patties packaged in modified atmosphere. *Meat Sci.* **58,** 421–429.

Silliker, J. H. and Wolfe, S. K. (1980). Microbiological safety considerations in controlled atmosphere storage of meats. *Food Technol.* **34(3)**, 59–63.

Silva, J. L. and White, T. D. (1994). Bacteriological and color changes in modified atmosphere-packaged refrigerated channel catfish. *J. Food Prot.* **57,** 715–719.

Sivertsvik, M., Rosnes, J. T., Vorre, A., Randell, K., Ahvenainen, R. and Bergslien, H. (1999). Quality of whole gutted salmon in various bulk packages. *J. Food Qual.* **22,** 387–401.

Sivertsvik, M., Jeksrud, W. K. and Rosnes, J. T. (2002). A review of modified atmosphere packaging of fish and fishery products – significance of microbial growth, activities and safety. *Intl J. Food Sci. Technol.* **37,** 107–127.

Skura, B. J. (1991). Modified atmosphere packaging of fish and fish products. In: *Modified Atmosphere Packaging of Foods* (B. Olraikul and M. E. Stiles, eds), pp. 148–168. Ellis Horwood, New York, NY.

Sørheim, O., Aune, T. and Nesbakken, T. (1977). Technological, hygienic and toxicological aspects of carbon monoxide use in modified-atmosphere packaging of meat. *Trends Food Sci. Technol.* **8,** 307–312.

Sørheim, O., Nissen, H. and Nesbakken, T. (1999). The storage life of beef and pork packaged in an atmosphere with low carbon monoxide and high carbon dioxide. *Meat Sci.* **52,** 157–164.

Stiles, M. E. (1991). Modified atmosphere packing of meat, poultry and their products. In: *Modified Atmosphere Packaging of Foods* (B. Olraikul and M. E. Stiles, eds), pp. 118–147. Ellis Horwood, New York, NY.

Taylor, A. A. (1985). Packaging fresh meat. In: *Developments in Meat Science*, Vol. 3 (R. Lawrie, ed.), pp. 89–113. Elsevier, Barking, UK.

Whitfield, F. B. (1998). Microbiology of food taints. *Intl J. Food Sci. Technol.* **33,** 31–51.

Young, L. L., Reviere, R. D. and Cole, A. B. (1988). Fresh red meats: a place to apply modified atmospheres. *Food Technol.* **42,** 65–69.

Centralized packaging systems for meats

Roberto A. Buffo and Richard A. Holley

14

Introduction

During processing and preservation, the texture, flavor, color, and fresh taste of food are affected. Useful storage life is the time a food remains acceptable from a sensory, nutritional, microbiological or safety perspective following its manufacture and/or packaging. Storage life can be extended when the most probable mechanism for its deterioration is known, and when techniques that will not affect sensory and nutritional characteristics can be applied (Labuza, 1996). A variety of effective techniques have been adopted by the food industry to extend the storage life of liquid foods (retort or aseptic processing), liquid foods with particulates (canning and retort treatments), dehydrated foods (freeze drying), cut fruits and vegetables (cryogenic freezing), and meat (refrigeration or freezing). However, most of these techniques can cause a loss of nutrients and 'freshness' (i.e. fresh taste). The value of freshness is measured by the fact that fresh fruits, vegetables, and meats are premium-priced products in both local and international markets (Tewari *et al.*, 1999). Chilling reduces the growth of micro-organisms and allows storage-life extension with minimal changes in the textural and nutritional qualities of foods. If properly used, it can be considered the most conven-ient and reliable means of providing fresh food products (Jeyamkondan *et al.*, 2000).

To date, modified atmosphere packaging (MAP) is the only technique available that consistently delays microbial growth, ensures freshness, and provides a long shelf-life for fresh meat (Gill and Harrison, 1989). Nevertheless, shelf-life goals can only be achieved with strict temperature control to achieve maximum microbial inhibition.

Innovations in Food Packaging
ISBN: 0-12-311632-5

Temperatures must be controlled as closely as possible to the freezing point of meat tissue ($-2°C$) in order to inhibit organisms like the pseudomonads, which can grow at $-3°C$, and still avoid ice crystallization damage to the meat tissue. The rate of spoilage in muscle foods is high because they are a protein-rich medium that encourages microbial growth. Muscle tissue continues to metabolize just after slaughter, using stored carbohydrates and fats. Without further processing, microbial growth occurs as the energy stored is depleted and metabolic products, typically low molecular weight organic compounds, accumulate in the tissues. This growth results in undesirable odors and flavors, and eventually microbiological growth produces signs of decay on the meat surface (Tewari et al., 1999).

Traditional meat distribution

Optimized storage of fresh meat can yield 9–15 weeks of useful product life if meats are held at $-1.5 \pm 0.5°C$ in atmospheres saturated with CO_2 and devoid of O_2. These systems are most useful for long-term storage and the intercontinental transport of primal cuts of meat (Tewari et al., 1999). The current meat distribution system in North America involves the transportation of primal and subprimal cuts in vacuum packages to retail stores, yielding a storage life of about 42 days for pork to 45 days for beef. Due to residual O_2 present in the package, some surface discoloration occurs due to the oxidation of myoglobin to form metmyoglobin (brown). Once the primal cuts reach the retail stores, they are trimmed to remove the surface-discolored tissue, which can carry a high microbial load. The primal cuts are then reduced to consumer-size cuts and packaged in disposable retail trays using O_2-permeable films (Jeyamkondan et al., 2000). Oxygen from the atmosphere combines with the free binding site of deoxymyoglobin to form oxymyoglobin (a process known as oxygenation), which yields a bright red color. This reaction, which takes place in less than 30 minutes due to the high affinity between deoxymyoglobin and O_2, is known as "blooming" in the meat industry. Because spoilage occurs primarily on exposed surfaces, the creation of new surfaces after extended storage sets the spoilage clock back to near zero (Seideman et al., 1984; Penny and Bell, 1993).

This distribution system is considered to be inefficient, based on the following premises (Jeyamkondan et al., 2000):

1. Packaging is carried out at both packer and retail levels
2. Quality control at the retail level is generally deficient
3. Considerable floor space in retail stores is used for making retail cuts
4. Fabrication costs of retail cuts are usually high because of the lack of specialized labor crews and appropriate machinery at the retail level
5. By-products and edible waste are underutilized at the retail level.

Centralized packaging systems

Moving the packaging of retail-ready cuts from retail stores to the packer level or, alternatively, to centralized packaging operations, eliminates the time-consuming labor required for cutting, trimming, and re-wrapping meat at retail stores. Higher efficiency can be achieved at a centralized location by the incorporation of robotics, which minimizes human handling, thus improving quality and safety (Scholtz et al., 1992a). Only the saleable meat cuts are transported from the centralized operation to the retail store; fat trimmings, bones, and other waste products remain at the packaging center, thus eliminating unnecessary refrigerated transport (Jeyamkondan et al., 2000). It should be noted that storage life of master-packaged retail-ready meat containing O_2 is considerably shorter, typically no more than a week. There are two main reasons for this (Shay and Egan, 1987):

1. As meat color must be red in the retail arena to attract consumers, meat must be packaged either in an O_2-permeable film or in packages containing high concentrations of O_2.
2. No further manipulation is possible and no new surfaces are created – in other words, no additional trimming is possible following initial packaging.

Nevertheless, there is presently an irreversible shift of meat distribution systems towards centralized packaging operations, especially by large supermarket chains. Advantages associated with the system are as follows (Quigley, 2002):

1. A case-ready program transfers a large portion of the labor component from the store to the packaging plant, reducing costs and adding storage space for other merchandise at the retail point
2. A case-ready program makes it logistically possible for a retailer to keep the meat case filled with specific cuts and concentrate on meat products that are most in demand throughout the day and over holiday periods, when other stores may run out of popular cuts
3. A case-ready program ensures that the cuts in inventory are the exact ones most in demand, as opposed to an in-store meat-cutting program, where a range of different cuts (some popular and some not) is produced and then have to be further merchandized
4. Case-ready programs provide uniformity in terms of product weight and quality throughout the retain chain. Customers can expect to receive the same quality and cut of meat or meat product, such as sausage, from every store throughout the chain without having variations introduced by different butchers with different meat-cutting techniques or different recipes at different stores
5. Since store employees do not handle unpacked meat, the case-ready program reduces the opportunity for contamination and enhances food safety assurance levels; this removes the responsibility from the store manager and frees up his or her time
6. Case-ready products have significantly reduced the amount of in-store shrinkage – according to some market analysts, by up to two-thirds
7. Since each case-ready package is completely traceable, stores can determine the exact quantities sold; also, for quality control purposes, packages can even be traced back to the farm level
8. Case-ready products lend themselves to the product-branding programs that are rapidly becoming an essential part of the North American meat business, offering consumers significantly more information about the meat they purchase.

Researchers at universities and industry scientists have developed a number of centralized packaging systems. Some of these are discussed below.

Individual Packages

High O_2 MAP system

This type of packaging system has been known to the meat industry for many years. It keeps the color of the meat bright red throughout the storage period. Aerobic bacteria tolerate high concentrations of O_2; thus, their growth can be reduced by including CO_2 (typically 80% O_2/20% CO_2) in the gas mixture (Gill and Molin, 1991). This system is able to extend the shelf life of beef by 2 weeks while maintaining an acceptable red color at $-1.5°C$ (Gill and Jones, 1994). Yet the shelf life is considerably less than that given by anoxic atmospheres due to the oxidative rancidity (lipid oxidation) induced by O_2 and the rapid growth of psychrotrophic bacteria (Cole, 1986).

There have been successful applications of this system. For example, Tesco, a large British supermarket chain, converted its entire fresh meat operation into a high O_2 MAP centralized operation (Brody, 1996). Another British chain, Marks & Spencer, has combined vacuum-skin packaging (VSP) and high O_2 MAP for the delivery of fresh meat cuts. Each cut is first vacuum-skin packaged in an O_2-permeable film, and then bagged in an O_2-impermeable nylon pouch with a gas mixture of 80% O_2 and 20% CO_2 (Anonymous, 1989).

Cryovac peeling

Cryovac (Sealed Air Corporation, Duncan, SC, USA) developed the Darfresh peelable VSP system for retail-ready meat cuts. In this system, cuts are first vacuum-packaged in a transparent O_2-permeable film and then vacuum-skin packaged in a transparent O_2-impermeable film. Just before the retail cut is displayed in the corresponding case, the outer impermeable film is removed and meat blooms on exposure to air. A similar technique can be applied to retail-ready cuts in flexible plastic trays. However, these double-film systems are expensive (Jeyamkondan *et al.*, 2000).

Flavaloc fresh

This system was developed by Garwood Packaging, Inc. (Indianapolis, IN, USA) and ASI Plastic Solutions (Moorabbin, Victoria, Australia). The package has three parts: a shallow, white base tray made of polyethylene terephthalate plastic, Estapac 9921, in which the retail cut is placed; a transparent polyvinyl chloride (PVC), O_2-permeable film being stretched across the cut; and a transparent O_2-impermeable lid or dome, made of polyethylene terephthalate G 6763 polyester (Eastman Chemical Company, Kingsport, TN, USA), which is sealed on top. Retail cuts are prepared at a centralized packaging operation, and placed in the preformed trays together with pads to absorb drip that may form. Air is then evacuated and back-flushed with a gas mixture at atmospheric pressure containing at least 30% CO_2 with the remainder N_2, which yields less than 300 ppm of residual O_2 (Zhao *et al.*, 1994). The meat is purple when the dome is in place. Once the package reaches the store, the dome is removed and the meat blooms within 20 minutes. The company claims a storage life of 21–40 days in the dome, and subsequently 4 days of retail display (Jeyamkondan *et al.*, 2000).

Windjammer case-ready packaging system

In this system, retail cuts are first individually packed in MAP with high O_2-barrier films to give a storage life of up to 21 days. Before retail display, a gas-exchange machine is used to evacuate the anoxic atmosphere. The package is then back-flushed with high O_2 modified atmosphere containing 80% O_2 and 20% CO_2, which allows blooming of the meat. This technology has been characterized as a "dynamic gas-exchange system" (Zhao *et al.*, 1994).

Master packaging (multiple packages)

In this technique, four to six conventional retail packs, wrapped in O_2-permeable films, are placed within a large pouch made of an aluminum laminate with negligible O_2 transmission, known as the master pack. This pack is evacuated to remove O_2 and back-flushed with a desired gas mixture, typically 100% CO_2. When CO_2 is introduced, meat absorbs it, reducing the atmospheric volume. Thus, a sufficient amount of CO_2 (about 2 litres/kg of meat) must be used in order to maintain a 100% concentration throughout the storage period. Needle holes are burned in at least two different locations of the O_2-permeable films to allow the meat rapidly to come in contact with the modified atmosphere. As needed, according to consumer demands, master packs are opened, the holes sealed with a label, and the retail cuts placed in the retail display case. Blooming takes place within 30 minutes (Penny and Bell, 1993; Gill and Jones, 1994).

Master packaging is more economical than the previous systems because of the grouping of the individual packs under a common master bag. It differs from the Flavaloc Fresh system in that it uses an opaque aluminum laminate film as the outer barrier rather than a transparent dome. The aluminum laminate has an O_2 transmission rate near zero, thereby substantially reducing the formation of metmyoglobin. Today there are transparent, non-metallized, laminated films with no measurable O_2 transmission rates (Winpak, Winnipeg, MB, Canada) that can be used instead. Master packaging differs from the Windjammer system in that it relies on atmospheric air for bloom development, which eliminates the need for expensive gas-exchange machines at retail stores (Jeyamkondan *et al.*, 2000).

There are three essential requirements for achieving the maximum storage life of master-packaged meat (Jeyamkondan *et al.*, 2000):

1. Superior initial microbial quality of the meat
2. A consistent storage temperature of $-1.5°C$ throughout the distribution period
3. Headspace gas composition consistently maintained at 100% CO_2 during the distribution period without any residual O_2.

By following good manufacturing practices, and adoption of an appropriate HACCP program for production and distribution, the first condition may be satisfied. With regard to the second, there have been a number of studies recommending a cryogenic fluid, liquid N_2, as a refrigeration system in portable containers for storage and distribution of retail-ready meats. It is reported that liquid N_2 can control the temperature of meat in a narrow range ($-1.5 \pm 0.5°C$) (Bailey et al., 1997; Habok, 1999; Jeyamkondan, 1999). The third condition is the most problematic, because during storage of meat the headspace gas composition changes dynamically. Change depends on the gas transmission rate of the packaging film, the respiration rate of meat tissues and microflora, the absorption of gases by meat tissues, and the initial gas composition. The respiration of meat tissues and microflora contributes in a negligible manner to the total CO_2 present, although the capacity of meat to absorb CO_2 has to be taken into account to prevent collapsing of the package after initial packaging. An important issue is that most commercial machines presently available are not capable of evacuating O_2 completely from the package. Even a small amount of residual O_2 is conducive to the formation of metmyoglobin, which is stable and undesirable. Because meat has only a limited capacity to reduce metmyoglobin to deoxymyoglobin, from which oxymyoglobin can form, a large amount of the former in the package may prevent development of the desirable red color when the retail-ready pack is displayed (Jeyamkondan et al., 2000).

There is a packaging system developed at the New Zealand Meat Research Institute (Hamilton, New Zealand) called CAPTECH (chilled atmosphere packaging technology) capable of creating a CO_2 atmosphere with <300 ppm of residual O_2. In this process, the meat is packaged in a triple-laminated pouch consisting of high-strength plastic film, aluminum foil, and a low-strength plastic film (SecureFresh Pacific Ltd, Auckland, New Zealand). The aluminum foil is a good gas barrier and, together with the high-strength plastic film, provides enough mechanical strength to avoid puncture during transportation and handling. The inner low-strength film is helpful in heat-sealing the package. The volume of CO_2 added to the package in relation to the mass of meat is metered by a packaging machine (CAPTRON, Secure Fresh Pacific, Auckland, New Zealand). As the packaging material has a negligible, non-measurable O_2 transmission rate, the O_2 remaining in the package is at a sufficiently low concentration to be converted to CO_2 through meat respiration. In addition, metmyoglobin formed due to these small amounts of residual O_2 is small enough to be reduced to deoxymyoglobin and later converted to oxymyoglobin during display (Zhao et al., 1994; Varnam and Sutherland, 1995; Jeyamkondan et al., 2000).

As mentioned previously, headspace composition within a package changes dynamically during the distribution period, and must be controlled continuously in order to achieve the desired food quality and safety. These packaging techniques are known as active or intelligent packaging (Labuza and Breene, 1989). Among them, O_2-scavenging systems to remove residual O_2 from the package atmosphere are of central importance with regard to successful master packaging. The Japanese were the forerunners in the development of this technology. O_2-scavenging ageless sachets, named Ageless FX-100 and manufactured by Mitsubishi Gas Chemical Co., Inc. (Tokyo, Japan), are now used in the food industry around the world. These sachets contain ferrous iron powder sealed in a moisture- and gas-permeable package. The ferrous iron absorbs any O_2 present in the package headspace and is oxidized to the ferric state; typically, this requires moisture as an activating agent. Ageless FX-100 sachets have been commercially successful for various food product applications, although the kinetics of this system at very low O_2 concentrations (300–600 ppm) and low temperature (*ca.* $-1.5°C$) in meats is not fully known (Jeyamkondan *et al.*, 2000).

Another O_2-scavenging system is the Cryovac OS1000, a multi-layer flexible film with a co-extruded sealant from the Sealed Air Corporation (Saddle Brook, NJ, USA). A proprietary O_2-scavenging polymer is incorporated within the sealant and therefore is invisible to consumers, unlike the sachets and labels. The polymer is dormant until it is activated by a UV light system (Cryovac 4100 UV triggering unit, Sealed Air Corporation) just before packaging. Unlike the Ageless FX-100 sachets, it does not require moisture as an activating agent; therefore, the product is unaffected. The manufacturing company claims that the system can reduce O_2 levels within the modified atmosphere of the package from 0.5–1% to a few parts per million, in 4 to 10 days (Jeyamkondan *et al.*, 2000).

Applied research

Scholtz *et al.* (1992b) compared the performance of a commercial bulk pre-packaging system, Cryovac GFII, manufactured by Darex Africa (Pty) Ltd. (Kemptonpark, South Africa), to that of a laboratory-based system in terms of quality attributes such as microbiology, color, odor, and acceptability of PVC-overwrapped pork retail cuts. A similar retail shelf life of 3 days was achieved after 0, 7 or 14 days bulk storage with both packaging systems. The color of the samples from both systems was pale to normal during the trial. After 14 days of storage, samples from both systems were still acceptable and had a fresh meat odor. These results showed that the commercial system tested may be applied successfully for 100% CO_2 bulk packaging of PVC-overwrapped pork retail cuts.

Gill and Jones (1994) studied the storage life of master-packaged beefsteak under various atmospheres at $-1.5°C$. Steaks stored under 100% CO_2 or 100% N_2 for less than 4 days were only slightly desirable because of the formation of metmyoglobin on the surface due to the presence of small amounts of residual O_2 in packages. By 4 days of storage, metmyoglobin was reduced to deoxymyoglobin by the meat tissue (action of the enzyme metmyoglobin reductase), and the product acquired the potential to develop the desirable red color upon final exposure to O_2 during retail display. This phenomenon is known as *transient discoloration*, and occurs with beef shortly after packaging in anoxic atmospheres where there are O_2 residuals $\leqslant 1000$ ppm. Master packaging under a CO_2–O_2 (1:2 v/v) atmosphere gave an initial acceptable appearance, but the color began to deteriorate after 12 days of storage, and it was concluded that the combination may be suitable for storage of less than 4 days. Master packaging in 100% N_2 gave a shelf life of 4 weeks, and 2 days of subsequent retail display, while master packaging in 100% CO_2 gave about 7 weeks of shelf life in the master pack and 2 days of subsequent retail display. This study also emphasized that these storage lives can only be obtained with products of high initial microbiological quality.

Jeremiah and Gibson (1997) investigated the effects of 100% CO_2 atmospheres on the flavor and texture profiles of master-packaged pork cuts at $-1.5°C$. Flavor became inappropriate, unbalanced and unblended after 12 days of storage, and it was concluded that off-flavor development, most likely caused by lactic acid bacteria, constitutes the limiting factor for shelf-life extension of fresh pork under modified atmosphere conditions. As in the previous study, the importance of using cuts with very low initial bacterial numbers was stressed. In a follow-up study, Nattress et al. (1998) found that off-flavor development coincides with a shift in the dominant microflora from non-aciduric to aciduric lactic acid bacteria.

Gill and McGinnis (1995) used FreshPax 200R from Multisorb Technologies, Inc. (Buffalo, NY, USA), an iron-based O_2-scavenger in a sachet, to prevent transient discoloration. This scavenging system is composed of activated iron-oxide powder mixed with acids, salts and humectants to promote oxidation of iron, which removes residual O_2 from the package atmosphere avoiding meat discoloration. These authors also reported that O_2 concentration in the pack atmosphere must be reduced to <10 ppm within 30 minutes at $2°C$, or within 2 hours at $-1.5°C$ for acceptable blooming of ground beef to occur.

Tewari (2000) studied the storage life of master-packaged beef tenderloins and pork loins in combination with other technologies such as O_2-scavenging systems, the CAPTECH process, and a N_2-based refrigeration unit able to maintain the temperature at $-1.5 \pm 0.5°C$. Preliminary results showed that Ageless FX-100 sachets needed to be placed inside the retail packs rather than in the master bag. Eight sachets were found to be optimal when they were placed below the absorption pads under each retail cut. Four packaged retail cuts were master-packaged, filled with 4.5 l of N_2, sealed using a CAPTRON system, and stored at $-1.5 \pm 0.5°C$ using a liquid N_2-fueled refrigeration system. The storage life of beef and pork cuts was about 10 weeks, with 3 days of subsequent retail display life.

Conclusions

A variety of centralized packaging systems for retail-ready meat is a commercial reality, and each of them satisfies different requirements. Thus, high O_2 MAP is suitable for local distribution due to the limited shelf-life, whereas master packaging in the absence of O_2, combined with appropriate temperature control, good manufacturing practices, and suitable packaging materials can achieve 6–10 weeks of storage life, besides being the most economical system due to the grouping of individual packs (Jeyamkondan et al., 2000).

The case-ready program has been a reactionary process built on market demand, satisfying cost-cutting requirements at specific segments of the supply and distribution channels. Communications between packer and retail buyer, and between retail buyer and store meat manager, are key to the success of the system, because they allow prediction of sales on a store-by-store basis and provide better pricing strategies and promotional activities (Pizzico, 2002). In the USA, a January 2001 survey by the Cryovac Division of Sealed Air

Corporation showed that of the 127 000 grocery stores in the country, about 25 000 carried case-ready poultry, 6000 carried case-ready pork, 10 000 carried case-ready ground beef, and 1000 carried case-ready total muscle cuts. The report also found that stores making the move to case-ready products have experienced, on average, a 3.8% growth in sales. Based on these numbers, it is clear that this sector of the industry has nowhere to go but up, pushed both by consumers and improved productivity (Quigley, 2002).

References

Anonymous (1989). Fresh meat package combines. VSP and MAP. *Food Eng.* **61,** 72–75.
Ailey, C. G., Jayas, D. S., Holley, R. A., Jeremiah, L. E. and Gill, C. O. (1997). Design, fabrication, and testing of a returnable, insulated, nitrogen refrigerated shipping container for distribution of fresh red meat under controlled CO_2 atmosphere. *Food Res. Intl* **30,** 743–753.
Brody, A. L. (1996). Integrating aseptic and modified atmosphere packaging to fulfill a vision of tomorrow. *Food Technol.* **50(4),** 56–66.
Cole, A. B. Jr (1986). Retail packaging systems for fresh red meat cuts. Reciprocal Meat Conference. *Proc. Am. Meat Sci. Assoc.* **39,** 106–111.
Gill, C. O. and Harrison, J. C. L. (1989). The storage life of chilled pork packaged under carbon dioxide. *Meat Sci.* **26,** 313–324.
Gill, C. O. and Jones, T. (1994). The display life of retail-packaged beef steaks after their storage in master packs under various atmospheres. *Meat Sci.* **38,** 385–396.
Gill, C. O. and McGinnis, J. C. (1995). The use of oxygen scavengers to prevent transient discolouration of ground beef packaged under controlled oxygen-depleted atmospheres. *Meat Sci.* **41,** 19–27.
Gill, C. O. and Molin, G. (1991). Modified atmospheres and vacuum packaging. In: *Food Preservatives* (N.J. Russell and G.W. Gould, eds), pp. 172-199. Blackie & Sons Ltd., London, UK.
Habok, M. N. N. (1999). Modification and testing of a nitrogen refrigerated, controlled atmosphere container for distribution of fresh red meats. MSc thesis, Department of Biosystems Engineering, University of Manitoba, Winnipeg, Canada.
Jeremiah, L. E. and Gibson, L. L. (1997). The influence of controlled atmosphere storage on the flavor and texture profiles on display-ready pork cuts. *Food Res. Intl* **30,** 117–129.
Jeyamkondan, W. (1999). Design and evaluation of a portable, nitrogen-refrigerated, jacketed container for storage and distribution of chilled meat. MSc thesis, Department of Biosystems Engineering, University of Manitoba, Winnipeg, Canada.
Jeyamkondan, W., Jayas, D. S. and Holley, R. A. (2000). Review of centralized packaging systems for distribution of retail-ready meat. *J. Food Prot.* **63,** 796–804.
Labuza, T. P. (1996). An introduction to active packaging for foods. *Food Technol.* **59(4),** 68–71.
Labuza, T. P. and Breene, W. M. (1989). Applications of "active packaging" for improvement of shelf-life and nutritional quality of fresh and extended shelf-life foods. *J. Food Proc. Pres.* **13,** 1–69.

Nattress, F. M., Worobo, R. J., Greer, G. C. and Jeremiah, L. E. (1998). Effect of lactic acid bacteria on beef flavour. *Proc. 44th Intl. Congress Meat Sci. Technol.* **44**, 321–327.

Penny, N. and Bell, R. G. (1993). Effect of residual oxygen on the colour, odour and taste of carbon-dioxide packaged beef, lamb and pork during short term storage at chill temperatures. *Meat Sci.* **33**, 245–252.

Pizzico, B. (2002). The key to case ready. *Meat & Poultry* **48(2)**, 38–41.

Quigley, L. (2002). Canadian style case ready. *Meat & Poultry* **48(2)**, 30–36.

Scholtz, E. M., Jordaan, E., Krüger, J. and Nortjé, R.T. (1992a). The influence of different centralized pre-packaging systems on the shelf-life of fresh pork. *Meat Sci.* **32**, 11–29.

Scholtz, E. M., Krüger, J., Nortjé, G. and Naudé, R. (1992b). Centralized bulk pre-packaging of pork retail cuts. *Intl. J. Food Sci. Technol.* **27**, 391–398.

Seideman, S. C., Cross, H. R., Smith, G. C. and Durland, P. R. (1984). Factors associated with fresh meat color: a review. *J. Food Qual.* **6**, 211–237.

Shay, B. J. and Egan, A. F. (1987). The packaging of chilled red meats. *Food Technol. Aust.* **39**, 283–285.

Tewari, G. (2000). Centralized packaging of retail meat cuts. PhD thesis, Department of Biosystems Engineering, University of Manitoba, Winnipeg, Canada.

Tewari, G., Jayas, D. S. and Holley, R. A. (1999). Centralized packaging of meat cuts: a review. *J. Food Prot.* **62**, 418–425.

Varnam, A. H. and Sutherland, J. P. (1995). *Meat and Meat Products: Technology, Chemistry and Microbiology,* 2nd edn. Chapman & Hall, London, UK.

Zhao, Y., Wells, J. H. and McMillen, K. W. (1994). Applications of dynamic modified atmosphere packaging systems for fresh red meats: a review. *J. Muscle Foods* **5**, 299–328.

Edible and biodegradable coatings and films

Edible films and coatings: a review

Jung H. Han and Aristippos Gennadios

Introduction

Films are generally defined as stand-alone thin layers of materials. They usually consist of polymers able to provide mechanical strength to the stand-alone thin structure. Sheets are thick films. Therefore, there is not an obvious difference in material composition between films and sheets other than their thickness. Films can form pouches, wraps, capsules, bags, or casings through further fabrication processes. Coatings are a particular form of films directly applied to the surface of materials. Removal of coating layers may be possible; however, coatings are typically not intended for disposal separately from the coated materials. Therefore, coatings are regarded as a part of the final product.

The use of edible films and coatings as carriers of active substances has been suggested as a promising application of active food packaging (Cuq *et al.*, 1995; Han, 2000, 2001). In fact, the use of edible films and coatings for food packaging without active substances is arguably also an application of active food packaging, since the edibility and biodegradability of the films are additional functions not offered by conventional packaging materials (Cuq *et al.*, 1995; Han, 2002).

Edible films and coatings

Edible films and coatings are produced from edible biopolymers and food-grade additives. Film-forming biopolymers can be proteins, polysaccharides (carbohydrates and

Innovations in Food Packaging
ISBN: 0-12-311632-5

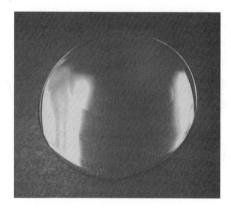

Figure 15.1 Edible film made from carboxymethyl cellulose.

gums) or lipids (Gennadios *et al.*, 1997; see Figure 15.1). Plasticizers and other additives are combined with the film-forming biopolymers to modify the physical properties or functionality of films.

The film-forming mechanisms of biopolymers include intermolecular forces such as covalent bonds (e.g., disulfide bonds and cross-linking) and/or electrostatic, hydrophobic or ionic interactions. Fabrication procedures initiate these film-forming mechanisms. For the resulting films or coatings to be edible, the film-forming mechanism involved in fabrication should be an appropriate food process, namely pH modification, salt addition, heating, enzymatic modification, drying, use of food-grade solvents, and addition of other food-grade chemicals. The control of fabrication process conditions is very important because changes in treatment conditions can alter kinetics and reaction mechanisms (Guilbert *et al.*, 1996, 1997).

Edible films and coatings enhance the quality of food products, protecting them from physical, chemical, and biological deterioration (Kester and Fennema, 1986). The application of edible films and coatings can readily improve the physical strength of food products, reduce particle clustering, and improve visual and tactile features on product surfaces (Cuq *et al.*, 1995; Cisneros-Zevallos *et al.*, 1997). It can also protect food products from moisture migration, microbial growth on the surface, light-induced chemical changes, oxidation of nutrients etc. (Kester and Fennema, 1986). Most commonly, edible films and coatings function as barriers against oils, gases or vapors, and as carriers of active substances, such as antioxidants, antimicrobials, colors, and flavors (Kester and Fennema, 1986; Gennadios and Weller, 1990; Guilbert and Gontard, 1995; Krochta and De Mulder-Johnston, 1997; Miller *et al.*, 1998; see Figure 15.2). These protective functions are aimed to enhance the quality of food products, resulting in shelf-life extension and safety improvement (Gennadios and Weller, 1990).

Biodegradable films and coatings

Inherently, edible film and coating materials are biodegradable (Krochta, 2002). In fact, biodegradability is one of the greatest benefits of edible films and coatings, along with edibility (Debeaufort *et al.*, 1998). There are several potential non-food applications

Figure 15.2 Whey protein concentrate edible coating protects the surface of nuts from physical damage and oxidation.

for edible films and coatings, such as films for agricultural uses (e.g. mulching, and tunnel and bale wrap), grocery bags, paper/paperboard coatings, and cushioning foams (Guilbert and Gontard, 1995; Han and Krochta, 1999, 2001). Many functions of edible films and coatings are similar to those of synthetic packaging films; however, edible film and coating materials must be chosen according to specific food applications, the types of food products, and the major mechanisms of quality deterioration (Peterson *et al.*, 1999; Guilbert, 2002). The use of edible films and coatings as primary packaging can potentially replace conventional packaging materials, partially or totally, thus reducing their overall utilization (Krochta and De Mulder-Johnston, 1997; Petersen *et al.*, 1999). Due to their protective functions, edible films and coatings may simplify the total packaging structure (Krochta and De Mulder-Johnston, 1997; Debeaufort *et al.*, 1998). For example, the packaging of cookies in a plastic primary bag in a secondary cardboard carton can be simplified to coated cookies in a carton.

Historical and current uses of edible films and coatings

The use of free-standing edible films may have a shorter history than edible coatings. As an example of edible films, yuba (soy-milk skin) has been traditionally used in Asian countries since the fifteenth century (Wu and Bates, 1972; Park *et al.*, 2002). Wax coatings were applied to citrus fruits in the twelfth and thirteenth centuries, but only commercially utilized on apples and pears as recently as the 1930s (Baldwin, 1994; Debeaufort *et al.*, 1998; Park, 2000). Wax coatings reduce moisture loss and slow down the respiration of coated fruits and vegetables, resulting in shelf-life extension (Baldwin, 1994). Various waxes have been sprayed on the surface of fruits and vegetables as forms of hot-melt wax or emulsions. Lipid coatings (larding) on meats and cheeses have been used since the Middle Ages for shrinkage prevention (Kester and Fennema, 1986; Donhowe and Fennema, 1994; Debeaufort *et al.*, 1998).

Currently, edible films and coatings are used with several food products, mostly fruits, vegetables, candies, and some nuts (Krochta and De Mulder-Johnston, 1997;

Petersen *et al.*, 1999). Collagen films are used for sausage casings, and some hydroxy-methyl cellulose films as soluble pouches for dried food ingredients. In general, there are more applications of coatings than of films. Shellac and wax coatings on fruits and vegetables, zein coatings on candies, and sugar coatings on nuts are the most common commercial practices of edible coatings (Krochta and De Mulder-Johnston, 1997). The pharmaceutical industry uses sugar coatings on drug pills and gelatin films for soft capsules (Gennadios, 2002; Krochta, 2002). The use of cellulose ethers (such as carboxymethyl cellulose, hydroxypropyl cellulose, and methylcellulose) as ingredients in coatings for fruits, vegetables, meats, fish, nuts, confectionery, bakery, grains, and other agricultural products is increasing (Nussinovitch, 2003).

Film composition

Film-forming materials

The main film-forming materials are biopolymers, such as proteins, polysaccharides, lipids, and resins. They can be used alone or in combinations. The physical and chemical characteristics of the biopolymers greatly influence the properties of resulting films and coatings (Sothornvit and Krochta, 2000). Film-forming materials can be hydrophilic or hydrophobic. However, in order to maintain edibility, solvents used are restricted to water and ethanol (Peyron, 1991).

Proteins are commonly used film-forming materials. They are macromolecules with specific amino acid sequences and molecular structures. The secondary, tertiary, and quaternary structures of proteins can be easily modified by heat denaturation, pressure, irradiation, mechanical treatment, acids, alkalis, metal ions, salts, chemical hydrolysis, enzymatic treatment, and chemical cross-linking. The most distinctive characteristics of proteins compared to other film-forming materials are conformational denaturation, electrostatic charges, and amphiphilic nature. Many factors can affect the conformation of proteins, such as charge density and hydrophilic–hydrophobic balance. These factors can ultimately control the physical and mechanical properties of prepared films and coatings. Protein film-forming materials are derived from many different animal and plant sources, such as animal tissues, milks, eggs, grains, and oilseeds (Krochta, 2002).

Polysaccharide film-forming materials include starch, non-starch carbohydrates, gums, and fibers. The sequence of polysaccharides is simple compared to proteins, which have 20 common amino acids. However, the conformation of polysaccharide structures is more complicated and unpredictable, resulting in much larger molecular weights than proteins. Most carbohydrates are neutral, while some gums are mostly negatively charged. Although this electrostatic neutrality of carbohydrates may not affect significantly the properties of formed films and coatings, the occurrence of relatively large numbers of hydroxyl groups or other hydrophilic moieties in the structure indicate that hydrogen bonds may play significant roles in film formation and characteristics. Some negatively charged gums, such as alginate, pectin, and carboxymethyl cellulose, show significantly different rheological properties in acidic than in neutral or alkaline conditions.

Table 15.1	Materials used for edible films and coatings
Functional compositions	**Materials**
Film-forming materials	*Proteins*: collagen, gelatin, casein, whey protein, corn zein, wheat gluten, soy protein, egg white protein, fish myofibrillar protein, sorghum protein, pea protein, rice bran protein, cottonseed protein, peanut protein, keratin *Polysaccharides*: starch, modified starch, modified cellulose (CMC, MC, HPC, HPMC), alginate, carrageenan, pectin, pullulan, chitosan, gellan gum, xanthan gum *Lipids*: waxes (beeswax, paraffin, carnauba wax, candelilla wax, rice bran wax), resins (shellac, terpene), acetoglycerides
Plasticizers	Glycerin, propylene glycol, sorbitol, sucrose, polyethylene glycol, corn syrup, water
Functional additives	Antioxidants, antimicrobials, nutrients, nutraceuticals, pharmaceuticals, flavors, colors
Other additives	Emulsifiers (lecithin, Tweens, Spans), lipid emulsions (edible waxes, fatty acids)

CMC, carboxy methylcellulose; MC, methylcellulose; HPC, hydroxypropyl cellulose; HPMC, hydroxypropyl methylcellulose.

Lipids and resins are also used as film-forming materials, but they are not polymers and, evidently, "biopolymers" is a misnomer for them. Nevertheless, they are edible, biodegradable, and cohesive biomaterials. Most lipids and edible resins are soft-solids at room temperature and possess characteristic phase transition temperatures. They can be fabricated to any shape by casting and molding systems after heat treatment, causing reversible phase transitions between fluid, soft-solid, and crystalline solid. Because of their hydrophobic nature, films or coatings made from lipid film-forming materials have very high water resistance and low surface energy. Lipids can be combined with other film-forming materials, such as proteins or polysaccharides, as emulsion particles or multi-layer coatings in order to increase the resistance to water penetration (Gennadios *et al.*, 1997; Pérez-Gago and Krochta, 2002).

Biopolymer composites can modify film properties and create desirable film structures for specific applications. Similar to multi-layered composite plastic films, biopolymer films can be produced as multiple composite layers, such as protein coatings (or film layers) on polysaccharide films, or lipid layers on protein/polysaccharide films. This multi-layered film structure optimizes the characteristics of the final film. Composite films can also be created by mixing two or more biopolymers, yielding one homogeneous film layer (Yildirim and Hettiarachchy, 1997; Debeaufort *et al.*, 1998; Were *et al.*, 1999). Various biopolymers can be mixed together to form a film with unique properties that combine the most desirable attributes of each component (Wu *et al.*, 2002; see Table 15.1).

Plasticizers

In most cases, plasticizers are required for edible films and coatings, especially for polysaccharides and proteins. These film structures are often brittle and stiff due to extensive interactions between polymer molecules (Krochta, 2002). Plasticizers are low molecular weight agents incorporated into the polymeric film-forming materials, which decrease the glass transition temperature of the polymers. They are able to position themselves between polymer molecules and to interfere with the polymer–polymer interaction to increase flexibility and processability (Guilbert and Gontard, 1995; Krochta, 2002). Plasticizers increase the free volume of polymer structures or the molecular mobility of polymer molecules (Sothornvit and Krochta, 2000). These properties imply that the plasticizers decrease the ratio of crystalline region to the amorphous region and lower the glass transition temperature (Guilbert et al., 1997; Krochta, 2002). The addition of plasticizers affects not only the elastic modulus and other mechanical properties, but also the resistance of edible films and coatings to permeation of vapors and gases (Sothornvit and Krochta, 2000, 2001). Most plasticizers are very hydrophilic and hygroscopic so that they can attract water molecules and form a large hydrodynamic plasticizer–water complex. For protein and polysaccharide edible films, plasticizers disrupt inter- and intra-molecular hydrogen bonds, increase the distance between polymer molecules, and reduce the proportion of crystalline to amorphous region (Krochta, 2002). Water molecules in the films function as plasticizers. Water is actually a very good plasticizer, but it can easily be lost by dehydration at a low relative humidity (Guilbert and Gontard, 1995). Therefore, the addition of hydrophilic chemical plasticizers to films can reduce water loss through dehydration, increase the amount of bound water, and maintain a high water activity. There are two main types of plasticizers (Sothernvit and Krochta, 2000, 2001):

1. Agents capable of forming many hydrogen bonds, thus interacting with polymers by interrupting polymer–polymer bonding and maintaining the farther distance between polymer chains
2. Agents capable of interacting with large amounts of water to retain more water molecules, thus resulting in higher moisture content and larger hydrodynamic radius.

However, owing to the hydrophilic nature of water, biopolymers, and plasticizers, and due to the abundantly existing hydrogen bonds in their structures, it is very difficult to separate these two mechanisms. Sothornvit and Krochta (2001) suggested that several factors of plasticizers affect plasticizing efficiency, including size/shape of plasticizer molecules, number of oxygen atoms and their spatial distance within the structure of the plasticizers, and water-binding capacity. Besides the effect of hydrogen bonding, repulsive forces between molecules of the same charge or between polar/non-polar polymers (e.g. acetylated starch) can increase the distance between polymers, thus achieving the function of plasticization in the case of charged polymeric film structures. Therefore, compared to neutral polymer films (e.g. starch films), the flexibility of charged polymer films (e.g. soy protein, carboxymethyl cellulose or alginate films) may be affected more significantly by altering pH and salt addition at the same water activity level.

Additives

Edible films and coatings can carry various active agents, such as emulsifiers, antioxidants, antimicrobials, nutraceuticals, flavors, and colorants, thus enhancing food quality and safety, up to the level where the additives interfere with physical and mechanical properties of the films (Kester and Fennema, 1986; Baldwin *et al.*, 1995, 1997; Guilbert *et al.*, 1996; Howard and Gonzalez, 2001; Han, 2002, 2003). Emulsifiers are surface-active agents of amphiphilic nature able to reduce the surface tension of the water–lipid interface or the water–air surface. Emulsifiers are essential for the formation of protein or polysaccharide films containing lipid emulsion particles. They also modify surface energy to control the adhesion and wettability of the film surface (Krochta, 2002). Although many biopolymers possess certain levels of emulsifying capacity, it is necessary to incorporate emulsifiers into film-forming solutions to produce lipid-emulsion films. In the case of protein films, some film-forming proteins have sufficient emulsifying capacity due to their amphiphilic structure. Antioxidants and antimicrobial agents can be incorporated into film-forming solutions to achieve active packaging or coating functions (Han, 2002, 2003). They provide additional active functions to the edible film and coating system to protect food products from oxidation and microbial spoilage, resulting in quality improvement and safety enhancement. When nutraceutical and pharmaceutical substances are incorporated into edible films and coatings, the system can be used for drug delivery purposes (Han, 2003). Incorporated flavors and colorants can improve the taste and the visual perception of quality, respectively. Because of the various chemical characteristics of these active additives, film composition should be modified to keep a homogeneous film structure when heterogeneous additives are incorporated into the film-forming materials (Debeaufort *et al.*, 1998).

Functions and advantages

Edibility and biodegradability

The most beneficial characteristics of edible films and coatings are their edibility and inherent biodegradability (Guilbert *et al.*, 1996; Krochta, 2002). To maintain edibility, all film components (i.e. biopolymers, plasticizers, and other additives) should be food-grade ingredients and all process facilities should be acceptable for food processing (Guilbert *et al.*, 1996). With regard to biodegradability, all components should be biodegradable and environmentally safe. Human toxicity and environmental safety should be evaluated by standard analytical protocols by authorized agencies.

Physical and mechanical protection

Edible films and coatings protect packaged or coated food products from physical damage caused by mechanical impact, pressure, vibrations, and other mechanical factors. Standardized mechanical examinations of commercial film structures are also applied to edible film and coating structures. Such tests may include tensile strength, elongation-at-break, elastic modulus, compression strength, puncture strength, stiffness, tearing strength, burst strength, abrasion resistance, adhesion force, folding endurance and others. Table 15.2 shows the tensile properties of various edible films and common

Table 15.2 Tensile and gas barrier characteristics of edible films and common plastics

Inferior	Marginal	Good	Superior
Tensile strength (MPa)			
<1	1–10	10–100	>100
PPC : Gly	Coll : Cell : Gly	WPI : Gly	Cellophane
	Na-caseinate : Gly	FMP : Gly	MC
	Ca-caseinate : Gly	CZ : Gly	HPMC
	WPI : Sor	HAPS : Gly	Amylose
	EWP : Gly	LDPE	OPP
	SPI : Gly	HDPE	PVDC
	CZ : PEG	PP	PET
	WG : Gly	PS	
	WPI : BW : Gly		
	SPI : FA : Gly		
	Pea protein : Gly		
Elongation (%)			
<1	1–10	10–100	>100
Ca-caseinate : Gly	MC	Coll : Cell : Gly	CZ : Gly
PS	HPMC	Na-caseinate : Gly	WG : Gly
	WPI : BW : Gly	WPI : Gly	SPI : FA : Gly
		WPI : Sor	HAPS : Gly
		FMP : Gly	LDPE
		EWP : Gly	HDPE
		EWP : PEG	
		SPI : Gly	
		CZ : PEG	
		Pea protein : Gly	
		Cellophane	
		Amylose	
		OPP	
		PET	
Oxygen permeability (cm^3 μm m^{-2} d^{-1} kPa^{-1})			
>1000	1000–100	100–10	<10
LDPE	HPMC	Collagen	WG : Gly
Starch : Gly	MC	CZ : Gly	SPI : Gly
	Shellac	WPI : Gly	WPI : Sor
	Beeswax	EWP : Gly	HAPS : Gly
	Most waxes	PPC : Gly	EVOH
	HDPE	Cellophane	PVDC
		Polyester	
Water vapor permeability (g mm m^{-2} d^{-1} kPa^{-1})			
>10	10–1	1–0.1	<0.1
Na-caseinate : Gly	WG : Gly	Shellac	HPMC : FA : PEG
Ca-caseinate : Gly	WG : BW : Gly	Chocolate	HPMC : BW : PEG
EWP : Gly	Ca-caseinate : BW		Beeswax
WPI : Sor	WPI : BW : Sor		Paraffin wax
WPI : Gly	WPI : BW : Gly		Most waxes
SPI : Gly	Cellophane		LDPE
PPC : Gly			HDPE
SPI : FA : Gly			PVDC

(Continued)

Table 15.2 (*Continued*)			
Inferior	Marginal	Good	Superior
CZ : Gly			EVOH
Pea protein : Gly			PVC
HAPS : Gly			PET

Gly, glycerol; Sor, sorbitol; PEG, polyethylene glycol; PPC, peanut protein concentrate; Coll, collagen; Cell, cellulose; WPI, whey protein isolate; EWP, egg white protein; SPI, soy protein isolate; CZ, corn zein; WG, wheat gluten; FMP, fish myofibrillar protein; MC, methylcellulose; HPMC, hydroxypropyl methylcellulose; HAPS, high-amylose pea starch; LDPE, low-density polyethylene; HDPE, high-density polyethylene; PP, polypropylene; PS, polystyrene; OPP, oriented polypropylene; EVOH, ethylene vinylalcohol; PVC, polyvinyl chloride; PVDC, polyvinylidene chloride; PET, polyethylene terephthalate; BW, beeswax; FA, fatty acids.
Test conditions for tensile test and water vapor permeability are at approximately 50% RH and 25°C.
Data collected from Gennadios *et al.* (1994), McHugh and Krochta (1994), Guilbert *et al.* (1996), Krochta (1997, 2002), Choi and Han (2001), Wu *et al.* (2002), Mehyar and Han (2003).

plastic films. Edible films have lower tensile strength than common plastic films, while their elongation-at-break varies widely. Some edible films have elongation values comparable to those of common plastic films. Many edible film and coating materials are very sensitive to moisture (Guilbert and Gontard, 1995; Guilbert *et al.*, 1996; Krochta, 2002). At higher relative humidity conditions, their physical strength is lower than that at lower relative humidity since absorbed moisture functions as a plasticizer. Temperature is also an important variable affecting the physical and mechanical properties of edible films and coatings (Guilbert *et al.*, 1997; Miller *et al.*, 1998; Wu *et al.*, 2002). The physical strength of materials dramatically decreases when temperature increases above the glass transition temperature. High relative humidity and large amounts of plasticizers lower the glass transition temperature of film-forming materials.

Migration, permeation, and barrier functions

The quality of most food products deteriorates via mass transfer phenomena, including moisture absorption, oxygen invasion, flavor loss, undesirable odor absorption, and the migration of packaging components into the food (Kester and Fennema, 1986; Debeaufort *et al.*, 1998; Miller *et al.*, 1998; Krochta, 2002). These phenomena can occur between food and the atmospheric environment, food and packaging materials, or among heterogeneous ingredients in the food product itself (Krochta, 1997). For example, atmospheric oxygen penetration into foods causes oxidation of food ingredients; inks, solvents and monomeric additives in packaging materials can migrate into foods; essential volatile flavors of beverages and confections may be absorbed into plastic packaging materials; and pie/pizza crusts absorb moisture from fillings/toppings, leading to the loss of crispness. Edible films and coatings may wrap these food products or be located between heterogeneous parts of food products to prevent these migration phenomena and preserve quality (Guilbert *et al.*, 1997; Krochta, 2002).

To characterize the barrier properties of edible films and coatings, the transmission rates of specific hazardous migrants should be determined using stand-alone edible

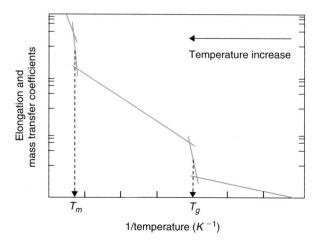

Figure 15.3 Effects of phase transition temperature on tensile properties and mass transfer. T_g and T_m are glass transition temperature and melting temperature, respectively. Y-axis has a logarithmic scale.

films. Most research has dealt with water vapor permeability, oxygen permeability, carbon dioxide permeability, flavor permeability, and oil resistance of edible films. Table 15.2 shows oxygen permeability and water vapor permeability values of edible films and common plastic films. Edible films possess a wide range of oxygen permeability values. Certain edible films are excellent oxygen barriers. Except for lipid-based materials, the water vapor permeability of most edible films is generally higher than that of common plastic films. All barrier properties of edible films and coatings are affected greatly by film composition and environmental conditions (relative humidity and temperature). Plasticizers in edible film-forming materials reduce glass transition temperatures and increase the permeability of most migrants. Oxygen permeability is very sensitive to relative humidity (Guilbert *et al.*, 1997; Maté and Krochta, 1998). At higher relative humidity conditions, oxygen permeability increases substantially. Therefore, it is very important to maintain low relative humidity environments to maximize the effectiveness of edible films as gas barriers.

Temperature is also an important factor of migration (Guilbert *et al.*, 1997; Amarante and Banks, 2001; Wu *et al.*, 2002). A temperature increase provides more energy to the migrating substances and increases the permeability. At temperatures far distant from the phase transition, changes of migration coefficients such as permeability and diffusivity follow the Arrhenius equation (Guilbert *et al.*, 1997; Miller *et al.*, 1998). At the glass transition and melting temperatures of film materials, most mass transfer coefficients change substantially (Figure 15.3).

Convenience and quality preservation

Edible films and coatings provide many benefits in terms of convenience. Reinforced surface strength of fragile products makes handling easier. Coated fruits and vegetables have much higher resistance against bruising and tissue damage caused by physical impact and vibration. Besides this protective function for foods, edible films and coatings are utilized in the food and pharmaceutical industries to develop single-dose

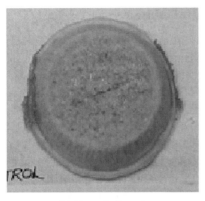

(a) Uncoated crust   (b) Coated crust

Figure 15.4 Bottom of apple pies after 7-day storage. (a) Regular apple pie; (b) Inside piecrust had been sprayed with edible coating materials before filling. The edible coating layer prevented the migration of moisture from the filling to the crust, and maintained crispiness of the crust. (Courtesy of BioEnvelop Agro Technologies, St. Hyacinth, Quebec, Canada.)

pre-measured pouches of food ingredients and drugs, as well as to mask the eventual undesirable taste of medicines (Gennadios and Weller, 1990).

Quality maintenance and enhancement are also very significant functions of edible films and coatings (Krochta, 1997). They can retard surface dehydration, moisture absorption, oxidation of ingredients, aroma loss, frying oil absorption, ripening/aging, and microbial deterioration of food products. In addition to the physical and chemical quality enhancement, edible films and coatings contribute to visual quality, surface smoothness, flavor carriage, edible color print, and other marketing related quality factors. Edible coatings maintain the quality of foods even after the package is opened (Krochta, 1997).

Edible films and coatings may be used to preserve the quality of several food commodities. The oxygen-barrier properties of film and coating layers can prevent oxidation of lipid ingredients, colorants, and flavors of food products such as nuts, confectionary, fried products, and colored produce (Baldwin *et al.*, 1997). The oxygen-barrier property is also quite useful for retarding the respiration rate of fresh produce (Baldwin *et al.*, 1995). Many climacteric fruits and vegetables can be coated with edible film-forming materials to slow down their respiration rate (Park, 2000; Amarante and Banks, 2001). High-fat meat and fish products, such as sausages, jerky, and fillets, can be protected from oxidation after an edible coating process. Moisture-barrier properties of edible films and coatings can protect fresh fruits and vegetables from dehydration. Moisture loss is the most critical quality degradation factor of fresh produce (Guilbert *et al.*, 1997). The moisture barrier property can also be utilized to prevent moisture migration between heterogeneous food product ingredients, for example between raisins and breakfast cereals (Kester and Fennema, 1986), pie fillings and crusts (Figure 15.4), fruit chips and baking dough (Buss, 1996), and the like. The active ingredient carrier function is very useful for the addition of quality preserving agents as well as nutrients and nutraceuticals, resulting in an upgraded quality level of products such as colored/flavored confectionary, glazed bakery, flavored nuts, and vitamin-enriched rice.

Shelf-life extension and safety enhancement

The enhancement and maintenance of quality are directly related to shelf-life extension and safety improvement. An increased protective function of food products extends shelf life (Kester and Fennema, 1986; Krochta, 1997) and reduces the possibility of contamination by foreign matter (Han, 2002, 2003). The market for minimally processed foods and fresh produce has recently seen a significant increase, and accordingly there is a requirement to secure the safety and extend the shelf life of the products involved (Baldwin *et al.*, 1995, 1997; Park, 2000). The massive scale of modern food manufacturing, distribution systems, and fast-food business also dictates the need for improved systemic procedures for maintaining safety and shelf life.

The use of biodegradable materials for food packaging in the food service business is attractive because it reduces the total amount of synthetic materials and appeals to environmentally conscious consumers (Krochta and De Mulder-Johnston, 1997; Amarante and Banks, 2001). However, it is obvious that the period of disintegration of edible films and coatings through biodegradation mechanisms should be longer than the expected shelf life of the packaged products.

Active substance carriers and controlled release

Edible films and coatings can carry food ingredients, pharmaceuticals, nutraceuticals, and agrochemicals in the form of hard capsules, soft gel capsules, microcapsules, soluble strips, flexible pouches, coatings on hard particles and others (Kester and Fennema, 1986; Gennadios and Weller, 1990; Baldwin *et al.*, 1996). The most important parameter to judge the effectiveness of these applications is the ability of controlled release (Han, 2003). Depending on the nature of the application, various release rates are required, which may include immediate release, slow release, specific release rate, or non-migration of active substances. Edible films and coatings should control the release rates of the incorporated active substances to the surrounding media as specifically as possible to achieve maximum effectiveness of the function(s) of the active substances. It is also important to consider chemical interactions between the active substances and the film-forming materials, and between the former and the environmental conditions.

Many different active substances can be incorporated into film-forming materials to design controlled release systems. Such active substances requiring specific migration rates are antimicrobials, antioxidants, bioactive nutraceuticals, pharmaceuticals, flavors, inks, fertilizers, pesticides, insect repellents, and medical/biotechnology diagnostic agents.

Non-edible product applications

Edible films and coatings may substitute for conventional synthetic packaging materials since they have comparable material properties (Krochta, 1997; Krochta and De Mulder-Johnston, 1997). Because edible films and coatings are biodegradable, their mechanical and physical properties do not last for a very long time. Their properties should remain effective during the whole period of shelf life. However, because of

Figure 15.5 SEM image of cross-sections of edible protein coated pulp-papers. Right-hand side, paper was coated by whey protein isolate at the level of 10 g of protein per m² of paper surface. Edible coating resulted in smoother surface as well as stronger resistance to moisture, oil, and oxygen transmissions.

environmental sensitivity (such as to humidity) and relatively inferior mechanical and optical properties compared to those of conventional packaging materials (e.g. paper, paperboard, plastics, glass, and metals), complete substitution of the latter with edible film and coating materials may not be suitable to maintain the functional requirements of packaging systems. Therefore, partial replacement is recommended to reduce the use of synthetic materials (Han and Krochta, 2001).

Whey protein coatings on paper and paperboard can improve the barrier properties of paper-based packaging materials by slowing even further the moisture and oil penetration, as well as improving surface smoothness (Han and Krochta, 1999, 2001; see Figure 15.5). Edible film layers on conventional packaging materials could improve their optical properties and modify their hydrophilic/hydrophobic nature, including surface energy and water/oil absorption of material surfaces.

Edible films and coatings may be used in non-food situations, such as for agricultural applications, paper manufacturing, plastic modification, consumer products, and in the painting industry. Edible films could replace synthetic films, maximizing biodegradation activity. Coating applications may modify the surface properties of any compatible materials. Both applications can carry active substances and prevent moisture and gas migration through the materials. In agricultural uses, they can be used for mulching, wrapping, seed coating, and agrochemical delivery (Guilbert and Gontard, 1995; Nussinovitch, 2003). Surface coating of low-grade papers with edible biopolymers can upgrade their quality for printing and other special purposes (Han and Krochta, 1999). Low-gauge thin paper can be transformed into heavy-gauge thick and strong paper by adding an edible coating. The modified paper surface may accept various inks and further lamination.

Edible coatings can also modify plastic surfaces. Surface modification of plastics has various benefits in many industrial applications. Hydrophilic edible coatings on hydrophobic packaging material surfaces increase the surface energy and provide great advantages in the printing and packaging processes (Hong et al., 2003). The coated packaging materials with increased surface energy can accept various types of inks and ink solvents, such as benzene-, alcohol-, and water-based (Han and Krochta, 1999, 2001; Hong et al., 2003), and may eliminate electrostatic problems through a fast and easy electrostatic discharge since the increased hydrophilicity raises the electroconductivity of the material. The latter application has been studied frequently in the clothing and textile area, but requires more experimental verification for the packaging application.

Besides the benefit of surface modification, common plastic grocery bags may be substituted with biopolymer bags. Many consumer plastic products may be replaced by the edible biopolymer materials, which may be used for disposable plastic bags, cups, plates, containers and utensils, and other plastic products. Edible biopolymers may be incorporated in water-based paints to control paint viscosity and the properties of the paint surface after drying.

Edible biopolymers may also contain chemical cross-linking agents, resulting in a stronger film structure with higher resistance and greater longevity. Even though the cross-linked film structures are no longer edible, they still preserve their biodegradability, minimizing environmental impact (Rhim, 1998; Micard *et al.*, 2000). Most cross-linking agents that are commonly used for cross-linking the reactions of various specific ligands in the column-packing matrix of affinity chromatography can be utilized for the cross-linking of biopolymers, as long as they do not impart significant toxicity to the corresponding non-food application.

Other process-aiding functions

Edible films and coatings can increase the effectiveness of some food processing unit operations. For example, edible coatings on potato slices/strips or fish can reduce oil absorption during frying (Balasubramaniam *et al.*, 1997; Sensidoni and Peressini, 1997). Edible coatings on fruits and vegetables can retard the oxidation of dried products during dehydration and improve the shelf-life extension imparted by irradiation (Lacroix and Ouattara, 2000). Edible coatings of plasticizers can also reduce the loss of color, flavor, or nutrients in particulate fluid foods during processing and distribution (Buss, 1996). Edible coatings improve the effectiveness of the popping process of popcorn (Wu and Schwartzberg, 1992), and act as adhesion agents between heterogeneous food ingredients (Anonymous, 1997).

Many advantages may result from the osmotic dehydration of fruits, vegetables, and functional foods. Since the osmotic dehydration utilizes the migration of water caused by osmotic pressure, many other water-soluble ingredients can be released from the food into the dehydrating fluids (which are generally specific sugar solutions). Edible coatings applied prior to the osmotic dehydration can prevent the migration of valuable ingredients into the dehydrating fluids during the dehydration process, as well as minimizing the invasion of dehydrating agents into the food itself (Dabrowska and Lenart, 2001). Therefore, edible coatings can broaden the selection of dehydrating agents and optimize operation conditions. To maximize the benefit of edible coatings for the osmotic dehydration process, the edible coating layer should have selective migration rates. Water vapor permeability should be high to enhance dehydration; however, the permeability of valuable ingredients or dehydrating agents should be very low to protect their migration during dehydration. Edible coatings on food products may be beneficial to freeze-drying processes, since the moisture-permeable coating can prevent the evaporation of volatile flavors. Evidently, extensive experimental studies are required to verify and validate the benefits of edible coatings as applied to food operations.

Scientific parameters

Chemistry of film-forming materials

It is essential to understand the chemical properties and structure of film-forming materials, biopolymers as well as additives, to tailor them to specific applications (Han, 2002; Nussinovitch, 2003). Solubility in water and ethanol is very important to select a solvent for wet casting or active agent mixing. Thermoplasticity of biopolymers, including phase transition, glass transition, and gelatinization characteristics, should be understood for dry casting or thermoforming (Guilbert *et al.*, 1997). Many hydrophilic properties of film-forming materials are also very important characteristics that should be identified. These may be determined indirectly from water-related properties, such as hydrophilic–lipophilic balance, hygroscopicity, water solubility, and solid surface energy of films; hydrodynamic radius of biopolymers and plasticizers; and surface tension and viscosity of film-forming solutions. The chemical characteristics of plasticizers and any other additives should also be identified to verify their chemical compatibility with biopolymers and to determine the changes in film structure caused by the addition of plasticizers and additives. These investigations are very important to obtain critical information related to film-forming mechanisms and film property modification.

Film-forming mechanisms

An edible film is essentially a dried and extensively interacting polymer network of a three-dimensional gel structure. Despite the film-forming process, whether it is wet casting or dry casting, film-forming materials should form a spatially rearranged gel structure with all incorporated film-forming agents, such as biopolymers, plasticizers, other additives, and solvents in the case of wet casting. Biopolymer film-forming materials are generally gelatinized to produce film-forming solutions. Further drying of the hydrogels eliminates excess solvents from the gel structure. For example, whey-protein films are produced from whey-protein gels by dehydration after heat-set or cold-set gel formation. This does not mean that the film-forming mechanism during the drying process is only the extension of the wet-gelation mechanism. The film-forming mechanism during the drying process may differ from the wet-gelation mechanism, though wet gelation is the initial stage of the film-forming process. There could be a critical stage of a transition from a wet gel to a dry film, which relates to a phase transition from a polymer-in-water (or other solvents) system to a water-in-polymer system. The complete film-forming mechanisms of most biopolymers after gelation are not clearly determined yet. Several polymer chemistry laboratory techniques are required to identify them, including X-ray diffraction, FTIR spectrometry, NMR spectrometry, electrophoresis, polarizing microscopy, and other polymer analysis methodologies. For extrusion casting (dry process), many thermoplastic properties,

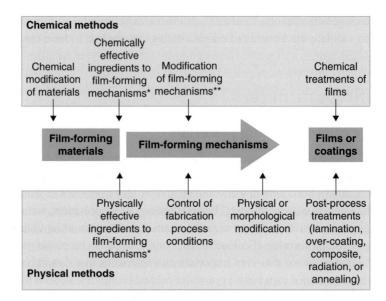

Figure 15.6 Various ways for modifying the characteristics of edible films and coatings.
* indicates the addition of chemically or physically active ingredients, which may enhance or interfere with the film-forming mechanisms; ** includes any chemical cross-linking, chemical substitution of side chains to create hydrophobic interactions or electrostatic interactions, and other extra mechanisms caused by chemical modifications.

such as gelatinization, polymer melting, flow profile, polymer rearrangement and others, should be investigated to predict the film-forming mechanisms. Figure 15.6 describes potential chemical and physical approaches to the modification of film-forming mechanisms by altering film-forming raw materials, varying film-forming processing conditions, and applying treatments on formed films.

As examples, potential chemical methods of modifying the film-forming mechanisms of protein-based films include pH changes, salt addition, heat denaturation, solvent changes, chemical modification of the side chains of peptides, cross-linking, and hydrolysis of peptides (Yildirim and Hettiarachchy, 1997; Rhim, 1988; Were *et al.*, 1999), irradiation of peptides (Lacroix and Ouattara, 2000), and the addition of foreign proteins. For polysaccharide-based films several chemical modifications are available, including salt addition, solvent changes, heat gelatinization, pH changes, chemical modification of hydroxyl groups, cross-linking of polysaccharides, hydrolysis of polysaccharides, and the addition of foreign polysaccharides. Physical modifications of edible films and coatings include lamination, formation of composites, addition of particles or emulsions, perforation, over-coating, annealing/heat curing (Gennadios *et al.*, 1996; Miller *et al.*, 1997; Micard *et al.*, 2000), orientation, radiation (Gennadios *et al.*, 1998; Micard *et al.*, 2000), and ultrasound treatment (Banerjee *et al.*, 1996).

Physical chemistry of films

While polymer structural chemistry is a very important tool for the determination of film-forming mechanisms, physical chemistry is essential to determine film characteristics.

The determination of most physical and mechanical characteristics of film structures is related to physical chemistry parameters, which include mechanical strength, elasticity, moisture and gas permeation, cohesion of polymers, film adhesion onto food surfaces, surface energy, surface roughness/smoothness, light transmittance, color (opaque/gloss), viscosity, thermoplastic characteristics and others. More research is needed to identify the relationship of these physical properties with the polymer structure chemistry (Sothornvit and Krochta, 2000).

Cohesion of film-forming materials is a very important parameter that influences the mechanical strength of films, especially homogeneously continuous film structures (Guilbert *et al.*, 1996). Cohesion is the attractive force between molecules of the same substance (Anonymous, 1992). If the film-forming materials contain heterogeneous ingredients that are not compatible with the main biopolymers, the cohesion of the film-forming materials decreases and the film strength weakens. When the use of new biopolymers or additives is investigated, the compatibility of all film-forming ingredients should be maintained to obtain strong cohesion. Plasticizers are the agents reducing the cohesion of film-forming polymers (Guilbert *et al.*, 1996).

Adhesion of film-forming materials is an important parameter, practically, for film casting and coating processes (Peyron, 1991; Guilbert *et al.*, 1996). Adhesion is the attractive force between the surface molecules of different substances, such as between coating materials and food surfaces (Anonymous, 1992). A low adhesion force results in incomplete coatings on the surface, or easy peel-off of the coating layers from the surface. The surface energy of film-forming materials (surface tension of the film-forming solution), the solid surface energy of uncoated product, and that of the dried film should be determined to achieve strong adhesion. A larger difference of the surface energy of a coating material from the uncoated product surface lowers the work of adhesion and results in a poor coating performance (Peyron, 1991). Surface-active agents, such as emulsifiers and other amphiphilic chemicals in the film-forming solution, reduce the surface tension of the coating solution, thus decreasing the difference between the solid surface energy and the surface tension of the coating solution and ultimately increasing the work of adhesion. From the following Young–Dupré equation (Good, 1993), the work of adhesion (W_{ad}) can be increased by decreasing γ_{SL} (interfacial energy between the solid surface and liquid), which means reducing the difference of energy between the solid food surface (γ_S) and liquid coating solution (γ_L) and lowering the contact angle θ (i.e., increasing $\cos\theta$).

$$W_{ad} = \gamma_s + \gamma_L - \gamma_{SL} = \gamma_L (1 + \cos\theta)$$

Practical parameters for commercialization

Film-production processes

There are two categories of film-production processes; dry and wet (Guilbert *et al.*, 1997). The dry process of edible film production does not use liquid solvents, such as water or alcohol. Molten casting, extrusion, and heat pressing are good examples of dry processes. For the dry process, heat is applied to the film-forming materials to increase the temperature to above the melting point of the film-forming materials, to cause them to flow. Therefore, the thermoplastic properties of the film-forming materials should be identified in order to design film-manufacturing processes. It is necessary to determine the effects of plasticizers and any other additives on the thermoplasticity of the film-forming materials. Plasticizers decrease the glass transition temperature (Guilbert *et al.*, 1997; Krochta, 2002).

The wet process uses solvents for the dispersion of film-forming materials, followed by drying to remove the solvent and form a film structure. For the wet process, the selection of solvents is one of the most important factors. Since the film-forming solution should be edible and biodegradable, only water, ethanol, and their mixtures are appropriate as solvents (Peyron, 1991). All the ingredients of film-forming materials should be dissolved or homogeneously dispersed in the solvents to produce film-forming solutions. The film-forming solution should be applied to flat surfaces using a sprayer, spreader or dipping roller, and dried to eliminate the solvent, forming a film structure. Phase separation of incompatible ingredients from the film-forming solution is not generally desirable unless the phase separation is intentionally designed for the formation of a bi-layer film structure. To produce a homogeneous film structure avoiding phase separation, various emulsifiers can be added to the film-forming solution (Peyron, 1991). This solvent compatibility of ingredients is very important to develop homogeneous edible film and coating systems carrying active agents. All ingredients, including active agents as well as biopolymers and plasticizers, should be homogeneously dissolved in the solvent to produce film-forming solutions. Most film-forming solutions possess much higher surface tension than the surface energy of dried films, since they contain excessive amounts of water or ethanol. Therefore, it is difficult evenly to coat a flat surface that has low surface energy with film-forming solutions that possess high surface energy using high-speed coating equipment in manufacturing. Nonetheless, during the solvent drying process the film-forming solution is concentrated and its surface energy is decreased due to the loss of solvent. The viscosity of the film-forming solution also affects the coating process. A lower viscosity accelerates the separation process (coacervation) of the film-forming solution from the flat surface and causes the uneven coating on the surface, followed by dripping down of the coating solution from the surface to the floor. Higher viscosity of the film-forming solution is desirable to reduce this coating phase separation, unless this creates an uncontrollably heavy coating thickness. If the film-forming solution has a lower surface tension and higher viscosity, the high-speed coating process is more likely to form a film layer on the flat surface. However, the lower surface energy of coated films after drying makes the peeling process of the film from the flat surfaces harder, since there is a very high adhesion between the film and the flat surface when there is a smaller difference of the surface energy between the film and the flat surface (Figure 15.7). Conversely, this is a desirable phenomenon for the direct coating process of the film-forming solution onto food surfaces to avoid the peeling problem of the coated layer.

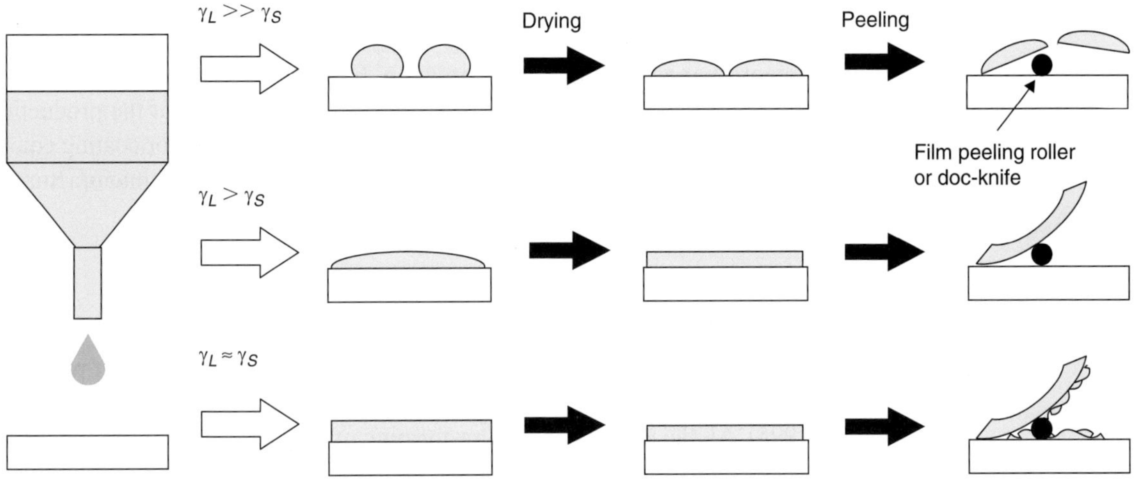

Figure 15.7 Effect of surface tension (γ_L) of coating solution on the adhesion of films during a film-forming process. γ_S is the surface energy of a flat-base solid, such as a drum surface or a conveyor belt.

Feasibility of commercialized system

The production of edible films and coatings may require new processing systems, which may include extrusion, roll orientation, conveyor drying, bath coating, pan coating, or other processes. These new production systems should be economically feasible, and compatible with current synthetic film production processes and food coating processes. They will be practicable when simple modifications of synthetic film and coating processes lead to the production of edible films and coatings without significant investments. To satisfy this feasibility of new production systems, the composition of film-forming materials should be carefully optimized and the film-forming mechanisms modified (Petersen *et al.*, 1999).

Consumer-related issues

Since edible films and coatings are consumable parts of food products, the potential uses of edible materials may be significantly affected by consumer acceptance (Petersen *et al.*, 1999). Consumer acceptance is an integrated index of the subjective preferences of consumers for the products, and includes organoleptic properties, safety, marketing, and cultural hesitation regarding the use of new materials. Organoleptic properties may include favorable flavor, tastelessness, sensory compatibility with coated/packaged foods, texture, and appearance (Han, 2002; Nussinovitch, 2003). Safety issues relate to the potential toxicity or allergenicity of the new edible film-forming materials, and microflora changes of the packaged/coated food products. Marketing factors include the price of the final products, consumers' reluctance to use of the new materials, and special attention of consumers to the use of the new packages if special instructions are required for opening the packages, consuming the packaged/coated foods, and disposing of the used packaging materials.

Besides consumer acceptance, there are many limiting factors for the commercial use of edible films and coatings. They may include the complexity of the production process, large investment for the installation of new film-production or coating equipment, potential conflict with conventional food packaging systems, manufacturers' resistance to the use of new materials, and regulatory issues.

Regulatory issues

Since edible films and coatings are an integral part of the edible portion of products, they should follow all required regulations regarding food products (Guilbert and Gontard, 1995). All the ingredients of film-forming materials, as well as any functional additives in the film-forming materials, should be food-grade, non-toxic materials (Guilbert and Gontard, 1995; Han, 2002; Nussinovitch, 2003). Most GRAS status ingredients have specific restrictions. For example, when the GRAS notification document describes the intended use of a certain material as a surface treatment for poultry meat, the use of the same material for red meats, cheese, or other food products would not be consistent with the GRAS notification.

From a regulatory standpoint, edible films and coatings could be classified as food products, food ingredients, food additives, food contact substances, or food packaging materials (Debeaufort et al., 1998; Petersen et al., 1999). In the case of pharmaceutical and nutraceutical applications, there may be other regulations regarding their use. If there is a different viewpoint of categorization between manufacturers and regulatory agencies, critical mislabeling problems may occur, resulting in mandatory product recall situations. When food manufacturers formulate film-forming materials and apply the materials on their food products, they should include all the film-forming ingredients on the labels of their final products. However, if they use edible film or coating materials that have been produced by different suppliers, there may be the opportunity to categorize them as food contact substances or food packaging materials. It is recommended that the edible film and coating material suppliers obtain no-objection notifications from related authorizing agencies for the use of their film and coating products as food ingredients, with careful considerations of proper labeling, including nutritional information and possible allergenicity (Han, 2001, 2002; Krochta, 2002).

Conclusions

Edible films and coatings are promising systems for the improvement of food quality, shelf life, safety and functionality. They can be used as individual packaging materials, food coating materials, active ingredient carriers, and to separate the compartments of heterogeneous ingredients within foods. The efficiency and functional properties of edible film and coating materials are highly dependent on the inherent characteristics of film-forming materials, namely biopolymers (such as proteins, carbohydrates, and lipids), plasticizers, and other additives. Most biopolymers are relatively hydrophilic compared to commercial plastic materials. For industrial use, it is necessary to conduct scientific research to identify the film-forming mechanisms of biopolymers in order to optimize their properties. It is also suggested that feasibility studies be performed regarding the commercial uses of edible films and coatings by extending the results of research and development studies to commercialization studies, such as new process evaluation, safety and toxicity determination, regulatory assessment, and consumer studies.

References

Amarante, C. and Banks, N. H. (2001). Post harvest physiology and quality of coated fruits and vegetables. *Hort. Rev.* **26,** 161–238.

Anonymous (1992). In: *Comprehensive Dictionary of Physical Chemistry* (L. Ulicky and T. J. Kemp, eds), pp. 15, 65. Ellis Horwood, Chichester, UK.

Anonymous (1997). Edible films solve problems. *Food Technol.* **51(2),** 60.

Balasubramaniam, V. M., Chinnan, M. S., Mallikarjunan, P. and Phillips, R. D. (1997). The effect of edible film on oil uptake and moisture retention of a deep-fat fried poultry product. *J. Food Process Eng.* **20,** 17–29.

Baldwin, E. A. (1994). Edible coatings for fresh fruits and vegetables: past, present and future. In: *Edible Coatings and Films to Improve Food Quality* (J. M. Krochta, E. A. Baldwin and M. Nisperos-Carriedo, eds), pp. 25–64. Technomic Publishing, Lancaster, PA.

Baldwin, E. A., Nispero-Carriedo, M. O. and Baker, R. A. (1995). Edible coatings for lightly processed fruits and vegetables. *Hort. Sci.* **30(1),** 35–38.

Baldwin, E. A., Nispero-Carriedo, M. O., Hagenmaier, R. D. and Baker, R. A. (1997). Use of lipids in coatings for food products. *Food Technol.* **51(6),** 56–64.

Banerjee, R., Chen, H. and Wu, J. (1996). Milk protein-based edible film mechanical strength changes due to ultrasound process. *J. Food Sci.* **61(4),** 824–828.

Buss, D. D. (1996). Stretching gum applications. *Food Prod. Des.* **December**, 79–94.

Choi, W. S. and Han, J. H. (2001). Physical and mechanical properties of pea-protein-based edible films. *J. Food Sci.* **66(2),** 319–322.

Cisneros-Zevallos, L., Saltveit, M. E. and Krochta, J. M. (1997). Hygroscopic coatings control surface white discoloration of peeled (minimally processes) carrots during storage. *J. Food Sci.* **62(2),** 363–366, 398.

Cuq, B., Gontard, N. and Guilbert, S. (1995). Edible films and coatings as active layers. In: *Active Food Packaging* (M. Rooney, ed.), pp. 111–142. Blackie, Glasgow, UK.

Dabrowska, R. and Lenart, A. (2001). Influence of edible coatings on osmotic treatment of apples. In: *Osmotic Dehydration and Vacuum Impregnation: Applications in Food Industries* (P. Fito, A. Chiralt, J. M. Barat, W. E. L. Spiess and D. Behsnilian, eds), pp. 43–49. Technomic Publishing, Lancaster, PA.

Debeaufort, F., Quezada-Gallo, J. A. and Voilley, A. (1998). Edible films and coatings: tomorrow's packaging: a review. *Crit. Rev. Food Sci. Nutr.* **38(4),** 299–313.

Donhowe, I. G. and Fennema, O. (1994). Edible films and coatings: characteristics, formation, definitions, and testing methods. In: *Edible Coatings and Films to Improve*

Food Quality (J. M. Krochta, E. A. Baldwin and M. Nisperos-Carriedo, eds), pp. 1–24. Technomic Publishing, Lancaster, PA.

Gennadios, A. (2002). Soft gelation capsules. In: *Protein-based Films and Coatings* (A. Gennadios, ed.), pp. 393–443. CRC Press, Boca Raton, FL.

Gennadios, A. and Weller, C. L. (1990). Edible films and coatings from wheat and corn proteins. *Food Technol.* **44(10),** 63–69.

Gennadios, A., McHugh, T. H., Weller, C. L. and Krochta, J. M. (1994). Edible coatings and film based proteins. In: *Edible Coatings and Films to Improve Food Quality* (J. M. Krochta, E. A. Baldwin and M. Nisperos-Carriedo, eds), pp. 201–277. Technomic Publishing, Lancaster, PA.

Gennadios, A., Ghorpade, V. M., Weller, C. L. and Hanna, M. A. (1996). Heat curing of soy protein films. *Trans. ASAE* **39(2),** 575–579.

Gennadios, A., Hanna, M. A. and Kurth, L. B. (1997). Application of edible coatings on meats, poultry and seafoods: a review. *Lebensm. Wiss. u. Technol.* **30(4),** 337–350.

Gennadios, A., Rhim, J. W., Handa, A., Weller, C. L. and Hanna, M. A. (1998). Ultraviolet radiation affects physical and molecular properties of soy protein films. *J. Food Sci.* **63(2),** 225–228.

Good, R. J. (1993). Contact angle, wetting, and adhesion: a critical review. In: *Contact Angle, Wettability and Adhesion* (K. L. Mittal, ed.), pp. 3–36. VSP, Utrecht, The Netherlands.

Guilbert, S. (2002). Edible and biodegradable coating/film systems. In: *Active Food Packaging* (J. H. Han, ed.), pp. 4–10. SCI Publication and Communication Services, Winnipeg, Canada.

Guilbert, S. and Gontard, N. (1995). Edible and biodegradable food packaging. In: *Foods and Packaging Materials – Chemical Interactions* (P. Ackermann, M. Jägerstad and T. Ohlsson, eds), pp. 159–168. The Royal Society of Chemistry, Cambridge, England.

Guilbert, S., Gontard, N. and Gorris, L. G. M. (1996). Prolongation of the shelf life of perishable food products using biodegradable films and coatings. *Lebensm. Wiss. u. Technol.* **29,** 10–17.

Guilbert, S., Cuq, B. and Gontard, N. (1997). Recent innovations in edible and/or biodegradable packaging materials. *Food Add. Contam.* **14(6–7),** 741–751.

Han, J. H. (2000). Antimicrobial food packaging. *Food Technol.* **54(3),** 56–65.

Han, J. H. (2001). Design of edible and biodegradable films/coatings containing active ingredients. In: *Active Biopolymer Films and Coatings for Food and Biotechnological Uses* (H. J. Park, R. F. Testin, M. S. Chinnan and J. W. Park, eds), pp. 187–198. Proceedings of Pre-congress Short Course of IUFoST, Seoul, Korea. 10th IUFoST World Congress Organization.

Han, J. H. (2002). Protein-based edible films and coatings carrying antimicrobial agents. In: *Protein-based Films and Coatings* (A. Gennadios, ed.), pp. 485–499. CRC Press, Boca Raton, FL.

Han, J. H. (2003). Antimicrobial food packaging. In: *Novel Food Packaging Techniques* (R. Ahvenainen, ed.), pp. 50–70. Woodhead Publishing Ltd., Cambridge, UK.

Han, J. H. and Krochta, J. M. (1999). Wetting properties and water vapor permeation of whey-protein-coated-paper. *Trans. ASAE* **42(5),** 1375–1382.

Han, J. H. and Krochta, J. M. (2001). Physical properties and oil absorption of whey-protein coated paper. *J. Food Sci.* **66(2),** 294–299.

Hong, S. I., Han, J. H. and Krochta, J. M. (2003). Optimal and surface properties of whey protein isolate coatings on plastic films as influenced by substrate, protein concentration, and plasticizer types. *J. Appl. Polym. Sci.* **92(1),** 335–348.

Howard, L. R. and Gonzales, A. R. (2001). Food safety and produce operations: what is the future? *Hort. Sci.* **36(1),** 33–39.

Kester, J. J. and Fennema, O. R. (1986). Edible films and coatings: a review. *Food Technol.* **48(12),** 47–59.

Krochta, J. M. (1997). Edible protein films and coatings. In: *Food Proteins and Their Applications* (S. Damodaran and A. Paraf, eds), pp. 529–549. Marcel Dekker, New York, NY.

Krochta, J. M. (2002). Proteins as raw materials for films and coatings: definitions, current status, and opportunities. In: *Protein-based Films and Coatings* (A. Gennadios, ed.), pp. 1–41. CRC Press, Boca Raton, FL.

Krochta, J. M. and De Mulder-Johnston, C. (1997). Edible and biodegradable polymer films: challenges and opportunities. *Food Technol.* **51(2),** 61–74.

Lacroix, M. and Ouattara, B. (2000). Combined industrial processes with irradiation to assure innocuity and preservation of food products – a review. *Food Res. Intl* **33(2),** 719–724.

Maté, J. I. and Krochta, J. M. (1998). Oxygen uptake model for uncoated and coated peanuts. *J. Food Eng.* **35,** 299–312.

McHugh, T. H. and Krochta, J. M. (1994). Permeability properties of edible films. In: *Edible Coatings and Films to Improve Food Quality* (J. M. Krochta, E. A. Baldwin and M. Nisperos-Carriedo, eds), pp. 139–187. Technomic Publishing, Lancaster, PA.

Mehyar, G. F. and Han, J. H. (2003). Physical and mechanical properties of edible films made from high-amylose rice and pea starches. In: *2003 IFT Annual Meeting: Book of Abstracts*, p. 111. Institute of Food Technologists, Chicago, IL.

Micard, V., Belamri, R., Morel, M. H. and Guilbert, S. (2000). Properties of chemically and physically treated wheat gluten films. *J. Agric. Food Chem.* **48,** 2948–2953.

Miller, K. S., Chiang, M. T. and Krochta, J. M. (1997). Heat curing of whey protein films. *J. Food Sci.* **62(6),** 1189–1193.

Miller, K. S., Upadhyaya, S. K. and Krochta, J. M. (1998). Permeability of d-limonene in whey protein films. *J. Food Sci.* **63(2),** 244–247.

Nussinovitch, A. (2003). *Water Soluble Polymer Applications in Foods*, pp. 29–69. Blackwell Science, Oxford, UK.

Park, H. J. (2000). Development of advanced edible coatings for fruits. *Trends Food Sci. Technol.* **10,** 254–260.

Park, S. K., Hettiarachchy, N. S., Ju, Z. Y. and Gennadios, A. (2002). Formation and properties of soy protein films and coatings. In: *Protein-based Films and Coatings* (A. Gennadios, ed.), pp. 123–137. CRC Press, Boca Raton, FL.

Pérez-Gago, M. B. and Krochta, J. M. (2002). Formation and properties of whey protein films and coatings. In: *Protein-based Films and Coatings* (A. Gennadios, ed.), pp. 159–180. CRC Press, Boca Raton, FL.

Petersen, K., Nielsen, P. V., Bertelsen, G. *et al.* (1999). Potential of biobased materials for food packaging. *Trends Food Sci. Technol.* **10,** 52–68.

Peyron, A. (1991). L'enrobage et les produits filmogenes: un nouveau mode démballage. *Viandes Prod. Cares* **12(2),** 41–46.

Rhim, J. M. (1998). Modification of soy protein film by formaldehyde. *Korean J. Food Sci. Technol.* **30(2),** 372–378.

Sensidoni, A. and Peressini, D. (1997). Edible films: potential innovation for fish products. *Industrie Alimentari* **36(356),** 129–133.

Sothornvit, R. and Krochta, J. M. (2000). Plasticizer effect on oxygen permeability of β-lactoglobulin films. *J. Agric. Food Chem.* **48,** 6298–6302.

Sothornvit, R. and Krochta, J. M. (2001). Plasticizer effect on mechanical properties of β-lactoglobulin films. *J. Food Eng.* **50,** 149–155.

Were, L., Hettiarachcky, N. S. and Coleman, M. (1999). Properties of cysteine-added soy protein-wheat gluten films. *J. Food Sci.* **64(3),** 514–518.

Wu, L. C. and Bates, R. P. (1972). Soy protein-lipid films. 1. Studies on the film formation phenomenon. *J. Food Sci.* **37(1),** 36–39.

Wu, P. J. and Schwartzberg, H. G. (1992). Popping behavior and zein coating of popcorn. *Cereal Chem.* **69(5),** 567–573.

Wu, Y., Weller, C. L., Hamouz, F., Cuppett, S. L. and Schnepf, M. (2002). Development and application of multicomponent edible coatings and films: a review. *Adv. Food Nutr. Res.* **44,** 347–394.

Yildirim, M. and Hettiarachchy, N. S. (1997). Biopolymers produced by cross-linking soybean 11S globulin with whey proteins using transglutaminase. *J. Food Sci.* **62(2),** 270–275.

Agro-polymers for edible and biodegradable films: review of agricultural polymeric materials, physical and mechanical characteristics

Stéphane Guilbert and Nathalie Gontard

Introduction

Various polymers obtained from products or by-products of agricultural origin are proposed for the formulation of biodegradable materials or edible films. These polymers (polysaccharides, proteins and lipids, or polyesters) can be used in various forms

Innovations in Food Packaging
ISBN: 0-12-311632-5

(coatings, simple or multi-layer films, three-dimensional items, simple materials, mixtures, blends, and composites). The materials obtained from these agro-polymers are fully renewable and biodegradable (except when severe chemical modifications are applied). They are non-toxic to the soil and the environment, and when food-grade ingredients are used to formulate the materials, they can be edible.

In addition, edible and biodegradable coatings or films produced from agro-polymers provide a supplementary and sometimes essential means of controlling physiological, microbiological, and physicochemical changes in the food products. This is achieved by controlling mass transfers between food products and ambient atmospheres, or between the components in heterogeneous food products. Furthermore, due to some original material properties, they can also be used as an "active bio-packaging" to modify and control food surface conditions (for example, gas-selective materials, controlled release of specific functional agents, of flavor compounds, etc.).

Agro-polymers

The formulation of bio-plastic or edible films implies the use of at least one component able to form a matrix, having sufficient cohesion and continuity. They are polymers which, under preparation conditions, have the property to form crystalline or amorphous continuous structures. Only polyesters, polysaccharides or proteins are used for making "materials". Natural lipid compounds are also used, primarily in the field of edible film and coating applications. Generally they are applied in thin layers, or as a composite with a polymeric matrix.

Three different techniques using agricultural raw materials (fully renewable raw materials) are possible to make agro-polymers (Guilbert, 1999):

1. Agricultural polymers (polysaccharides or proteins) can be extracted and eventually purified. They can be used alone or in a mixture with synthetic biodegradable polymers such as polycaprolactone or other synthetic biodegradable polyesters.
2. Agricultural products can be used as fermentation substrates to produce microbial polymers (*e.g.* polyhydroxyalcanoates).
3. Agricultural products (or by-products) can be used as fermentation substrates to produce monomers or oligomers which will be polymerized by conventional chemical processes (*e.g.* polylactic acid obtained by polymerization of natural lactic acid produced by fermentation of corn).

Edible films or coatings

Agro-polymers that have been proposed to formulate edible films or coatings are numerous (Cuq *et al.*, 1995; Guilbert and Cuq, 1998). Polysaccharide, protein or lipid materials are used in various forms (simple or composite materials, single-layer or multi-layer films).

Polysaccharides used for material formulations are generally the same ones as those used as stabilizing, thickening, and gelling agents. These polysaccharides are of various origin: plant polysaccharides such as cellulose and derivatives; starches and derivatives; pectin or arabinoxylanes; algae gums such as alginates or carrageenans; and microbial gums, pullulan, xanthan and gellan.

Many plant and animal proteins have been studied as raw materials for films and coatings. These proteins are generally characterized by having interesting functional properties (Guilbert and Cuq, 2002).

Lipids and derivatives are used due to their good water-barrier properties. The use of lipids (or derivatives) with a polysaccharide or protein-based matrix or support is generally advised.

A few examples of applications of edible films or coatings, used to improve product appearance or conservation, include sugar and chocolate coatings for candies or icecreams, wax coatings for fresh fruits, and oil or fat coatings for raisins and dry fruits. Edible films become a part of the whole food product; their composition is dependent on the product nature, and must conform to regulations that apply to the respective food product. Therefore, edible film and coating technologies are regarded as a "problem" for food formulations (Kester and Fennema, 1986; Krochta *et al.*, 1994; Guilbert and Cuq, 1998).

Biodegradable materials

As far as biodegradable materials are concerned, starch is the most commonly used agricultural raw material. Starch is inexpensive, widely available, and relatively easy to handle. "All-starch" bio-plastics are made from thermoplastic starches, which are produced by standard techniques of synthetic polymer films, such as extrusion or injection molding (Colonna, 1992). The use of thermoplastic proteins has also been investigated (Gontard and Guilbert, 1994; Guilbert and Cuq, 2002), but its commercial applications are still expected. Among the proteins, milk proteins (casein, whey proteins), soy proteins and cereal proteins (wheat gluten, zein) have been more extensively studied (Genadios *et al.*, 1994; Redl *et al.*, 1996; Guilbert *et al.*, 2001).

Materials based on hydrocolloids constitute effective oil or fat barriers, but they are generally not very resistant to water and their moisture-barrier properties are poor. However, in some cases, water solubility or the sensitivity to water is a functional advantage – for example, in the formulation of soluble sachets of edible films or coatings (less perceptible in the mouth) or in the formulation of "active" materials, where the water swelling is used to induce a drastic change in the properties. However, in general, improving water resistance and water-barrier properties is of most importance. Therefore, chemical modifications of the biopolymers and the development of specific additives (cross-linking agents or plasticizers) adapted to the polymer structures are proposed. Regarding these developments, proteins that have a "non-monotonous" complex structure and many potential functional properties are promising (Cuq *et al.*, 1998).

Commercial water-resistant starch-based bio-plastics (for non-edible applications) are produced by using fine molecular blends of biodegradable synthetic polymers and starches. These composite materials are made with gelatinized starch (up to 60–85%), hydrophilic synthetic polymers (e.g. ethylene vinyl alcohol copolymer) or hydrophobic synthetic polymers (e.g. polycaprolactone), and with compatibility agents (Fritz *et al.*, 1994). The most important starch-based material on the market is Materbi®, proposed by Novamont.

Microbial polymers (e.g. poly(3)-hydroxybutyrate-hydroxyvalerate) are excreted or stored by micro-organisms cultivated on starch hydrolysates or lipid mediums. The isolation and purification costs of these products, which are obtained from complex mixtures, can be high. Monsanto stopped the commercialization of their product "Biopol®" in 1999. Coopeazucar, in Brazil, has built new facilities for pilot plant production of these polyhydroxyalkanoates.

Polylactic and polyglycolic acids are mainly produced by chemical polymerization of lactic acid and glycolic acid, which are obtained by Lactobacillus fermentation. Commercial applications of these materials are rapidly increasing under the trademarks of Ecopla® from Cargill/Dow Chemical, or Lacea® from Mitsui.

Processing

Two general process pathways for agro-materials are distinguished:

1. The "dry process", such as thermoplastic extrusion, is based on the biopolymers' thermoplastic properties when plasticized and heated above their glass transition temperature under low water content conditions
2. The "solvent process" or "casting" is based on the drying or dispersion of the film-forming solution.

The casting process is used to form edible preformed films, or to apply coatings directly onto the products (Guilbert and Cuq, 1998). This process is generally adapted for coating seeds and foods, for making cosmetic masks or varnishes, and for making pharmaceutical capsules. Heat processing of agro-polymer based materials, using techniques usually used for synthetic thermoplastic polymers (e.g. extrusion, injection, molding etc.) is more cost effective. This process is often used for making flexible films (e.g. films for agricultural applications, packaging films, and cardboard coatings) or objects (e.g. biodegradable materials) that are sometimes reinforced with fibers (composite bio-plastics for construction, automobile parts etc.).

The material characteristics (e.g. polysaccharide, protein, polyester or lipid; plasticized or not; chemically modified or not; used alone or in combination) and the fabrication procedures (e.g. casting of a film-forming solution, thermoforming) must be adapted to each specific food product and the conditions in which it will be used (relative humidity, temperature).

Properties and applications of edible and biodegradable films

Edible and biodegradable films must meet a number of specific functional requirements (e.g. moisture barrier; solute and/or gas barrier; water or lipid solubility; color and appearance; mechanical and rheological characteristics; non-toxicity, etc.). These properties are dependent on the type of material used, its formation, and its application. Plasticizers, cross-linking agents, antimicrobials, anti-oxygen agents, texture agents etc. can be added to enhance the functional properties of the films.

The properties of edible or biodegradable films depend on the type of film-forming materials used, and especially on their structural cohesion characteristics. Cohesion depends on the structure of the polymer, its molecular length, geometry, and molecular weight distribution, and on the type and position of its lateral groups. Film properties are also linked to the film-forming conditions (e.g. type of process and process parameters). The properties of amorphous or semi-crystalline materials are seriously modified when the temperature of the compounds rises above the glass transition temperature (T_g).

The glass transition phenomenon separates materials into two domains, according to clear structural and property differences, thus dictating their processing conditions and potential applications (temperature and water resistance). Generally, fully amorphous bio-plastic applications are limited by the fact that a polymer's T_g is highly affected by the relative humidity (especially for hydrophilic polymers). Below the T_g the material is rigid, and above the T_g it becomes visco-elastic or even liquid. Below this critical threshold, only weak, non-cooperative local vibration and rotation movements are possible. Film relaxation in relation to temperature follows an Arrhenius time course. Above the T_g threshold, strong, cooperative movements of whole molecules and polymer segments can be observed.

Organoleptic properties

Edible films and coatings must have organoleptic properties that are as neutral as possible (clear, transparent, odorless, tasteless etc.) so as not to be detected when eaten. Enhancing the surface appearances (e.g. brilliance) and the tactile characteristics (e.g. reduced stickiness) can be required. Hydrocolloid-based films are generally more neutral than those formed from lipids or derivatives and waxes, which are often opaque, slippery, and waxy tasting. It is possible to obtain materials with ideal organoleptic properties, but they must also be compatible with the food's filling – for example, sugar coatings, chocolate layers (or chocolate analogs), and starch films for candies, biscuits, some cakes and icecream products (wafer coatings) etc.

Films and coatings can also help to maintain desirable concentrations of coloring, flavor, spiciness, acidity, sweetness, saltiness etc. Some commercial films, especially Japanese pullulan-based films, are available in several colors, or with spices and seasonings included. This procedure could be used to provide nutritional improvement without destroying the integrity of the food product – for example, by using edible films and coatings enriched with vitamins and various nutrients.

The optical properties of films depend on the film formulation and fabrication procedures.

Biodegradability

Biodegradation kinetics are dependent on the type of polymer used (molecular weight, structure, crystallinity) and on the additives used (e.g. plasticizers, fillers etc.). The methods used to evaluate biodegradability are generally based on Sturm's procedure (Sturm, 1973) – i.e. international standard ISO 14852 – which measures the ultimate aerobic biodegradability of materials. Another approach consists of the evaluation of the biodegradability in soil.

The agro-polymer based materials are generally fully biodegradable (apart from when some very severe chemical modifications are applied). They are non-ecotoxic. Polylactic acid is biodegradable, but at temperatures above its T_g (>45–$55°C$) – that is, in a composting medium.

It is often interesting to adjust biodegradation rates to each desired type of application. This can be achieved by applying formulation and/or chemical modifications.

Mechanical properties

Mechanical properties are also dependent on the type of polymer and the additives used. For many biodegradable polyesters, the tensile strength is often similar to that of polyethylene or PET, but the elongation is often much lower (without external plasticization). Hydrocolloids, which generally require plasticization (e.g. using low molecular weight hydrophilic molecules such as glycerol or with amphipolar molecules such as fatty acid derivatives) show lower tensile strength, but their elongation is mainly dependent on the plasticizer's content. The mechanical properties of various biodegradable, edible or conventional materials are illustrated in Figure 16.1. It is clear from this figure that biodegradable synthetic and conventional materials have very similar properties (high tensile strength and high elongation), while microbial bio-plastics and agricultural bio-plastics are either highly deformable or highly resistant, but not both simultaneously.

Water vapor transfer

Owing to their relatively low water-vapor barrier properties, agro-polymer based materials and edible films can only be used as protective barriers to limit moisture exchange for short-term applications. However, they can be of considerable interest for numerous applications where very high water vapor permeability is required, such as in the case of modified atmosphere packaging of fresh, minimally processed or fermented foods (fish, meat, fruits, vegetables, and cheeses). Biodegradable polyesters (lipids or waxes for edible applications) show better water-vapor barrier properties compared to starch or protein-based materials, but still have significantly lower water-vapor barrier properties than most conventional materials. As far as edible applications

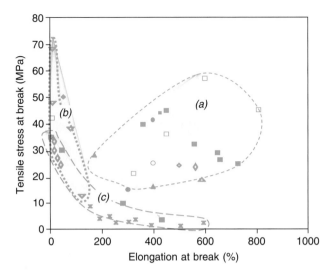

Figure 16.1 Mechanical properties of various biodegradable, edible and conventional materials (adapted from Guilbert *et al.*, 2001).
(a) *Synthetic materials (closed symbols):* thermoplastic polyurethane elastomer (Dow Chemical); Polyvinylchloride; PVC plasticized with Di-2-ethylheylphtalate; Polypropylene; low-density polyethylene and *biodegradable synthetic materials (open symbols):* Bak 1095, polyester amide (Bayer, Germany); Ecoflex: 1,4 Butandiol adipinic-dicarbonic and terephtalate copolyester (Basf, Germany); Eastar 14766: poly (tetramethylene adipate-coteréphtalate) (Eastman, USA); Bionolle 3000: polybutylene succinate/adipate (Showa, Japan).
(b) *Biodegradable materials from microbial origin:* Biopol: polyhydroxybutyrate (Monsanto, Italy); Lacea: Polylactic acid (Mitsui, Japan); Ecopla, Polylactic acid (Cargill/Dow, USA).
(c) *Biodegradable materials from agricultural origin:* Biotec: starch/polyester (Biotec, Germany); Materbi: starch/polycaprolactone (Novamont, Italy) and *edible films:* wheat gluten films; molded soy protein isolate films, alginate based films; chitosan based film; pullulan based films.

are concerned, lipid compounds such as animal and vegetable fats (natural waxes and derivatives, acetoglycerides, surfactants, etc.) are generally proposed for their excellent moisture-barrier properties, but can cause textural and organoleptical problems due to oxidation and a waxy taste.

Control of gas exchange

Agro-polymer based materials have impressive gas-barrier properties in dry conditions, especially against oxygen. For instance, the oxygen permeability of a wheat gluten film was 800 times lower than that of low-density polyethylene, and two times lower than that of polyamide 6, a well-known high oxygen-barrier polymer. Increasing the water activity promotes both the gas diffusivity (due to the increased mobility in the hydrophilic macromolecule chains) and the gas solubility (due to the water swelling in the matrix), leading to a sharp increase in the gas permeability. With carbon dioxide, a sharp increase in the permeability is more important than with the oxygen permeability. The selectivity coefficient between carbon dioxide and oxygen (Gontard *et al.*, 1996; Mujica Paz and Gontard, 1997) is very sensitive to moisture and temperature (for example, from 4 up to 35 for gluten films), whereas the selectivity coefficient for synthetic polymers remains relatively constant, at 4–6. This could be explained by the differences in the water solubility of these gases (e.g. carbon dioxide is very soluble), but also by specific interactions between carbon dioxide and the water-plasticized polymer.

Very high gas selectivity is particularly interesting for modified atmosphere packaging of cheeses (to control the proliferation of the microflora) and fresh fruit and vegetables (to control the respiration rates). Some selective gluten-based films have been shown to lead to the production of new modified atmospheres (e.g. low oxygen and carbon dioxide concentrations at equilibrium) when used to wrap fresh vegetables (Guilbert *et al.*, 1996, 1997; Barron *et al.*, 2001).

Modification of surface conditions and controlled release of additives

"Active" edible or biodegradable films can be applied to foods to modify and control the atmosphere of food surface conditions (Cuq *et al.*, 1995). Improvement in food microbial stability can be obtained using edible active layers as surface retention agents to limit the diffusion of food additives into the food core, or they can be used in combination with treatments such as refrigeration and a controlled atmosphere. Maintaining a high concentration of an effective preservative in a local area may allow a reduction in the total amount of the preservative while sustaining the same effectiveness. Improvement of food microbial stability can also be obtained by reducing the surface pH. This can be achieved using films that immobilize specific acids or charged macromolecules (Torres and Karel, 1985). It is important to be able to predict and control preservative release (Torres *et al.*, 1985; Redl *et al.*, 1996). Microbiological analysis generally confirms the efficiency of the preservative retention within surface coatings that are able to increase the shelf life of intermediate moisture foods (Torres and Karel, 1985; Guilbert, 1988).

Applications of agro-polymer based materials

Applications of biodegradable plastics

Biodegradable plastics can be classified into three categories (Guilbert, 1999):

1. Plastics to be composted or recycled (i.e. where reuse or fine recovery is difficult)
2. Plastics to be used in the natural environment (i.e. where recovery is not economically or practically feasible)
3. Specialty plastics (with specific features where bio-plastics possess preferential properties).

Since agro-polymers are relatively costly to produce, actual applications are limited to special niches with environmental considerations. Loose-fill packaging and compost bags are the two major end uses, constituting nearly 90% of the demand in 1996 (Bohlmann and Takei, 1998). However, new applications in agriculture and food packaging are emerging. In the domain of plastics for agriculture, the demand for biodegradable mulching plastics is increasing because biodegradability is a real functional advantage. For food packaging products such as plastic films, nets and small containers or trays are of concern, even if they are more expensive than the traditional products. The principal applications of agro-polymers are listed in Table 16.1.

Table 16.1 Applications of agro-polymers (adapted from Guilbert, 1999)

Agro-polymers	Applications
Plastics to be composted or recycled	
Loose-fill packaging	Shock absorbers
Waste and carrier bags	Compost bags
Dishes and cutlery	Trays, spoons, cups
Hygiene disposables	Diapers, sanitary napkins, sticks for cotton swabs, razors, toothbrushes
Miscellaneous short-life goods	Pens, tees, toys, gadgets, keyrings
Food packaging	Dried foods, short lifecycle food packaging (e.g. foast-food packaging, egg boxes
	Fresh or minimally processed fruits and vegetables, dairy products, organically grown products
Paper or cardboard	Accessories or windows for paper envelopes or cardboard packaging, coating for paper or cardboard
Plastics used in natural environments (no recovery)	
Biodegradable/soluble/controlled-release materials for agriculture and fisheries	Mulching plastic, films for banana culture, twine, nursery pots
	Materials for controlled-release fertilizers or agrochemicals
	High water-retention materials for planting
	Soluble sachets, biodegradable containers for fertilizers or agrochemicals
	Fishing lines and nets
Civil engineering, car industry and construction materials	Heat insulators, noise insulators, form wares
	Car interior door casings
	Retaining walls or bags for mountain areas or sea, protective sheets and nets for tree planting
Disposable leisure goods	Golf tees, miscellaneous goods for marine or mountain sports
Speciality ingredients or materials	
Functionalized nanoparticles	Nanoparticles for rubber reinforcement
Medical goods	Bone fixation, sutures, films, non-woven tissues
Edible films/coatings	Barrier internal layers, surface coatings, "active" superficial layers
	Soluble sachets for instant dry food and beverages, for food additives
	Component of simple or multi-layer packaging
O_2 barrier, selective O_2/CO_2 barrier, aroma barrier	
Matrix for controlled-release systems	Slow release of fertilizers, agrochemicals, pharmaceuticals, food additives
Super-absorbents	
Adhesives	
Paints	
Dyes and pigments	

Application of edible films

Edible films and coatings provide an easy means of structurally strengthening certain foods, reducing particle clustering and improving visual and tactile features on product surfaces (Cuq et al., 1995). Edible films can also be used to package components or additives that are to be dissolved in hot water or food mixes, and also act as an additional parameter for improving overall food quality and stability. They represent one way to apply hurdle technology to solid foods without affecting their structural integrity (Guilbert, 1986; Guilbert et al., 1997). Many functions of edible films are the same as those of synthetic packaging, but they must be chosen according to a specific application and type of food product, and its main deterioration mechanisms.

Films that have a significant gas- and moisture-barrier properties are required for many applications. For example, they are used to control gas exchange for fresh or oxidable foods, and to reduce moisture exchange with the external atmosphere, etc. The retention of specific additives in edible films can lead to a functional response. This is generally confined to the surface of the product, where they are used to modify and control "surface conditions". Edible coatings can also limit oil and solute penetration into foods during processing. Figure 16.2 is a schematic representation of a

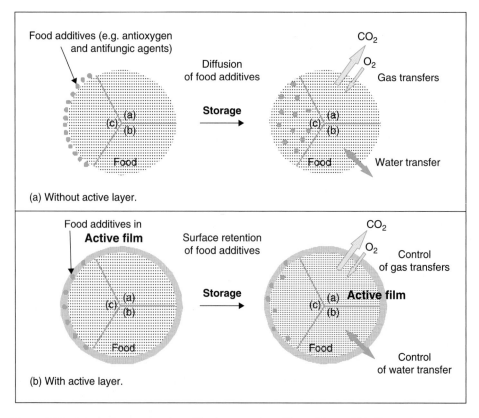

Figure 16.2 Schematic representation of food preservation with or without edible films and coatings as active layers, when the first mode of deterioration results from respiration (a), from dehydration or moisture uptake (b), or from surface microbial development or oxidation (c), adapted from Cuq et al. (1995).

food preservation method using agro-polymers as edible/biodegradable active layers when the food product undergoes three types of deterioration, due to (a) respiration, (b) dehydration or moisture uptake, or (c) surface microbial development or oxidation. The protective feature of the film is dependent on gas- and water-vapor barrier properties, on surface condition modifications, and on its own antimicrobial properties.

Market opportunities

Edible films and coating are generally not commercialized as preformed self-supporting materials, but as pre-mix ingredients. This makes it very difficult to estimate the demand in the market, especially when this is not significant compared to demand for other specialty ingredients and additives. Companies in this type of market are usually suppliers of cellulose and derivatives, gums, food-texturing agents, food-grade surfactants, waxes, and lipid derivatives. Recently, lipid materials specifically designed to be used as edible water barriers have been proposed for commercial use. Most of them are composed of specifically designed acetoglycerides, either alone or mixed with waxes.

The market for biodegradable plastic materials is currently expanding, but it remains a niche market. The total world production capacity of "modern thermoplastic bio-plastics" is approximately 300 000 tons (estimation for 2003). However, the 2003 world consumption is estimated to be only between 70 000 and 90 000 tons.

Eighty-seven percent of biodegradable materials will result from renewable resources. This market still has great diversity in producers and materials, but attempts are being made to decrease this variety. The major companies involved are Cargill Dow, Novamont, Basf, Eastman, Mitsui, and Solvay. These corporations have a double function, acting both as purchasers and as suppliers.

The total target market is 500 000–900 000 tons in 2005–2007. Today, these film applications account for about 30% of the sales, and this percentage is increasing, with estimates in the range of 30–50% by 2010. They are not meant to replace the traditional plastics. The applications that require biodegradability as a functional characteristic currently account for about 75% of sales, and it is expected that this proportion will decrease to 60–50% in the future. Currently, the largest market share for bio-plastics is provided by blends of thermoplastic starch and synthetic biodegradable polyesters, representing about 65% of the total market. They contain an average starch content of about 60–70%. Novamont is the leader in this market, with the Materbi® products. Polylactic acid is the other major bio-plastic produced commercially at the moment (mainly by Cargill Dow with Ecopla®).

The industrial future for biodegradable plastics is still under discussion, owing to numerous problems. One of the main obstacles to worldwide use of bio-plastics is the cost of commercially available products (which is similar to that of speciality plastics, e.g. 1.5 to 10 euro/kg). Bio-plastics are still unknown to the processing industry, and their performances (e.g. the water sensitivity of agricultural-based plastics) and processibility often remain a problem. In terms of the marketing point of view, the major hurdles are the absence of well-identified demands, the lack of eligibility of the offer, the risk in terms of the image, consumer perception, and the complexity of the whole system. Other major obstacles are the absence of a recovery chain, and of the infrastructure of nationwide disposal in most developed countries.

New legislations in Western Europe and in Japan, which are in favor of using biodegradable plastics, has helped to increase the demand, but developing international standards for biodegradable polymers is difficult and the competition regarding the recycling of plastics is still an obstacle. In addition to being biodegradable, the plastics must now also be non-ecotoxic. However, with the development of a new range of functionalities and the benefits to the environment, topics that are in line with international objectives, a dynamic market is predicted.

Conclusion

After a long period of latency, biodegradable plastics have now become credible. Major polymer manufacturers are entering the market, material costs are quickly falling, and performances and processibility are improving significantly. Important niche markets are opening due to consumer demands, which are considered important. In addition, new legislations and standards in favor of the use of biodegradable plastic are helping to encourage market growth. More significant markets are expected to be located in Western Europe and in Japan.

Among the different categories of biodegradable plastics obtained from agro-polymers, the starch/polyester blends and the "microbial" biodegradable plastics satisfy the majority of the requirements proposed by plastic packaging industries (material qualities, processibility, performance, etc.). Interest has been shown in other bio-plastics based on natural polysaccharides or proteins, due to their low costs; however, their non-reproductive quality and their lower performance are still problems.

Edible films and coatings provide an added and at times necessary means to control physiological, microbiological, and physicochemical changes in food products. Recently, the concept of "edible active layers" (Cuq *et al.*, 1995) has been introduced. These films or coatings contribute to food preservation, for instance by controlling mass transfer of water vapor, oxygen, carbon dioxide, ethylene, etc., or by modifying and controlling the food surface conditions (e.g. pH, level of functional agents, slow release of additives etc.).

Edible and/or biodegradable packages formed with several compounds (composite or complex materials) have been developed to take advantage of the complementary functional properties of the different constituent materials, and to overcome their respective drawbacks. Most composite films studied to date combine one or several polyester (or lipid) compounds with a hydrocolloid-based one. The future of edible and/or biodegradable materials is therefore probably dependent on the development of applications where some of their preferential properties (such as gas selectivity) are enhanced, or on the development of composite materials.

References

Barron, C., Varoquaux, P., Guilbert, S., Gontard, N. and Gouble, B. (2001). Modified atmosphere of cultivated mushroom (*Agaricus Bisporus L.*) with hydrophilic films. *J. Food Sci.* **67(1),** 251–255.

Bohlmann and Takei (1998). CEH marketing research report on biodegradable polymers, plastics and tesins. *Chemical Economics Handbook*, 580.0280 A. SRI International.

Colonna, P. (1992). Biodégradabilité et bioassimilabilité des matériaux d'emballage. In: *Conditionnement Alimentaire: Innovation et Environnement* (ISECA, ed.), pp. 203–218. ISECA, Pouzauges, France.

Cuq, B., Gontard, N. and Guilbert, S. (1995). Edible films and coatings as active layers. In: *Active Food Packagings* (M. L. Rooney, ed.), pp. 111–142. Blackie, Glasgow, UK.

Cuq, B., Gontard, N. and Guilbert, S. (1998). Proteins as agro-polymer for packaging production. *Cereal Chem.* **75(1),** 1–9.

Fritz, H. G., Seidenstücker, T., Bölz, U., Juza, M., Schroeter, J. and Gendres, H. J. (1994). *Study on Production of Thermoplastics and Fibers Based Mainly on Biological Materials.* Science Research Development, European Commission, EUR 16102 EN.

Gennadios, A., McHugh, T. H., Weller, C. L. and Krochta, J. M. (1994). Edible coatings and films based on proteins. In: *Edible Coatings and Films to Improve Food Quality* (J. M. Krochta, E. A. Baldwin and M. O. Nisperos-Carriedo, eds), pp. 201–278. Technomic Publishing, Lancaster, PA.

Gontard, N. and Guilbert, S. (1994). Bio-packaging: technology and properties of edible and/or biodegradable material of agricultural origin. In: *Food Packaging and Preservation* (M. Mathlouthi, ed.), pp. 159–181. Blackie, Glasgow, UK.

Gontard, N., Thibault, R., Cuq, B. and Guilbert, S. (1996). Influence of relative humidity and film composition on oxygen and carbon dioxide permeabilities of edible films. *J. Agric. Food Chem.* **44(4),** 1064–1069.

Guilbert, S. (1986). Technology and application of edible protective films. In: *Food Packaging and Preservation* (M. Mathlouthi, ed.), pp. 371–394. Elsevier Applied Science Publishers, New York, NY.

Guilbert, S. (1988). Use of superficial edible layer to protect intermediate moisture foods: application to the protection of tropical fruit dehydrated by osmosis. In: *Food Preservation by Moisture Control* (C. C. Seow, T. T. Teng and C. H. Quah, eds), pp. 199–219. Elsevier Applied Science Publishers, London, UK.

Guilbert, S. (1999). Biomaterials for food packaging: applications and future prospects. In: *Trends in Food Engineering* (G. Barbosa, ed.). Aspen, New York, NY (in press).

Guilbert, S. and Cuq, B. (1998). Les films et enrobages comestibles. In: *L'emballage des Denrées Alimentaires de Grande Consommation*, pp. 471–530, Technique et Documentation, Lavoisier, Paris, France.

Guilbert, S. and Cuq, B. (2002). Protein as raw material for biodegradable products. In: *Handbook of Biodegradable Polymers* (C. Bastioli, ed.) Rapra Tech., London, UK.

Guilbert, S., Gontard, N. and Gorris, L. G. M. (1996). Prolongation of the shelf-life of perishable food products using biodegradable films and coatings. *Lebensm. Wiss. u. Technol.* **29,** 10–17.

Guilbert, S., Cuq, B. and Gontard, N. (1997). Recent innovations in edible and/or biodegradable packagings. *Food Add. Contam.* **14(6–7),** 741–751.

Guilbert, S., Gontard, N., Morel, M. H., Chalier, P., Micard, V. and Redl, A. (2001). Formation and properties of wheat gluten films and coatings. In: *Protein-based Films and Coatings* (A. Gennadios, ed.), pp. 69–122. CRC Press, Boca Raton, FL.

Kester, J. J. and Fennema, O. (1986). Edible films and coatings: a review. *Food Technol.* **40,** 47–59.

Krochta, J. M., Baldwin, E. A. and Nisperos-Carriedo, M. (1994). *Edible Films and Coatings to Improve Food Quality.* Technomic Publishing, Lancaster, PA.

Mujica Paz, H. and Gontard, N. (1997). Oxygen and carbon dioxide permeability of wheat gluten film: effect of relative humidity and temperature. *J. Agric. Food Chem.* **45(10),** 4101–4105.

Redl, A., Gontard, N. and Guilbert, S. (1996). Determination of sorbic acid diffusivity in edible wheat gluten and lipid based films. *J. Food Sci.* **61,** 116–120.

Sturm, R. N. (1973). Biodegradability of nonionic surfactants: screening test for predicting rate and ultimate biodegradation. *J. Oil Chem. Soc.* **50,** 159–167.

Torres, J. A. and Karel, M. (1985). Microbial stabilization of intermediate moisture food surfaces. III. Effects of surface preservative concentration and surface pH control on microbial stability of an intermediate moisture cheese analog. *J. Food Process. Preserv.* **9,** 107–119.

Torres, J. A., Bouzas, J. O. and Karel, M. (1985). Microbial stabilization of intermediate moisture food surfaces. II. Control of surface pH. *J. Food Process. Preserv.* **9,** 93–106.

Edible films and coatings from plant origin proteins

Roberto A. Buffo and Jung H. Han

Introduction

The 1990s saw a remarkable increase in research efforts for the development of biopolymer films and coatings from protein, polysaccharide, and lipid materials. The qualities of renewability, degradability, and edibility make such films particularly appealing for food and non-food packaging applications. Moreover, wide commercialization of biopolymer films would provide a value-added innovative use for traditional agricultural commodities as sources of film-forming materials. This chapter discusses the film formation ability and associated functional properties of a number of plant proteins, namely those extractable from corn, soybean, wheat, cottonseed, and other crops.

Zein films and coatings

Zein comprises a group of prolamins found in corn endosperm. It accounts for 50% or more of total endosperm protein, occurring in small and compact bodies embedded in the glutelin protein matrix and distributed mainly in the outer layers of the endosperm (Dombrink-Kurtzman and Bietz, 1993).

Innovations in Food Packaging
ISBN: 0-12-311632-5

Commercial zein is a by-product of the corn wet-milling industry. After cleaning, corn is steeped in water containing sulfur dioxide to soften the grains. Coarse milling separates the hulls and germ from the endosperm, which is further milled into a fine slurry. Centrifugal separation of starch from the endosperm slurry leaves a protein-rich mass, or corn gluten meal, from which zein is extracted with aqueous alcohol and dried to a granular powder (Padua and Wang, 2002).

Zein occurs as aggregates linked by disulfide bonds in whole corn. Those bonds may be cleaved by reducing agents during extraction or wet-milling operations (Reiners *et al.*, 1973). Based on solubility differences, zein consists of three protein fractions, α-zein, β-zein and γ-zein, α-zein being predominant (75–85% of total zein) (Esen, 1987).

Coatings

Zein is one of a few proteins, such as collagen and gelatin, used commercially as an edible coating. Zein coatings are used as oxygen, lipid, and moisture barriers for nuts, candies, confectionery, and other foods (Andres, 1984), and for controlled ingredient release and protection in pharmaceutical tablets (Gennadios and Weller, 1990).

Zein coatings are prepared in three steps. Zein powder is dissolved in warm, aqueous ethyl alcohol or isopropanol. Approved plasticizers such as propylene glycol or glycerin are added to increase coating flexibility. The products to be coated are dipped into, sprayed with or brushed with the zein–plasticizer solution. Antioxidants such as BHT or BHA are added to prevent lipid oxidation, and vegetable oils to enhance film shine. The coating is formed on the product surface upon solvent evaporation (Andres, 1984; Gennadios and Weller, 1990; Park *et al.*, 1994a, 1994b; Wong *et al.*, 1996).

Thickness is a critical property of coatings that becomes a quality-control bottleneck in manufacturing, as it decisively influences physical and barrier properties. To measure thickness, a direct observation of a thin section of the coated product under a light microscope is preferred over an indirect estimation (Park *et al.*, 1994b).

Zein coatings have oxygen and carbon dioxide transmission rates that are lower by one or two orders of magnitude than those of low-density polyethylene (LDPE), methylcellulose (MC), and hydroxypropyl cellulose (HPC) films, and similar to those of polyester films at the same relative humidity (RH) and testing temperature (Krochta, 1992). Water vapor permeability (WVP) is satisfactory at low relative humidity, but increases rapidly as the RH increases. The importance of zein coatings in vegetables to enhance their shelf life is the way this limits their exposure to ambient oxygen and increases the internal carbon dioxide concentration (Park *et al.*, 1994a, 1994b).

Films

Zein films are easily cast from alcohol solutions. Zein is dissolved in warm (65–85°C) aqueous ethanol or isopropanol with added plasticizers. The solution is cooled to 40–50°C, allowing bubbling to cease prior to casting, and is then poured over a glass plate where a film is formed as the alcohol evaporates from the surface. The dried film is finally peeled off the plate (Gennadios *et al.*, 1993a).

Plasticizers are added to reduce film brittleness. They decrease the intermolecular forces between polymer chains, thus increasing flexibility and extendibility. The most common plasticizer for zein films is glycerol. However, it tends to migrate from the bulk of the film matrix to the food surface because its interaction with protein molecules is weak (Park *et al.*, 1994c). Cast zein films are initially transparent, but the glycerol "sweating out" causes a characteristic cloudiness, with a consequent loss in film flexibility (Park *et al.*, 1992). Mixtures of glycerol and polyethylene glycol (PEG) present slower migration rates of plasticizers than glycerol alone, but neither is effective at temperatures below freezing or in low relative humidity environments. Unsaturated fatty acids remain effective plasticizers in those conditions, but then lipid oxidation becomes an issue (Park *et al.*, 1994d; Lai and Padua, 1998).

The tensile properties of zein films are generally evaluated by two parameters; tensile strength (TS) and elongation at break (E). The TS of zein films is similar to that of wheat gluten films, and two- to threefold lower than that of MC and HPC films (Gennadios *et al.*, 1993a). In general, adding plasticizers decreases TS and increases E (Park *et al.*, 1994c). The TS of zein films varies substantially with RH and temperature conditioning, decreasing linearly with increase in RH and quadratically with temperature increase (Gennadios *et al.*, 1993a). TS increases up to twofold with the addition of cross-linking agents such as formaldehyde, glutaraldehyde or citric acid to the alcoholic film-forming solution (Yang *et al.*, 1996; Parris and Coffin, 1997).

Zein films have WVP values lower than or similar to those of other protein films, cellulose ethers, and cellophane (Krochta, 1992), but much higher than that of LPDE (Smith, 1986). Addition of plasticizers increases the WVP because of greater mobility of chain segments and subsequent greater diffusion coefficients of permeants (Seymour and Carraher, 1984). The WVP also increases with RH due to moisture plasticization and increasing vapor pressure differences between the two sides of the film (Lai and Padua, 1998). However, the low vapor-barrier ability of zein films could be considered an advantage, as it would allow excessive water vapor movement across the film, thus preventing water condensation inside the package leading to microbial spoilage (Ben-Yehoshua, 1985). Low WVP films require either treatment with cross-linking agents such as polymeric dialdehyde starch (PDS) (Parris and Coffin, 1997), or preparation of bilayer films (i.e. pouring, spraying or rolling a second film directly on top of a previously prepared one). Zein/soy protein and zein/wheat gluten films have lower permeability than single zein films (Foulk and Bunn, 1994).

The oxygen and carbon dioxide permeabilities of zein films are lower than those of other polysaccharide and composite polysaccharide/lipid films (Greener and Fennema, 1989) but higher than those of wheat gluten films (Gennadios *et al.*, 1993b). Apparently, oxygen molecules can more readily permeate through the helical conformation of zein than through the highly cross-linked gluten structure. The oxygen permeability (OP) of zein films is also lower than that of common plastic films such as LPDE, propylene, polystyrene and polyvinyl chloride (Billing, 1989). The addition of plasticizers and increase in RH raise the OP (Yamada *et al.*, 1995) because moisture causes swelling of hydrophilic polymers, resulting in higher gas permeability (Ashley, 1985). Temperature affects the OP according to the characteristic Arrhenius activation energy model (Gennadios *et al.*, 1993b).

Zein may be plasticized in solution with long chain fatty acids, and collected as a soft solid mass after precipitation with cold water. Zein films may be drawn from wet, freshly precipitated resins after a kneading step, or formed by extrusion of dry resin pellets. Dry films are translucent, flexible, ductile, and heat-sealable. Under the microscope, they show a fiber network structure with pinholes and structural gaps (Padua and Wang, 2002). Fusion lamination, performed by heating several sheets together under pressure, produces transparent zein films that are more ductile and pliable than the original ones, filling pinholes and gaps in the structure and thus reducing macroscopic defects and increasing film uniformity (Rakotonirainy and Padua, 1999).

The TS and E values of drawn zein films are similar to those of cast films (Lai and Padua, 1997). Ductility increases at high RH due to the plasticization by moisture, resulting in higher E (Lai and Padua, 1998). Tensile properties of zein/fatty acid films are affected by fatty acid content. For example, increasing oleic acid from 0.5 to 1.0 g/g zein decreases the TS and increases E (Santosa and Padua, 1999).

Drawn zein films have a lower equilibrium moisture content (7 g water/100 g solids) than cast zein films of identical composition (12 g water/100 g solids) at a water activity value of 0.95, owing to more extensive zein–fatty acid interactions, which cause a reduction on zein exposure to water (Lai and Padua, 1998). The WVP is relatively low compared to that of hydrophilic biopolymers such as wheat gluten (Park and Chinnan, 1990) and cellophane (Taylor, 1986), but is at least one order of magnitude higher than that of LPDE films (Smith, 1986). Expectedly, WVP increases at higher RH and temperatures (Lai and Padua, 1998). Coating drawn-zein films with tung or linseed oils reduces the WVP by up to tenfold (Rakotonirainy and Padua, 1999).

Morphological differences are clear between cast and drawn films plasticized with oleic acid. Cast films show a different finish on their two sides; the side in contact with the casting surface is smooth and glossy, while the one in contact with air during drying is irregular and opaque. In contrast, both drawn-film surfaces are smooth and low gloss. Scanning electron microscopy reveals that the glossy sides of cast films are featureless, whereas the opaque sides show globular deposits due to fatty acids separated from the zein matrix by the evaporating solvent. Thermograms also suggest a more extensive interaction between zein and fatty acids in drawn films than in cast films (Lai and Padua, 1997).

Applications

The potential increase in zein availability has activated research on its properties and utilization. Although presently pharmaceutical coating and confectionery glazing form the bulk of zein commercial applications, intensive research on zein utilization for edible and biodegradable film formulations has resulted in improved performance. Plasticization with polyols and fatty acids reduces stiffness and increases film flexibility, while heat treatment under high pressure improves clarity and texture. Coating with water-resistant lipids reduces WVP and permits exposure to high RH environments. There are even attempts to scale film production by extrusion. The main challenge is to modify the mechanical properties of films produced by the present methods, which increase TS, resulting in higher stiffness and reduced flexibility (Padua and Wang, 2002). As zein films present very important advantages as biodegradable packaging materials for their resistance to rodent and insect attack and freezing and thawing cycles, it is expected that there will be advances in improving their properties, thus allowing commercial applications in the near future.

Soy protein films

Soy protein used in the food industry is classified as soy flour, concentrate or isolate, based on the protein content. Defatted soy flour (DSF) contains 50–59% protein, and is obtained by grinding defatted soy flakes. Soy protein concentrate (SPC) contains 65–72% protein, and is obtained by aqueous liquid extraction or acid leaching processes. Soy protein isolate (SPI) contains more than 90% protein, and is obtained by aqueous or mild alkali extraction followed by isoelectric precipitation (S. K. Park et al., 2002). Soy protein is a viable and renewable resource for producing edible and environmentally friendly biodegradable films. The use of soy protein as a film-forming agent can add value to soybeans by creating new channels for marketing soy protein, in particular SPI, which is of low cost and plentiful supply and is actually an under-utilized product despite its high protein quality (Raynes et al., 2000).

Formation of soy protein films

Soy protein films are typically prepared from SPI by drying thin layers of cast film-forming solutions (Subirade et al., 1998). The drying temperature and RH that determine the drying rate of cast solutions can also affect the film structure and properties. Rapid drying at 95°C and 30% RH yields films that are stronger, thinner, stiffer, and less extendable than films dried more slowly at 21°C and 50% RH. Also, the former have lower WVP than the latter (Alcantara et al., 1998). Soy protein films need the addition of plasticizers for texture and flexibility, and glycerol and sorbitol are the ones most commonly used (Gennadios et al., 1994).

Extruding soy protein formulations into films is also possible, using SPI, plasticizer, and water as the main formulation ingredients (Naga et al., 1996). Films have also been extruded from blends of SPI, polyethylene oxide, and LDPE (Ghorpade and Hanna, 1996). Yuba, a Japanese word for soymilk film, is the dried film formed on the surface of heated soymilk at temperatures close to the boiling point (Wu and Bates, 1972; Sian and Ishak, 1990; Gennadios and Weller, 1991). Soy protein polymerization is attributed to disulfide bonds and, to a lesser degree, hydrophobic interactions and hydrogen bonds. Dried yuba is quite brittle, has a relatively long shelf life, and is used as a wrapper for other foods, is cooked with other foods, or is used in soups (Liu, 1997).

Functional properties of soy protein films

In general, soy protein films have poor moisture resistance and water vapor barrier ability due to the inherent hydrophilicity of the protein and the substantial amounts of hydrophilic plasticizers used to impart film flexibility (Rhim *et al.*, 2000). Compared to LDPE films, cast SPI films have greater WVP values by roughly four orders of magnitude (Gennadios *et al.*, 1993c). In contrast, soy protein films are potent oxygen barriers, at least at low RH environments (Gennadios *et al.*, 1993b; Ghorpade *et al.*, 1995).

In the case of SPI films, mechanical properties depend on the development of disulfide bonds, hydrophobic interactions and hydrogen bonds within the protein matrix. Interactions between proteins and small molecules, including water, plasticizers, lipids, and other additives dispersed in the space of the matrix, also contribute to the mechanical behavior of protein films (Chen, 1995). In general, soy protein films, similar to films from other proteins, have only moderate mechanical properties compared to commonly used plastic films such as polyethylene, polypropylene, and polyvinylidene chloride (Gennadios *et al.*, 1994; Krochta and De Mulder-Johnston, 1997).

Heating above 60°C and alkaline conditions below pH 10.5 promote soy protein polymerization by altering the three-dimensional structure through the unfolding of polypeptide chains, thus exposing sulfhydryl and hydrophobic groups (Kelley and Pressey, 1966). The alkaline conditions also favor thiol-disulfide interchange reactions by deprotonating sulfhydryl groups and allowing them to act as nucleophiles. Upon drying the cast solutions, unfolded macromolecules approach each other and are linked by disulfide bonds and hydrophobic interactions. SPI films from solutions heated at 95°C have greater TS and E values than films from solutions heated at 75°C (Park and Hettiarachchy, 2000).

Protein-film formation is also pH dependent. SPI films can be formed in both alkaline and acidic conditions. In a study of interfacial film formation, lipids (corn oil or glycerol trioleate) were added to SPI solutions and satisfactory films were produced at pH values lower than 4.6 (as low as pH 1), the isoelectric region of soy protein, and as high as pH 6.5. Above pH 7.4, the increasing solubility of SPI limited film formation (Flint and Johnson, 1981). Cast films formed in alkaline conditions have been reported to have greater TS and E and lower WVP than films formed in acidic conditions. Similar to heating, alkaline conditions facilitate soy protein denaturation, thus promoting the formation of disulfide bonds within the structure of the dried films (Gennadios *et al.*, 1993c).

Besides heating soy protein film-forming solutions prior to casting, cast SPI films have also been subjected to heat treatments (heat curing) to modify their properties. For example, heating SPI films at 80 or 95°C for 2, 6, 14 or 24 hours increased TS and reduced E, WVP and water solubility (Gennadios *et al.*, 1996). In general, thermal treatments of proteins promote the formation of intra and intermolecular cross-links, mainly involving lysine and cystine amino acid residues (Cheftel *et al.*, 1985). The increased cross-linking within the protein network directly affects film properties (e.g. increased TS and reduced E). Moreover, this heat-induced cross-linking probably involves the reactions among polar protein groups, thus enhancing film hydrophobicity. For example, SPI films heat-cured in an air-circulating oven at 90°C for 24 hours had a lower moisture content than untreated SPI films after conditioning at 50% RH and 25°C for 2 days (Rhim *et al.*, 2000). Because water has a plasticizing effect on protein-based films, the decreased moisture content of heat-cured films likely contributes further to their increased mechanical toughness (Gontard *et al.*, 1993).

The use of enzymes that promote protein cross-linking to improve protein film properties appears feasible. Horseradish peroxidase has been used for such purposes. This enzyme catalyzes the oxidation of tyrosine residues, forming di-, tri- and tetra-tyrosine. Incubating SPI film-forming solutions with this enzyme at 37°C for 24 hours does not affect WVP but increases TS and protein solubility, and decreases E. Because of the increase in protein solubility, it is assumed that horseradish peroxidase causes protein degradation in addition to cross-linking (Stuchell and Krochta, 1994).

The ability of low molecular weight aldehydes such as formaldehyde to react with primary amino groups and sulfhydryl groups in proteins, thereby forming intra- and intermolecular crosslinks, is well documented (Feeney *et al.*, 1975). Formaldehyde can be directly added to SPI film-forming solutions, or it can be applied by immersing dried SPI films into formaldehyde solutions (Rhim *et al.*, 2000). Treatment with formaldehyde results in larger than twofold increases in TS and puncture strength (PS), while also reducing WVP by about 6% and water solubility by about 42% (Ghorpade *et al.*, 1995). However, aldehydes are generally toxic and non-edible, which limits their possible use to non-edible applications (S. K. Park *et al.*, 2002).

Reportedly, aromatic amino acids such as tyrosine and phenylalanine are able to absorb UV light and recombine to form covalent cross-links in proteins (Tomihata *et al.*, 1992). Subjecting cast SPI films to UV irradiation increases TS and decreases E, although WVP is not affected (Gennadios *et al.*, 1998a).

The water-barrier capability of protein films can be improved through combination with hydrophobic lipid materials (Gennadios *et al.*, 1994; Krochta and De Mulder-Johnston, 1997). One approach is to add the lipids to protein film-forming solutions, which are then cast to prepare emulsified, bi-layered or multi-component films. Alternatively, molten lipids can be deposited (laminated) onto preformed protein films to prepare multi-layered composite films. For example, the addition of fatty acids (i.e. lauric, myristic, palmitic, oleic) to SPI film-forming solutions caused a substantial decrease in WVP as well as a plasticizing effect translated into a much higher E (228% versus 70% in the control). However, because lipids lack the structural integrity of proteins or polysaccharides, the addition of fatty acids greatly reduces TS (Gennadios *et al.*, 1998b). Another concern regarding protein–lipid films is their potential susceptibility to lipid oxidation.

The formation of covalent cross-links between proteins and propylene glycol alginate (PGA) has been documented (McKay *et al.*, 1985). The ε-amino groups of lysine are involved in soy protein interaction with PGA. SPI–PGA films prepared with low levels of PGA (10% w/w of protein) have increased TS and water resistance (Shih, 1994; Rhim *et al.*, 1999). Although PGA is edible, its use as a cross-linker for protein films would reduce the nutritional value of protein by impairing the availability of lysine. Nevertheless, an edible film or coating is unlikely to contribute notably to the nutritional content of foods (S. K. Park *et al.*, 2002).

As aforementioned, soy protein films are formed through polymerization of heat-denatured proteins, disulfide bonds and hydrophobic interactions being the main forces that maintain the film network (Shimada and Cheftel, 1998). In sulfur-containing protein systems such as SPI, cysteine groups can undergo polymerization via sulfhydryl–disulfide interchange reactions during heating to form a continuous covalent network upon cooling (Arntfield *et al.*, 1991). For example, addition of cysteine (1% w/w) to SPI film-forming solutions at pH 7 increased the number of disulfide bonds in the solutions by 576.6%, and also increased the TS of cast films (Were *et al.*, 1999).

Applications

The high oxygen-barrier capability of SPI films could be utilized in the manufacture of multi-layer packaging where protein films would function as the oxygen barrier-providing layer. SPI coating on precooked meat products could control lipid oxidation and limit surface moisture loss. Incorporation of antioxidants and flavoring agents in SPI coatings could improve the overall quality characteristics of food products (Kunte *et al.*, 1997). Also, SPI films may find applications as microencapsulating agents of flavors and pharmaceuticals, or in coatings of fruits, vegetables, and cheese (Petersen *et al.*, 1999). Protective SPI coatings could also be used on certain food products, such as meat pies and high-moisture–low-sugar cakes, which require films that are highly permeable to water vapor (Gennadios *et al.*, 1993b).

Wheat gluten films

The very high molecular weight, markedly apolar character, complexity, and diversity of their fractions are features of wheat gluten (WG) proteins that can be utilized to make films with novel functional properties, such as selective gas-barrier properties and rubber-like mechanical properties. WG-based materials are homogeneous, transparent, mechanically strong, and relatively water resistant. They are biodegradable, biocompatible, and edible when food-grade additives are used (Guilbert *et al.*, 2002).

Vital WG is the cohesive and elastic mass that is left over after starch is washed away from wheat flour dough. Commercially, it is an industrial by-product of wheat starch production via wet milling. Whereas dry wheat flour comprises 9–13% protein and 75–80% starch, WG consists mainly of wheat storage protein (70–80% dry matter basis) with traces of starch and non-starch polysaccharides (10–14%), lipids (6–8%), and minerals (0.8–1.4%). WG is suitable for numerous food and non-food uses. Its main application is in the bakery industry, where it is used to strengthen weak flours, rendering them suitable for bread baking. It is widely accepted that the unique rheological properties of WG are derived from its protein composition, with lipids and carbohydrates being contaminants entangled in the protein matrix (Guilbert *et al.*, 2002).

There are four wheat-protein classes, based on solubility in different solvents, namely albumins, globulins, gliadins, and glutenins. The albumins and globulins (15–22% of total protein), which are water and saline soluble respectively, are removed with starch granules during gluten separation processing. In contrast, gliadins (prolamins), which are soluble in alcohol, and the glutenins, which are soluble in dilute acid or alkali solutions, are collected into gluten. Gliadins and glutenins amount to up to 85% (w/w) of total wheat flour protein, and are evenly distributed into gluten (Wrigley and Bietz,

1988). Gliadin is named prolamin due to its large proline and glutamine content. The solubility of gliadin in 60–70% alcohol results from its peculiar amino acid composition (Lafiandra and Kasarda, 1985). In the fully hydrated state, gliadin exhibits viscous flow properties without significant elasticity. For cereal technologists, gliadin accounts for the extensibility of wheat flour dough, also acting as a filler by diluting glutenin inter- actions (Guilbert *et al.*, 2002). Glutenin has a similar amino acid composition to gliadin, with a slightly lower content of hydrophobic amino acids (Wrigley and Bietz, 1988). Contrary to gliadin, which is comprised of distinct polypeptide chains, glutenin consists of polymers made from polypeptide chains (also named subunits) linked end- to-tail by disulfide bonds. Glutenin is believed to be one of the largest natural poly- meric molecules, with an estimated molecular mass of well over 10^7 (Kasarda, 1999).

Formation of wheat gluten films

Film formation from WG solutions or dispersions has been extensively studied (Gennadios and Weller, 1990; Gennadios *et al.*, 1993b, 1993c, 1993d, 1993e; Park and Chinnan, 1995). WG films can be obtained by casting from solution or by extrusion, the latter being more cost effective. WG films have also been produced by collecting the surface skin formed during the heating of WG solutions to temperatures near boil- ing (Watanabe and Okamoto, 1976).

WG proteins are insoluble in water, and require a complex solvent system with basic or acidic conditions in the presence of alcohol and disulfide bond-reducing agents. Generally, changing the pH of the medium disrupts hydrogen and ionic interactions, while ethanol disrupts hydrophobic interactions (mainly occurring among gliadin mol- ecules). Intermolecular (for glutenins) and intramolecular (for gliadins and glutenins) covalent disulfide bonds are cleaved and reduced to sulfhydryl groups when dispers- ing WG in alkaline environments (Okamoto, 1978). Reducing agents, such as sodium sulfite, cysteine or β-mercaptoethanol, should be used in acidic environments to destroy disulfide bonds. A wide variety of acids can be used, including citric, acetic, phos- phoric, propionic, and lactic acids. On the other hand, bases such as sodium hydrox- ide, potassium hydroxide, and (mainly for its volatility) ammonium hydroxide are used to reduce disulfide bonds (Guilbert *et al.*, 2002).

During drying, all these volatile disruptive agents are progressively eliminated. Solvent removal increases the concentration of WG proteins. Consequently, active sites for bond formation become free, and are close enough to each other to create new interactions. New hydrogen bonds, hydrophobic interactions, and disulfide bonds contribute to the formation of a three-dimensional network. Solvent removal can be fractioned to obtain film-forming solutions with a very low alcohol concentration (Guilbert *et al.*, 2002).

The addition of plasticizers is essential to avoid film brittleness as a consequence of extensive intermolecular associations. Sorbitol, manitol, diglycerol, propylene glycol, triethylene glycol, maltitol, polyvinyl alcohol, and PEG are used as plasticizers for WG films, but glycerol and ethanolamine are preferred. These molecules are small enough to insert themselves between the protein chains, and to form hydrogen bonds through their hydroxyl groups. In terms of level of use, plasticizers are employed in amounts sufficient to lower the glass transition temperature to the ambient temperature. Typical plasticizer levels are 15–40% of protein weight (preferably 20–30%). Polar plasticizers increase the hygroscopicity of WG-based materials. In addition, they diffuse out easily if the material is in contact with a moisture-rich product. For these reasons, the research for non-polar, low-diffusivity plasticizers is essential for the development of WG-based plastics. The use of amphiphilic plasticizers, such as fatty acids and derivatives, should be recommended (Gennadios *et al.*, 1993b, 1993c, 1993d, 1993e; Park and Chinnan, 1995; Guilbert *et al.*, 2002).

Extrusion for WG-film formation is a thermoplastic process involving a thermal treatment with glass transition followed by a thermoforming step. Extrusion of plasticized WG has been performed using a co-rotating, intermeshing twin-screw extruder (Redl *et al.*, 1999).

Functional properties of wheat gluten-based films

Cohesion of WG-based materials depends mainly on the type and density of intra- and intermolecular interactions, but also results from interactions with other constituents. WG films show low TS but high E compared to other films (Gennadios *et al.*, 1993c). The mechanical properties of WG films depend greatly on processing conditions, the addition of plasticizers, lipids, and cross-linking agents, and external conditions such as RH and temperature. Glycerol and water act as plasticizers of WG films, generally reducing mechanical resistance and increasing extensibility. However, at a very low water content and thus low RH (below 15 g/100 g dry matter and 60%, respectively), hydration has a positive effect on film mechanical resistance, probably due to formation of supplementary hydrogen bonds among protein chains (Gontard *et al.*, 1993). Increasing the processing temperature from 80 to 135°C induces an increase in mechanical resistance of the WG network and a decrease in elongation (Cuq *et al.*, 2000).

The WVP of WG films is equivalent to that of other protein- or polysaccharide-based films, but is relatively high compared to that of synthetic polymer films. The WVP through WG films increases with increasing RH and temperature – for example, at 30°C the WVP is tenfold lower at 50% RH than at 93% RH. The WVP of WG films is also highly dependent on the nature and amount of the added plasticizer. A linear relationship exists between the glycerol concentration and WVP of WG films (Gontard *et al.*, 1993).

Similar to WVP, the oxygen, carbon dioxide, and ethylene permeabilities of WG films are highly dependent on RH and temperature. At low RH and temperature a WG film has very low oxygen and carbon dioxide permeabilities compared to other biopolymer or synthetic polymer films. For example, at 0% RH the oxygen permeability of a WG film was found 800-fold lower than that of LDPE and 9.6-fold lower than that of polyamide-6 (a well-known potent oxygen-barrier polymer at 0% RH). Also, the WG film was 2.6- and 570-fold less permeable to carbon dioxide than are polyamide-6 and LPDE, respectively. Interesting applications of WG films as oxygen and carbon dioxide barriers in dry conditions could thus be considered (e.g. for low water activity products, or sandwiched between two layers of a highly water-resistant material) (Billing, 1989). However, the gas permeability values of WG films increase substantially with increasing RH and temperature in an almost exponential fashion above 50% RH (Mujica-Paz and Gontard, 1997). Increased RH promotes both gas diffusivity (due to the increased mobility of hydrophilic macromolecular chains) and gas solubility (due to swelling of the matrix by absorbed water), leading to a sharp increase in gas permeability. With carbon dioxide, the sharp increase of permeability is more pronounced than with oxygen. The effect of RH on gas permeability also varies with temperature, following (approximately) the Arrhenius model (Gontard *et al.*, 1996).

It is also possible to combine WG with other materials in order to optimize film properties. Films made from WG and lipids can combine the water vapor resistance of the lipids with the relatively good mechanical properties of the WG. Lipids can either be incorporated in WG-film-forming solutions to obtain composite films (emulsion technique) or deposited as a layer onto the surface of preformed WG films to obtain bi-layer films (coating technique). The effects of lipids incorporated into WG film-forming solutions can vary from destruction of the protein network to improvement of the mechanical and barrier properties of the cast films. These effects are complex, and are dependent on the nature and structure of the lipids and on the chemical interactions between the lipids and the protein. For example, combining WG with a diacetyl tartaric ester of monoglycerides at 20% w/w dry matter produced films that, when compared to the control glycerol-plasticized WG films, were as transparent, had higher mechanical resistance, and had about 50% lower WVP. However, improving the moisture-barrier properties of such composite films by increasing lipid concentration has proved to be of limited value because the hydrophilic protein matrix alters the resistance of the lipid components to WVP (Gontard, 1991). The coating technique for preparing bi- or multi-layer WG–lipid films has shown opposite results in the literature, with some studies reporting the process to be more effective than the emulsion technique (Gontard *et al.*, 1995) and others reporting it to be less effective (Koelsch, 1994).

There have been studies in which bi-layer films composed of a WG film and a modified PE film were prepared, thus combining the gas selectivity of the WG films with the excellent mechanical properties and moisture resistance of the PE films. WG films were prepared first using a casting procedure and then hot-pressed with the PE films at different temperatures. The TS and E values of bi-layer films made of polyethylene maleic anhydride (PGMA) and WG were not affected by RH, unlike those of WG films. Also, the gas and water vapor permeability values of the PGMA–WG films were considerably reduced as compared to those of control WG films (Pérez-Pérez, 1997).

Post-production treatments of wheat gluten films

WG films are relatively water resistant and water insoluble, but their high WVP due to their inherent hydrophilic nature remains the greatest limitation for their use (Gennadios *et al.*, 1993d). Furthermore, their mechanical and water vapor barrier properties are strongly affected by water or other plasticizers (Gontard *et al.*, 1993). The functional properties of WG-based films can also be improved by subjecting the film-forming solutions to various treatments prior to casting. Such treatments can be enzymatic, chemical, thermal or radiative (Gennadios and Weller, 1990; Gontard *et al.*, 1993; Herald *et al.*, 1995).

Heat treatments promote protein aggregation through disulfide bonding and hydrophobic interactions (Gennadios *et al.*, 1996; Ali *et al.*, 1997; Cuq *et al.*, 2000; Micard and Guilbert, 2000). This may be responsible for the observed changes in the mechanical and barrier properties of heat-treated protein films. Heat curing of WG films increases TS and decreases E. For example, the TS and E of cast films heated at 140°C reached up to 300% and 43%, respectively, of the value obtained when the films were heated at 80°C. The WVP of WG films at 25°C and 60% RH decreases as the exposure time increases at a given temperature. WG films appear to be rather sensitive to color changes following heat curing, particularly with regard to yellowness.

Formaldehyde is the main chemical that has been used to treat preformed protein films. It leads to a drastic insolubilization of proteins by reinforcing the protein network through covalent linkages. Both soaking (Guéguen *et al.*, 1998) and vapor exposure (Micard and Guilbert, 2000) have been used to apply formaldehyde to protein films, the latter approach offering more reproducible results. Such exposure increased TS and stiffness by 400% and decreased E by 62%, although no significant changes in WVP were observed (Micard and Guilbert, 2000). Immersion of cast WG films in calcium chloride or buffer solutions at the isoelectric point of WG (pH 7.5) increased TS by 46% and WVP by about 15%. No changes were observed with regard to OP (Gennadios *et al.*, 1993e).

Gamma radiation increases TS (43%) and stiffness (79%) and decreases E (32%) of WG films. This effect is explained by the formation of dityrosine cross-linkages within polypeptide macromolecules. The WVP increases as well (19%), probably because of glutenin degradation (Micard and Guilbert, 2000).

Aging WG films at 20°C and 60% RH for 360 hours increases TS (75%) and stiffness (314%) and decreases E (36%) in comparison to similar conditions of aging over 48 hours. No significant changes are observed with respect to WVP (Morel *et al.*, 2000).

Applications

Application of WG-based materials can be envisioned for the coating of seeds, pills, and foodstuffs, and for making cosmetic masks, polishes or drug capsules. It is important to note that the allergenic character of WG-based products for people suffering from celiac disease could limit such applications to cereal-based products in which the presence of WG or similar cereal proteins is known and expected. The moisture-, gas-, and solute-barrier properties of WG-based films could be particularly interesting for applications in the fields of active coatings, active packaging, drug delivery systems, and modified atmosphere packaging (Guilbert *et al.*, 2002).

Cottonseed protein films

Cotton is a plant that has been cultivated in 70 different countries for centuries as a fiber crop. Cottonseed accounts for about 13% of world oilseed production, although oil production accounts only for about 15% of the commercial value of the cottonseed crop. To optimize the usage of this low-cost, oil-rich, and protein-rich commodity, the film-forming properties of cottonseed proteins could be used to produce biodegradable materials of economic and environmental interests (Wu and Bates, 1973; Marquié *et al.*, 1996, 1997a, 1997b).

Cottonseed is crushed to extract oil and to produce cake that is chiefly used for ruminant livestock (Tacher *et al.*, 1985; Zongo and Coulibaly, 1993). As it contains gossypol, a toxic compound, this cake cannot be used for human consumption. Cottonseed proteins make up 30–40% w/w of the cottonseed kernel. Other important cottonseed components include lipids, soluble carbohydrates, cellulose, minerals, phytates, and polyphenolic pigments. The protein components are mainly globulins (60%) and albumins (30%), with much lower proportions of prolamins (8.6%) and glutelins (0.5%) (Saroso, 1989).

Cottonseed proteins have a high content of ionizable amino acids (aspartic and glutamic acids, arginine, histidine and lysine) and a low content of sulfur amino acids. These proteins are more soluble at a basic pH than at an acidic pH, and have an isoelectric point of around 5 (Bérot and Guéguen, 1985; Dumay *et al.*, 1986). Cottonseed proteins are readily denatured by thermal treatment above 80°C, leading to loss of solubility and nutritional value (Besançon *et al.*, 1985). The temperature increase during the crushing of cottonseed induces chemical interactions between gossypol and lysine, and also Maillard reactions (Marquié, 1987; Bourély, 1990).

Formation of cottonseed films

Initially, cottonseed films were prepared by soaking cottonseed kernels in hot water to prepare an "oilseed milk" from which films were formed on the surface of the heated liquid, similarly to traditional yuba (soy protein) films in East Asia (Wu and Bates, 1973). Owing to their poor mechanical properties, films made with this process may be used as edible items but not as packaging materials.

Cottonseed protein-based films are obtained directly from cottonseed flour using a casting process (Marquié, 1995, 1996; Marquié *et al.*, 1995, 1997a, 1997b, 1998). The first step involves solubilizing cottonseed flour proteins under appropriate conditions (solvent, pH, and temperature, and the addition of salts, plasticizers, and dissociating agents) to minimize protein–protein interactions. Changes in the protein structure prevent "remelting" of macromolecules and reveal potentially interactive zones. The dispersions are then centrifuged to remove insoluble substances, and the supernatants are homogenized. Glycerol is added as a plasticizer to the film-forming solution. Drying allows solvent removal and the formation of three-dimensional protein networks through new inter- and intramolecular linkages.

In contrast to protein isolates, the conditions required to obtain films from cottonseed flour-based solutions are difficult to determine because of the complex nature of the raw material, which contains proteins, lipids, ash, cellulose, and carbohydrates. The following general ranges have been established for cottonseed flour film formation: pH 8–12, temperature 20–60°C, solid/solvent ratio 10–50% w/v, use of dispersing agents and plasticizer content (10–50% w/w, dry basis). Films containing lipids are weaker than those made from defatted flour; this is explained by the protein content, which is evidently lower in lipoprotein films than in films made from delipidated flour. The presence of gossypol leads to less soluble films. The film color varies from light yellow (in the absence of gossypol) to brown (in the presence of gossypol). Films from defatted cottonseed flour are smooth and transparent.

Cottonseed flour films are excessively brittle when the glycerol content is below 10% w/w dry basis, and become sticky when glycerol content is higher than 30% (Marquié et al., 1995). Glycerol decreases the intermolecular forces between polymer chains, with a decrease in cohesion, tensile strength, and glass transition temperature (Guilbert, 1986). Gossypol is characteristically oxidized into quinones at an alkaline pH, forming Schiff bases with ε-amino groups of lysine. In protein films, gossypol strengthens the protein network (Tchiegang and Bourély, 1990; Marquié et al., 1995).

Films can be strengthened by cross-linking agents able to modify proteins chemically during the preparation of film-forming solutions. The main cross-linking agent for cottonseed protein is formaldehyde. In alkaline solution, formaldehyde forms short methylene cross-links between lysine amino groups (Means and Feeney, 1968; Bizzini and Raymund, 1974). Chemical treatment of cottonseed proteins with formaldehyde, using a formaldehyde–lysine reactive mix in the film-forming solution, produces films with a greater puncture strength (PS) than when using other cross-linking agents such as glyoxal or glutaraldehyde (Marquié, 1996).

It is also possible to incorporate carded cotton fibers into cottonseed flour films and plastic-like materials. The resulting films resemble bi-layer films in combining the properties of each polymer component (cellulose and protein). For example, plastic-like materials with proteins that were cross-linked by glutaraldehyde and contained 8% carded cotton fibers (w/w, dry basis) were five times more water resistant (Marquié et al., 1996).

Functional properties of cottonseed flour-based films

Films from cottonseed flour are highly hydrophilic owing to their protein content and protein composition (Gontard et al., 1993; Gennadios and Weller, 1994). Their mechanical properties are affected by both temperature and RH. Water acts as a plasticizer in cottonseed flour films by increasing the molecular mobility in the protein network and decreasing the apparent glass transition temperature at which film mechanical properties change (Marquié, 1996).

The PS of cottonseed flour films varies as a function of film moisture content and water activity. At low water activity (0.1), films with low moisture content (2–3% w/w, dry basis) are very brittle. PS then increases sharply for film moisture contents within 3–7% (w/w, dry basis). The increase in film PS at low water content may be interpreted as an antiplasticizing effect of water (Guo, 1994a, 1994b). At film moisture contents up to 7% (w/w, dry basis), water plasticizes the film structure through extensive hydrogen bonding among protein chains, thus increasing molecular mobility with a concomitant decrease in PS. The TS of cottonseed flour films, even cross-linked with formaldehyde, is five- to tenfold weaker than films made from synthetic materials;

however, cottonseed flour films that contain cotton fibers are as strong as them. The WVP of cottonseed flour films compares to that of other protein films such as SPI or zein, but is about 100-fold higher than that of polyethylene (Marquié, 1996).

Applications

Cottonseed flour films may be suitable for certain applications in non-food packaging (e.g. compostable waste bags) where good mechanical resistance and insolubility in water are required. They can also be used for agricultural packaging and mulching films to protect crops and fixate seeds, and other applications where film color, porosity, and biodegradability are beneficial (Marquié and Guilbert, 2002). They could be used in certain medical areas where biodegradability is necessary (e.g. prostheses and resorbable dressings). The ability of these films to absorb considerable amounts of moisture is a further advantage for exudate absorption (Marquié, 1995). It should be also possible during film preparation to introduce active compounds (e.g. insecticides, fungicides and bactericides) that could migrate out of the films and offer a specific environmental action. Film hydration, accompanied by increased WVP and diffusion of small molecules, would make these materials suitable vectors for active substances in delicate environments (Marquié and Guilbert, 2002).

Other protein films

Corn, soybean, wheat, and cottonseed are the prevalent types of plant from which proteins with film-forming ability are obtained. Nevertheless there are other plant proteins, of limited availability, that may be of interest, due to a unique property they provide to films or an advantage with regard to film formation. Limited availability may be because of relatively low production of the protein source or limitations in recovering the protein as a co-product from a process. The nature, recovery, film formation and film properties of protein obtained from peanut, rice, pea, pistachio, and grain sorghum are discussed here (H. J. Park *et al.*, 2002).

Peanut protein

Peanut protein concentrates (PPC) and isolates (PPI) are commercially produced from defatted peanut flour using several methods. Hydraulic pressing, screw pressing, solvent extraction, and pre-pressing followed by solvent extraction have been used for defatting (Natarajan, 1980; Woodroof, 1983). After oil removal, protein isolation and purification are necessary. Peanut concentrates (about 70% protein) are produced from dehulled kernels by removing most of the oil and the water-soluble, non-protein components. Techniques for removing water-soluble components include water leaching at the isoelectric pH, aqueous alcohol leaching, air classification, liquid cyclone fractionation, and moist heat denaturation followed by water leaching. PPI (about

90–95% protein) are produced by further processing of PPC to remove water-insoluble oligosaccharides and polysaccharides, and other minor constituents (Natarajan, 1980). Peanut proteins are deficient in methionine, lysine, and, possibly, threonine and tryptophan. In addition, there is a proven allergenicity associated with peanut proteins. These limitations must be considered when peanut proteins are added to foods for nutritional purposes (Natarajan, 1980).

Two different methods have been used to prepare peanut-protein films. The first method involves the formation of peanut-protein–lipid films on the surface of heated peanut milk (Wu and Bates, 1973; Aboagye and Stanley, 1985; Del Rosario et al., 1992). This film production is similar to that of the traditional Asian *yuba*. The second method involves casting of PPC or PPI solutions. For example, films were prepared from PPC solutions adjusted to pH 6.0, 7.5 or 9.0, and dried at 70, 80 or 90°C. Glycerin, sorbitol, PEG or propylene glycol were added as plasticizers. Films formed at a higher pH (7.5 or 9.0) and higher drying temperature (80 or 90°C) were less moist and sticky at the surface than films formed at pH 6.0 and a drying temperature of 70°C. Glycerin was determined to be the best performing plasticizer. Increasing the film-drying temperature decreased the WVP, but increasing the pH had no effect, and it was hypothesized that the reduction of WVP with increased drying temperature was due to greater cross-linking, resulting in a tight and compact protein film structure. The effects of pH and drying temperature on OP were similar to those on WVP. TS and E increased when film-drying temperature increased from 70 to 90°C. This was attributed to increased protein denaturation at the higher temperature, which also likely resulted in a tighter, more compact film structure (Jangchud and Chinnan, 1999a, 1999b).

Rice protein

Abrasive milling removes the outer tissue material from rice kernels, producing polished rice and the side products of bran and polish. Usually, 10% by weight of brown rice is removed during milling. Bran accounts for 5–8% of the rough rice weight, and polish accounts for a further 2–3%. Rice bran and polish contain 12.0–15.6% and 11.8–13.0% protein, respectively (Houston, 1972). Rice bran protein concentrates are prepared from commercially available unstabilized or heat-stabilized rice bran by alkaline extraction and isoelectric precipitation (Gnanasambandam and Hettiarachchy, 1995). Rice protein concentrate can be prepared by alkaline extraction of rice flour or of broken rice kernels, a side-product of rice milling. Alternatively, rice flour or kernels may be treated with enzymes to partially remove the starch component (Chen and Chang, 1984).

Films from rice bran protein solutions (70% protein, dry basis) have been prepared using glycerol as a plasticizer (2% w/v), adjusting the pH to either 9.5 or 3.0, heated to 80°C, poured onto polyethylene plates and dried at 60°C (Gnanasambandam et al., 1997). Films prepared at pH 9.5 resulted in darker (a more intense reddish-yellow tan color) and more transparent films than those prepared at pH 3.0. No substantial differences were observed with regard to WVP between films prepared at pH 3.0 and 9.5. Rice bran films have higher WVP than synthetic films because of the hydrophilic nature of the protein – the majority of proteins in rice bran are albumins and globulins.

OP was affected by pH, with films prepared at pH 3.0 showing lower values than films prepared at pH 9.5. The lower solubility of rice bran protein at pH 3.0 may result in a tighter film structure and, thus, reduced OP. PS was greater for rice bran protein films prepared at pH 9.5 than for those prepared at pH 3.0. Rice bran proteins were less soluble at pH 3.0, resulting in fewer protein–protein interactions and hence lower PS than films prepared at pH 9.5. The TS showed a similar trend with pH changing from the basic to the acid region (Gnanasambandam *et al.*, 1997).

Pea protein

Pea proteins are separated from starch and fiber by multi-step solubilization at pH 2.5–3.0, followed by centrifugation (Nickel, 1981). These protein isolates have a mean crude protein content of 85.3% and an ash content of 4.1–5.0%. They show a lower fat absorption than SPI, suggesting the presence of more numerous hydrophilic than hydrophobic groups on the surface of protein molecules (Nackz *et al.*, 1986). Pea protein consists primarily of globulins (>80% of total protein) and a small fraction of albumins. The globulins are storage proteins located in the cotyledons, whereas albumins are cytoplasmic proteins consisting of many different subunits (Boulter, 1983; Mosse and Pernollet, 1983). Pea protein possesses good nutritional quality (i.e. protein efficiency ratio and essential amino acid content) and has potential as a dietary protein fortifier (Linden, 1985), although a relatively poor functionality has been reported with respect to protein solubility, emulsion stability, and gelation characteristics (Sumner *et al.*, 1981; Jackman and Yada, 1989).

Owing to their globular structure, pea proteins need unfolding under alkaline conditions prior to film formation. An industrial pea protein concentrate (70.6% protein) has been utilized to cast films plasticized with various polyols (e.g. mono-, di-, tri- and tetra-ethylene glycol, glycerol and propane diol) and dried at 60°C. Glycerol-plasticized pea-protein films are sticky, while propane diol-plasticized films become brittle rapidly. Ethylene glycol gives films of greater TS and E, and lower WVP, than di-, tri- or tetra-ethylene glycol. Overall, pea-protein isolate films show lower TS and E, and greater WVP (by one or two orders of magnitude), than LPDE. Cross-linking with formaldehyde substantially reduces the water solubility and increases the TS of these films – an expected effect due to the high lysine content of pea protein (Guéguen *et al.*, 1995, 1998).

Films made from denatured pea-protein isolate (*ca.* 85% protein) show physical and mechanical properties similar to those of soy protein and whey protein films, and possess the strength and elasticity to resist handling. Increasing the concentration of glycerol as a plasticizer in the film decreases the TS and elastic modulus, but increases the E and WVP. Very strong and stretchable films can be obtained from 70/30 and 60/40 PPC/glycerol compositions, respectively. Film solubility is not affected by the amount of plasticizer (Choi and Han, 2001). Heat treatment of pea-protein isolate solutions for 5 minutes at 90°C to induce denaturation increases the TS and E, and decreases WVP and the elastic modulus. FTIR spectroscopy shows more water in the non-heated than in the heat-denatured ones, while electrophoresis suggests that intermolecular disulfide bonds are created during heat denaturation, which most feasibly results in the observed greater film integrity with respect to the non-heated control films (Choi and Han, 2002).

Table 17.1	References listed by type of protein

Protein	References
Zein	Andres (1984), Ashley (1985), Ben-Yehoshua (1985), Billing (1989), Dombrink-Kurtzman and Bietz (1993), Esen (1987), Foulk and Bunn (1994), Gennadios and Weller (1990), Gennadios *et al.* (1993a, 1993b), Grenner and Fennema (1989), Krochta (1992), Lai and Padua (1997, 1998), Padua and Wang (2002), Park and Chinnan (1990), Park *et al.* (1992, 1994a, 1994b, 1994c, 1994d), Parris and Coffin (1997), Rakotonirainy and Padua (1999), Reiners *et al.* (1973), Santosa and Padua (1999), Seymour and Carraher (1984), Smith (1986), Taylor (1986), Wong *et al.* (1996), Yamada *et al.* (1995), Yang *et al.* (1996)
Soybean	Alcantara *et al.* (1998), Artnfield *et al.* (1991), Cheftel *et al.* (1985), Chen (1995), Feeney *et al.* (1975), Flint and Johnson (1981), Gennadios and Weller (1993), Gennadios *et al.* (1993b, 1993c, 1994, 1996, 1998a, 1998b), Ghorpade and Hanna (1996), Ghorpade *et al.* (1995), Gontard *et al.* (1993), Kelley and Pressey (1966), Krochta and De Mulder-Johnston (1997), Kunte *et al.* (1997), Liu (1997), McKay *et al.* (1985), Naga *et al.* (1996), Park and Hetiarachchy (2000), S. K. Park *et al.* (2002), Petersen *et al.* (1999), Raynes *et al.* (2000), Rhim *et al.* (1999, 2000), Shih (1994), Shimada and Cheftel (1988), Sian and Ishak (1990), Stuchell and Krochta (1994), Subirade *et al.* (1998), Tomihata *et al.* (1992), Were *et al.* (1999), Wu and Bates (1972)
Wheat gluten	Ali *et al.* (1997), Billing (1989), Cuq *et al.* (2000), Gennadios and Weller (1990), Gennadios *et al.* (1993b, 1993c, 1993d, 1993e, 1996), Guilbert *et al.* (2002), Gontard (1991), Gontard *et al.* (1993, 1995, 1996), Guéguen *et al.* (1998), Herald *et al.* (1995), Kasarda (1999), Koelsch (1994), Lafiandra and Kasarda (1985), Micard and Guilbert (2000), Morel *et al.* (2000), Mujica-Paz and Gontard (1997), Okamoto (1978), Park and Chinnan (1995), Pérez-Pérez (1997), Redl *et al.* (1999), Watanabe and Okamoto (1976), Wrigley and Bietz (1998)
Cottonseed	Bérot and Guéguen (1985), Besançon *et al.* (1985), Bizzini and Raynaud (1974), Bourély (1990), Dumay *et al.* (1986), Gennadios and Weller (1994), Gontard *et al.* (1993), Guilbert (1986), Guo (1994a, 1994b), Marquié (1987, 1995, 1996), Marquié and Guilbert (2002), Marquié *et al.* (1995, 1996, 1997a, 1997b, 1998), Means and Feeney (1968), Saroso (1989), Tacher *et al.* (1990), Tchiegang and Bourély (1990), Wu and Bates (1973), Zongo and Coulibaly (1993)
Other proteins	H. J. Park *et al.* (2002)
• Peanut	Abogaye and Stanley (1985), Del Rosario *et al.* (1992), Jangchud and Chinnan (1999a, 1999b), Natarajan (1980), Woodroof (1983), Wu and Bates (1973)
• Rice	Chen and Chang (1984), Gnanasambandam and Hettiarachchy (1995), Gnanasambandam *et al.* (1997), Houston (1972)
• Pea	Boulter (1983), Choi and Han (2001, 2002), Guéguen *et al.* (1995, 1998), Jackman and Hada (1989), Linden (1985), Mosse and Pernollet (1983), Nackz *et al.* (1986), Nickel (1981)
• Pistachio	Ayranci and Çetin (1995), Ayranci and Dalgiç (1992), Crane (1979)
• Grain sorghum	Buffo (1995), Buffo *et al.* (1997), Dendy (1995), Hamaker *et al.* (1995), Shull *et al.* (1991)

Pistachio protein

Pistacia is mainly grown in hot, dry climates, such as in Western Asia, Asia Minor, Southern Europe, Northern Africa, and California. Its fruits (kernels) are used as edible nuts, for making a coffee-like drink, and as a source of oil and coloring (Crane, 1979). Between 35 and 40% of the raw kernel consists of protein, in which the hydrophilic amino acids predominate (e.g. glutamic and aspartic acids) (Ayranci and Dalgiç, 1992). A protein isolate (about 95% protein) can be prepared by extraction from ground kernels, followed by a bleaching process to change the color from brownish to white using hydrogen peroxide, and a coagulation step to obtain the isolate (Ayranci and Dalgiç, 1992; Ayranci and Çetin, 1995).

The protein isolate is combined with hydroxypropyl methylcellulose (HPMC) and PEG as plasticizers to cast films. Film WVP increases with increasing amounts of protein isolate due to the high hydrophilic amino acid content. In addition, films swell by absorbing water, showing a greater thickness with respect to the control HPMC-only film (Ayranci and Çetin, 1995).

Grain sorghum protein

Grain sorghum ranks fifth among cereals produced in the world (Dendy, 1995). The main protein fraction in the sorghum kernel is prolamin, known as kafirin. Kafirin is similar to corn zein in molecular weight, solubility, structure, and amino acid composition (Shull *et al.*, 1991). According to its prolamin nature, kafirin can be readily extracted with alcohol and/or a reducing agent such as β-mercaptoethanol. The addition of β-mercaptoethanol greatly enhances extraction yield, but its toxicity makes it useless if edibility is to be maintained. In sorghum wholegrain flour and sorghum endosperm, the kafirin content is 68–73% and 77–82% of total protein, respectively (Hamaker *et al.*, 1995).

Films have been prepared from laboratory-extracted kafirin (89% protein) (Buffo, 1995) dissolved in ethanol, with glycerin and PEG added as plasticizers. The dry films have similar TS, E and WVP to those of films prepared from commercial corn zein. However, kafirin films are notably darker than zein films, presumably due to a higher content of inherent pigments (Buffo *et al.*, 1997).

Appendix

Table 17.1 lists the references relevant to each type of protein.

References

Abogaye, Y. and Stanley, D. W. (1985). *Can. Inst. Food Sci. Technol. J.* **18,** 12–20.

Alcantara, C. R., Rumsey, T. R. and Krochta, J. M. (1998). *J. Food Process Eng.* **21,** 387–405.

Ali, Y., Ghorpade, V. M. and Hanna, M. A. (1997). *Ind. Crops Prod.* **6,** 177–184.

Andres, C. (1984). *Food Process.* **45(13),** 48–49.

Arntfield, S. D., Murray, E. D. and Ismond, M. A. H. (1991). *J. Agric. Food Chem.* **39,** 1376–1385.

Ashley, R. J. (1985). In: *Polymer Permeability* (J. Comyn, ed.), pp. 269–308. Elsevier Applied Science Publishers, London, UK.

Ayranci, E. and Çetin, E. (1995). *Lebensm. Wiss. u. Technol.* **28,** 241–244.

Ayranci, E. and Dalgiç, A. C. (1992). *Lebensm. Wiss. u. Technol.* **25,** 442–444.

Ben-Yehoshua, S. (1985). *Hort. Sci.* **20,** 32–37.

Bérot, S. and Guéguen, J. (1985). In: *Proceedings of the Colloque IDESSA-CIDT TRITU-RAF: Le Cottonier sans Gossypol: Une Nouvelle Ressource Alimentaire, 26–27 November, Abidjan, Côte d'Ivoire*, pp. 186–210.

Besançon, P., Henri, O. and Rouanet, J. M. (1985). In: *Proceedings of the Colloque IDESSA-CIDT TRITURAF: Le Cottonier sans Gossypol: Une Nouvelle Ressource Alimentaire, 26–27 November, Abidjan, Côte d'Ivoire*, pp. 63–79.

Billing, O. (1989). *Flexible Packaging*. Akerlund & Rausing, Lund, Sweden.

Bizzini, B. and Raynaud, M. (1974). *Biochim*. **56,** 297–303.

Boulter, D. (1983). In: *Encyclopedia of Food Science* (M. S. Peterson and A. H. Johnson, eds), pp. 221–224. Martinus Nijhoff Publishers, The Hague, The Netherlands.

Bourély, J. (1990). *Sci. des Aliments* **10,** 485–514.

Buffo, R. A. (1995). MS thesis, Department of Food Science and Technology, University of Nebraska, Lincoln, NE.

Buffo, R. A., Weller, C. L. and Gennadios, A. (1997). *Cereal Chem*. **74,** 473–475.

Cheftel, J. C., Cuq, J. L. and Lorient, D. (1985). In: *Food Chemistry*, 2nd edn. (O. R. Fennema, ed.), pp. 245–369. Marcel Dekker, New York, NY.

Chen, H. (1995). *J. Diary Sci*. **78,** 2563–2583.

Chen, W. P. and Chang, Y. C. (1984). *J. Sci. Food Agric*. **35,** 1128–1135.

Choi, W. S. and Han, J. H. (2001). *J. Food Sci*. **67(4),** 1399–1406.

Choi, W. S. and Han, J. H. (2002). *J. Food Sci*. **66(2),** 319–322.

Crane, J. C. (1979). In: *Tree Nuts: Production, Processing, Products* (J. G. Woodroof, ed.), pp. 572–603. The AVI Publishing Co., Westport, CT.

Cuq, B., Boutrot, F., Redl, A. and Lullien-Pellerin, V. (2000). *J. Agric. Food Chem*. **48,** 2954–2959.

Del Rosario, R. R., Rubio, M. R., Maldo, O. M., Sabiniano, N. S., Real, M. P. N. and Alcantara, V. A. (1992). *Philippine Agriculturist* **75,** 93–98.

Dendy, D. A. V. (1995). In: *Sorghum and Millets: Chemistry and Technology* (D. A. V. Dendy, ed.), pp. 11–26. American Association of Cereal Chemists, St Paul, MN.

Dombrink-Kurtzman, M. A. and Bietz, J. A. (1993). *Cereal Chem*. **70,** 105–108.

Dumay, E., Condet, F. and Cheftel, J. C. (1986). *Sci. des Aliments* **6,** 623–656.

Esen, A. (1987). *J. Cereal Sci*. **5,** 117–128.

Feeney, R. E., Blankenhorn, G. and Dixon, H. B. F. (1975). *Adv. Protein Chem*. **29,** 135–203.

Flint, F. O. and Johnson, R. F. P. (1981). *J. Food Sci*. **46,** 1351–1353.

Foulk, J. A. and Bunn, J. M. (1994). *ASAE Paper No. 94-6017*. American Society of Agricultural Engineers, St Joseph, MO.

Gennadios, A. and Weller, C. L. (1990). *Food Technol*. **44(10),** 63–69.

Gennadios, A. and Weller, C. L. (1991). *Cereal Foods World* **36,** 1004–1009.

Gennadios, A. and Weller, C. L. (1994). *Trans. ASAE* **37,** 535–539.

Gennadios, A., Park, H. J. and Weller, C. L. (1993a). *Trans. ASAE* **36,** 1867–1872.

Gennadios, A., Weller, C. L. and Testin, R. F. (1993b). *J. Food Sci*. **58,** 212–214, 219.

Gennadios, A., Brandenburg, A. H., Weller, C. L. and Testin, R. F. (1993c). *J. Agric. Food Chem*. **41,** 1835–1839.

Gennadios, A., Weller, C. L. and Testin, R. F. (1993d). *Trans. ASAE* **36,** 465–470.

Gennadios, A., Weller, C. L. and Testin, R. F. (1993e). *Cereal Chem*. **70,** 426–429.

Gennadios, A., McHugh, T. H., Weller, C. L. and Krochta, J. M. (1994). In: *Edible Coatings and Films to Improve Food Quality* (J. M. Krochta, E. A. Baldwin and M. O. Nisperos-Carriedo, eds), pp. 201–277. Technomic Publishing Company, Lancaster, PA.

Gennadios, A., Ghorpade, V. M., Weller, C. L. and Hanna, M. A. (1996). *Trans. ASAE* **39,** 575–579.

Gennadios, A., Rhim, J. W., Handa, A., Weller, C. L. and Hanna, M. A. (1998a). *J. Food Sci.* **63,** 225–228.

Gennadios, A., Cezeirat, C., Weller, C. L. and Hanna, M. A. (1998b). In: *Paradigm for Successful Utilization of Renewable Resources* (D. J. Sessa and J. L. Willett, eds), pp. 213–226. AOCS Press, Champaign, IL.

Ghorpade, V. M., and Hanna, M. A. (1996). *Trans. ASAE* **39,** 611–615.

Ghorpade, V. M., Li, H., Gennadios, A. and Hanna, M. A. (1995). *Trans. ASAE* **38,** 1805–1808.

Gnanasambandam, R. and Hettiarachchy, N. S. (1995). *J. Food Sci.* **60,** 1066–1069, 1074.

Gnanasambandam, R., Hettiarachchy, N. S. and Coleman, M. (1997). *J. Food Sci.* **62,** 395–398.

Gontard, N. (1991). Thèse de Doctorat en Sciences des Aliments, Université Montpellier II, Montpellier, France.

Gontard, N., Guilbert, S. and Cuq, J.-L. (1993). *J. Food Sci.* **58,** 206–211.

Gontard, N., Marchesseau, S., Cuq, J.-L. and Guilbert, S. (1995). *Intl J. Food Sci. Technol.* **30,** 49–56.

Gontard, N., Thobault, R., Cuq, B. and Guilbert, S. (1996). *J. Agric. Food Chem.* **44,** 1064–1069.

Greener, I. K. and Fennema, O. (1989). *J. Food Sci.* **54,** 1393–1399.

Guéguen, J., Viroben, G. and Barbot, J. (1995). In: *Proceedings of 2nd European Conference on Grain Legumes, July 9–13, Copenhagen, Denmark,* pp. 358–359.

Guéguen, J., Viroben, G., Noireaux, P. and Subirade, M. (1998). *Ind. Crops Products* **7,** 149–157.

Guilbert, S. (1986). In: *Food Packaging and Preservation* (M. Mathlouthi, ed.), pp. 371–394. Elsevier Applied Science Publishers, New York, NY.

Guilbert, S., Gontard, N., Morel, M. H., Chalier, P., Micard, V. and Redl, A. (2002). In: *Protein-Based Films and Coatings* (A. Gennadios, ed.), pp. 69–121. CRC Press, Boca Raton, FL.

Guo, J. (1994a). *Drug Develop. Ind. Pharm.* **20,** 1883–1893.

Guo, J. (1994b). *J. Pharm. Sci.* **83,** 447–449.

Hamaker, B. R., Mohamed, A. A., Habben, J. E., Huang, C. P. and Larkins, B. A. (1995). *Cereal Chem.* **72,** 583–588.

Herald, T. J., Gnanasambandam, R., McGuire, B. H. and Hachmeister, K. A. (1995). *J. Food Sci.* **60,** 1147–1150.

Houston, D. F. (1972). In: *Rice Chemistry and Technology*, 1st edn. (D. F. Houston, ed.), pp. 272–300. American Association of Cereal Chemists, St Paul, MN.

Jackman, R. L. and Yada, R. Y. (1989). *J. Food Sci.* **54(5),** 1287–1292.

Jangchud, A. and Chinnan, M. S. (1999a). *J. Food Sci.* **64,** 153–157.

Jangchud, A. and Chinnan, M. S. (1999b). *Lebensm. Wiss. u. Technol.* **32,** 89–94.

Kasarda, D. D. (1999). *Cereal Foods World* **44,** 566–572.

Kelley, J. J. and Pressey, R. (1966). *Cereal Chem.* **43,** 195–206.

Koelsch, C. (1994). *Trends Food Sci. Technol.* **5,** 76–81.

Krochta, J. M. (1992). In: *Advances in Food Engineering* (R. P. Singh and M. A. Wirakartakasumah, eds), pp. 517–538. CRC Press, Boca Raton, FL.

Krochta, J. M. and De Mulder-Johnston, C. (1997). *Food Technol.* **51(2),** 61–74.

Kunte, L. A., Gennadios, A., Cuppett, S. L., Hanna, M. A. and Weller, C. L. (1997). *Cereal Chem.* **74,** 115–118.

Lafiandra, D. and Kasarda, D. D. (1985). *Cereal Chem.* **62,** 314–319.

Lai, H.-M. and Padua, G. W. (1997). *Cereal Chem.* **74,** 771–775.

Lai, H.-M. and Padua, G. W. (1998). *Cereal Chem.* **75,** 194–199.

Linden, M. C. (1985). In: *Nutritional Biochemistry and Metabolism, with Clinical Applications* (M. C. Linden, ed.), pp. 51–52. Elsevier Publishing Co., New York, NY.

Liu, K. (1997). *Soybeans: Chemistry, Technology and Utilization*, pp. 198–202. Chapman and Hall, New York, NY.

Marquié, C. (1987). *Cot. Fib. Trop.* **42,** 65–73.

Marquié, C. (1995). Breve deposé au nom du Cirad du 07 Mars, No. 9502640.

Marquié, C. (1996). Thèse de Doctorat, Université Montpellier II, Montpellier, France.

Marquié, C. and Guilbert, S. (2002). In: *Protein-Based Films and Coatings* (A. Gennadios, ed.), pp. 139–157. CRC Press, Boca Raton, FL.

Marquié, C., Aymard, C., Cuq, J. L. and Guilbert, S. (1995). *J. Agric. Food Chem.* **43,** 2762–2767.

Marquié, C., Héquet, E., Guilbert, S., Tessier, A. M. and Vialettes, V. (1996). In: *Proceedings of the 23rd International Cotton Conference, Faserinstitut, Bremen* (H. Harig and S. A. Heap, eds), pp. 145–156.

Marquié, C., Héquet, E., Vialettes, V. and Tessier, A. M. (1997a). In: *Beltwide Cotton Conferences, Cotton Improvement Conference, New Orleans*, pp. 469–473.

Marquié, C., Tessier, A. M., Aymard, C. and Guilbert, S. (1997b). *J. Agric. Food Chem.* **45,** 922–926.

Marquié, C., Tessier, A. M., Aymard, C. and Guilbert, S. (1998). *Nahrung* **42,** 264–265.

McKay, J. E., Stainsby, G. and Wilson, E. L. (1985). *Carbohydr. Polym.* **5,** 223–236.

Means, G. E. and Feeney, R. E. (1968). *Biochem.* **7,** 2192–2201.

Micard, V. and Guilbert, S. (2000). *Intl J. Biol. Macromol.* **27,** 229–236.

Morel, M. H., Bonicel, J., Micard, V. and Guilbert, S. (2000). *J. Agric. Food Chem.* **48,** 186–192.

Mosse, J. and Pernollet, J. C. (1983). In: *Chemistry and Biochemistry of Legumes* (S. K. Akora, ed.), pp. 111–193. Oxford & IBH Publishing, New Delhi, India.

Mujica-Paz, H. and Gontard, N. (1997). *J. Agric. Food Chem.* **45,** 4101–4105.

Nackz, M., Rubin, L. J. and Shahidi, F. (1986). *J. Food Sci.* **51,** 1245–1247.

Naga, M., Kirihara, S., Tokugawa, Y., Tsuda, F. and Hirotsuka, M. (1996). US Patent 5,569,482.

Natarajan, K. R. (1980). In: *Advances in Food Research*, Vol. 26 (C. O. Chichester, E. M. Mrak and G. F. Stewart, eds), pp. 215–273. Academic Press, New York, NY.

Nickel, G. B. (1981). Canadian Patent 1,104,871.

Okamoto, S. (1978). *Cereal Foods World* **23,** 256–262.

Padua, G. W. and Wang, Q. (2002). In: *Protein-Based Films and Coatings* (A. Gennadios, ed.), pp. 43–67. CRC Press, Boca Raton, FL.

Park, H. J. and Chinnan, M. S. (1990). *ASAE Paper No. 90-6510.* American Society of Agricultural Engineers, St Joseph, MO.

Park, H. J. and Chinnan, M. S. (1995). *J. Food Engr.* **25,** 497–507.

Park, S. K. and Hettiarachchy, N. S. (2000). Unpublished data. University of Arkansas, Fayeteville, AR.

Park, H. J., Weller, C. L., Vergano, P. J. and Testin, R. F. (1992). Paper No. 428, presented at the Annual Meeting of the Institute of Food Technologists, June 20–24, New Orleans, LA.

Park, H. J., Chinnan, M. S. and Shewfelt, R. L. (1994a). *J. Food Sci.* **59,** 568–570.

Park, H. J., Chinnan, M. S. and Shewfelt, R. L. (1994b). *J. Food Process. Preserv.* **18,** 317–331.

Park, H. J., Bunn, J. M., Weller, C. L., Vergano, P. J. and Testin, R. F. (1994c). *Trans. ASAE* **37,** 1281–1285.

Park, H. J., Testin, R. F., Park, J. W., Vergano, P. J. and Weller, C. L. (1994d). *J. Food Sci.* **59,** 916–919.

Park, H. J., Rhim, J. W., Weller, C. L., Gennadios, A. and Hanna, M. A. (2002). In: *Protein Based Films and Coatings* (A. Gennadios, ed.), pp. 305–327. CRC Press, Boca Raton, FL.

Park, S. K., Hettiarachchy, N. S., Ju, Z. Y. and Gennadios, A. (2002). In: *Protein-Based Films and Coatings* (A. Gennadios, ed.), pp. 123–137. CRC Press, Boca Raton, FL.

Parris, N. and Coffin, D. R. (1997). *J. Agric. Food Chem.* **45,** 1596–1599.

Pérez-Pérez, C. (1997). Diplôme d'Etude Approfondie en Sciences des Aliments, Université Montpellier II, Montpellier, France.

Petersen, K., Nielsen, V. P., Bertelsen, G. *et al.* (1999). *Trends Food Sci. Technol.* **10,** 52–68.

Rakotonirainy, A. M. and Padua, G. W. (1999). Paper No. 11B-25, presented at the Annual Meeting of the Institute of Food Technologists, 24–28 July, Chicago, IL.

Raynes, M., Ciolfi, V., Maves, B., Stedman, P. and Mittal, G. S. (2000). *J. Sci. Food Agric.* **80,** 777–782.

Redl, A., Morel, M. H., Bonicel, J., Vergnes, B. and Guilbert, S. (1999). *Cereal Chem.* **76,** 361–370.

Reiners, R. A., Wall, J. S. and Inglett, G. E. (1973). In: *Industrial Uses of Cereals* (Y. Pomeranz, ed.), pp. 285–298. American Association of Cereal Chemists, St Paul, MN.

Rhim, J. W., Gennadios, A., Weller, C. L. and Schnepf, M. (1999). *J. Food Sci.* **64,** 149–152.

Rhim, J. W., Gennadios, A., Handa, A., Weller, C. L. and Hanna, M. A. (2000). *J. Agric. Food Chem.* **48,** 4937–4941.

Santosa, F. X. B. and Padua, G. W. (1999). *J. Agric. Food Chem.* **47,** 2070–2074.

Saroso, B. (1989). *Ind. Crop Res. J.* **1,** 60–65.

Seymour, R. B. and Carraher, C. E. (1984). *Structure–Property Relationships in Polymers*, pp. 107–111. Plenum Press, New York, NY.

Shih, F. F. (1994). *JAOCS* **71,** 1281–1285.

Shimada, K. and Cheftel, J. C. (1988). *J. Agric Food Chem.* **36,** 147–153.

Shull, J. M., Watterson, J. J. and Kirlies, A. W. (1991). *J. Agric. Food Chem.* **39,** 83–87.

Sian, N. K. and Ishak, S. (1990). *Cereal Foods World* **35,** 748, 750, 752.

Smith, S. A. (1986). In: *The Wiley Encyclopedia of Packaging Technology* (M. Bakker, ed.), pp. 514–523. John Wiley & Sons, New York, NY.

Stuchell, Y. M. and Krochta, J .M. (1994). *J. Food Sci.* **59,** 1332–1337.

Subirade, M., Kelly, I., Guéguen, J. and Pézolet, M. (1998). *Biol. Macromol.* **23,** 241–249.

Sumner, A. K., Nielsen, M. A. and Youngs, C. G. (1981). *J. Food Sci.* **46(2),** 364–366.

Tacher, G., Rivière, R. and Landry, C. (1985). In: *Proceedings of the Colloque Alimentaire, 26–27 November, Abidjan, Côte d'Ivoire*, pp. 97–112.

Taylor, C. C. (1986). In: *The Wiley Encyclopedia of Packaging Technology* (M. Bakker, ed.), pp. 159–163. John Wiley & Sons, New York, NY.

Tchiegang, C. and Bourély, J. (1990). *Cot. Fib. Trop.* **45,** 115–117.

Tomihata, K., Burczak, K., Shiraki, K. and Ikada, Y. (1992). *Polym. Preprints* **33,** 534–535.

Watanabe, K. and Okamoto, S. (1976). *New Food Ind.* **18(4),** 65–77.

Were, L., Hettiarachchy, N. S. and Coleman, M. (1999). *J. Food Sci.* **64,** 514–518.

Wong, Y. C., Herald, T. J. and Hacmeister, K. A. (1996). *Poultry Sci.* **75,** 417–422.

Woodroof, J. G. (1983). In: *Peanuts: Production, Processing, Products*, 3rd edn. (J. G. Woodroof, ed.), pp. 165–179. The AVI Publishing Company, Westport, CT.

Wrigley, C. W. and Bietz, J. A. (1988). In: *Wheat Chemistry and Technology* (Y. Pomeranz, ed.), pp. 159–275. American Association of Cereal Chemists, St Paul, MN.

Wu, L. C. and Bates, R. P. (1972). *J. Food Sci.* **37,** 36–44.

Yamada, K., Takahashi, H. and Noguchi, A. (1995). *J. Food Sci. Technol.* **30,** 599–608.

Yang, Y., Wang, L. and Li, S. (1996). *J. Appl. Polym. Sci.* **59,** 433–441.

Zongo, D. and Coulibaly, M. (1993). *Tropicultura* **11,** 95–98.

Edible films and coatings from animal-origin proteins

Monique Lacroix and Kay Cooksey

Introduction

Edible coatings can potentially extend the shelf life and improve the quality of food systems by controlling mass transfer, moisture and oil diffusion, gas permeability (O_2, CO_2), and flavor and aroma losses. Coating formulations can be used to serve as adhesive for seasonings or to improve the appearance of foods. For example, edible coatings can be sprayed or dipped onto the surface of snack foods and crackers to serve as a foundation or adhesive for flavorings (Druchat and De Mulder Johnston, 1997). Candies are often coated with edible films to improve their texture by reducing the stickiness (McHugh, 2000).

Active compounds (antimicrobials, antioxidants, nutrients) can be added to food coatings in order to extend the shelf life, preserve the color, and improve the nutritional value of foods. Edible coatings also have the potential for maintaining the quality of foods after the packaging is opened by protecting it against moisture change, oxygen uptake, and aroma losses (Krochta, 1997). The application of edible coatings to meat and fish products may be produced by dipping, spraying, casting, rolling, brushing, and foaming (Donhowe and Fennema, 1994). Coatings must have barrier properties with regard to water vapor, oxygen, carbon dioxide, and lipid transfer, while maintaining good color, appearance, mechanical, and rheological characteristics (Guilbert *et al.*, 1996).

Composite coatings improve gas exchange, the adherence to coated products and moisture vapor permeability properties (Baldwin *et al.*, 1995). The addition of glycerol, polyethylene glycol, and sorbitol can reduce film brittleness. A composite film containing antimicrobial or antioxidant compounds can prevent rancidity and improve shelf life by controlling bacterial proliferation. Also, the addition of vegetable oils can act as a moisture barrier. Sealing meats with a cross-linked sodium caseinate gel produced a juicier product by reducing drip loss; it also reduced the use of absorbent pads and protected the color of the meat (Ben and Kurth, 1995).

Edible films as a solid sheet can be applied between the food components or on the surface of the food system in order to inhibit moisture, oxygen, CO_2, aroma, and lipid migration. Edible films with adequate mechanical properties could conceivably also serve as edible packaging for selected foods (Krochta and De Mulder-Johnston, 1997).

The benefits of edible films can be numerous, but some barriers to commercial implementation have not been overcome. The raw material for much of the edible films comes from underutilized sources, but the cost of purification can be financially impractical. According to Krochta and De Mulder-Johnston (1997), the cost of edible films can range from as low as $1.30/lb for soy protein isolate to $6–12/lb for whey protein isolate. Collagen films were cited as $49–54/lb. In comparison, synthetic polymers traditionally cost less than a $1/lb.

Another barrier is the concept of biodegradability. According to Hunt et al. (1990), inert and non-biodegradable plastic materials represent 30% of municipal solid waste. Although, the idea of biodegradable films continues to be of research interest, such films are not practical under the current solid-waste handling conditions. Biodegradable films could work in a compost environment, but it is well known that even biodegradable materials do not degrade well in landfills, where the majority of all packaging waste material is disposed.

To be accepted, an edible film should be generally recognized as safe (GRAS), and used within any limitations specified by the Food and Drug Administration (FDA). Many edible films can be produced from food-grade materials, but many require solvents to become soluble. Ultimately, any material that is used for direct food contact will face regulatory scrutiny. In particular, biopolymers that act as carriers of additives intended to migrate to the food for preservative effects.

In addition, another challenge for the successful use of biodegradable films such as protein films is the stabilization of the functional properties of the films during storage, and the improvement of the insolubility and mechanical properties. Sensory and quality characteristics of films also need further improvement for successful implementation for commercial applications.

Animal-origin proteins

Proteins are good film formers exhibiting excellent oxygen, carbon dioxide and lipid barrier properties, particularly at low relative humidity. Edible films based on proteins were found to possess satisfactory mechanical properties (Kester and Fennema, 1986; Peyron, 1991; Cuq et al., 1998a). However, their predominantly hydrophilic character results in poor water-barrier characteristics (Kester and Fennema, 1986; Peyron, 1991; McHugh, 2000). The increase of cohesion between protein polypeptide chains was thought to be effective for the improvement of the barrier properties of the films.

For instance, the cross-linking of proteins by means of chemical, enzymatic (transglutaminase) or physical treatments (heating, irradiation) was reported to improve film water-vapor barriers as well as the mechanical properties and the resistance to proteolysis (Brault *et al.*, 1997; Ressouany *et al.*, 1998, 2000; Sabato *et al.*, 2001; Ouattara *et al.*, 2002a).

When cross-linked whey proteins are entrapped in cellulose, this generates insoluble films with good mechanical properties, high resistance to attack by proteolytic enzymes, and with decreased water-vapor permeability properties (Le Tein *et al.*, 2000). The addition of lipids in the film formulations can act as a good moisture barrier (McHugh, 2000). Although interesting, protein–lipid films are often difficult to obtain. For example, bi-layer film formation requires the use of solvents or high temperatures, making production more costly. Furthermore, the separation of the layers may occur with time. For films cast from aqueous lipid emulsion solutions, the process is complex and the incorporation of the lipids is limited. However, the addition of emulsifying agents or surfactants can improve the emulsion stability (Everett, 1989) and decrease the mean particle diameters of the emulsion, which results in a linear decrease in water-vapor permeability values (McHugh and Krochta, 1994).

Protein films are brittle and susceptible to cracking due to the strong cohesive energy density of the polymer (Lim *et al.*, 2002). The addition of compatible plasticizers improves the extensibility and viscoelasticity of the films (Brault *et al.*, 1997). According to Ressouany *et al.* (1998), sorbitol is a good plasticizer and significantly increased puncture resistance. However, polyethylene glycol and mannitol had lower plasticizer effects. The addition of glycerol in the film formulation produced an important loss of protein–water interactions (Letendre *et al.*, 2002a). The presence of plasticizers can enhance the formation of cross-links. It was observed by Brault *et al.* (1997), and by Mezgheni *et al.* (1998), that the presence of glycerol, propylene glycol, and triethylene glycol in film formulations enhanced the formation of cross-links during irradiation treatments.

Composite films may be of heterogenic nature and be formed via a mixture of a protein/polysaccharide/plasticizer and a lipid. This approach allows better exploitation of the functional properties of each of a film's components. The addition of a polysaccharide in a film formulation could improve the moisture barrier, the resistance, and the mechanical properties of the films (Ressouany *et al.*, 1998; Letendre *et al.*, 2002b). It is believed that some polysaccharides, such as carboxymethylcellulose, alginate, and pectin, form charge–charge electrostatic complexes with proteins (Imeson *et al.*, 1977; Shih, 1994; Thakur *et al.*, 1997; Sabato *et al.*, 2001; Letendre *et al.*, 2002b). Under certain conditions, polysaccharides like pectin may form cross-links with proteins (Thakur *et al.*, 1997). Heat treatments may enhance protein–polysaccharide interactions, resulting in a three-dimensional network with improved mechanical properties. Untreated polysaccharides keep their orderly structure, thus preventing any favorable interactions between their functional groups and those of the proteins (Letendre *et al.*, 2002b). The addition of starch in film formulations can improve oxygen and oil barriers (Kroger and Igoe, 1971; Morgan, 1971; Sacharow, 1972). Alginate can reduce dehydration and retard oxidative off-flavours in meats (Wanstedt *et al.*, 1981). Pectin and chitin can reduce bacterial growth in foods (Baldwin *et al.*, 1995; Chen *et al.*, 1998).

Milk proteins, such as whey and casein proteins, were extensively studied due to their excellent nutritional value and their numerous functional properties, which are important for the formation of edible films (McHugh and Krochta, 1994; Chen, 1995). Caseinates can easily form films from aqueous solutions due to their random-coil nature and their ability to form extensive intermolecular hydrogen, electrostatic, and hydrophobic bonds, resulting in an increase in the inter-chain cohesion (McHugh and Krochta, 1994). Moreover, edible films based on milk proteins were reported to be flavorless, tasteless, and flexible, and, depending on the formulation, they varied from transparent to translucent (Chen, 1995). Whey and caseins are the main milk-protein fractions. Caseins represent 80% of the total composition of milk proteins, with a mean concentration of 3% in milk (Brunner, 1977; Dalgleish, 1989).

Caseins

Commercial caseinates are produced by adjusting acid-coagulated casein to pH 6.7 using calcium or sodium hydroxide. Caseins are predominantly phosphoproteins that precipitate at pH 4.6 and at 20°C (Gennadios et al., 1994). Four principal components, α_{s1}, α_{s2}, β-, and κ-caseins are identified. The amino acid composition in casein is characterized by low levels of cysteine. Consequently, disulfide cross-linkages cannot form water-insoluble films (Gennadios et al., 1994; Chen, 2002). Among the protein fractions of casein, β-casein is the most interesting fraction to produce films of weak permeability to water vapor. The α-casein fraction contains more charged residues and fewer hydrophobic residues than the β-casein fraction (Dalgleish, 1989). In the presence of calcium, κ-casein associates with α_{s1} casein and β-casein to form thermodynamically stable micelles (Brunner, 1977). Calcium acts by reducing the electrostatic repulsions of casein phosphate groups, consequently facilitating hydrophobic interactions and the formation of micelles (Kinsella, 1984).

Caseins form films from aqueous solutions without further treatment due to their random-coil nature and their ability to hydrogen bond extensively. It is believed that electrostatic interactions also play an important role in the formation of casein-based edible films (Gennadios et al., 1994). According to Vachon et al. (2000), pure caseinate/glycerol/CMC films are highly soluble in water (88%). Film solubility in water decreased with buffer treatments at the isoelectric point of the films (Chen, 2002), with cross-linking of the protein using irradiation (Vachon et al., 2000), and with the transglutaminase enzyme (Nielson, 1995). A direct correlation between mechanical properties and film porosity based on caseinate–whey proteins was also observed by Vachon et al. (2000). This study demonstrated that the structure of cross-linked films and caseinate-based films containing whey-protein isolated (WPI) were generally more dense and homogenous, and comparable to caseinate films without WPI (Figure 18.1).

A protein ratio up to 50–50 (2.5% WPI and 2.5% caseinate) did not significantly affect the puncture strength of the films. At higher WPI concentrations, the puncture strength of the films was significantly reduced (Figure 18.2). The film based on whey-protein concentrate (WPC)–caseinate showed a granular structure and a large content of impurities such as fats, salts and lactose. An increase of WPC concentration in the WPC–caseinate based films significantly decreased the puncture strength of the films (Figure 18.3).

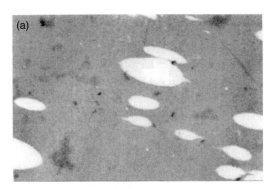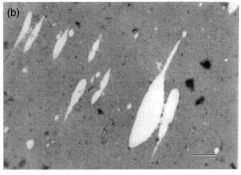

Figure 18.1 Cross-sections of WPI–caseinate films: (a) heated at 90°C for 30 minutes; (b) heated at 90°C for 30 minutes and irradiated at 32 kGy (9 mm bar = 3 μm). Reprinted with permission from Vachon *et al.* (2000). Copyright (2000) American Chemical Society.

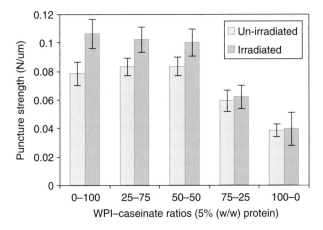

Figure 18.2 Puncture strength of un-irradiated and irradiated (32 kGy) whey-protein isolate (WPI)–caseinate films. Reprinted with permission from Vachon *et al.* (2000). Copyright (2000) American Chemical Society.

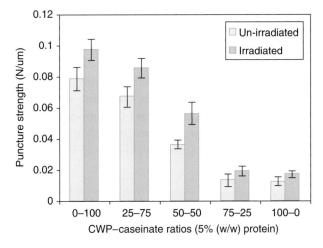

Figure 18.3 Puncture strength of un-irradiated and irradiated (32 kGy) whey-protein concentrate (WPC)–caseinate films. Reprinted with permission from Vachon *et al.* (2000). Copyright (2000) American Chemical Society.

Whey proteins

Whey proteins are soluble proteins present in milk serum after caseinate is coagulated during the cheese processing. Whey proteins represent around 20% of total milk proteins (Brunner, 1977), and contain five main proteins: α-lactalbumin, β-lactoglobulin, bovine serum albumin (BSA), immunoglobulin, and proteose-peptones. The β-lactoglobulin monomer makes up approximately 57% of the protein in whey (Dybing and Smith, 1991). Monomeric β-lactoglobulin contains one free S–H group and two S–S bonds.

The α-lactalbumin is the second most abundant whey protein (20%), and contains four S–S bonds. However, bound calcium and the four S–S bridges maintain the globular structure of the protein and stabilize it against denaturation (Kinsella and Whitehead, 1989).

BSA represents around 7% of the protein in whey. This protein contains fourteen S–S bonds and one free thiol, which makes it highly structured (Morr and Ha, 1993). The industrial processes used for whey-protein recovery are ultrafiltration, reverse osmosis, gel filtration, electrodialysis, ion exchange chromatography, and diafiltration (De Boer *et al.*, 1977; Matthews, 1984; Glover, 1985; St-Gelais *et al.*, 1992; Letendre *et al.*, 2002b). These processes are generally used to produce whey-protein concentrate (WPC: 25–80% protein) or whey-protein isolate (WPI: >90% protein).

Formation of intact and insoluble whey-protein films can be produced by heat denaturation of the proteins. Heating modifies the three-dimensional structure of the protein, exposing internal S–H and hydrophobic groups (Shimada and Cheftel, 1998), which promote intermolecular S–S bonds and hydrophobic interactions upon drying (McHugh and Krochta, 1994). McHugh and Krochta (1994) produced WPI films by heat treatment using a protein concentration from 8–12% and heating temperatures between 75 and 100°C.

Le Tein *et al.* (2000) have observed that a film formulation solution containing a protein concentration of ⩾10% of native WPI (protein obtained by diafiltration without any heat treatment) at pH 7 gelled during cross-linking treatment when heated at 90°C. An optimal temperature and concentration was obtained at 80°C for 30 minutes with a protein concentration of 5%. Banerjee and Chen (1995) reported that heating WPC solutions at 75°C for 30 minutes and at pH 6.6 produced uniform films. Contrary to casein films, insoluble whey-protein films can be produced by heat treatment due to the formation of covalent disulfide bonds (Gennadios *et al.*, 1994). Transglutaminase has also been developed to cross-link whey proteins (Mahmoud and Savello, 1993). Transglutaminase catalyzes the introduction of ε–(γ-glutamyl)-lysine cross-links into proteins via an acyl transfer reaction between whey proteins, as well as between peptides and primary amines (Mahmoud and Savello, 1993).

Collagen and gelatin

Collagen is a constituent of skin, tendon, and connective tissues. It is a fibrous protein, and represents about 30% of the total mass of the body (Gustavon, 1956). Collagen fibrils are produced by self-assembly of collagen molecules in the extracellular matrix, and provide tensile strength to animal tissues (Trotter *et al.*, 2000). Collagen can be dissolved in dilute acid or alkali solutions, and in neutral solutions. Two major components are identified; α (MW 100 000 Da) and β (MW 200 000 Da), and consist of two different types of covalent cross-linked chain pairs α_1–α_1 and α_2–α_2 (Harrington, 1966; Piez et al., 1968).

Hydrolysis of collagen results in gelatin. The molecular weight of gelatin covers a broad range, from 3000–200 000 Da, depending on the raw material employed during gelatin production and handling conditions (Young, 1967). Edible coatings made with gelatin reduce the migration of moisture, oxygen and oil.

Collagen is the most commercially successful edible protein film. Film-forming collagen has been traditionally used in the meat industry, for the production of edible sausage casings. This protein has largely replaced natural gut casings for sausages. Collagen films are not as strong and tough as cellophane, but have good mechanical properties (Hood, 1987). Collagen films have an excellent oxygen barrier at 0% relative humidity, but the oxygen permeability increases rapidly with increasing relative humidity in a manner similar to cellophane (Lieberman and Gilbert, 1973). Different cross-linking chemical agents have been used to improve the mechanical properties, to reduce the solubility, and to improve the thermal stability of these films. Carbodimide, microbial transglutaminase, and glutaraldehyde are usually used as cross-linking agents (Jones and Whitmore, 1972; Takahashi et al., 1999; Taylor et al., 2002).

Gelatin is known to form clear, flexible, strong, and oxygen-impermeable films when cast from aqueous solutions in the presence of plasticizers (Gennadios et al., 1994). Industrial applications of gelatin include capsule coating and microencapsulation. Edible coatings with gelatin reduce oxygen, moisture and oil migration, or can carry antioxidant or antimicrobial agents (Krochta and De Mulder-Johnson, 1997). They can be good gas barriers but poor water barriers due to their hydrophilic nature.

According to Arvanitoyannis (2002), several researchers have studied the glass transition temperature (Tg) and melting point temperature (Tm) of gelatin from various sources. The Tg of gelatin from cow tendon was the lowest at 95°C and the highest was from limed ossein with a Tg of 220°C. The lowest Tm was from gelatin from cow tendon (145°C) and the highest was from gelatin from pigskin (237°C).

The tensile strength (TS) and percentage elongation of a variety of gelatin-based films was measured by Arvanitoyannis et al. (1997, 1998a, 1998b). A combination of low and high temperature processes was used in the production of films containing different types of gelatin/carbohydrate combinations with sorbitol and water ratios. The film that produced the highest tensile strength (130 MPa) was made from a combination of gelatin/chitosan/sorbitol and water using a low-temperature process. The lowest tensile-strength (34.4 MPa) film was produced using a high-temperature process and gelatin combined with soluble starch, sorbitol, and water; however, it also had the highest elongation at 44.7%.

Myofibrillar proteins

Edible protein films have been formed from myofibrillar proteins from a variety of animal sources, but most of the work has been focused on fish as the main source of protein. The probability of forming intermolecular bonds mainly depends on the protein shape (fibrous or globular), and on the physicochemical conditions during processing. High molecular weight proteins (e.g. myosin) and fibrous proteins (e.g. myosin and F-actin) can generally form films with good mechanical properties, while globular or pseudo-globular proteins (e.g. G-actin) need to be unfolded first, before film formation. Myofibrillar proteins are found in the muscle, and are mainly composed of myosin and actin, regardless of whether the source of protein is mammalian or fish (Table 18.1).

Proteins obtained from mammal or fish sources must be washed and purified to obtain the myofibrillar portion. Components such as blood, lipids, myoglobin, and collagen must be removed through a series of washing treatments. According to Cuq (2002), a suitable method for obtaining myofibrillar protein from fish is the process known as surimi. It produces a product that is odorless, colorless, forms a gel upon heating, and is stable under freezing conditions. Beef, pork, and chicken proteins have also been processed using the surimi process, but it is not commercially commonly performed as it is with fish (Stanley *et al.*, 1994; Sousa *et al.*, 1997).

In order to form a film, the proteins must be soluble in solution. According to Steffansson and Hultin (1994), cod myofibrillar protein was soluble in water at neutral pH with 0.0003 ionic strength or less. There was a strong relationship between pH and salt concentration. Increasing the sodium chloride concentration or decreasing the pH to 5.5 reduced the solubility, possibly due to the denaturation of the protein. Aggregation and precipitation of proteins can occur when there is a loss of stable hydrophilic surface charges, thus exposing the hydrophobic areas of the protein to interact. This phenomenon is commonly referred to as "salting out". Steffansson and Hultin's (1994) research established important factors for the separation and solubilization of fish myofibrillar proteins to be used in gel or film formations.

Orban (1992) stated that the isoelectric point of myofibrillar proteins is pH 5, and that they are not soluble at that pH but are soluble at pH < 4.5 and > 6.5. Below pH 4, myofibrillar proteins gelate in water–acetic acid solutions at room temperatures. According to Cuq (2002), there are three stages of film formation:

1. Destabilization of intermolecular protein bonds through the use of solvents or heat
2. Polymerization

Table 18.1 Protein composition of myofibrillar proteins (adapted from Cuq, 2002)

Myofibrillar protein	Mammal source (g/100 g protein)	Fish source (g/100 g protein)
Myosin	42	40–48
Actin	21	15–20
Tropomyosin	5	4–6
Troponins	8	4–6
M-proteins	4	3–5
C-proteins	2	2.5–3.0
F-proteins	0.1	0.1
α-Actinin	2	2–3
β-Actinin	1	<1
Nebulin	4	4
Titin	10	10

3. Formation and stabilization of the three-dimensional protein network, upon the removal of the initial destabilization component (removal of solvent or removal of heat).

As stated above, a method for destabilizing intermolecular protein bonds can be obtained by using solvents such as acetic acid, lactic acid or ammonium hydroxide. A low pH (2–3) provides optimal film-forming conditions (Cuq *et al.*, 1995, 1998b; Cuq, 2002). The viscosity of the film is also affected by the protein content and the pH. As the protein content decreased, the film solution viscosity decreased (Cuq *et al.*, 1995). Plasticizers such as glycerol, sorbitol, and sucrose increase the flexibility of myofibrillar protein films. Lactic acid also has a plasticizing effect on films, in addition to lowering the pH (Sousa *et al.*, 1997).

The second method of destabilizing proteins for film formation involves heat. According to Cuq (2002), the glass transition temperature (Tg) of fish myofibrillar protein is between 215°C and 250°C, which is similar to that of collagen, gelatin, and starch. Above 250°C thermal degradation occurs, and this is too low for most polymer-processing methods, such as extrusion (Cuq *et al.*, 1997a). The addition of sucrose and sorbitol as plasticizers decreased the Tg, as did water. Cuq *et al.* (1997b) found that high thermal degradation occurred with low moisture content, and even caused the material to turn brown in color due to caramelization. However, glassy translucent materials were produced with moisture content of 2.2% and at temperatures less than 200°C.

Myofibrillar proteins are not considered to be soluble in water (Cuq *et al.*, 1997b, 1998c). Films immersed in water for 24 hours did not lose their integrity. Reasons for their water insolubility have been studied, but it is yet not fully understood. However, it is known that the addition of plasticizers increases the water solubility of most polymers.

Myofibrillar proteins have a higher tensile strength than other protein-based films shown in Table 18.2. The only synthetic film that compares to fish myofibrillar protein

Table 18.2 Properties of selected synthetic and biopolymer films for comparison to myofibrillar protein films (Krochta *et al.*, 1994; Cuq, 2002)

Film	Tensile strength (MPa)	Elongation (%)	WVP*	O$_2$P**
Polyester	178	85	0.02	12
Polyvinyl chloride	93	30	1.2	23
Low-density polyethylene	13	500	0.04–0.05	1003
High-density polyethylene	26	300	0.014	224
Hydroxypropyl cellulose	15	33	6.2	300
Wheat gluten	3.3	192	5.1	1290
Soy protein	3.6	160	194	14
Corn zein	3.9	213	6.5	35
Fish myofibrillar protein	17	23	3.9–3.8	1–873***

* WVP (water-vapor permeability) ($\times\ 10^{-12}$ mol $\cdot$ m $\cdot$/m$^2\cdot$ s $\cdot$ Pa)
** O$_2$P (oxygen permeability) ($\times\ 10^{-18}$ mol $\cdot$ m $\cdot$/m$^2\cdot$ s $\cdot$ Pa)
*** 1 measured under dry conditions, 873 measured under high relatively humidity.

films in tensile strength is low-density polyethylene, while polyester is approximately ten times stronger. Fish myofibrillar protein film has a relatively low percentage elongation, but comparison with other films is difficult without knowing the relative thickness and whether other films have plasticizers. However, it can be generally concluded that myofibrillar protein films are stronger than most biopolymer films, but less elastic.

Cuq et al. (1996) developed a rheological model to predict the mechanical properties of fish myofibrillar proteins. The model included factors such as elastic modulus, viscosity, force-deformation, relation, and creep. They determined that the model could be used successfully to characterize the molecular interactions in the films. Furthermore, their research compared the elastic modulus of fish myofibrillar protein films to that of cellulose and of low-density polyethylene films. The fish myofibrillar protein films were found to have more relaxation than the cellulose-based films, but less relaxation than the polyethylene-based films.

Permeability rates of synthetic and edible films are also shown in Table 18.2. The water-vapor permeability (WVP) of fish myofibrillar protein is relatively high due to its hydrophilic nature (Cuq, 2002). However, it has a lower WVP compared to most other edible films. The oxygen permeability of the myofibrillar protein is highly dependent upon the relative humidity and the water activity. Under dry conditions the film is an excellent oxygen barrier, but when it is exposed to high moisture conditions it is not as good. This is an important consideration, since edible films are usually used under moist conditions.

The use of plasticizers has a significant effect on the properties of the films. For example, the addition of glycerol, sorbitol or sucrose decreased film strength and elastic modulus; however, it increased the percentage elongation and the water-vapor permeability (Cuq et al., 1997a). Strain at failure increased 100-fold when the ambient relative humidity increased from 11% to 95% (20°C), and 12-fold when the glycerol content increased (Cuq, 2002). The role of water as a plasticizer of amorphous materials in relation to the T_g was studied by Cuq et al. (1997d). Using differential scanning calorimetry (DSC), myofibrillar films exhibited a non-linear decrease in T_g with increase in water content.

Very little work has been done on the sensory properties of myofibrillar protein films; however, Cuq et al. (1996) included measurements of mechanical resistance of sliced fish containing a fish myofribrillar protein coating. Films were applied to fish specimens and tested using a strength-deformation rheogram. The coating was found to affect the texture of the meat slightly by increasing the mechanical resistance when sliced. It was determined that the film should be reduced in strength for it to have acceptable sensory properties as an edible coating on fish surfaces. No studies have reported the effects of protein films on taste; however, Cuq (2002) indicated that the films had a slight fishy odor.

The majority of the research on myofibrillar proteins has focused its attention on coating fish to reduce the oxidative effects. A driving force for further development of these films appears to be the underutilization of fish and fish by-products. In addition, other sources of myofibrillar protein films (beef, pork and chicken) are too well utilized to find a cost-effective purpose for them to be used to produce edible films.

Egg white

The surplus of egg-white products has created a reason to produce more edible films and coatings, and has led to development of new applications in the food industry. Egg white is a mixture of eight globular proteins; ovalbumine, ovotransferrin, ovomucoid, ovomucin, lysozyme, G2 globulin, G3 globulin, and avidin. Ovalbumin, ovotransferin, and ovomucoid constitutes 54%, 12% and 11% of the protein weight, respectively. In addition, ovalbumin is the only protein containing free S–H groups. Ovotransferrin, ovomucoid, and lysozyme contain S–S bonds (Mine, 1995). The S–S bonds are considered important in film formation (Gennadios et al., 1994). The use of egg white in film and coating production represents a nutritional interest, since it is an effective antioxidant (Negbenebor and Chen, 1985). Edible packaging made from egg white is clear and transparent, and has properties similar to other proteins (Gennadios et al., 1996). It has been reported that a pH range from 10 to 12 is necessary to obtain homogeneous films (Okamoto, 1978; Gennadios et al., 1996; Handa, 1999). In proteins, at alkaline pH, the S–S bonds are reduced to S–H groups. The S–H groups are converted to inter- and intramolecular S–S covalent cross-links during the heating treatment (Gennadios et al., 1996). The formation of cross-linked films was also observed when proteins were treated under ultraviolet light, resulting in lower total soluble matter and better mechanical properties (Rhim et al., 1999).

Egg albumen has been studied for its potential to retain moisture inside raisins in cereal/raisin mixtures (Bolin, 1976) and inside meat products (Reutimann et al., 1996). Egg-white coatings have also been evaluated for their effectiveness in making pizza components impermeable (Berberat and Wissgott, 1993).

Bioactive protein-based coatings and films

Active compounds such as antioxidants and antimicrobials can be incorporated into edible films and coatings. The resulting bioactive films or coatings provide more inhibitory effects against spoilage and pathogenic bacteria by lowering the bacterias' diffusion through the food product and maintaining high concentrations of the active compounds on the food surfaces (Gennadios and Kurth, 1997). Organic acids, essential oils, salts, lipids, spices, and bacteriocins (nisin) have been largely studied for their efficiency in controlling the growth of spoilage and pathogenic bacteria in foods (Siragusa and Dickson, 1992; Ming et al., 1997; Padgett et al., 1998; Ouattara et al., 2001).

Ouattara et al. (2001) demonstrated that the use of an edible coating based on milk proteins containing natural antimicrobial compounds (e.g. thyme oil and trans-cinnamaldehyde) that was applied by immersion on shrimp was able to extend the shelf life by 12 days. It was also demonstrated that a synergistic inhibitory effect was obtained by using a combination of a coating with irradiation (Lacroix and Ouattara, 2000). The incorporation of antimicrobial compounds extracted from spices and incorporated in milk-protein film reduced lipid oxidation and –S–H radical production during the post-irradiation storage of ground beef (Ouattara et al., 2002b). Whey proteins are effective antioxidants (Le Tein et al., 2001), and are effective in protecting salmon against oil oxidation. The application of active compounds in whey-protein coatings can also reduce the oxidation of cut fruits and sliced mushrooms (Nisperos-Carriedo *et al.*, 1991; Le Tein *et al.*, 2001). Cysteine and aromatic amino acids are potent free-radical targets (Berlett and Stadtman, 1997), and can reduce the polyphenol oxidase activity in vegetables and thus reduce browning reactions. Coatings based on proteins rich in this amino acid can have good antioxidant properties (Le Tein *et al.*, 2001).

References

Arvanitoyannis, I. S. (2002). Formation and properties of collagen and gelatin films and coatings. In: *Protein Based Films and Coatings* (A. Gennadios, ed.), pp. 275–304. CRC Press, Boca Raton, FL.

Arvanitoyannis, I., Psomiadou, E., Nakayama, A., Aiba, S. and Yamamoto, N. (1997). Edible films made from gelatin, soluble starch and polyols, Part 3. *Food Chem.* **90**, 593–604.

Arvanitoyannis, I., Nakayama, A. and Aiba, S. (1998a). Chitosan and gelatin based edible films: state diagram, mechanical and permeation properties. *Carbohydr. Polym.* **37**, 371–382.

Arvanitoyannis, I., Nakayama, A. and Aiba, S. (1998b). Edible films made from hydroxypropyl starch and gelatin and plasticized by polyols and water. *Carbohydr. Polym.* **36**, 105–119.

Baldwin, E. A., Nisperos, N. O. and Baker, R. A. (1995). Use of edible coatings to preserve quality of lightly and slightly processed products. *Crit. Rev. Food Sci. Nutr.* **35**, 509–524.

Banerjee, R. and Chen, H. (1995). Functional properties of edible films using whey protein concentrates. *J. Dairy Sci.* **78**, 1673–1683.

Ben, A. and Kurth, L. B. (1995). Edible film coatings for meat cuts and primal meat 95. The Australian Meat Industry Research Conference, CSIRO, 10–12 September.

Berberat, A. and Wissgott, U. (1993). US Patent 5,248,512.

Berlett, B. S. and Stadtman, E. R. (1997). Protein oxidation in aging disease and oxidative stress. *J. Biol. Chem.* **272**, 20313–20316.

Bolin, H. R. (1976). Texture and crystallization control in raisins. *J. Food Sci.* **41**, 1316–1319.

Brault, D., D'Aprano, G. and Lacroix, M. (1997). Formation of free-standing sterilized edible films from irradiated caseinates. *J. Agric. Food Chem.* **45**, 2964–2969.

Brunner, J. R. (1977). Milk proteins. In: *Food Proteins* (J. R. Witaker and S. R. Tannenbaum, eds), pp. 175–208. AVI Publishers, Westport, CT.

Chen, H. (1995). Functional properties and applications of edible films made of milk proteins. *J. Dairy Sci.* **78**, 2563–2583.

Chen, H. (2002). Formation and properties of casein films and coatings. In: *Protein-Based Films and Coatings* (A. Gennadios, ed.), pp. 181–211. CRC Press, Boca Raton, FL.

Chen, C. S., Liau, W. Y. and Tsai, G. J. (1998). Antibacterial effects of N-sulfonated and N-sulfobenzoyl chitosan and application to oyster preservation. *J. Food Prot.* **61**, 1124–1128.

Cuq, B. (2002). Formation and properties of fish myofibrillar protein films coatings. In: *Protein-Based Films and Coatings* (A. Gennadios, ed.), pp. 213–232. CRC Press, Boca Raton, FL.

Cuq, B., Aymard, C., Cuq, J.-L. and Guilbert, S. (1995). Edible packaging films based on fish myofibrillar proteins: formation and function properties. *J. Food Sci.* **60(6)**, 1369–1374.

Cuq, B., Gontard, N., Cuq, J.-L. and Guilbert, S. (1996). Stability of myofibrillar protein-based biopackagings during storage. *Lebensm. Wiss. u. Technol.* **29**, 344–348.

Cuq, B., Gontard, N. and Guilbert, S. (1997a). Thermoplastic properties of fish myofibrillar proteins: application to biopackaging fabrication. *Polymer* **38**, 4071–4078.

Cuq, B., Gontard, N. and Guilbert, S. (1997b). Thermal properties of fish myofibrillar protein-based films as affected by moisture content. *Polymer* **38**, 2399–2405.

Cuq, B., Gontard, N., Aymard, C. and Guilbert, S. (1997c). Relative humidity and temperature effects on mechanical and water vapor barrier properties of myofibrillar protein-based films. *Polymer Gels Networks* **5**, 1–15.

Cuq, B., Gontard, N., Cuq, J.-L. and Guilbert, S. (1997d). Selected functional properties of fish myofibrillar protein-based films as affected by hydrophilic plasticizers. *J. Agric. Food Chem.* **45**, 622–626.

Cuq, B., Gontard, N. and Guilbert, S. (1998a). Proteins as agricultural polymers for packaging production. *Cereal Chem.* **75**, 1–9.

Cuq, B., Gontard, N., Cuq, J.-L. and Guilbert, S. (1998b). Packaging films based on myofibrillar proteins: fabrication, properties and applications. *Nahrung* **42**, 260–263.

Dalgleish, D. G. (1989). Milk proteins – chemistry and physics. In: *Food Proteins* (J. E. Kinsella and W. G. Soucie, eds), pp. 155–178. American Oil Chemists Society, Champaign, IL.

De Boer, R., de Wit, J. N. and Hiddink, J. (1977). Processing of whey by means of membranes and some applications of whey protein concentrate. *J. Soc. Dairy Tech.* **30**, 112–120.

Donhowe, I. G. and Fennema, O. (1994). Edible films and coatings: characteristics, formation, definitions and testing methods. In: *Edible Coatings and Films to Improve Food Quality* (J. M. Krochta, E. A. Baldwin and M. O. Nisperos-Carriedo, eds), pp. 1–24. Technomic Publishing Company, Lancaster, PA.

Druchta, J. and De Mulder Johnston, C. (1997). Edible films solve problems. *Food Technol.* **51**, 61–74.

Dybing, S. T. and Smith, D. E. (1991). Relation of chemistry and processing procedures to whey protein functionality: a review. *Cult. Dairy Prod. J.* **26**, 4–12.

Everett, D. H. (1989). *Basic Principles of Colloidal Science*, pp. 167–182. Royal Society of Piccadilly, London, UK.

Gennadios, A. and Kurth, L. B. (1997). Application of edible coatings on meats, poultry and seafoods: a review. *Lebensm. Wiss. u. Technol.* **30**, 337–350.

Gennadios, A., McHugh, T., Weller, C. and Krochta, J. M. (1994). Edible coatings and films based on proteins. In: *Edible Coatings and Films to Improve Food Quality* (J. M. Krochta, E. A. Baldwin and M. Nisperos-Carriedo, eds), pp. 231–247. Technomic Publishing, Lancaster, PA.

Gennadios, A., Weller, C. L., Hanna, M. A. and Froning, G. W. (1996). Mechanical and barrier properties of egg albumen films. *J. Food Sci.* **61**, 585–589.

Glover, F. A. (1985). Ultrafiltration and reverse osmosis for the dairy industry. *Tech. Bull. 5.* NIRD, Reading, UK.

Guilbert, S., Gontard, N. and Gorris, L. G. M. (1996). Prolongation of the shelf-life of perishable food products using biodegradable films and coatings. *Lebensm. Wiss. u. Technol.* **29,** 10–17.

Gustavson, K. H. (1956). *The Chemistry and Reactivity of Collagen.* Academic Press, New York, NY.

Handa, A., Gennadios, A., Froning, G. W., Kuroda, N. and Hanna, M. A. (1999). Tensile, solubility and electrophoretic properties of egg white films as affected by surface sulfhydryl groups. *J. Food Sci.* **64,** 82–85.

Harrington, W. F. (1966). Collagen. In: *Encyclopedia of Polymer Science and Technology,* (H. F. Mark, N. G. Gaylord and N. M. Bikales, eds), Vol. 4, *Plastics, Resins, Rubbers, Fibers,* pp. 1–16. Interscience Publishers, New York, NY.

Hood, L. L. (1987). Collagen in sausage casting. In: *Advances in Meat Research* (A. M. Pearson, T. R. Dutson and A. J. Bailey, eds), pp. 109–129. Van Nostrand Reinhold, New York, NY.

Hunt, R. G., Sellers, V. R., Franklin, W. E. *et al.* (1990). *Estimation of the Volume of Municipal Solid Waste and Selected Components in Trash Cans and Landfills.* Franklin Associates Ltd., Prairie Village, KS, and Garbage Project, Tucson, AZ.

Imeson, A. P., Ledward, D. A. and Mitchell, J. R. (1977). On the nature of the interaction between some anionic polysaccharide acid proteins. *J. Sci. Food Agric.* **28,** 661.

Jones, H. W. and Whitmore, A. (1972). US Patent 3,694,234.

Kester, J. J. and Fennema, O. (1986). Edible films and coatings: a review. *Food Technol.* **December,** 47–59.

Kinsella, J. E. (1984). Milk proteins: physiological and functional properties. *Crit. Rev. Food Sci. Nutr.* **21(3),** 197–262.

Kinsella, J. E. and Whitehead, D. M. (1989). Proteins in whey: chemical, physical and functional properties. *Adv. Food Nutr. Res.* **33,** 343–438.

Krochta, J. M. (1997). Edible protein films and coatings. In: *Food Proteins and Their Applications* (S. Damodaran and A. Paraf, eds), pp. 529–549. Marcel Dekker, New York, NY.

Krochta, J. M. and De Mulder-Johnston, C. (1997). Edible and biodegradable polymer films: challenges and opportunities. *Food Technol.* **51,** 60–74.

Kroger, M. and Igoe, R. S. (1971). Edible containers. *Food Prod. Dev.* **5,** 74, 76, 78–79, 82.

Lacroix, M. and Ouattara, B. (2000). Combined industrial processes with irradiation to assure innocuity and preservation of food products: a review. *Food Res. Intl* **33,** 719–724.

Le Tien, C., Letendre, M., Ispas-Szabo, P. *et al.* (2000). Development of biodegradable films from whey proteins by cross-linking and entrapment in cellulose. *J. Agric. Food Chem.* **48,** 5566–5575.

Le Tien, C., Vachon, C., Mateescu, M. A. and Lacroix, M. (2001). Milk protein coatings prevent oxidative browning of apples and potatoes. *J. Food Sci.* **66,** 512–516.

Letendre, M., D'Aprano, G., St-Gelais, D., Delmas-Patterson, G. and Lacroix, M. (2002a). Isothermal calorimetry study of calcium caseinate and whey protein isolate edible films cross-linked by heating and gamma irradiation. *J. Agric. Food Chem.* **50,** 6053–6057.

Letendre, M., D'Aprano, G., Lacroix, M., Salmieri, S. and St-Gelais, D. (2002b). Physicochemical properties and bacterial resistance of biodegradable milk protein films containing agar and pectin. *J. Agric. Food Chem.* **50,** 6017–6022.

Lieberman, E. R. and Guilbert, S. G. (1973). Gas permeation of collagen films as affected by cross-linkage, moisture and plasticizer content. *J. Polym. Sci. Symp.* **41**, 33–43.

Lim, L. T., Mine, Y., Britt, I. J. and Tung, M. A. (2002). Formation and properties of egg white protein films and coatings. In: *Protein-Based Films and Coatings* (A. Gennadios, ed.), pp. 233–252. CRC Press, Boca Raton, FL.

Mahmoud, R. and Savello, P. A. (1993). Mechanical properties of and water vapor transferability through whey protein films. *J. Dairy Sci.* **75**, 942–946.

Matthews, M. E. (1984). Whey protein recovery processes and products. *J. Dairy Sci.* **67**, 2680–2692.

McHugh, T. H. (2000). Protein–lipid interactions in edible films and coatings. *Nahrung* **44**, 148–151.

McHugh, T. H. and Krochta, J. M. (1994). Water vapor permeability properties of edible whey protein-lipid emulsion films. *J. Am. Oil Chem. Soc.* **71**, 307–312.

Mezgheni, E., D'Aprano, G. and Lacroix, M. (1998). Formation of sterilized edible films based on caseinates: effects of calcium and plasticizers. *J. Agric. Food Chem.* **46**, 318–324.

Mine, Y. (1995). Recent advances in the understanding of egg white protein functionality. *Trends Food Sci. Technol.* **6**, 225–232.

Ming, X., Weber, G. H., Ayres, J. W. and Sandine, W. E. (1997). Bacteriocins applied to food packaging materials to inhibit *Listeria monocytogenes* on meats. *J. Food Sci.* **62**, 413–415.

Morgan, B. H. (1971). Edible packaging update. *Food Prod. Dev.* **5**, 75–77, 108.

Morr, C. V. and Ha, E. Y. W. (1993). Whey protein concentrates and isolates: processing and functional properties. *Crit. Rev. Food Sci. Nutr.* **33**, 431–476.

Negbenebor, C. A. and Chen, T. C. (1985). Effect of ovalbumen on TBA values of comminuted poultry meat. *J. Food Sci.* **50**, 270–271.

Nielson, P. M. (1995). Reactions and potential industrial applications of transglutaminase. *Rev. Lit. Patents, Food Biotechnol.* **9**, 119–156.

Nisperos-Carriedo, M. O., Baldwin, E. A. and Shaw, P. E. (1991). Development of an edible coating for extending postharvest life of selected fruits and vegetables. *Proc. Fl. State Hort. Soc.* **104**, 122–125.

Okamoto, S. (1978). Factors affecting protein film formations. *Cereal Foods World* **23**, 256–262.

Orban, E., Quaglia, G. B., Casini, I., Caproni, I. and Moneta, E. (1992). Composition and functionality of sardine native proteins. *Lebensm. Wiss. u. Technol.* **25**, 371–373.

Ouattara, B., Sabato, S. F. and Lacroix, M. (2001).Combined effect of antimicrobial coating and gamma irradiation on shelf life extension of pre-cooked shrimp (*Penaeus* spp.). *Intl J. Food Microbiol.* **68**, 1–9.

Ouattara, B., Le Tien, C., Vachon, C., Mateescu, M. A. and Lacroix, M. (2002a). Use of gamma irradiation cross-linking to improve the water vapor permeability and the chemical stability of milk protein films. *Rad. Physics Chem.* **63**, 821–825.

Ouattara, B., Giroux, M., Smoragiewicz, W., Saucier, L. and Lacroix, M. (2002b). Combined effect of gamma irradiation, ascorbic acid and edible film on the improvement of microbial and biochemical characteristics of ground beef. *J. Food Protect.* **6**, 981–987.

Padgett, T., Han, I. Y. and Dawson, P. L. (1998). Incorporation of food-grade antimicrobial compounds into biodegradable packaging films. *J. Food Protect.* **61**, 1330–1335.

Peyron, A. (1991). L'enrobage et les produits filmogènes: un nouveau mode d'emballage. *Viandes Prod. Carnes* **12**, 41–46.

Piez, K. A., Bornstein, P. and Kang, A. H. (1968). The chemistry and biosynthesis of inter-chain crosslinks in collagen. In: *Symposium on Fibrous Proteins* (W. G. Crewther, ed.), pp. 205–211. Plenum Press, New York, NY.

Ressouany, M., Vachon, C. and Lacroix, M. (1998). Irradiation dose and calcium effect on the mechanical properties of cross-linked caseinate films. *J. Agric. Food Chem.* **46**, 1618–1623.

Ressouany, M., Vachon, C. and Lacroix, M. (2000). Microbial resistance of caseinate films crosslinked by gamma irradiation. *J. Dairy Res.* **67**, 119–124.

Reutimann, E. J., Vadehra, D. V. and Wedral, E. R. (1996). US Patent 5,567,453.

Rhim, J. W., Gennadios, A., Fu, D., Weller, C. L. and Hanna, M. A. (1999). Properties of ultraviolet irradiated protein films. *Lebensm. Wiss. u. Technol.* **32**, 129–133.

Sabato, S. F., Ouattara, B., Yu, H., D'Aprano, G. and Lacroix, M. (2001). Mechanical and barrier properties of cross-linked soy and whey protein based films. *J. Agric. Food Chem.* **49**, 1397–1403.

Sacharow, S. (1972). Edible films. *Packaging* **43**, 6, 9.

Shih, F. F. (1994). Interaction of soy isolate with polysaccharide and its effect on film properties. *J. Am. Oil Chem. Soc.* **71**, 1281–1285.

Shimada, K. and Cheftel, J. C. (1998). Sulfhydryl group disulfide bond interchange during heat induced gelation of whey protein isolate. *J. Agric. Food Chem.* **37**, 161–168.

Siragusa, G. A. and Dickson, J. S. (1992). Inhibition of *Listeria monocytogenes* on beef tissue by application of organic acids immobilized in a calcium alginate gel. *J. Food Sci.* **57**, 293–296.

Sousa, S. A., Sobral, P. J. A. and Menegalli, F. C. M. (1997). Desenvolvimento de filmes comestiveis a base de proteinas miofibrilares extraidas de carne bovina. In: *Proceedings of Workshop Sobre Biopolimeros, Pirrassununga, Brazil* (P. J. A. Sobral and G. Chuzel, eds), pp. 102–106.

Stanley, D. W., Stone, A. P. and Hultin, H. O. (1994). Solubility of beef and chicken myofibrillar proteins in low ionic strength media. *J. Agric. Food Chem.* **42**, 863–867.

Stefansson, G. and Hultin, H. O. (1994). On the solubility of cod muscle proteins in water. *J. Agric. Food Chem.* **42**, 2656–2664.

St-Gelais, D., Hache, S. and Gros-Louis, M. (1992). Combined effects of temperature, acidification and diafiltration on composition of skim milk retentate and permeate. *J. Dairy Sci.* **75**, 1167–1172.

Takahashi, K., Nakata, Y., Someya, K. and Hattori, M. (1999). Improvement of the physical properties of pepsin-solubilized elastin collagen film by crosslinking. *Biosci. Biotechnol. Biochem.* **63**, 2144–2149.

Taylor, M. M., Liu, C. K., Latona, N., Marmer, W. N. and Brown, E. M. (2002). Enzymatic modification of hydrolysis products from collagen using a microbial transglutaminase. II. Preparation of films. *J. ALCA* **97**, 225–234.

Thakur, B. R., Singh, R. K. and Handa, A. K. (1997). Chemistry and uses of pectin – a review. *Crit. Rev. Food Sci. Nutr.* **37**, 47–73.

Trotter, J. A., Kadler, K. E. and Holmes, D. F. (2000). Echinoderm collagen fibrils grow by surface-nucleation and propagation from both centers and ends. *J. Mol. Biol.* **300,** 531–540.

Vachon, C., Yu, H.-L., Yefsah, R., Alain, R., St-Gelais, D. and Lacroix, M. (2000). Mechanical and structural properties of milk protein edible films cross-linked by heating and gamma irradiation. *J. Agric. Food Chem.* **48,** 3202–3209.

Wanstedt, K. G., Seideman, S. C., Connelly, L. S. and Quenzer, N. M. (1981). Sensory attributes of precooked, calcium alginate-coated pork patties. *J. Food Protect.* **44,** 732–735.

Young, H. H. (1967). Gelatin. In: *Encyclopedia of Polymer Science and Technology* (H. F. Mark, N. G. Gaylord and N. M. Bikales, eds), Vol. 7, *Plastics, Resins, Rubbers, Fibers*, pp. 456–460. Interscience Publishers, New York, NY.

19 Edible films and coatings from starches

Zhiqiang Liu

Introduction

Starch is one of the most abundant natural polysaccharides (Narayan, 1994). The world starch production in 2000 was 49 million tons, of which over 80% was from corn starch (LMC International, 2002). Other main agricultural sources for starch production include wheat, potato, and cassava (Ellis *et al.*, 1998; LMC International, 2002).

As major ingredients or additives, native or modified starches have been playing important roles in the food industry. Nutritionally, starches supply more than 50% of the calories consumed by the human population (Whistler, 1984). In addition, starches and their derivatives have been used to modify physical properties of food products such as soups, sauces, snacks, batters, and meat products (Thomas and Atwell, 1997), contributing mainly to texture, viscosity, gel formation, adhesion, binding, moisture retention, product homogeneity, and film formation (Thomas and Atwell, 1997).

Starch films are often transparent (Wolff *et al.*, 1951; Lourdin *et al.*, 1997a; Myllärinen *et al.*, 2002a) or translucent (Rindlav *et al.*, 1997), odorless, tasteless, and colorless (Wolff *et al.*, 1951). They have been utilized in the packaging and coating of food products, because of their edibility and low permeability to oxygen (Mark *et al.*, 1966; Roth and Mehltretter, 1970). Compared to non-starch edible films described in this book, starch films are usually cost effective. The overall performance of starch films and coatings is highly likely to be customizable, because of the availability of a wide variety of starches and their capacity for physical and/or chemical modifications (Ellis *et al.*, 1998; Liu, 2002). In fact, edible starch films are perhaps the most

Innovations in Food Packaging
ISBN: 0-12-311632-5

effective forms of soluble packaging in terms of performance, adaptability to products, and production operations and costs (Daniels, 1973). Edible starch films and coatings are commonly used in bakery, confectionery, batters, and meat products (Thomas and Atwell, 1997). A series of starch-based film formers has been commercialized by National Starch and Chemical Company under the brand of CRYSTAL, which can form thin, clear, and non-tacky films over bakery glazes, and reduce moisture migration into icings (National Starch and Chemical Company, 2003). High-amylose starch films are potential coatings for oxygen-sensitive foods under normal and refrigerator storage (Senti and Dimler, 1959).

This chapter is dedicated to edible films and coatings from starches, with emphasis on the film formation process and the relationship between the structure and properties of self-support starch films.

Starch fundamentals

Starch is normally a mixture of amylose and amylopectin polymers. Amylose is a nearly linear polymer of α-1,4 anhydroglucose units, with a molecular weight of 10^5–10^6 (Galliard and Bowler, 1987; Durrani and Donald, 1995). Amylopectin is a highly branched polymer consisting of short α-1,4 chains linked by α-1,6 glucosidic branching points occurring every 25–30 glucose units, with a molecular weight of 10^7–10^9 (Gaillard and Bowler, 1987; Durrani and Donald, 1995). The content of amylose in starch varies from 0 to 100%, depending on the botanic origins (Shannon and Garwood, 1984; Yoshimoto et al., 2000; Gerard et al., 2001). Most starches, such as those from wheat, corn and potato, contain 20–30% amylose. However, the amylose content can be as high as 50–80% in high-amylose starches such as amylomaize and Hylon VII, or as low as less than 5% in waxy starches such as waxy corn, sorghum and rice. In addition, native cereal starches often contain a small amount (<2%) of non-carbohydrate components such as lipids, phosphates, and proteins (Galliard and Bowler, 1987; van Soest, 1996).

Native starches are in the form of granules, in which amylose and amylopectin are structured by hydrogen-bonding in a manner of alternative semi-crystalline and amorphous rings (Cameron and Donald, 1992; Jenkins et al., 1993). Due to the orderly arrangement of the crystalline areas, starch granules are birefringent (French, 1984; Paris et al., 1999), showing an interference pattern of the Maltese cross under polarized light. Since amylopectin accounts for the crystallinity in native starch granules (Zobel, 1988; Jenkins et al., 1993), higher crystallinity is expected for starches containing more amylopectin (Cheetham and Tao, 1998). Basically, starch granules are semicrystalline. The degree of crystallinity even for waxy corn starch (almost 100% amylopectin) is lower than 50% (Table 19.1).

In general, native starch granules exhibit three X-ray diffraction (XRD) patterns: A, B and C (Figure 19.1). In the A-type crystal, the double helices pack in an anti-parallel manner into an orthorhombic unit-cell (Figure 19.2a), resulting in nearly hexagonal close-packing (Wu and Sarko, 1978a). In the B-type crystal, the double helices pack also in an anti-parallel manner but into a hexagonal unit cell with two helices per cell

Table 19.1 Physical properties of some starches

Starch	Amylose (%)	Polymorph	X_c (%)	T_{Gel} (°C)
Corn/maize				
Waxy (Amioca)[a]	<1	A	40	65–70[b]
Regular[a]	30	A	27	75–80[b]
Amylomaize V[a]	50	B	19	63–92[c]
Amylomaize VII[a]	70	B	16	63–92[c]
Pea (*rug4*)[d]	33	C	23	58–75
Tapioca[b]	17	A	38	65–70
Potato[b]	21	B	28	60–65
Wheat[b]	28	A	36	80–85

X_c, degree of crystallinity; T_{Gel}, gelatinization temperature; [a]Mani and Bhattacharya (2001); [b]Ellis *et al.* (1998); [c]Bogracheva *et al.* (1999); [d]High amylose starches are not completely gelatinized in boiling water (Murphy, 2000).

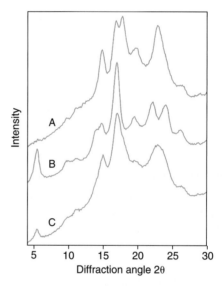

Figure 19.1 X-ray diffraction spectrum for A-type (corn), B-type (potato) and C-type (pea) starch. Reprinted from Bogracheva *et al.* (1999), with permission from Elsevier.

(Figure 19.2b), leaving an open channel that is filled with water molecules (Wu and Sarko, 1978b). Although the B-type crystal has less denser packing than the A-type (Figure 19.2), the similarity in their packing modes suggest that A- and B-type crystals are inter-convertible (Wu and Sarko, 1978a). Regular cereal (e.g. wheat, corn) starches show the A-type polymorph, whereas root (e.g. potato) and high-amylose (>50%) starches show the B-type polymorph (Table 19.1). The C-type polymorph (Figure 19.1), which is actually a mixture of A- and B-type polymorphs (Bogracheva *et al.*, 1998, 2002), is usually found with low-amylose pea starches (Table 19.1). Some starch granules show a trace of V-type polymorph, attributed to complexes of amylose with lipids (Fukui and Nikuni, 1969). The V-type crystal resembles a single amylose helix with small lipid molecules aligned with the helical center (Buléon *et al.*, 1998).

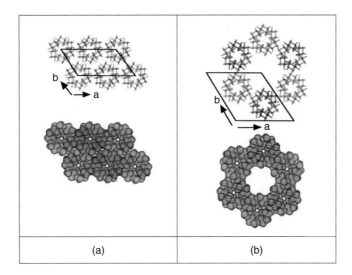

Figure 19.2 Packing of starch double helices. (a) A-type and (b) B-type polymorphs. Each circle represents a view down the z-axis of a double helix. Reprinted from Buléon *et al.* (1998), with permission from Elsevier.

Film formation

Starch solution

Preparing a clear starch solution is normally the first step when making edible starch films. While it is important to minimize or avoid hydrolysis and oxidation of starch, a solution-making process must ensure that starch polymers are completely gelatinized, disintegrated, and solubilized (Protzman *et al.*, 1967; Bader and Göritz, 1994a; Lourdin *et al.*, 1995, 1997b).

It is important to prepare starch solutions with a proper solid concentration, in order to insure both the continuity of films and the ease of casting. When the solid concentration is very low (e.g. <4%), the solution of completely solubilized starch polymers hardly gels, causing problems in forming a continuous film of sufficient thickness (Protzman *et al.*, 1967). However, when the concentration is very high (e.g. >30%), the solution is too viscous to be cast evenly (Protzman *et al.*, 1967). A solid concentration of 10–15% is therefore suggested for casting conventional starch films.

To completely dissolve high-amylose (>50%) starches in water it requires elevated temperatures (>120°C) and super-atmospheric pressures, which can be fulfilled by autoclaving. Bader and Göritz (1994a) found that 155°C is the lowest possible temperature at which high-amylose (55%) starch granules will have completely disintegrated so as to obtain a homogeneous solution. Even higher temperatures are needed for fractionated or retrograded amylose (Muetgeert *et al.*, 1962; Sarko *et al.*, 1963). Although low-amylose (<40%) starches, including waxy starches and fractionated amylopectin, are literally soluble in hot water (>50°C) (Daniels, 1973; Ritzl *et al.*, 1998), high temperatures and super-atmospheric pressures are still necessary to obtain a clear starch solution (Lourdin *et al.*, 1997a, 1997b). To minimize or avoid thermal and oxidative degradation of starch in water at high temperatures, it is often necessary to heat up the solution rapidly (e.g. 2.5°C/min) (Bader and Göritz, 1994a) and keep the solubilization process under the protection of nitrogen atmosphere (Lourdin *et al.*, 1997a, 1997b).

Complete solubilization of high-amylose starches in water at relatively low temperatures can be facilitated by using amylose-complexing agents, which have been utilized to assist isolating amylose from native starches (Wolff *et al.*, 1952). Normally, amylose is readily dissolvable in hot water (e.g. at 95°C) saturated with a complexing agent (Wolff *et al.*, 1951, 1952; Davis *et al.*, 1953; Lourdin *et al.*, 1995), which can be low molecular weight aliphatic alcohols (e.g. ethanol, propanol, butanol, pentanol), glycol ethers (e.g. diethyl Cellosolve), or organic bases (e.g. pyridine) (Davis *et al.*, 1953). Among those agents, butanol is the most commonly used in making amylose films. It is recommended that the complexing agent should be driven off prior to film casting (Wolff *et al.*, 1952; Lourdin *et al.*, 1995), otherwise the film may appear spotted (Wolff *et al.*, 1951).

Aqueous starch solutions are normally unstable. In particular, those with a high solid concentration and/or high-amylose starch tend to gel immediately upon cooling (Young, 1984), which makes casting more difficult (Daniels, 1973). It is therefore necessary to keep the starch solutions at a temperature above their gelation temperatures prior to casting (Davis *et al.*, 1953). For amylose solutions, the gelation temperature is basically a linear function of the concentration of amylose (Muetgeert *et al.*, 1962). When the amylose solution has a solid concentration of 10–15%, the gelation temperature is 60–74°C (Muetgeert *et al.*, 1962). To be on the safe side, the amylose solutions should be kept at about 80°C before casting.

Film formation

Self-supporting starch films are commonly produced by casting the starch solution upon a support with a smooth surface, on which it is dried until the film can be taken off the support (Wolff *et al.*, 1951; Daniels, 1973). Instead of a static support, a rotating drum or a moving belt can be used for the continuous production of starch films. Sometimes, applying a release agent (e.g. lecithin) to the surfaces is necessary to make the film-stripping process easier (Protzman *et al.*, 1967).

Having been cast onto a surface at room temperature, the aqueous solution of high-amylose starch or amylose starts gelling upon cooling, due to the association of glucose chains (Protzman *et al.*, 1967; Wu and Sarko, 1978a, 1978b; Leloup *et al.*, 1992). The concurrent evaporation of water leads to the concentration of the solution and accelerates the gelling process. When all possible chain associations have been completed, further evaporation of water reduces the bulk of the gel until the majority of the free water is removed. Often, the surface upon which the solution is cast serves as a sufficient restraint to prevent the gel or film from shrinking (Protzman *et al.*, 1967). Structurally, starch films resemble condensed starch gels. For example, amylose films show an open network composed of nanoscale rod-like strands (Unbehend and Sarko,

1974; Rindlav-Westling *et al.*, 1998), which presumably consist of contiguously associated blocks of amylose double helices (Leloup *et al.*, 1992).

Drying is profoundly crucial to the appearance and performance of starch films
(Davis *et al.*, 1953; Protzman *et al.*, 1967). Rapid drying is normally undesirable,
because it can induce cracks, warping, and concentric drying marks on the films (Davis
et al., 1953), or even granulate the films (Sarko *et al.*, 1963; Germino *et al.*, 1964;
Unbehend and Sarko, 1974). Over-drying should also be avoided to prevent the films
from becoming too brittle (Protzman *et al.*, 1967). Film thickness is roughly controlled
by the solid concentration and the amount of solution cast on the support (Wolff *et al.*,
1951). For example, amylose films with thickness ranging from 11 to 170 μm can be
prepared by varying the concentration of the casting solution from 7 to 15%, and the
clearance of the casting blade from 380 to 1270 μm to the support (Wolff *et al.*, 1951).

Structures in starch films

Similar to starch granules, starch films are often semicrystalline, containing both
amorphous and crystalline phases. The amorphous and crystalline phases are characterized by the glass transition temperature and the degree of crystallinity, respectively,
both of which will affect the major properties of starch films such as mechanical and
gas barrier properties.

Amorphous structure and glass transition temperature

Upon complete gelatinization and solubilization, native starch loses its crystallinity,
which can rarely be restored upon immediate drying (Ritzl *et al.*, 1998) because the retrogradation or recrystallization of starch is often much slower than the rate of drying
(Kalichevsky *et al.*, 1992). Therefore, freshly prepared starch films are basically amorphous (Rindlav *et al.*, 1997; Rindlav-Westling *et al.*, 1998; García *et al.*, 2000; Myllärinen
et al., 2002a), even though some fresh amylose films may contain a little crystallinity
(<10%) (Myllärinen *et al.*, 2002a). Because of its high molecular weight and large number of branches, amylopectin retrogrades or recrystallizes more slowly than amylose
(Goodfellow and Wilson, 1990), and its major presence in starch could interfere with the
recrystallization of amylose. Even though aged starch films are semicrystalline, the
majority (>70%) is still amorphous (Rindlav *et al.*, 1997; Rindlav-Westling *et al.*, 1998).

Glass transition is by far the most important transition among the many transitions
observed in amorphous and semicrystalline polymers, and has a drastic effect on the
performance of such polymers (Bicerano, 2003). Glass transition temperature (T_g) is
defined as the temperature at which the forces holding the principle components (e.g.
amylose and/or amylopectin) of an amorphous solid together are overcome, so that
these components are able to undergo large-scale molecular motions (Bicerano, 2003).
The glass transition temperature of starch films is commonly measured by differential
scanning calorimetry (DSC) (van Soest, 1996; Bizot *et al.*, 1997; Lourdin *et al.*, 1997a,
1997b) and dynamic thermal mechanical analysis (DMTA) (Lourdin *et al.*, 1997a;
Gaudin *et al.*, 1999, 2000; Moates *et al.*, 2001). Even though amylose and amylopectin

Table 19.2 Glass transition temperature T_g and coefficient ΔC_p for the Couchman–Karasz equation

	T_g (K)	ΔC_p (J·kg^{-1}·K^{-1})
Pea amylose (cast 100°C)[a]	605	265
Potato starch (cast 90°C)[a]	589	265
Waxy maize starch (cast 90°C)[a]	558	295
Glycerol[b]	187	970
Water[b]	134	1830
Sodium lactate[b]	246	1960
Sorbitol[b]	271	2450

[a] Bizot et al. (1997); [b] Lourdin et al. (1997c).

are incompatible in aqueous solutions (Miles et al., 1985; Kalichevsky and Ring, 1987), their blends apparently behave compatibly, with a single calorimetric glass transition spanning a similar range on their DSC curves (Bizot et al., 1997).

The T_g of completely anhydrous starch cannot be directly measured by DSC, because the material tends to degrade before the occurrence of the glass transition. However, it can be obtained by extrapolation on a plot of T_g versus water content for a starch–water system (Bizot et al., 1997). In general, low molecular weight polymers show low T_g. For example, the T_g (410 K) of lintnerized potato starch (M_w 2.3 $\times$ 10^3) is lower than that (605 K) of pea amylose (M_w 5.9 $\times$ 10^5) (Table 19.2).

The amorphous phase of starch, because of its less ordered structure, is more prone to attack by small molecules such as water and other plasticizers like glycerol. These small molecules tend to interact through hydrogen bonding with glucose chains, thereby loosening up the very strong inter- and intra-molecular interactions (also through hydrogen bonding) of starch polymers, and apparently facilitating the large-scale motion of glucose chains. In other words, the presence of plasticizers depresses the T_g of starch (Biliaderis, 1991). As the content of plasticizer increases, the T_g of starch decreases. As a function of water content, the T_g for a starch-water binary system can be estimated by using the Gordon and Taylor (1952) equation

$$T_g = \frac{(1 - m_w)T_{gs} + km_w T_{gw}}{1 - m_w + km_w} \tag{19.1}$$

where m_w is the water content, T_{gw} and T_{gs} are the glass transition temperatures of water and starch, respectively, and k is a constant. The glass transition temperature for a multi-component system (e.g. starch–water–glycerol) can be predicted using the Couchman–Karasz equation (Couchman, 1978, 1987):

$$T_g = \frac{\sum m_i \Delta C_{pi} T_{gi}}{\sum m_i \Delta C_{pi}} \tag{19.2}$$

where m_i, T_{gi}, and ΔC_{pi} are the mass fraction, the glass transition temperature, and the heat capacity change at T_{gi} of the ith component, respectively. For a starch–water system, equation (19.2) is equivalent to equation (19.1) with $k = \Delta C_{pw}/\Delta C_{ps}$. Table 19.2 lists the values of T_g and ΔC_p for some starches and plasticizers. The glass transition temperatures predicted by the Couchman–Karasz equation agree well with their experimental counterparts for a starch–water–glycerol system (Lourdin *et al.*, 1997b).

Since amylose and amylopectin are polar polymers, with a large number of $-OH$ groups, the effect of plasticizers on the T_g depends on how many functional groups (e.g. $-OH$) the plasticizer can provide to interact with the polar polymers. Consequently, the decrease in the T_g should be proportional to the mole of functional groups in the plasticizer, instead of the mass fraction of the plasticizer. For example, using 100 g of glycerol or water as the plasticizer for 200 g of starch, glycerol can supply 3.26 moles of $-OH$ groups, whereas water can provide 5.56 moles of $-OH$ groups. Consequently, water is a better plasticizer than glycerol to decrease the glass transition temperature of starch. For example, the T_g of starch films containing 21% water is close to room temperature, whereas that of films containing the same amount of glycerol is 93°C (Myllärinen *et al.*, 2002b).

Crystalline structure and degree of crystallinity

Starch films normally show a B-type XRD pattern, regardless of the original polymorphs of starch from which the films are made (Bader and Göritz, 1994a; Rindlav *et al.*, 1997; Rindlav-Westling *et al.*, 1998; Mali *et al.*, 2002; Myllärinen *et al.*, 2002a). However, the A-type crystalline structure can be formed with excessive heating and/or moisture, which promotes the mobility of macromolecular chains and thus allows the rearrangement of the crystalline structure. For example, high-amylose (55%) corn starch films show a B-type (Figure 19.2b) crystalline polymorph when the drying temperature is not higher than 60°C at ambient relative humidity, but show an A-type (Figure 19.2a) when the drying temperature is between 80°C and 100°C, and a C-type when the drying temperature is between 60 and 80°C (Bader and Göritz, 1994a). However, the critical drying temperature at which the B to A polymorph transition occurs is lower for low-amylose starch such as wheat starch (Zobel, 1964). In fact, treating starch granules at high temperatures (95–130°C) and at proper water content has been known to induce a polymorph transition from B to A (Sair, 1967; Kulp and Lorenz, 1981; Colonna *et al.*, 1987; Stute, 1992; Kawabata *et al.*, 1994) or from C to A (Lorenz and Kulp, 1982).

Since amylose is nearly linear whereas amylopectin is highly branched, the crystallinity of starch films is primarily associated with amylose (García *et al.*, 2000), even though amylopectin accounts for the crystalline regions in native starch granules (Jenkins *et al.*, 1993). Consequently, starch films containing more amylose show higher crystallinity (García *et al.*, 2000), which is contrary to the fact that native corn starch containing more amylose has a lower degree of crystallinity. However, amylopectin by itself can recrystallize under some circumstances (Myllärinen *et al.*, 2002a). To favor recrystallization, amylopectin should have an average chain length of 15 or longer (Ring *et al.*, 1987). In this sense, potato amylopectin is a better candidate than cereal amylopectin to form semicrystalline films.

Initial crystallinity of starch films depends on the drying temperature, the air humidity, and the time that elapses during the drying from gel to film (Rindlav *et al.*, 1997). Although the degree of crystallinity of high-amylose (55%) starch films seems independent of the drying temperature up to 60°C, a significantly higher degree of crystallinity is obtained by drying the films at 90°C in ambient relative humidity (Bader and Göritz, 1994a). However, this drying temperature effect diminishes for low-amylose starch films (Rindlav et al., 1997). As the air humidity increases, the crystallinity of low-amylose starch films dried at room temperature slightly increases (Rindlav et al., 1997). The highest crystallinity of 23% can be obtained by drying potato starch films at 20°C in 92% RH (Rindlav et al., 1997). However, this positive effect of air humidity does not hold for amylose and amylopectin films dried at room temperature (Rindlav-Westling et al., 1998). Irrespective of the relative humidity, the degree of crystallinity remains at about 34% for amylose films and zero for amylopectin films (Rindlav-Westling et al., 1998). Once again, the positive effect of the drying temperature and humidity comes from their capability to control the mobility of glucose chains. Since the initial crystallinity of amylose films stems from the association of amylose chains during gelling, and the crystallinity is fully developed during the early stages of drying when the water content is still high (Rindlav-Westling et al., 1998), the air humidity has little effect on the crystallinity of amylose films.

Even though fresh starch films are almost amorphous (Rindlav et al., 1997; Rindlav-Westling et al., 1998; García et al., 2000; Myllärinen et al., 2002a), they develop crystallinity with time (Bader and Göritz, 1994a; García et al., 2000). Normally, the degree of crystallinity of amylose and high-amylose starch films increases and levels off within the first one or two weeks of storage (van Soest and Essers, 1997; Myllärinen et al., 2002a), whereas low-amylose starch films need much longer time (>1 month) to develop detectable crystallinity (Forssell et al., 1999; Myllärinen et al., 2002a). Regardless of the relative humidity, amylopectin films stored at room temperature are basically amorphous (Myllärinen et al., 2002a).

The degree of crystallinity of starch films also depends on air humidity, temperature during storage, and the content of the plasticizer. To increase the crystallinity requires the mobility of macromolecular chains, so that the movement of amorphous molecular chain segments towards crystalline areas is possible (Protzman et al., 1967). An approach often used for this purpose is hydrothermal treatment, or annealing, in which a starch film is conditioned at a temperature well above the T_g but below its melting temperature. In general, the addition of plasticizers such as water and glycerol tends to decrease the T_g of starch films. If the storage temperature is higher than the T_g, starch tends to recrystallize. It is therefore not surprising that amylose films containing other plasticizers (e.g. glycerol) show appreciably higher crystallinity (Myllärinen et al., 2002a). Even amorphous amylopectin films containing 30% glycerol develop 19% crystallinity after one month's storage at 91% RH (Myllärinen et al., 2002a). Sugars can also be used to change the crystallinity of starch films. For example, incorporating fructose and glucose into starch films leads to an increased rate of starch crystallization, whereas adding sucrose will retard crystallization (Arvanitoyannis et al., 1994).

Water sorption

Water is the most important plasticizer for starch films. As discussed above, both the glass transition temperature and the degree of crystallinity are affected by the water content, which usually varies with the environment where the films are stored.

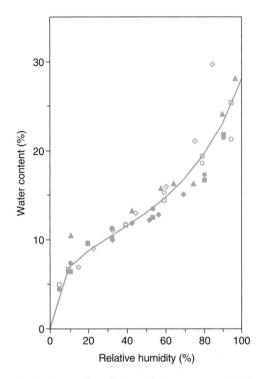

Figure 19.3 Water sorption isotherms of starch films. ■,Waxy corn starch (2% amylose), 20°C (data from Myllärinen *et al.*, 2002b); ◆, potato starch (25% amylose), 25°C (data from Lourdin *et al.*, 1997c); ▲, yam starch (30% amylose), 25°C (data from Mali *et al.*, 2002); ●, high-amylose corn starch (55% amylose), and □, defatted high-amylose corn starch (55% amylose), 20°C (data from Bader and Göritz, 1994b); ◇,wrinkled pea starch (65% amylose), 23°C (data from Funke and Lindhauer, 1994); and ○, potato amylose, 20°C (data from Myllärinen *et al.*, 2002b). The solid line represents the moisture contents predicted by the Guggenheim–Anderson–De Boer (GAB) model with $W_1 = 9.1\%$, $c = 32.2$, and $K = 0.68$.

Water exists in both crystalline and amorphous phases (Rindlav *et al.*, 1997). The water molecules in the crystalline phase are attached to the crystalline structure (Wu and Sarko, 1978a, 1978b). The B-type crystals can hold 25% water, whereas the A-type crystals can hold 12.5% water (Wu and Sarko, 1978b). There is a strong positive correlation between the water content and the crystallinity of starch films (Cleven *et al.*, 1978; Buléon *et al.*, 1982; Rindlav *et al.*, 1997; Cheetham and Tao, 1998). Cheetham and Tao (1998) stated that the driving force for increased water uptake is the formation of crystallites.

Water sorption isotherms are commonly used to study the effect of temperature and relative humidity on the water content of starch films. Normally, starch films possess higher water sorption than native starches (Mali *et al.*, 2002), because the latter have a compact granular structure. However, the type of starch and the amylose content seem to show little effect on the water sorption of starch films (Figure 19.3), particularly at intermediate relative humidity (20–70%) (Bizot *et al.*, 1997). A nearly linear region in the range of intermediate humidity (20–70%) on the sigmoidal curve (Figure 19.3) suggests that a slight fluctuation in the environmental relative humidity could cause a significant change in the water content of starch films, which could consequently affect the T_g and the degree of crystallinity of the films.

Table 19.3 Values of the fitting parameters for the Guggenheim–Anderson–De Boer (GAB) water sorption model

Starch	% Amylose	W_1 (%)	c	K
Waxy maize (cast 90°C)[a]	0	7.7	21.2	0.77
Potato (cast 90°C)[a]	23	8.1	13.6	0.76
High-amylose corn (cast 60°C)[b]	55	9.1	32.2	0.68
Pea amylose (cast 100°C)[a]	100	8.3	18.0	0.75

[a]25°C (Bizot et al., 1997); [b]20°C (Bader and Göritz, 1994b).

The water sorption behavior fits well with the Guggenheim–Anderson–De Boer (GAB) equation (Strauss et al., 1990; Bizot et al., 1997), which is a modification after the Brunauer–Emmet–Teller (BET) model (Brunauer et al., 1938):

$$W = \frac{cW_1 Ka}{(1 - Ka)[1 + (c - 1)Ka]} \qquad (19.3)$$

where W is the water content in the equilibrated starch, W_1 is the amount of water present when every sorption site on the glucose units is covered with one water molecule, c is a temperature dependent adsorption constant, K is a constant, and a is the relative humidity or water activity. The GAB parameters for high-amylose starch (Figure 19.3) are $W_1 = 9.1\%$, $c = 32.2$, and $K = 0.68$ (Bader and Göritz, 1994b). The values of the parameters for other starch films are shown in Table 19.3.

The effect of plasticizers on the water sorption of starch films is complicated. In the case of glycerol, the effect is not a simple superposition (Myllärinen et al., 2002b). At low relative humidity (<60%), the water content of glycerol-plasticized amylose and amylopectin films is significantly lower than that of water-plasticized films (Myllärinen et al., 2002b). However, Mali et al. (2002) observed the opposite for yam starch films – the glycerol-plasticized films are more hydrated than the water-plasticized films. The content of plasticizers exerts marginal effect on the water sorption behavior of starch films (Lourdin et al., 1997b; Mali et al., 2002; Myllärinen et al., 2002b).

Mechanical properties

Internally, the mechanical properties of starch films primarily depend on the mobility of macromolecular chains in the amorphous phase, and on the degree of crystallinity. When serving at a temperature below the T_g, the starch films behave much like a brittle or glassy material, because very little free space is available for the movement of glucose chains as a response to an applied force. In contrast, when working at a temperature higher than the T_g, the films appear flexible and extendible. Crystallites, although less in quantity than the amorphous phase in starch films, behave much like hard particles or physical cross-linkers, which tend to strengthen and stiffen the films. Externally, the ratio of amylose/amylopectin, plasticizer and/or water content, and

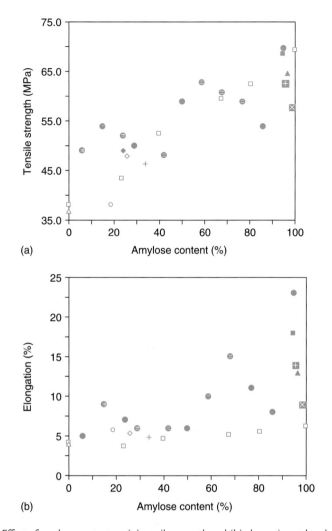

Figure 19.4 Effect of amylose content on (a) tensile strength and (b) elongation at break of starch films. Films (20–30 μm thick) of native corn starch (◆), fractionated amylose from tapioca (■), wheat (▲), sweet potato (⊞), and white potato (⊠), and blends of fractionated amylose and amylopectin from corn starch (●), conditioned at 22°C and 50% RH for more than 4 days (data from Wolff *et al.*, 1951); films (30–60 μm thick) of native waxy maize starch (△), potato starch (○), wheat starch (◇), smooth pea starch (+), and blends of fractionated amylose from smooth pea starch and amylopectin from waxy maize (□), conditioned at 23°C and 57% RH for 48 hours (data from Lourdin *et al.*, 1995).

storage, can affect the mechanical properties of starch films, through their effect on the T_g and the degree of crystallinity of the starch films.

Basically, amylose films exhibit improved mechanical properties over amylopectin films (Myllärinen *et al.*, 2002b). However, it seems that the origin of starch has little effect on the mechanical properties of amylose films (Wolff *et al.*, 1951). Increasing amylose content normally leads to improved mechanical properties (Wolff *et al.*, 1951; Lourdin *et al.*, 1995; Palviainen *et al.*, 2001), including tensile strength and elongation (Figure 19.4). This is probably due to the higher degree of crystallinity and flexibility of amylose. However, Muetgeert and Hiemstra (1958) reported that the starch film containing 10–20% amylopectin exhibits the highest tensile strength.

Plasticizers are primarily utilized to improve the workability and flexibility of plastics (Sears and Darby, 1982), often at the sacrifice of strength and stiffness. The content of plasticizer can vary over a wide range (15–60%), depending on the particular plasticizer used and the overall properties (e.g. flexibility, strength) desired for the starch films or coatings (Wolff et al., 1951; Moore and Robinson, 1968). Although many chemicals can serve as plasticizers for starch (Mark et al., 1964, 1966; Nakamura and Tobolsky, 1967; Moore and Robinson, 1968), only a few are frequently used, including glycerol and sorbitol (Lourdin et al., 1997a, 1997c; Gaudin et al., 1999, 2000; Hulleman et al., 1999; García et al., 2000; Moates et al., 2001; Myllärinen et al., 2002b).

The effect of plasticizer on the mechanical properties of starch films is complicated by its effect on the water content. To be more specific, the films with different amounts of plasticizer also contain different amounts of water, which gives rise to differences in both the crystallinity and the glass transition temperatures. Therefore, it is not surprising that the tensile strength or elastic modulus does not linearly decrease, and the elongation does not linearly increase with the plasticizer content (Lourdin et al., 1997a; Gaudin et al., 1999; Chang et al., 2000; Myllärinen et al., 2002b). An anti-plasticization effect may come into play when the plasticizer content is low (Lourdin et al., 1997a; Gaudin et al., 1999; Chang et al., 2000; Myllärinen et al., 2002b). For example, increasing glycerol content up to 15% leads to brittleness of amylose films (Myllärinen et al., 2002b) and potato starch films (Lourdin et al., 1997a). In the case of sorbitol, the critical content is 27% (Gaudin et al., 1999). Although the cause of anti-plasticization has not been clearly elucidated, it is reportedly associated with the disappearance of local molecular mobility in the amorphous phase (Gaudin et al., 2000), due to the very strong hydrogen bonding between the plasticizer and glucose chains.

Upon aging, starchy materials usually become stronger and stiffer but less flexible (van Soest and Knooren, 1997; Forssell et al., 1999), primarily because of the increased crystallinity. Although the elongation and elastic modulus remain almost unchanged after 2 weeks' storage (van Soest and Knooren, 1997), the strength will probably not be stabilized until 2 months (van Soest and Knooren, 1997; Forssell et al., 1999). Even the presence of a plasticizer like glycerol would not make the aged films flexible, because its plasticization effect varies greatly with climatic conditions (Young, 1967). In fact, aged glycerol-plasticized amylose films have about the same relatively poor elongation and brittleness as water-plasticized amylose films (Young, 1967). There is a continuing need for a plasticizer that will permanently plasticize starch films and/or does not adversely affect the strength and water solubility or water sensitivity of the films (Muetgeert and Hiemstra, 1958).

Oxygen and carbon dioxide barrier

According to the free volume theory, gas molecules pass or diffuse through amorphous phases where the free volume is present (Benczédi, 1999). The rate of transport through a barrier is determined by the size of the gas molecules and by the amount of free volume in the barrier (Benczédi, 1999). Since the free volume is adversely proportional to the degree of crystallinity, any additives or processes that promote the degree of crystallinity in starch films will lead to a better barrier to gases (Arvanitoyannis *et al.*, 1994, 1996). Another mechanism of gas permeability through starch films is related to the absorption and solubility of gases by the film (Rankin *et al.*, 1958). Molecules with similar structure to the starch films (e.g. $-OH$ groups) are more likely to be absorbed or dissolved by starch. These absorbed molecules cause weakening or the disruption of the intra- and inter-molecular hydrogen bonds of starch, and leave on the other side of the weakened film through evaporation (Rankin *et al.*, 1958).

Normally starch films are good barriers to oxygen, since oxygen is a non-polar gas and cannot be dissolved in starch films (Rankin *et al.*, 1958). In ambient environments (e.g. 20°C, 50–60% RH), amylose and amylopectin films have very low oxygen permeability comparable to ethylene vinyl alcohol copolymer (EVOH), which is a good commercial oxygen barrier film (Forssell *et al.*, 2002). Amylose films are better oxygen barriers than amylopectin films, irrespective of the air humidity during film formation (Rindlav-Westling *et al.*, 1998). High-amylose starch films have no detectable oxygen permeability at RH below 100% at 25°C (Mark *et al.*, 1966). However, the oxygen permeability of starch films is greatly affected by the water content in the films (Figure 19.5) (Gaudin *et al.*, 2000; Forssell *et al.*, 2002). The films containing

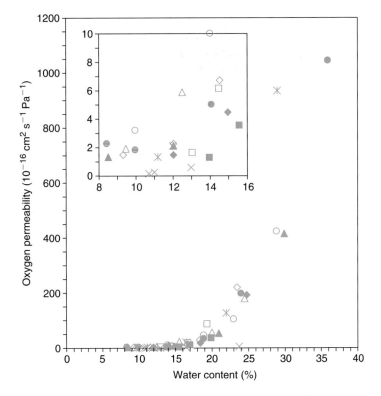

Figure 19.5 Effect of water content on oxygen permeability of starch films at 20°C. Amylose (■); amylose with 10% (◆), 20% (▲), and 30% (●) glycerol; amylopectin (□); amylopectin with 10% (◇), 20% (△), and 30% (○) glycerol (data from Forssell *et al.*, 2002). Wheat starch (27% amylose) with 8.8% (×) and 28% (∗) sorbitol (data from Gaudin *et al.*, 2000).

less than 15% water are good oxygen barriers, whereas those containing more than 20% water lose the oxygen-barrier property (Forssell *et al.*, 2002).

As expected, starch films are less resistant to CO_2 (polar gas) than to O_2 (nonpolar gas), because of the higher solubility of CO_2 in starch films (Arvanitoyannis *et al.*, 1994). The addition of plasticizers significantly reduces the CO_2 permeability (García *et al.*, 2000). For example, high-amylose starch (65% amylose) films (20°C, 63.8% RH, pressure difference of CO_2 101.3 kPa) have an initial CO_2 permeability of $28.05 \pm 7.37 \times 10^{-11}$ $cm^2 s^{-1} Pa^{-1}$, whereas those containing 50% glycerol and 50% sorbitol have values of $3.85 \pm 1.28 \times 10^{-11}$ $cm^2 s^{-1} Pa^{-1}$ and $2.96 \pm 0.46 \times 10^{-11}$ $cm^2 s^{-1} Pa^{-1}$, respectively. Due to the increased crystallinity, aged starch films are expected to be more resistant to CO_2 permeability. After 20 days' storage at 20°C and 63.8% RH, the high-amylose starch films and those with 50% glycerol and 50% sorbitol show decreased CO_2 permeability of $7.71 \pm 1.40 \times 10^{-11}$ $cm^2 s^{-1} Pa^{-1}$, $2.28 \pm 0.33 \times 10^{-11}$ $cm^2 s^{-1} Pa^{-1}$, and $1.82 \pm 0.15 \times 10^{-11}$ $cm^2 s^{-1} Pa^{-1}$, respectively (García *et al.*, 2000). These results also demonstrate the stabilizing effect of the plasticizer on the gas permeability of starch films – plasticized films show less change.

Summary and conclusions

From the perspective of polymer science, starch films and coatings are multicomponent and multiphase systems. Starch films usually contain amylose, amylopectin, water, and/or other plasticizers, and are often semicrystalline. The film composition, in particular the contents of amylose, water, and/or plasticizer, and the film formation conditions have significant effects on the glass transition temperature and the crystallinity of starch films, which, as structurally-related factors, affect the overall properties of the films. Normally, films containing more amylose show better performance in terms of tensile strength, elongation, and gas barrier properties. Addition of plasticizers such as glycerol and sorbitol tends to make the films weaker, but softer and more flexible. The water content, which could be estimated by using the Guggenheim–Anderson–De Boer (GAB) model, greatly affects the oxygen permeability of starch films. Starch films containing more than 15% water lose the oxygen-barrier property. In order to obtain the optimum performance of edible starch films, more research is still needed to comprehensively understand the interactions between starch film composition (e.g. amylose/amylopectin ratio, plasticizer content, water content), film formation conditions (e.g. drying temperature and air humidity), structural factors (e.g. the crystallinity and glass transition temperature) and other film properties (e.g. mechanical, gas barrier).

References

Arvanitoyannis, I., Kalichevsky, M., Blanshard, J. M. V. and Psomiadou, E. (1994). Study of diffusion and permeation of gases in undrawn and uniaxially drawn films made from potato and rice starch conditioned at different relative humidities. *Carbohydr. Polym.* **24**, 1–15.

Arvanitoyannis, I., Psomiadou, E. and Nakamura, S. (1996). Edible films made from sodium caseinate, starches, sugars or glycerol. Part 1. *Carbohydr. Polym.* **31,** 179–192.

Bader, H. G. and Göritz, D. (1994a). Investigations on high amylose corn starch films. Part 1: Wide angle X-ray scattering (WAXS). *Starch,* **46,** 229–232.

Bader, H. G. and Göritz, D. (1994b). Investigations on high amylose corn starch films. Part 2: Water vapor sorption. *Starch* **46,** 249–252.

Benczédi, D. (1999). Estimation of the free volume starch-water barriers. *Trends Food Sci. Technol.* **10,** 21–24.

Bicerano, J. (2003). Glass transition. In: *Encyclopedia of Polymer Science and Technology,* Vols 1–4, Part 1 (H. F. Mark, ed.). John Wiley & Sons, New York, NY.

Biliaderis, C. G. (1991). Non-equilibrium phase transitions of aqueous starch systems. In: *Water Relationships in Foods* (H. Levine and L. Slade, eds), pp. 251–273. Plenum Publishers, New York, NY.

Bizot, H., Le Bail, P., Leroux, B., Roger, P. and Buléon, A. (1997). Calorimetric evaluation of the glass transition in hydrated, linear and branched polyanhydroglucose compounds. *Carbohydr. Polym.* **32,** 33–50.

Bogracheva, T. Y., Morris, V. J., Ring, S. G. and Hedley, C. L. (1998). The granular structure of C-type pea starch and its role in gelatinization. *Biopolymers* **45,** 323–332.

Bogracheva, T. Y., Cairns, P., Noel, T. R. *et al.* (1999). The effect of mutant genes at the r, rb, rug3, rug4, rug5 and lam loci on the granular structure and physico-chemical properties of pea seed starch. *Carbohydr. Polym.* **39,** 303–314.

Bogracheva, T. Y., Wang, Y. L., Wang, T. L. and Hedley, C. L. (2002). Structural studies of starches with different water contents. *Biopolymers* **64,** 268–281.

Brunauer, S., Emmett, P. H. and Teller, E. (1938). Adsorption of gases in multimolecular layers. *J. Am. Chem. Soc.* **60,** 309–319.

Buléon, A., Bizot, H., Delage, M. M. and Multon, J. L. (1982). Evolution of crystallinity and specific gravity of potato starch versus ad- and desorption. *Starch* **34,** 361–366.

Buléon, A., Colonna, P., Planchot, V. and Ball, S. (1998). Starch granules: structure and biosynthesis. *Intl J. Biol. Macromol.* **23,** 85–112.

Cameron, R. E. and Donald, A. M. (1992). A SAXS study of the annealing and gelatinisation of starch. *Polymer* **33,** 2628–2635.

Chang, Y. P., Cheah, P. B. and Seow, C. C. (2000). Plasticizing-antiplasticizing effects of water on physical properties of tapioca starch films in the glassy state. *J. Food Sci.* **65,** 445–451.

Cheetham, N. W. H. and Tao, L. (1998). Variation in crystalline type with amylose content in maize starch granules: an X-ray powder diffraction study. *Carbohydr. Polym.* **36,** 277–284.

Cleven, R., van den Berg, C. and van der Plas, L. (1978). Crystal structure of hydrated potato starch. *Starch* **30,** 223–228.

Colonna, P., Buléon, A. and Mercier, C. (1987). Physically modified starches. In: *Starch: Properties and Potential* (T. Galliard, ed.), pp. 79–114. John Wiley & Sons, Chichester, UK.

Couchman, P. R. (1978). A classical thermodynamic discussion of the effect of composition on glass-transition temperatures. *Macromolecules* **11,** 117–119.

Couchman, P. R. (1987). Glass transition of compatible blends. *Polym. Eng. Sci.* **27,** 618–621.

Daniels, R. (1973). *Edible Coatings and Soluble Packaging*. Noyes Data Corporation, Park Ridge, NJ.

Davis, H. A., Wolff, I. A. and Cluskey, J. E. (1953). US Patent 2656571.

Durrani, C. M. and Donald, A. M. (1995). Physical characterisation of amylopectin gels. *Polym. Gels Networks* **3**, 1–27.

Ellis, R. P., Cochrane, M. P., Dale, M. F. B. *et al.* (1998). Starch production and industrial use. *J. Sci. Food Agric.* **77**, 289–311.

Forssell, P., Hulleman, S., Myllärinen, P., Moates, G. and Parker, R. (1999). Ageing of rubbery thermoplastic barley and oat starches. *Carbohydr. Polym.* **39**, 43–51.

Forssell, P., Lahtinen, R. and Myllärinen, P. (2002). Oxygen permeability of amylose and amylopectin films. *Carbohydr. Polym.* **47**, 125–129.

French, D. (1984). Organization of starch granules. In: *Starch: Chemistry and Technology* (R. L. Whistler, J. N. BeMiller and E. F. Paschall, eds), pp. 183–247. Academic Press, New York, NY.

Fukui, T. and Nikuni, Z. (1969). Heat-moisture treatment of cereal starch observed by X-ray diffraction. *Agric. Biol. Chem.* **33**, 460–462.

Funke, U. and Lindhauer, M. G. (1994). Eigenschaften von Gißßfilmen aus nativen und chemisch modifizierten Stärken. *Starch* **46**, 384–388.

Galliard, T. and Bowler, P. (1987). Morphology and composition of starch. In: *Starch: Properties and Potential* (T. Galliard, ed.), pp. 55–78. John Wiley & Sons, New York, NY.

García, M. A., Martino, M. N. and Zaritzky, N. Z. (2000). Microstructural characterization of plasticized starch-based films. *Starch* **52**, 118–124.

Gaudin, S., Lourdin, D., Le Botlan, D., Ilari, J. L. and Colonna, P. (1999). Plasticisation and mobility in starch-sorbitol films. *J. Cereal Sci.* **29**, 273–284.

Gaudin, S., Lourdin, D., Forssell, P. M. and Colonna, P. (2000). Antiplasticisation and oxygen permeability of starch-sorbitol films. *Carbohydr. Polym.* **43**, 33–37.

Gerard, C., Barron, C., Colonna, P. and Planchot, V. (2001). Amylose determination in genetically modified starches. *Carbohydr. Polym.* **44**, 19–27.

Germino, F. L., Zeitlin, B. R. and Sarko, A. (1964). US Patent 3128209.

Goodfellow, B. J. and Wilson, R. H. (1990). A Fourier transform infrared study of the gelation of amylose and amylopectin. *Biopolymers* **30**, 1183–1189.

Gordon, M. and Taylor, J. S. (1952). Ideal copolymers and the second-order transitions of synthetic rubbers. I. Non-crystalline copolymers. *J. Appl. Chem.* **2**, 493–500.

Hulleman, S. H. D., Kalisvaart, M. G., Janssen, F. H. P., Feil, H. and Vliegenthart, J. F. G. (1999). Origins of B-type crystallinity in glycerol-plasticised, compression-moulded potato starches. *Carbohydr. Polym.* **39**, 351–360.

Jenkins, P. J., Cameron, R. E. and Donald, A. M. (1993). A universal feature in the structure of starch granules from different botanical sources. *Starch* **45**, 417–420.

Kalichevsky, M. T. and Ring, S. G. (1987). Incompatibility of amylose and amylopectin in aqueous solution. *Carbohydr. Res.* **162**, 323–328.

Kalichevsky, M. T., Jaroszkiewicz, E. M., Ablett, S., Blanshard, J. M. V. and Lillford, P. J. (1992). The glass transition of amylopectin measured by DSC, DMTA and NMR. *Carbohydr. Polym.* **18**, 77–88.

Kawabata, A., Takase, N., Miyoshi, E., Sawayma, S. and Kimura, T. (1994). Microscopic observation and X-ray diffractometry of heat/moisture-treated starch granules. *Starch* **46**, 463–469.

Kulp, K. and Lorenz, K. (1981). Heat-moisture treatment of starches I: Physicochemical properties. *Cereal Chem.* **58**, 46–48.

Leloup, V. M., Colonna, P., Ring, S. G., Roberts, K. and Wells, B. (1992). Microstructure of amylose gels. *Carbohydr. Polym.* **18**, 189–197.

Liu, Z. (2002). Starch: from granules to biomaterials. In: *Recent Research Developments in Applied Polymer Science*, Vol. 1, Part 1 (S. G. Pandalai, ed.), pp. 189–219. Research Signpost, Trivandrum, India.

LMC International (2002). *Evaluation of the Community Policy for Starch and Starch Products*. LMC International, Oxford, UK.

Lorenz, K. and Kulp, K. (1982). Cereal- and root starch modification by heat-moisture treatment. I. Physico-chemical properties. *Starch* **34**, 50–54.

Lourdin, D., Della Valle, G. and Colonna, P. (1995). Influence of amylose content on starch films and foams. *Carbohydr. Polym.* **27**, 261–270.

Lourdin, D., Bizot, H. and Colonna, P. (1997a). "Antiplasticization" in starch–glycerol films? *J. Appl. Polym. Sci.* **63**, 1047–1053.

Lourdin, D., Coignard, L., Bizot, H. and Colonna, P. (1997b). Influence of equilibrium relative humidity and plasticizer concentration on the water content and glass transition of starch materials. *Polymer* **38**, 5401–5406.

Lourdin, D., Bizot, H. and Colonna, P. (1997c). Correlation between static mechanical properties of starch-glycerol materials and low-temperature relaxation. *Macromol. Symp.* **114**, 179–185.

Mali, S., Grossmann, M. V. E., García, M. A., Martino, M. N. and Zaritzky, N. E. (2002). Microstructural characterization of yam starch films. *Carbohydr. Polym.* **50**, 379–386.

Mani, R. and Bhattacharya, M. (2001). Properties of injection moulded blends of starch and modified biodegradable polyesters. *Eur. Polym. J.* **37**, 515–526.

Mark, A. M., Roth, W. B., Mehltretter, C. L. and Rist, C. E. (1964). Physical properties of films from dimethyl sulfoxide pretreated-amylomaize starches. *Cereal Chem.* **41**, 197–199.

Mark, A. M., Roth, W. B., Mehltretter, C. L. and Rist, C. E. (1966). Oxygen permeability of amylomaize starch films. *Food Technol.* **20**, 75–77.

Miles, M. J., Morris, V. J., Orford, P. D. and Ring, S. G. (1985). The roles of amylose and amylopectin in the gelation and retrogradation of starch. *Carbohydr. Res.* **135**, 271–281.

Moates, G. K., Noel, T. R., Parker, R. and Ring, S. G. (2001). Dynamic mechanical and dielectric characterisation of amylose–glycerol films. *Carbohydr. Polym.* **44**, 247–253.

Moore, C. O. and Robinson, J. W. (1968). US Patent 3368909.

Muetgeert, J. and Hiemstra, P. (1958). US Patent 2822581.

Muetgeert, J., Bus, W. C. and Hiemstra, P. (1962). Influence of formaldehyde on viscosity stability of concentrated aqueous amylose solutions. *J. Chem. Eng. Data* **7**, 272–276.

Murphy, P. (2000). *Handbook of Hydrocolloids* (G. O. Phillips and P. A. Williams, eds), pp. 41–66. Woodhead Publishing Ltd, Cambridge, UK.

Myllärinen, P., Buléon, A., Lahtinen, R. and Forssell, P. (2002a). The crystallinity of amylose and amylopectin films. *Carbohydr. Polym.* **48**, 41–48.

Myllärinen, P., Partanen, R., Seppälä, J. and Forssell, P. (2002b). Effect of glycerol on behaviour of amylose and amylopectin films. *Carbohydr. Polym.* **50**, 355–361.

Nakamura, S. and Tobolsky, A. V. (1967). Viscoelastic properties of plasticized amylose films. *J. Appl. Polym. Sci.* **11**, 1371–1386.

Narayan, R. (1994). Polymeric materials from agricultural feedstocks. In: *Polymers from Agricultural Coproducts* (M. L. Fishman, R. B. Friedman and S. J. Huang, eds), pp. 2–28. American Chemical Society, Washington, DC.

National Starch and Chemical Company (2003). *Food Starch*. http://www.foodstarch.com

Palviainen, P., Heinämäki, J., Myllärinen, P., Lahtinen, R., Yliruusi, J. and Forssell, P. (2001). Corn starches as film formers in aqueous-based film coating. *Pharm. Develop. Technol.* **6**, 353–361.

Paris, M., Bizot, H., Emery, J., Buzaré, J. Y. and Buléon, A. (1999). Crystallinity and structuring role of water in native and recrystallized starches by ^{13}C CP-MAS NMR spectroscopy. *Carbohydr. Polym.* **39**, 327–339.

Protzman, T. F., Wagoner, J. A. and Young, A. H. (1967). US Patent 3344216.

Rankin, J. C., Wolff, I. A., Davis, H. A. and Rist, C. E. (1958). Permeability of amylose film to moisture vapor, selected organic vapors, and the common gasses. *Ind. Eng. Chem.* **3**, 120–123.

Rindlav, A., Hulleman, S. H. D. and Gatenholm, P. (1997). Formation of starch films with varying crystallinity. *Carbohydr. Polym.* **34**, 25–30.

Rindlav-Westling, A., Stading, M., Hermansson, A.-M. and Gatenholm, P. (1998). Structure, mechanical and barrier properties of amylose and amylopectin films. *Carbohydr. Polym.* **36**, 217–224.

Ring, S. G., Colonna, P., I'Anson, K. J. *et al.* (1987). The gelation and crystallisation of amylopectin. *Carbohydr. Res.* **162**, 277–293.

Ritzl, A., Regev, O. and Yerushalmi-Rozen, R. (1998). Structure and interfacial interactions of thin films of amylopectin. *Acta Polym.* **49**, 566–573.

Roth, W. B. and Mehltretter, C. L. (1970). Films from mixture of viscose and alkali high-amylose corn starch. *J. Appl. Polym. Sci.* **14**, 1387–1389.

Sair, L. (1967). Heat-moisture treatment of starch. *Cereal Chem.* **44**, 8–26.

Sarko, A., Zeitlin, B. R. and Germino, F. J. (1963). US Patent 3086890.

Sears, J. K. and Darby, J. R. (1982). *The Technology of Plasticizers*. John Wiley & Sons, New York, NY.

Senti, F. R. and Dimler, R. J. (1959). High-amylose corn properties and prospects. *Food Technol.* **13**, 663.

Shannon, J. C. and Garwood, D. L. (1984). Genetics and physiology of starch development. In: *Starch: Chemistry and Technology* (R. L. Whistler, J. N. BeMiller and E. F. Paschall, eds), pp. 25–86. Academic Press, New York, NY.

Strauss, U. P., Porcja, R. J. and Chen, S. Y. (1990). Volume effects of amylose-water interaction. *Macromolecules* **23**, 172–175.

Stute, R. (1992). Hydrothermal modification of starches: the difference between annealing and heat/moisture-treatment. *Starch* **44**, 205–214.

Thomas, D. J. and Atwell, W. A. (1997). *Starches*. Eagan Press, St Paul, MN.

Unbehend, J. E. and Sarko, A. (1974). Light scattering and x-ray characterization of amylose films. *J. Polym. Sci. Polym. Phys. Ed.* **12**, 545–554.

van Soest, J. J. G. (1996). *Starch Plastics: Structure–Property Relationships*. Utrecht University, Utrecht, The Netherlands.

van Soest, J. J. G. and Essers, P. (1997). Influence of amylose-amylopectin ratio on properties of extruded starch plastic sheets *J. Macromol. Sci. – Pure Appl. Chem.* **A34**, 1665–1689.

van Soest, J. J. G. and Knooren, N. (1997). Influence of glycerol and water content on the structure and properties of extruded starch plastic sheets during aging. *J. Appl. Polym. Sci.* **64**, 1411–1422.

Whistler, R. L. (1984). History and future expectation of starch use. In: *Starch: Chemistry and Technology* (R. L. Whistler, J. N. BeMiller and E. F. Paschall, eds), pp. 1–9. Academic Press, New York, NY.

Wolff, I. A., Davis, H. A., Cluskey, J. E., Gundrum, L. J. and Rist, C. E. (1951). Preparation of films from amylose. *Ind. Eng. Chem.* **43**, 915–919.

Wolff, I. A., Davis, H. A., Cluskey, J. E. and Gundrum, L. J. (1952). US Patent 2608723.

Wu, H.-C. H. and Sarko, A. (1978a). The double-helical molecular structure of crystalline B-amylose. *Carbohydr. Res.* **61**, 27–40.

Wu, H.-C. H. and Sarko, A. (1978b). The double-helical molecular structure of crystalline B-amylose. *Carbohydr. Res.* **61**, 7–25.

Yoshimoto, Y., Tashiro, J., Takenouchi, T. and Takeda, Y. (2000). Molecular structure and some physicochemical properties of high-amylose barley starches. *Cereal Chem.* **77**, 279–285.

Young, A. H. (1967). US Patent 3312641.

Young, A. H. (1984). Fractionation of starch. In: *Starch: Chemistry and Technology* (R. L. Whistler, J. N. BeMiller and E. F. Paschall, eds), pp. 249–284. Academic Press, New York, NY.

Zobel, H. F. (1964). X-ray analysis of starch granules. In: *Methods in Carbohydrate Chemistry* (R. L. Whistler, ed.), pp. 109–112. Academic Press, New York, NY.

Zobel, H. F. (1988). Starch crystal transformations and their industrial importance. *Starch* **40**, 1–7.

20 Edible films and coatings from non-starch polysaccharides

Monique Lacroix and Canh Le Tien

Introduction

The use of polysaccharides as coating materials for food protection has grown extensively in recent years. These natural polymers can prevent the product's deterioration, extending the shelf life and maintaining the sensory quality and safety of several food products (Robertson, 1993). Generally, these systems are designed by taking advantage of their barrier properties against physical/mechanical impacts, chemical reactions, and microbiological invasion. In addition, the use of polysaccharides presents advantages due to their availability, low cost, and biodegradability. The latter in particular is of great interest, as it leads to a reduction in the large quantities of non-biodegradable synthetic packaging materials. Furthermore, polysaccharides can be easily modified in order to improve their physiochemical properties.

Water-soluble polysaccharides are long-chain polymers that dissolve or disperse in water to give a thickening or viscosity-building effect (Glicksman, 1982). These compounds serve numerous functions, such as providing crispness, hardness, adhesiveness etc. Non-starch polysaccharides are obtained from a variety of sources, such as cellulose derivatives (carboxymethylcellulose, methylcellulose, hydroxypropyl cellulose, hydroxypropyl methyl cellulose, microcrystalline cellulose), seaweed extracts (agar, alginates, carrageenans, furcellaran), various plant and microbial gums (arabic, ghatti, karaya, tragacanth, guar, locust bean, xanthan, gellan, pullulan, levan, elsinan),

Innovations in Food Packaging
ISBN: 0-12-311632-5

Table 20.1 Classification of non-starch polysaccharides*	
Polysaccharide type	**Food use**
Cellulose and derivatives	
• Carboxymethylcellulose	Multipurpose
• Methylcellulose	Multipurpose
• Hydroxypropyl cellulose	Emulsifier, film-former, stabilizer, thickener, suspending agent, protective colloid
• Hydroxypropyl methylcellulose	Emulsifier, film-former, stabilizer, thickener, suspending agent, protective colloid
• Microcrystalline cellulose	Multipurpose
*Chitosan***	Antimicrobial activity, food wraps, controls moisture transfer, enzymatic browning and respiration rate, antioxidant, flavoring
• Gellan gum	Stabilizer
Seaweed extracts	
• Agar	Drying and flavoring agent, stabilizer, thickener, surface-finisher, formulation aid, humectant
• Alginates	Stabilizer, thickener, humectant, texturizer, formulation aid, firming agent, flavor adjuvant, flavor enhancer, processing aid, surface active agent
• Carrageenans	Emulsifier, stabilizer, thickener
• Furcellaran	Emulsifier, stabilizer, thickener
Pectins	Emulsifier, stabilizer, thickener
Microbial fermentation gums	
• Xanthan	Stabilizer, emulsifier, thickener, suspending agent, bodying agent, foam enhancer
Exudate gums	
• Gum arabic	Emulsifier, formulation aid, stabilizer, thickener, humectant, texturizer, surface-finishing agent, flavoring agent, adjuvant, emulsifier
• Gum ghatti	Emulsifier
• Gum karaya	Formulation aid, stabilizer, thickener
• Gum tragacanth	Emulsifier, thickener, stabilizer, formulation aid
Seed gums	
• Guar gum	Emulsifier, formulation aid, stabilizer, thickener, firming agent
• Locust bean gum	Stabilizer, thickener

*Nisperos-Carriedo (1994); **Rudrapatnam and Farooqahmed (2003).

and connective tissue extracts of crustaceans (chitosan) (Nisperos-Carriedo, 1994). The non-starch polysaccharides and their functions are summarized in Table 20.1.

Coatings are defined as thin layers of material applied to the surface of a food product. Edible coatings represent a unique category of packaging materials that differ from other conventional packaging materials in being edible. Edible coatings are applied and formed directly on the food product either by the addition of a liquid film-forming solution, or by being applied with a paintbrush, by spraying, dipping or fluidizing (Cuq *et al.*, 1995). Edible coatings form an integral part of the food product and hence should not interfere with the sensorial characteristics of the food (Guilbert *et al.*, 1997). The edible coating compositions generally contain a polymer as a matrix and a plasticizer or, in certain cases, a stabilizing agent to improve the rheological properties. An edible polymer should be generally recognized as safe (GRAS), and the materials should be used in accordance with current good manufacturing practices (food grade, prepared and handled as a food ingredient) and within any limitations specified by the Food and Drug Administration (Krochta and De Mulder-Johnston, 1997). These considerations are also valid for additives in edible coating formulations.

The ability of some water-soluble polysaccharides to form thermally induced gelatinous coatings can be used for the reduction of oil absorption during frying. Polysaccharide-based formulations may also retard moisture loss from meat products during short-term storage. They do this by acting as a sacrificial moisture barrier to the atmosphere, so that the moisture content of the coated food can be maintained (Kester and Fennema, 1986). Certain polysaccharide films may provide effective protection against surface browning, and oxidation of lipids and other food components (Kester and Fennema, 1986; Nisperos-Carriedo, 1994). In addition to preventing moisture loss, some types of polysaccharide films are less permeable to oxygen. Decreasing the oxygen permeability of coatings can help to preserve certain foods. The following is an overview of several types of polysaccharide coatings and their functions.

On the other hand, edible films are freestanding structures, first formed and then applied to foods. They are formed by casting and drying film-forming solutions on a levelled surface, or by using a drum drier or other traditional plastic processing techniques such as extrusion. Edible films can also be formed by applying the coating solutions directly on a food product. Edible films and coatings may be used to separate different components in multicomponent foods, thereby improving the quality of the product (Krochta and De Mulder-Johnston, 1997).

The film requirements for food products are complex. Unlike inert packaged commodities, foods are often dynamic systems with a limited shelf life and very specific packaging needs. In addition, since foods are consumed to sustain life, the need to guarantee safety is a critical factor in the packaging requirements. While the issue of food quality and safety is the priority in the minds of food scientists, a range of other issues surrounding the development of any food package must be considered before a particular packaging system becomes a reality.

Secondary packaging is often used for physical protection of the food product. It may be a box, surrounding a food packaged in a flexible plastic bag. It may be a corrugated box that contains a number of primary packages in order to ease handling during storage and distribution, improve stackability, or protect the primary packages from mechanical damage. Owing to the barrier properties of edible films and coatings, the coated foods may not require high-barrier packaging materials. Therefore, the entire packaging structure can be simplified, by satisfying the barrier requirements. The barrier functions concern oxygen, moisture, and aroma blocks, as well as physical damage prevention.

In addition to the increased barrier properties, edible films and coatings control adhesion, cohesion, and durability, and improve the appearance of coated foods (Krochta, 1997). They can incorporate active ingredients such as antioxidants, antimicrobial agents, colorants, flavorings, fortified nutrients, and/or spices (Floros *et al.*,

1997). Edible films and coatings must also meet the criteria that apply to conventional packaging materials associated with foods. These relate to barrier properties (water, gases, light, aroma), optical properties (transparency), strength, welding and molding properties, printing properties, migration/scalping requirements, chemical and temperature resistance properties, disposal requirements, and issues such as the user-friendly nature of the materials and whether the materials are price competitive. Edible film materials must also comply with food and packaging legislations, and the interactions between the food and packaging materials must not compromise food quality or safety. In addition, the intrinsic characteristics of edible film materials (for example, whether or not they are biodegradable) can place constraints on their use for foods.

Due to the hydrophilic nature of polysaccharides, polysaccharide-based films exhibit limited water vapor barrier ability (Gennadios *et al.*, 1997). However, films based on polysaccharides such as like alginate, cellulose ethers, chitosan, carrageenan or pectins exhibit good gas-barrier properties (Baldwin *et al.*, 1995; Ben and Kurth, 1995). The gas-permeability properties of such films result in desirable modified atmospheres, thereby increasing the product shelf life without creating anaerobic conditions (Baldwin *et al.*, 1995). Polysaccharide films are used in Japan for meat products, ham and poultry packaging, before smoking and steaming processes. The film is dissolved during the process, and the coated meat exhibits improved yield, structure, and texture, and reduced moisture loss (Labell, 1991; Stollman *et al.*, 1994).

Non-starch polysaccharides used for films and coatings

Cellulosic derivatives

Cellulose is present in all land plants, being the structural material of cell walls (Nisperos-Carriedo, 1994). Cellulose is the most abundantly occurring natural polymer on earth, and is an almost linear polymer of anhydroglucose. At a molecular level, cellulose is a relatively simple polymer consisting of β-[1,4] linked D-glucose molecules in a linear chain. Because of its regular structure and array of hydroxyl groups, it tends to form strong hydrogen-bonded crystalline microfibers, which are insoluble in several solvents. Cellulose is a cheap raw material, and is highly crystalline and insoluble. For film production, cellulose is dissolved in an aggressively toxic mixture of sodium hydroxide and carbon disulfide (xanthation) and then recast into sulfuric acid to produce cellophane films (Petersen *et al.*, 1999). The use of cellulose xanthate as a matrix for entrapment of cross-linked whey protein can produce insoluble films with good mechanical properties, good stability, and increased resistance to enzymatic attack (Le Tien *et al.*, 2000). The application potential of these films as packaging and wrapping materials can be of interest for various materials, and is probably compatible with several types of foods.

However, the usefulness of cellulose as a starting material for edible coatings can be extended by chemical modification to cellulosic derivatives. Generally, cellulose derivatives possess excellent film-forming properties, but they are simply too expensive for bulk use. This is a direct consequence of the crystalline structure of cellulose, making the initial steps of derivatization difficult and costly. Research is required to develop efficient processing technologies for the production of cellulose derivatives, if this situation is to change. Cellulose-based films and coatings reduce moisture loss and reduce the amount of oil absorbed by fried foods (Sanderson, 1981; Dziezak, 1991).

Edible coatings based on cellulose gums have also been used to delay ripening in some climatic fruits like mangoes, papayas and bananas. Furthermore, the application of the same formulation on sliced mushrooms has been shown significantly to reduce enzymatic browning (Nisperos-Carriedo et al., 1991).

Several cellulose derivatives are widely produced commercially, most commonly carboxymethyl cellulose (CMC), methyl cellulose (MC), hydroxypropyl cellulose (HPC), and hydroxypropylmethyl cellulose (HPMC). Edible coatings, which include CMC, MC, HPC, and HPMC, have been applied to a variety of foods to provide moisture, oxygen or oil barriers, and to improve batter adhesion. CMC, MC, HPC or HPMC can be solubilized in aqueous or aqueous–ethanol solutions, producing films with good film-forming properties, but are soluble in water (Gennadios et al., 1997). These cellulose ether films are generally transparent, flexible, odorless, tasteless, water-soluble, and resistant to oils and fats (Kester and Fennema, 1986; Hagenmaier and Shaw, 1990; Hanlon, 1992; Nisperos-Carriedo, 1994). HPC is the least hydrophilic of the cellulose ethers (Gennadios et al., 1997); it is also thermoplastic and capable of injection molding and extrusion. Cellulose can also be chemically modified to give ethyl cellulose (EC), which is biodegradable but not edible.

Non-ionic cellulose ethers have good film-forming properties. However, MC films still do not have good moisture barriers, but do provide an excellent barrier against migration of fats and oils (Nelson and Fennema, 1991). For example, MC and HPMC can be used to reduce oil absorption in fried products such as potato (Sanderson, 1981), fried poultry (Holownia et al., 2000) etc. MC and HPMC can reduce moisture loss during cooking of poultry and seafood (Baker et al., 1994). Balasubramaniam et al. (1997) have demonstrated that meatballs prepared from ground chicken breast and coated with HPMC absorbed 33.7% less fat and retained up to 16.4% more moisture than uncoated controls during deep-fat frying in peanut oil. HPC can retard spoilage and moisture absorption (Ganz, 1969) and retard spoilage and moisture absorption in coated nuts and candies (Krumel and Lindsay, 1976). CMC is an anionic cellulose ether and forms a complex in the presence of casein, increasing the coating formulation viscosity (Keller, 1984); it retains the firmness of fruits and vegetables (Mason, 1969), preserves important flavor components of some fresh commodities (Nisperos-Carriedo and Baldwin, 1990), reduces oxygen uptake without causing carbon dioxide increase in internal fruits and vegetables (Lowings and Cutts, 1982), and improves the puncture strength of films based on caseinate (Ressouany et al., 1998).

Chitosan

Chitosan is a polysaccharide derived from chitin, and is found in abundance in the shells of crustaceans. Chitosan is mainly composed of 2-amino-2-deoxy-β-D-glucopyranose repeating units, but still retains a small amount of 2-acetamido-2-deoxy-β-D-glucopyranose residues. Chitosan with a high amino content (pKa $\approx$ 6.2–7.0) is water-soluble in aqueous acids (Rinaudo and Domard, 1989). In general, chitosan has numerous uses; it can be used as a flocculent, clarifier, thickener, gas-selective membrane, promoter of plant disease resistance, wound-healing factor agent, and antimicrobial agent (Brine *et al.*, 1991). Chitosan also readily forms films, and in general produces materials with a high gas barrier. Chitosan may also be used as coatings for other bio-based polymers that lack gas-barrier properties. Moreover, chitosan has been widely used for the production of edible coatings (Krochta and De Mulder-Johnston, 1997). However, as with other polysaccharide-based polymers, care must be taken for moist conditions. The cationic properties of chitosan offer good opportunities to take advantage of electron interactions with numerous compounds during processing and incorporate specific properties into the material. The cationic property may further be used for the incorporation and/or slow release of active components, adding to the possibilities for the manufacturer in tailoring the properties (Hoagland and Parris, 1996). Further interesting properties of chitosan and chitin in relation to food packaging are their antimicrobial properties (Dawson *et al.*, 1998; Coma *et al.*, 2002) and their ability to absorb heavy metal ions (Chandra and Rustgi, 1998). The former could be valuable in relation to the microbial shelf life and safety of the food products, and the latter could be used to diminish oxidation processes in foods catalyzed by free metals.

Chitin and chitosan are natural antimicrobial compounds against different groups of micro-organisms, such as, bacteria, yeast, and fungi (Yalpani *et al.*, 1992; Ouattara *et al.*, 2000). Because of the positive charge on the C-2 of the glucosamine monomer at pH 6 and below, chitosan is more soluble and has better antimicrobial activity compared to chitin (Yalpani *et al.*, 1992). Also, the efficiency of the antimicrobial properties of chitosan depends on the degree of polymerization and of acetylation (Chen *et al.*, 1998). Sulfonated chitosans (DS = 0.63) have a minimal inhibitory concentration (MIC) of 100 ppm for *Staphylococcus aureus*, *Listeria monocytogenes*, *Escherichia coli*, and *Vibrio parahaemolyticus*. A MIC of 200 ppm was observed for *Pseudomonas aeruginosa* and *Shigella dysenteriae* (Chen *et al.*, 1998). Chitosan can also reduce the growth of numerous fungi. Chitosan with an –NH$_2$ content of 7.5% markedly reduced the radial growth of *Botrytis cinerea* and *Rhizopus stolonifer* in strawberries (El Ghaouth *et al.*, 1991a). A chitosan coating was shown to inhibit sclerotina rot in carrots (Cuero *et al.*, 1991). Treatment with chitosan also induces activity of chitinase enzymes, which are metabolized by the plant for the purpose of self-protection against pathogens (Mauch *et al.*, 1984).

Coatings and films based on chitosan and its *N,O*-carboxymethyl derivatives were used to reduce water loss, respiration, and fungal infection in peaches, Japanese pears, kiwi fruits, strawberries, tomatoes, bell peppers, cucumbers, bananas, and mangoes (El Ghaouth *et al.*, 1991a, 1991b; Kittur *et al.*, 1998). *N,O*-carboxymethyl chitin films have been approved in several countries to preserve fruits over a long period of time (Rudrapatnam and Farooqahmed, 2003).

Due to their good ability to form semipermeable films, chitosan-based films can be expected to modify the internal atmosphere as well as decrease the transpiration losses and delay the ripening in fruits. Chitosan films have good mechanical properties, being flexible and difficult to tear. Chitosan films also have moderate water vapor permeability properties, and exhibit good barriers to the permeation of oxygen (Rudrapatnam and Farooqahmed, 2003). They can also control enzymatic browning in fruits. For example, the application of a chitosan film onto lychee fruit delayed changes in the content of anthocyanins, flavonoids, and total phenolics, delayed an increase in polyphenol oxidase activity, and partially inhibited the increase in peroxidase activity. Browning reactions are related to the oxidation of phenolic compounds (Sapers and Douglass, 1987). The preparation of chitosan films and chitosan-laminated films with other polysaccharides has been reported (Kester and Fennema, 1986; Lobuza and Breene, 1986; Chen *et al.*, 1996; Kittur *et al.*, 1998). Hoagland and Parris (1996) have shown that the storage and loss moduli of chitosan-/pectin-laminated films were significantly greater than the respective moduli of chitosan films alone. However, the water vapor permeation of laminated chitosan/pectin films was not changed as compared to films based on chitosan alone. Makino and Hirata (1997) have shown that a biodegradable laminate consisting of chitosan cellulose and polycaprolactone can be used in modified atmosphere packaging of fresh produce.

Seaweed extracts

Alginate

Alginate is a polysaccharide derived from brown seaweed known as Phaeophyceae, considered to be a (1–>4) linked polyuronic, containing three types of block structure: M block (β-D-mannuronic acid), G block (poly α-L-guluronic acid), and MG block (containing both polyuronic acids) (Whistler and BeMiller, 1973). The relative amounts of the two uronic acid monomers and their sequential arrangement along the polymer chain vary widely, depending on the origin of the alginate. Alginates produce uniform, transparent and water-soluble films. Divalent cations are used as gelling agents in alginate film formation to induce ionic interactions followed by hydrogen bonding (Kester and Fennema, 1986). Films and coatings can be made from a sodium alginate solution by a rapid reaction with calcium in the cold, forming intermolecular associations involving the G-block regions (Nisperos-Carriedo, 1994). The treatment of an alginate film with multivalent cation (i.e. calcium) solutions converts the film to insoluble (Pavlath *et al.*, 1999). Alginates possess good film-forming properties, but tend to be quite brittle when dry; however, they may be plasticized with the inclusion of glycerol (Glicksman, 1983).

Alginate-based films are impervious to oils and fats, but are poor moisture barriers (Cottrell and Kovacs, 1980). However, alginate gel coatings can significantly reduce moisture loss from foods by acting sacrificially – i.e. moisture is lost from the coating before the food significantly dehydrates. Also, alginate coatings are good oxygen barriers (Conca and Yang, 1993), can retard lipid oxidation in foods (Kester and Fennema, 1986), and can improve flavor, texture and batter adhesion. Alginate provides beneficial properties when applied as an edible coating to precooked pork patties (Wanstedt *et al.*, 1981; Amanatidou *et al.*, 2000). Such coatings increase moisture and reduce flavors due to lipid oxidation. Alginate coatings have been found to reduce weight loss (Lazarus *et al.*, 1976) as well as the natural microflora counts in minimally processed carrots, stored lamb carcasses (Lazarus *et al.*, 1976; Amanatidou *et al.*, 2000), fish, shrimps, and sausages (Earle and Snyder, 1966; Earle, 1968; Daniels, 1973), and coliform bacteria on beef cuts (Williams *et al.*, 1978; Cuq *et al.*, 1995), which was attributed to the presence of calcium chloride (Williams *et al.*, 1978; Cuq *et al.*, 1995). Nisperos-Carriedo (1994) and Williams *et al.* (1978) reported that alginate coatings did not affect cooking losses, flavor, odor or overall acceptability of coated beef and poultry products. Alginate coatings have been used to retard oxidation off-flavors in re-heated pork patties (Wanstedt *et al.*, 1981), precooked beef patties (Wu *et al.*, 2001), and cooked pork chops (Hargens-Madsen, 1995). Coated pork patties and pork chops exhibited improved flavor and juiciness, with no lipid oxidation. Coated pre-cooked beef patties with an alginate-based coating containing stearic acid and toco-pherol were effective in controlling moisture loss and lipid oxidation. Coatings made from pectin have similar beneficial properties (Kester and Fennema, 1986).

Alginate coatings also improve adhesion between batter and product, thus reducing the loss of batter from the surface of meats and fish (Fischer and Wong, 1972). Calcium-alginate films can also have beneficial effects on the quality of fruits and vegetables – for example, coated mushrooms with calcium alginate were found to have a better appearance and color, and an advantage in weight, in comparison with uncoated ones (Hershko and Nussinovitch, 1998).

Carrageenan

Carrageenan is derived from red seaweed, and is a complex mixture of several poly-saccharides. The three principal carrageenan fractions are kappa (κ), iota (ι) and lambda (λ), which differ in the sulfate ester and 3,6-anhydro α-D-galactopyranosyl content (Morris *et al.*, 1980; Glicksman, 1982). κ-carrageenan fractions contain the lowest number of sulfate groups and the highest concentrations of the 3,6-anhydro-α-D-galactopyranosyl units. The ι-carrageenan fractions have an additional sulfate group at the 2-position.

Gelation of both fractions occurs in the presence of both monovalent and divalent ions (Nisperos-Carriedo, 1994). Gels prepared with alginates rich in L-glucopyranosyl-uronic acid tend to be stronger, more brittle, and less elastic than those prepared with alginates rich in D-mannopyranosyluronic acid (Nisperos-Carriedo, 1994). Commercial carrageenans are mixtures of these three fractions. In the presence of locust bean gum an association with the double helix structure is formed, increasing the elasticity of the gels. Pectin, guar, CMC, and starches could also be used to modify the gel structure (Nisperos-Carriedo, 1994).

Carrageenan-based coatings have been applied for a long time to a variety of foods to carry antimicrobials or antioxidants, and to reduce moisture loss, oxidation, or dis-integration. Defrosted coated whole chickens have a better resistance against off-flavor development during storage at 4°C (Pearce and Lavers, 1949). The incorporation of gallic acid or ascorbic acid as an antioxidant in a carrageenan-based coating was able to keep the sensorial qualities of frozen mackerel fillets over 8 months of storage at −18°C (Stoloff *et al.*, 1949). Food-grade antimicrobial agents (lysozyme, nisin, grape fruit seed extract, and EDTA) incorporated with Na-alginate and κ-carrageenan films can be used effectively in a wide range of food packagings (Choi *et al.*, 2001). Carrageenan protects against moisture loss by acting as a sacrificial moisture layer (Kester and Fennema, 1986). A carrageenan-based coating applied on cut grapefruit halves resulted in less shrinkage, leakage or deterioration of taste after 2 weeks of storage at 4°C (Bryan, 1972).

Agar

Agar is a gum that is derived from a variety of red seaweeds, and, like carrageenan, it is a galactose polymer (Sanderson, 1981). It forms strong gels characterized by melting points far above the initial gelation temperature (Whistler and Daniel, 1985). Like carrageenan, antibiotics, bacteriocins or natural antimicrobial compounds can be incorporated in agar-based films. These films can be used in order to improve shelf life and to control pathogenic bacterial growth. Natrajan and Sheldon (1995) observed a reduction in *Salmonella typhimurium* population of 1.8 to 4.6 log cycles after 96 hours of storage at 4°C on coated fresh poultry where nisin was incorporated.

Pectin

Pectin is a complex anionic polysaccharide composed of β-1, 4-linked D-galacturonic acid residues, where the uronic acid carboxyls are either fully (high-methoxy pectin) or partially (low-methoxy pectin) methyl esterified. High-methoxy pectin forms excellent films. Low-methoxyl pectin, derived by controlled de-esterification, forms gels in the presence of calcium ions and can also be used for edible film developments. Blends of pectin and starch can be used to make strong, self-supporting films. Pectin has been used in making biodegradable drinking straws in which coloring and flavoring substances incorporated in a pectin layer are released when liquids pass through the straw (Kasten, 1989). Plasticized blends of citrus pectin give strong, flexible films, which are thermally stable up to 180°C (Tharanathan and Kittur, 2003).

Pectin can also form cross-links with proteins under certain conditions (Thakur *et al.*, 1997). Autoclaving enhances pectin–protein interactions, resulting in a three-dimensional network with improved mechanical and barrier properties. Films obtained from irradiated solutions generated films with improved puncture strength (Letendre *et al.*, 2002a). Pectin is also miscible with poly(vinylalcohol) in all proportions. These films are solution-produced by air-drying after casting at ambient temperatures. Laminated films formed from pectin and chitosan with either glycerol or lactic acid as a plasticizer have been prepared (Fishman *et al.*, 1994). Although pectinate coatings have poor moisture barriers, they can retard water loss from food by acting as a sacrificial agent. Pectin coatings have been investigated for their ability to retard moisture loss and lipid migration (Brake and Fennema, 1993), and improve handling and appearance of foods. However, the addition of lipids may increase their resistance to water vapor transmission (Schultz *et al.*, 1949). Calcium pectinate gel coatings reduced shrinkage and bacterial growth on beef patties (Stubbs and Cornforth, 1980).

Microbial polysaccharides

Pullulan, levan, and elsinan are extracellular microbial polysaccharides that are edible and biodegradable. Pullulan films cast from aqueous solutions are clear, odorless, and tasteless, and have good oxygen barriers (Yuen, 1974; Conca and Yang, 1993). Pullulan coatings have been used successfully as oxygen barriers to prolong food shelf life. The polymorphic *Aureobasidium pullulans* secretes a polysaccharide pullulan, which is commercially a useful hydrocolloid (Seviour *et al.*, 1992). It is an α-glucan consisting of repeating maltotriose residues joined by 1,6-linkages. It can be extruded as films that are biodegradable, resistant to oils and grease, have excellent oxygen permeability rates, and are non-toxic. Pullulan is of use in the biodegradable packaging films industry.

Levan and elsinan can also be used as edible coating materials for foods and pharmaceuticals due to their low oxygen permeability properties (Kaplan *et al.*, 1993).

Xanthan gum is produced by fermentation from the organism *Xanthomonas campestris*. It contains five sugar residues: two β-D-glucopyranosyl, two β-D-mannopyranosyl, and one β-glucopyranosyluronic acid residue (Jansson *et al.*, 1975; Melton *et al.*, 1976). Xanthan gum is soluble in both cold and hot water and has a high viscosity. It is used for its thickening, suspending and stabilizing effects (Nisperos-Carriedo, 1994).

Gellan gum is produced by fermentation of a pure culture of *Pseudomonas elodea*. It is a linear tetrasaccharide with repeating units of (1–>3)-β-D-glucopyranosyl, (1–>4)-β-D-glucopyranosyluronic acid, (1–>4)-β-D-glucopyranosyl, and (1–>4)-α-L-rhamnopyranosyl units. A gel is formed after it is heated in the presence of cations and then allowed to cool. It can be used to improve the texture and the stability of foods.

Dextrans are microbial gums composed solely of α-D-glucopyranosyl units, but with varying types and amounts of glycosidic linkages (Whistler and Daniel, 1985). Dextran coatings have been proposed to preserve the flavor, color, and freshness of shrimp (Toulmin, 1956a, 1956b), fish (Novak, 1957), and meat products (Toulmin, 1957).

Exudate gums

Gum arabic or acacia is the dried, gummy exudate from the stems or branches of *Acacia Senegal* and other related species of *Acacia* (Balke, 1984). Gum arabic is a neutral or slightly acidic salt, and is a complex polysaccharide containing calcium, magnesium, and potassium ions. It is a D-galactopyranosyl, L-rhamnopyranosyl, L-arabinopyranosyl, L-arabinofuranosyl and D-glucopyranosyluronic acid containing residues (Prakash *et al.*, 1990; Whistler and Daniel, 1990).

Seed gums

Guar gum and locust bean gum are extracted from leguminous seeds (Glicksman, 1969; Whistler and BeMiller, 1973). Locust gum consists of β-D-mannopyranosyl and α-D-galactopyranosyl units in a ratio of 4 : 1. This gum has a high viscosity, is soluble in hot water (95°C), and is compatible with other hydrocolloids, carbohydrates, and proteins (Nisperos-Carriedo, 1994). It is used as a thickener, a binder of free water, and a suspending or stabilizing agent. Guar gum is soluble in cold and hot water, and has a high viscosity. It is also compatible with other plant hydrocolloids, chemical modified starches or cellulose, and water-soluble proteins.

Applications of edible films and coatings

The successful development of edible coatings and films for food applications is based on specific functional properties that the polymer should meet for each type of food, due to the different properties of the various components (plasticizers, stabilizers, lipids, proteins etc.). The addition of these compounds in the optimal concentrations and in the correct order after the appropriate treatment (cross-linking with proteins, extrusion, coacervation, functionalization, bi-layer coatings etc.) can be beneficial in developing films with superior functional properties.

Composition

The addition of plasticizers can improve the flexibility and reduce the brittleness of films (Brault *et al.*, 1997; Mezgheni *et al.*, 1998a; Ressouany *et al.*, 1998; Le Tien *et al.*, 2000; Letendre *et al.*, 2002a). Plasticizers such as glycerol, sorbitol, propylene glycol, triethylene glycol, polyethylene glycol, and ethylene glycol are commonly used. Certain plasticizers can inhibit fungal growth. For example, lactic acid was substituted for glycerol in pectin films and no further fungal growth was observed. The pectin films containing lactic acid had properties similar to those made with pectin, glycerol and lactic acid. No significant differences in the storage moduli (a measurement of stiffness) were observed in the films containing lactic acid instead of glycerol (Hoagland and Parris, 1996).

The addition of stabilizing agents in coating solutions has been shown to preserve the film integrity and, in certain cases, improve the mechanical properties and the transparency of the films. Stabilizing agents are generally proteins such as gelatin, collagen, prolamine etc. There are other known stabilizing agents, but the preferred ones are those of a proteinaceous nature. These can be associated with polysaccharide matrices. Furthermore, stabilizing agents that are included in edible coating formulations can in certain cases increase the biofilm stability (at a certain pH limit) and the water resistance (at elevated temperatures), and can limit film shrinkage (at a low or intermediate relative humidity) (Lacroix *et al.*, 2001).

Edible lipid components can be incorporated into coating formulations (Greener and Fennema, 1989; Martin-Polo *et al.*, 1992; Debeaufort and Voilley, 1994; Lacroix *et al.*, 2001). Such lipids include oils (canola oil), fatty acids (palmitic acid), fatty alcohols (stearyl alcohol), monoglycerides, and triglycerides having long carbon chains from 10 to 20 carbon atoms, which are either saturated or unsaturated. Composite films of proteins and lipids have also been formulated with the aim of decreasing their solubility. However, protein–lipid films are often difficult to obtain. For example, bi-layer film formation requires the use of solvents or high temperatures, making the production of such films more costly. Furthermore, with time, separation of the layers may occur. For films cast from aqueous–lipid emulsion solutions, the process is more complex and the incorporation of the lipids is limited.

Composite and bi-layer coatings are the edible coatings of the future. These two types of coatings have combined the beneficial properties of the various coating ingredients to create a superior film. Composite films of the future may consist of hydrophobic particles within a hydrophilic matrix. Such a configuration could give a water-soluble coating with good water vapor barrier properties (Baldwin, 1991). Bi-layer coatings (which have already been used to a limited extent) contain the beneficial water-barrier properties of the lipid coatings with the addition of the greaseless feel and good gas permeability characteristics of polysaccharide coatings. Such a bi-layer coating has been shown to reduce gas exchange in cut apple pieces (Wong *et al.*, 1994). However, few or no data exist showing the effects of bi-layer coatings on whole fruits or vegetables.

It may be desirable to include agents such as emulsifiers, lubricants, binders or defoaming agents to influence the spreading characteristics of coating solutions. For example, polysaccharide-based films can be applied as an emulsion to improve the moisture-barrier properties of the films (Kester and Fennema, 1989; Hernandez, 1994; Baldwin, 1995) and reduce the water loss from foods (Baldwin *et al.*, 1997).

Emulsions are colloidal systems containing two partially miscible fluids and, almost invariably, a surfactant. Emulsions are comprised of fat droplets that are generally greater than 0.2 μm and less than 5 μm in diameter. These droplets are enveloped by polysaccharide-based films of surfactant material that stabilize the droplets, preventing their association and coalescence. Emulsifying agents, such as proteins, lower the interfacial tension between the dispersed and the continuous phases. Emulsion stability depends on the delicate balance between the attractive intermolecular forces between molecules in the film and the repulsive forces on the outer surface of the film. Emulsion formation involves an increase in free energy and in the interfacial area between the two phases. Mechanical work is necessary to overcome the free energy barrier. Emulsion systems are generally unstable, and therefore emulsifying agents or surfactants are often necessary to improve the stability of the emulsions (Everett, 1989).

Composite edible films or coatings can be produced by coacervation. Coacervation is based on the separation of a macromolecular solution into two immiscible liquid phases; a dense coacervate phase, which is relatively concentrated in macromolecules, and a dilute equilibrium phase. This separation occurs when macromolecules have an increased tendency to interact with one another, which may be caused by a reduction in their ability to interact with the solvent (simple coacervation) or by an ionic interaction between oppositely charged macromolecules (complex coacervation) (Burgess and Carless, 1984; Burgess *et al.*, 1991; Burgess and Singh, 1993). Normally, different polymers with different charges are used to form an aggregative complex. This interaction is electrostatic in nature (Bungenberg de Jong, 1949). Chitosan–alginate films were prepared by Yan *et al.* (2000, 2001) using coacervation reactions. These films had good tensile properties, and were resistant to water and most solvents. However, the films exhibited microscopic heterogeneity, which led to the formation of porous films with high permeability to water vapor.

Property modification

Generally, polymers in their native state are sensitive to humidity and are soluble to water; consequently, certain modifications seem necessary to improve their properties. In certain formulations of films or coatings, a functionalization agent is included to increase the hydrophobicity of the polymer in order to improve its moisture-barrier properties (gas and water vapor). It is also possible that this functionalization agent may increase the mechanical properties of the films or coatings. This agent can be defined as a substance that is covalently linked on the polysaccharide matrix, with or without the help of a coupling agent. It generally includes glyceraldehyde, acyl chlorides, fatty acids, and anhydrides. Covalent modifications of the polysaccharide for the formation of coating formulations can be obtained by esterification, etherification, carboxymethylation etc. (Rutenberg and Solarek, 1984; Brode, 1991; Brode *et al.*, 1991; Lacroix *et al.*, 2001; Mulhacher *et al.*, 2001).

Cross-linking between proteins and polysaccharides, or treatments used to improve interactions between proteins and polysaccharides, may be used to improve the film's functional properties and its resistance (Mezgheni *et al.*, 1998b; Le Tien *et al.*, 2000; Ressouany *et al.*, 2000; Letendre *et al.*, 2002a). Pectin may form cross-links with proteins under certain conditions (Thakur *et al.*, 1997). Autoclaving enhances protein–polysaccharide, pectin–protein or agar–protein interactions, resulting in a three-dimensional network with improved mechanical properties. Letendre *et al.* (2002b) analyzed the interaction of pectin with calcium and whey proteins after certain heating treatments.

Extrusion processes can also be used to preserve flavors during food processing. This process consists of initially making a low-moisture carbohydrate melt (15% moisture), adding flavor (10–20% dried weight basis), forming an emulsion, and then extruding the melt under pressure. Gum arabic is normally used for the encapsulation of food flavors (Reineccius, 1994). Encapsulation could also be performed using functionalized polysaccharides. Le Tien *et al.* (2003) have demonstrated that lactic acid bacteria viability could be preserved when encapsulated in functionalized alginate or chitosan by succinylation, or in *N*-palmitoylaminoethyl alginate (Figure 20.1).

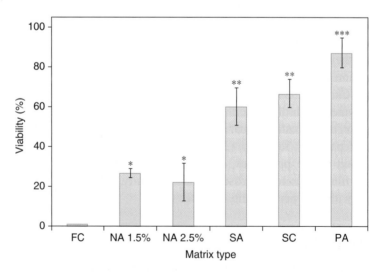

Figure 20.1 Viability (%) of *L. rhamnosus*, free and immobilized, on different functionalized matrices and in simulated gastric fluid (pH 1.5) for 30 minutes (FC, free cell; NA, native alginate; SA, succinylated alginate; SC, succinylated chitosan and PA, *N*-palmitoylaminoethyl alginate). (Reproduced with permission from Le Tien *et al.* (2004) ©Portland Press on behalf of the IUBMB)

Edible films as food packaging materials

Carbohydrate-based biopolymers may be used to modify conventional packaging systems by controlling gas transmission rates. Many foods require specific atmospheric conditions to sustain their freshness and overall quality during storage. Packaging foods under a specific mixture of gases can ensure the food product's optimum quality and safety. To ensure a constant gas composition inside the package, the packaging materials need to have certain gas-barrier specificities. In most packaging applications the gas mixture inside the package consists of carbon dioxide, oxygen, and nitrogen, or a combination of these gases. There is a vast amount of information in the literature on the barrier properties of biomaterials. However, comparison of biomaterials is complicated and at times impossible, due to the use of different types of equipment and dissimilar conditions of measurement.

The conventional approach to producing high-barrier edible films or coatings for food purposes in protective atmospheres is to incorporate multi-layers of different films in order to obtain the required properties. For example, an equivalent biobased laminate would consist of an outer layer of plasticized chitosan film, combined with polyhydroxyalkanoate (PHA) or alginate to obtain a required gas-barrier material. In the same fashion, PHA will protect the moisture-sensitive gas barrier made of polysaccharide.

Laminant films can be formed by applying another polymer (i.e. a lipid) as a laminant over a base polysaccharide film. Bi-layer films are laminant films that tend, over time, to delaminate, develop pinholes or cracks, and exhibit poor strength, non-uniform surfaces, and poor cohesion characteristics (Sherwin *et al.* 1998). Bi-layer films are often brittle and not very practical in many applications (Fairley *et al.*, 1997). In addition, they require multiple drying steps, whereas emulsion films require only a single dehydration step (Debeaufort and Voilley, 1995). The application of the lipid layer often requires the use of solvents or high temperatures, making its production more costly. However, bi-layer films are desirable owing to the fact that they generally exhibit better barrier properties compared to emulsion films (Fairley *et al.*, 1997; Sherwin *et al.*, 1998). Hoagland and Parris (1996) demonstrated that chitosan/pectin laminated films were transparent and more resistant to water dissolution. Yan *et al.* (2001) obtained similar results with chitosan/alginate laminated films. Additionally, these authors demonstrated that films prepared with low molecular weight chitosan were twice as thin and transparent, as well as 55% less permeable to water vapor, compared to those prepared with higher molecular weight chitosan. The improvement of the mechanical properties could have been due to the electrostatic interaction between the carboxylate groups of pectin or alginate and the amino groups of chitosan (Dutkiewicz *et al.*, 1992).

Park *et al.* (1994) reported better water vapor permeability and mechanical properties of laminated methylcellulose/corn zein–fatty acid films. Previous studies on bi-layer films made from methylcellulose and fatty acids exhibited distortion, cracking, and pinhole formation during drying. The addition of corn zein to the fatty acid component of these films eliminated all of these defects. In addition, the water-barrier properties of these films were very good. Makino and Hirata (1997) have also shown that a biodegradable laminate consisting of chitosan–cellulose and polycaprolactone can be used in modified atmosphere packaging of fresh produce.

Controlled emulsion destabilization offers another alternative method for the formation of bi-layer lipid–protein films. De-emulsification techniques are known in the food industry. Agents capable of displacing the stabilizing surfactant have been developed, leaving an unstable surfactant (Everett, 1989). These technologies may be applied in the future for the formation of effective edible films and coatings.

Carbohydrate chemistry

Structural analysis of carbohydrates

The structure of modified polymers can be characterized by several techniques such as Fourier transform infrared (FTIR), nuclear magnetic resonance (NMR), X-ray diffraction (XRD), and scanning electron microscopy (SEM). FTIR and NMR analysis are frequently used to characterize the polymer because they are rapid and give ample information. XRD and SEM are also good techniques, but are rarely used because they are expensive.

The FTIR analysis could be useful to identify the functional groups of the polymers after modifications. The important spectral regions at $3300–3000\ cm^{-1}$ are generally for the elucidation of the free or the interacting hydroxyl (–OH) groups; at $2950–2850\ cm^{-1}$ for the alkyl (–CH_2–) chains; and at $1750–1650\ cm^{-1}$, the bands represent the carbonyl groups (–C$=$O) and the amides bands (–C–N–). For example, Le Tien et al. (2004) functionalized the chitosan by acylation in order to improve the hydrophobicity of the polymer. FTIR analysis permitted the visualization of N-acyl chitosan (Figure 20.2a). For the native chitosan (non-modified chitosan), the absorption peaks at $1655\ cm^{-1}$ were assigned to the carbonyl stretching of secondary amides (amide I band), at $1570\ cm^{-1}$ to the N–H bending vibrations of non-acylated 2-aminoglucose primary amines, and at $1555\ cm^{-1}$ to the N–H bending vibrations of the amide II band (Xu et al., 1996). After N-acylation, the vibration band corresponding to the primary amino groups at $1570\ cm^{-1}$ disappeared (Figure 20.2), while prominent bands at 1655 and $1555\ cm^{-1}$ were observed. In addition, peaks at $2950–2850\ cm^{-1}$ represented (–CH_2–), and their intensity was proportional to the acyl chain length. These analyses clearly confirmed that the chitosan was substituted.

In NMR, the structure of the polymer can be determined by ^{1}H- or ^{13}C-nuclear magnetic resonance spectroscopy (NMR). Using the same example (Figure 20.2b), the ^{1}H NMR spectra of native chitosan shows that peaks at 2.0–2.1 ppm represent the three N-acetyl protons of N-acetylglucosamine (GlcNAc), and peaks at 3.1–3.2 ppm represent a H-2 proton of glucosamine (GlcN) residues. The ring protons (H-3, 4, 5, 6, 6′ are considered to resonate at 3.6–4.0 ppm, and the peaks at 4.6 and 4.8 ppm were assigned to the H-1 protons of the GlcN and GlcNAc residues, respectively (Xu et al., 1996; Signini et al., 1999; Sashiwa et al., 2002). However, for the N-acyl chitosan, new peaks at 0.7, 1.2 and 1.9 ppm were assigned to protons of CH_3–, –CH_2– and –CH_2–(CO), respectively, of the alkyl residue (Le Tien et al., 2004). It is worth noting that this method is not very useful for certain polymers due their high viscosity (i.e.

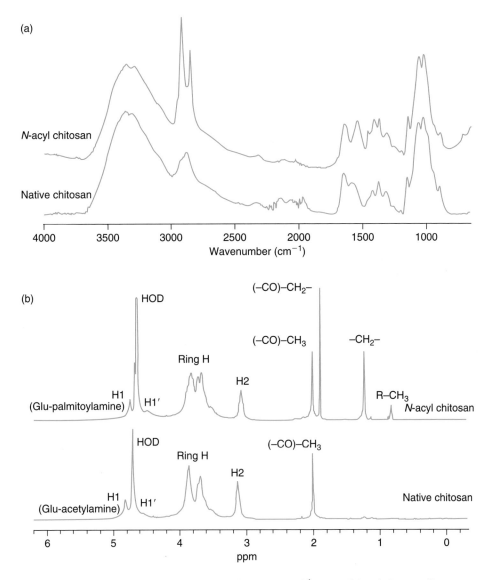

Figure 20.2 Characterization of *N*-acyl chitosan by FTIR (a) and ^{1}H NMR (b) analysis, according to Le Tien *et al.* (2003).

alginate). Generally, it is then therefore necessary to reduce the viscosity and increase the mobility of the molecules by depolymerization to obtain a degree of polymerization of about 20–50%. The NMR methodology and the peak identifications were based on data published by Grasdalen *et al.* (1979, 1981, 1983).

XRD and SEM are rarely used to characterize the polymers. XRD is mostly used to determine the degree of crystallinity of the polymer. Generally, polymers possess a high degree of crystallinity, forming a structure that is more organized and has good mechanical properties (Le Tien *et al.*, 2004). For SEM, the analysis is essentially based on the aspects of the polymer, such as uniformity and porosity.

Thermal and physicochemical properties

The thermal and mechanical properties of the polymeric materials are important for processing, and for products derived from these materials. Most biobased polymer materials perform in a similar fashion to conventional polymers. This indicates that both polystyrene-like polymers (relatively stiff materials with intermediate service temperatures) and polyethylene-like polymers (relatively flexible polymers with intermediate service temperatures) can be found among the available biobased polymers.

The mechanical properties (modulus and stiffness) of biopolymers are not very different compared to the conventional polymers. The modulus of biomaterials ranges from 2500–3000 MPa, and decreases for stiffer polymers such as thermoplastic starches (50 MPa) and even more for rubbery materials like medium-chain polyhydroxyalkanoates. Furthermore, the modulus of most biobased and petroleum-derived polymers can be tailored to meet the required mechanical properties by means of plasticizing, blending with other polymers or fillers, cross-linking, or the addition of fibers. Polymer such as bacterial cellulose could, for instance, be used in materials that require special mechanical properties. In theory, biomaterials can be made having similar strength to the ones that are used today (Iguchi *et al.*, 2000).

A major challenge for the material manufacturers is the improvement of the water vapor barriers of films and coatings for food applications. For example, the water vapor transmittance of various polysaccharide films, when compared to materials based on mineral oil, is lower. It becomes clear that it is difficult to produce an edible film or coating with water vapor permeability rates comparable to the ones provided by some conventional plastics. Consequently, research is focusing on this problem in order to produce future biomaterials that are able to mimic the water vapor barriers of the conventional materials known today (Butler *et al.*, 1996).

Conclusion

Polysaccharide-based films have the potential to be used in a variety of applications. These films can be used to extend the shelf life of fruits, vegetables, seafood, meats, and confectionery products by preventing dehydration, oxidation rancidity, surface browning, and oil diffusion, as well as by their ability to modify the internal atmosphere in fruit and vegetable packaging. Moreover, when applied to fruits and vegetables, polysaccharide-based coatings can improve the physicochemical, nutritional, and sensorial properties of the products. Several studies have demonstrated that the incorporation of active compounds like antimicrobials, antioxidants or bacteriocins in edible films and coatings can improve the microbial safety and the shelf life, and can be used to stabilize the quality of the products during storage. The development of new technologies (functionalization, cross-linking etc.) to improve the film properties (control release, bioactivity protection, resistance to water etc.) of active packaging and coatings is a major focus for future research.

References

Amanatidou, A., Slump, R. A., Gorris, L. G. M. and Smid, E. J. (2000). High oxygen and high carbon dioxide modified atmospheres for shelf life extension of minimally processed carrots. *J. Food Sci.* **65,** 61–66.

Baker, R. A., Baldwin, E. A. and Nisperos-Carriedo, M. O. (1994). Edible coatings and films for processed foods. In: *Edible Coatings and Films to Improve Food Quality* (J. M. Krochta, E. A. Baldwin and M. O. Nisperos-Carriedo, eds), pp. 89–105. Technomic Publishing, Lancaster, PA.

Balasubramaniam, V. M., Chinnan, M. S., Malikarjunan, P. and Phillips, R. D. (1997). The effect of edible film on oil uptake and moisture retention of deep-fat fried poultry product. *J. Food Process. Eng.* **20,** 17–29.

Baldwin, E. A. (1991). Edible coatings for fresh fruits and vegetables: past, present, and future. In: *Edible Coatings and Films to Improve Food Quality* (J. M. Krochta, E. A. Baldwin and M. O. Nisperos-Carriedo, eds), pp. 25–64. Technomic Publishing, Lancaster, PA.

Baldwin, E. A., Nisperos, M. O. and Baker, R. A. (1995). Use of edible coating to preserve quality of lightly and slightly processed products. *Crit. Rev. Food Sci. Nutr.* **35,** 509–524.

Baldwin, E. A., Nisperos, M. O., Hagenmaier, R. D. and Baker, R. A. (1997). Use of lipids in coatings for food products. *Food Technol.* **51,** 56–64.

Balke, W. J. (1984). The use of natural hydrocolloids as thickeners, stabilizers and emulsifiers. In: *Gum Starch and Technology, 18th Annual Symposium* (D. L. Downing, ed.), pp. 31–35. Institute of Food Science, Cornell University, NY.

Ben, A. and Kurth, L. B. (1995). Edible film coating for meat cuts and primal. Meat 95, The Australian Meat Industry Research Conference, CSIRO, September 10–12.

Brake, N. C. and Fennema, O. R. (1993). Edible coatings to inhibit lipid migration in a confectionery product. *J. Food Sci.* **58,** 1422–1425.

Brault, D., D'Aprano, G. and Lacroix, M. (1997). Formation of free-standing sterilised edible films from irradiated caseinates. *J. Agric. Food Chem.* **45,** 2964–2969.

Brine, C. J., Sandford, P. A. and Zikakis, J. P. (1991). *Advances in Chitin and Chitosan.* pp. 1–491. Elsevier Applied Science, London, UK.

Brode, G. L. (1991). Polysaccharides: natural for cosmetic and pharmaceuticals. In: *Cosmetic and Pharmaceutical Applications of Polymers* (C. G. Gebelein, T. C. Cheng and V. C. Yang, eds), pp. 105–115. Plenum Press, New York, NY.

Brode, G. L., Goddard, E. D., Harris, W. C. and Sale, G. A. (1991). Cationic polysaccharides for cosmetics and therapeutics. In: *Cosmetic and Pharmaceutical Applications of Polymers* (C. G. Gebelein, T. C. Cheng and V. C. Yang, eds), pp. 117–128. Plenum Press, New York, NY.

Bryan, D. S. (1972) US Patent 3,707,383.

Bungenberg de Jong, H. G. (1949). Crystallisation–coacervation–flocculation. In: *Colloid Science*, Vol. II (H. R. Kruyt, ed.), pp. 232–258. Elsevier, New York, NY.

Burgess, D. J. and Carless, J. E. (1984). Microelectrophoretic studies of gelatine and acacia for the prediction of complex coacervation. *J. Colloid. Interface Sci.* **98,** 1–8.

Burgess, D. J. and Singh, O. N. (1993). Spontaneous formation of small sized albumin-acacia coacervation. *J. Pharm. Pharmacol.* **45,** 586–591.

Burgess, D. J., Kwok, K. and Megremis, P. T. (1991). Characterization of albumin-acacia coacervation. *J. Pharm. Pharmacol.* **43,** 232–236.

Butler, B. L., Vergano, P. J., Testin, R. F., Bunn, J. M. and Wiles, J. L. (1996). Mechanical and barrier properties of edible chitosan films as affected by composition and storage. *J. Food Sci.* **61,** 953–956.

Chandra, R. and Rustgi, R. (1998). Biodegradable polymers. *Progr. Polym. Sci.* **23,** 1273–1335.

Chen, M., Yeh, G. H. and Chiang, B. (1996). Antimicrobial and physicochemical properties of methylcellulose and chitosan films containing a preservatives. *J. Food Process. Preserv.* **20,** 379–390.

Chen, C., Lian, W. and Isai, G. (1998). Antibacterial effects of *N*-sulfonated and *N*-sulfobenzoyl. *J. Food Protect.* **61,** 1124–1128.

Choi, J. H., Cha, D. S. and Park, H. J. (2001). The antimicrobial films based on Na-alginate and κ-carrageenan. *Food Packaging*, 73D. IFT Annual Meeting, New Orleans, LA.

Coma, V., Martial-Gros, A., Garreau, S., Copinet, A., Salin, F. and Deschamps, A. (2002). Edible antimicrobial films based on chitosan matrix. *J. Food Sci.* **67,** 1162–1169.

Conca, K. R. and Yang, T. C. S. (1993). Edible food barrier coatings. In: *Biodegradable Polymers and Packaging* (C. Ching, D. Kaplan and E. Thomas, eds), pp. 357–369. Technomic Publishing, Lancaster, PA.

Cottrell, I. W. and Kovacks, P. (1980). Alginates. In: *Handbook of Water-soluble Gums and Resins* (R. L. Davidson, ed.), pp. 143. McGraw-Hill, New York, NY.

Cuero, R. G., Osufi, G. and Washington, A. (1991). *N*-Carboxymethylchitosan inhibition of aflatoxin production: role of zinc. *Biotechnol. Lett.* **13,** 441–444.

Cuq, B., Gontard, N. and Guilbert, S. (1995). Edible films and coatings as active layers. In: *Active Food Packaging* (M. L. Rooney, ed.), pp. 111–142. Blackie Academic and Professional, Glasgow, UK.

Daniels, R. (1973). *Edible Coatings and Soluble Packaging*. Noyes Data Corp., Park Ridge, NJ.

Dawson, P. L., Han, I. Y., Orr, R. V. and Acton, J. C. (1998). Chitosan coatings to inhibit bacterial growth on chicken drumsticks. In: Proceedings of the 44th International Conference of Meat Science and Technology, Barcelona, Spain, pp. 458–462. Food Technology Publisher, Chicago, IL.

Debeaufort, F. and Voilley, A. (1994). Aroma compound and water vapor permeability of edible films and polymeric packaging. *J. Agric. Food Chem.* **42,** 2871–2875.

Debeaufort, F. and Voilley, A. (1995). Effect of surfactants and drying rate on barrier properties of emulsified edible films. *J. Food Sci. Technol.* **30,** 183–190.

Dutkiewicz, J., Tuora, M., Judkiewicz, L. and Ciszewski, R. (1992). New forms of chitosan polyelectrolyte complexes. In: *Advances in Chitin and Chitosan* (C. J. Brine, P. A. Sanford and J. P. Zikakis, eds), pp. 496–505. Elsevier Applied Science, Oxford, UK.

Dziezak, J. D. (1991). Special report: A focus on gums. *Food Technol.* **45,** 116–132.

Earle, R. D. (1968). US Patent 3,395,024.

Earle, R. D. and Snyder, C. E. (1966). US Patent 3,255,021.

El Ghaouth, A., Arul, J., Ponnampalam, R. and Boulet, A. (1991a). Special report: a focus on gums. *J. Food Sci.* **56,** 1618–1620.

El Ghaouth, A. E., Arul, J., Ponnampalam, R. and Boulet, M. (1991b). Use of chitosan coating to reduce weight loss and maintain quality of cucumber and bell pepper fruits. *J. Food Process. Preserv.* **15,** 359–368.

Everett, D. H. (1989). *Basic Principles of Colloidal Science*, pp. 167–182. Royal Society of Chemistry, London, UK.

Fairley, P., Krochta, J. M. and German, J. B. (1997). Interfacial interactions in edible emulsion films from whey protein isolate. *Food Hydrocolloids* **11,** 245–252.

Fischer, L. G. and Wong, P. (1972). US Patent 3,676,158.

Fishman, M. L., Friedman, R. B. and Huang, S. J. (1994). *Polymers from Agricultural Coproducts*, ACS Symposium, Series 575. American Chemical Society, Washington, DC.

Floros, J. D., Dock, L. L. and Han, J. H. (1997). *Food Cosmetics and Drug Packaging* **20,** 10–16.

Ganz, A. J. (1969). CMC and hydroxypropylcellulose-versatile gums for food use. *Food Prod. Dev.* **3,** 65.

Gennadios, A., Hanna, M. A. and Kurth, B. (1997). Application of edible coatings on meats, poultry and seafoods: a review. *Lebensm. Wiss. u. Technol.* **30,** 337–350.

Glicksman, M. (1969). *Gum Technology in the Food Industry*. Academic Press, New York, NY.

Glicksman, M. (1982). *Food Hydrocolloids*, Vol. I. CRC Press, Boca Raton, FL.

Glicksman, M. (1983). *Food Hydrocolloids*, Vol. II. CRC Press, Boca Raton, FL.

Grasdalen, H. (1983). High-field ^{1}H-NMR spectroscopy of alginate: sequential structure and linkage conformations. *Carbohydr. Res.* **118,** 255–260.

Grasdalen, H., Larsen, B. and Smidsrød, O. (1979). A PMR study of the composition and sequence of uronate residues in alginates. *Carbohydr. Res.* **68,** 23–31.

Grasdalen, H., Larsen, B. and Smidsrød, O. (1981). ^{13}C-NMR studies of monomeric composition and sequence in alginate. *Carbohydr. Res.* **89,** 179–191.

Greener, I. K. and Fennema, O. (1989). Evaluation of edible, bilayer films for use as moisture barriers for food. *J. Food Sci.* **54,** 1400–1406.

Guilbert, S., Cuq, B. and Gontard, N. (1997). Recent innovations in edible and/or biodegradable packaging materials. *Food Add. Contam.* **14,** 741–751.

Hagenmaier, R. D. and Shaw, P. E. (1990). Moisture permeability of edible films made with fatty acid and (hydroxypropyl)methyl cellulose. *J. Agric. Food Chem.* **38,** 1799–1803.

Hanlon, J. F. (1992). In: *Handbook of Package Engineering*, pp. 1–59. Technomic Publishing, Lancaster, PA.

Hargens-Madsen, M. R. (1995). Use of edible coatings and tocopherols in the control of warmed-over flavor. MS thesis, University of Nebraska, Lincoln, NE.

Hernandez, E. (1994). Edible coatings from lipids and resins. In: *Edible Coatings and Films to Improve Food Quality* (J. M. Krochta, E. A. Baldwin and M. O. Nisperos-Carriedo, eds), pp. 279–303. Technomic Publishing, Lancaster, PA.

Hershko, V. and Nussinovitch, A. (1998). Relationships between hydrocolloid coating and mushroom structure. *J. Agric. Food Chem.* **46,** 2988–2997.

Hoagland, P. D. and Parris, N. (1996). Chitosan/pectin laminated films. *J. Agric. Food Chem.* **44,** 1915–1919.

Holownia, K. I., Chinnan, M. S., Erickson, M. C. and Mallikarjunan, P. (2000). Quality evaluation of edible film-coated chicken strips and frying oils. *J. Food Sci.* **65,** 1087–1090.

Iguchi, M., Yamanaka, S. and Budhioni, A. (2000). Bacterial cellulose – a masterpiece of nature's arts. *J. Materials Sci.* **35,** 1–10.

Jansson, P. E., Kenne, L. and Lindberg, B. (1975). Structure of the extracellular polysaccharide from *Xanthomonas campestris. Carbohydr. Res.* **45,** 275–282.

Kaplan, D. L., Mayer, J. M., Ball, D., McCassie, J., Allen, A. L. and Stenhouse, P. (1993). Fundamentals of biodegradable polymers. In: *Biodegradable Polymers and Packaging* (C. Ching, D. Kaplan and E. Thomas. eds), pp. 1–42. Technomic Publishing, Lancaster, PA.

Kasten, H. (1989). German Patent DE 37 31058.

Keller, J. (1984). Sodium carboxymethylcellulose (CMC). In: *Gum Starch and Technology, 18th Annual Symposium* (D. L. Downing, ed.), pp. 9–19. Institute of Food Science, Cornell University, NY.

Kester, J. J. and Fennema, O. R. (1986). Edible films and coatings: a review. *Food Technol.* **12,** 47–59.

Kester, J. J. and Fennema, O. R. (1989). An edible film of lipids and cellulose ethers: barrier properties to moisture vapor transmission and structural evaluation. *J. Food Sci.* **54,** 1383–1389.

Kittur, F. S., Kuman, K. R. and Tharanathan, R. N. (1998). Functional packaging properties of chitosan films. *Z. Lebensmittel Untersuch. Forschung A* **206(1),** 44–47.

Krochta, J. M. (1997). Edible protein films and coatings. In: *Food Proteins and Their Applications in Foods* (S. Damodaran and A. Paraf, eds), pp. 529–549. Marcel Dekker, New York, NY.

Krochta, J. M. and De Mulder-Johnston, C. (1997). Edible and biodegradable polymer films. Challenges and opportunities. *Food Technol.* **51,** 61–74.

Krumel, K. L. and Lindsay, T. A. (1976). Nonionic cellulose ethers. *Food Technol.* **30,** 36–43.

Labell, F. (1991). Edible packaging. *Food Process. Eng.* **52,** 24.

Lacroix, M., Mateescu, M. A., Le Tien, C. and Patterson, G. (2001). Biocompatible composition as carriers or excipients for pharmaceutical formulations and for food protection. PCT/CA00/01386.

Lazarus, C. R., West, R. L., Oblinger, J. L. and Palmer, A. Z. (1976). Evaluation of calcium alginate coating and a protective plastic wrapping for the control of lamb carcass shrinkage. *J. Food Sci.* **41,** 639–641.

Letendre, M., D'Aprano, G., Lacroix, M., Salmieri, S. and St-Gelais, D. (2002a). Physicochemical properties and bacterial resistance of biodegradable milk protein films containing agar and pectin. *J. Agric. Food Chem.* **50,** 6017–6022.

Letendre, M., D'Aprano, G., Delmas-Patterson, G. and Lacroix, M. (2002b). Isothermal calorimetry study of calcium caseinate and whey protein isolate edible films cross-linked by heating and γ-irradiation. *J. Agric. Food Chem.* **50,** 6053–6057.

Le Tien, C., Letendre, M., Ispas-Szabo, P. *et al.* (2000). Development of biodegradable films from whey proteins by cross-linking and entrapment in cellulose. *J. Agric. Food Chem.* **48,** 5566–5575.

Le Tien, C., Lacroix, M., Ispas-Szabo, P. and Mateescu, M. A. (2003). Modified alginate and chitosan for lactic acid bacteria immobilization. *J. Contr. Rel.* **93,** 1–13.

Le Tien, C., Millette, M., Mateescu, M. A. and Lacroix, M. (2004). Modified alginate and chitosan for lactic acid bacteria immobilization. *Biotechnol. Appl. Biochem.* **39,** 347–354.

Lobuza, T. P. and Breene, W. M. (1989). Application of active packaging for improvement of shelf life and nutritional availability of fresh and extended shelf life in foods. *J. Food Process. Preserv.* **13**, 1–69.

Lowings, P. H. and Cutts, D. F. (1982). The preservation of fresh fruits and vegetables. In: *Proceedings of the Institute of Food Science and Technology Annual Symposium*, July 1981, p. 52. Nottingham, UK.

Makino, Y. and Hirata, T. (1997). Modified atmosphere packaging of fresh produce with a biodegradable laminate of chitosan-cellulose and polycaprolactone. *Postharv. Biol. Technol.* **10**, 247–254.

Martin-Polo, M., Mauguin, C. and Voilley, A. (1992). Hydrophobic films and their efficiency against moisture transfer. 1-Influence of the film preparation technique. *J. Agric. Food Chem.* **40**, 407–412.

Mason, D. F. (1969). U.S Patent 3,472,662.

Mauch, F., Hadwiger, L. A. and Bolier, T. (1984). Ethylene: symptom, not signal for the induction of chitinase and beta 1,3 glucanase in pea pods by pathogens and elicitors. *Plant Physiol.* **76**, 607–611.

Melton, L. D., Mindt, L., Rees, A. and Sanderson, G. R. (1976). Covalent structure of the polysaccharide from *Xanthomonas campestris*: Evidence from partial hydrolysis studies. *Carbohydr. Res.* **46**, 245–257.

Mezgheni, E., D'Aprano, G. and Lacroix, M. (1998a). Covalent structure of the polysaccharide from *Xanthomonas campestris*: Evidence from partial hydrolysis studies. *J. Agric. Food Chem.* **46**, 318–324.

Mezgheni, E., Vachon, C. and Lacroix, M. (1998b). Biodegradability behaviour of cross-linked calcium caseinates films. *Biotechnol. Prog.* **14**, 534–536.

Morris, E. R., Rees, D. A. and Robinson, G. (1980). Order-disorder transition for a bacterial polysaccharide in solution: A role for polysaccharide conformation in recognition between Xanthomonas as pathogen and its plant host. *J. Mol. Biol.* **138**, 349.

Mulhbacher, J., Ispas-Szabo, P., Lenaerts, V. and Mateescu, M. A. (2001). Cross-linked high amylose starch derivatives as matrices for controlled release of high drug loadings. *J. Contr. Rel.* **76**, 51–58.

Natrajan, N. and Sheldon, B. W. (1995). Evaluation of bacteriocin-based packaging and edible film delivery systems to reduce Salmonella in fresh poultry. *Poultry Sci.* **74(Suppl. 1)**, 31.

Nelson, K. L. and Fennema, O. R. (1991). Methylcellulose films to prevent lipid migration in confectionery products. *J. Food Sci.* **56**, 504–509.

Nisperos-Carriedo, M. O. (1994). Edible coating and films based on polysaccharide. In: *Edible Coatings and Films to Improve Food Quality* (J. M. Krochta, E. A. Baldwin and M. O. Nisperos-Carriedo, eds), pp. 305–335. Technomic Publishing, Lancaster, PA.

Nisperos-Carriedo, M. O. and Baldwin, E. A. (1990). Edible coating for fresh fruits and vegetables. Subtropical Technology Conference Proceedings, 18 October, Lake Alfred, Florida.

Nisperos-Carriedo, M. O., Baldwin, E. A. and Shaw, P. E. (1991). Development of an edible coating for extending postharvest life of selected fruits and vegetables. *Proc. Fla State Hort. Soc.* **104**, 122–125.

Novak, L. J. (1957). US Patent 2,790,720.

Ouattara, B., Simard, R. E., Piette, G., Begin, A. and Holley, R. A. (2000). Inhibition of surface spoilage bacteria in processed meats by application of antimicrobial films prepared with chitosan. *Intl J. Food Microbiol.* **62**, 139–148.

Park, J. W., Testin, R. F., Park, H. J., Vergano, P. J. and Weller, C. L. (1994). Fatty acid concentration effect on tensile strength, elongation, and water vapor permeability of laminated edible film. *J. Food Sci.* **59**, 916–919.

Pavlath, A. E., Gossett, C., Camirand, W. and Robertson, H. (1999). Ionomeric films of alginic acid. *J. Food. Sci.* **64**, 61–63.

Pearce, J. A. and Lavers, C. G. (1949). Frozen storage of poultry. V. Effects of some processing factors on quality. *Can. J. Food Res.* **27**, 253–265.

Peterson, K., Per Vaeggemose, N., Bertelsen, G. *et al.* (1999). Potential of biobased materials for food packaging. *Trends Food Sci. Technol.* **10**, 52–68.

Prakash, A., Joseph, M. and Mangino, M. E. (1990). The effects of added proteins on the functionality of gum arabic in soft drink emulsion systems. *Food Hydrocolloids* **4**, 177.

Reineccius, G. A. (1994). Flavor encapsulation. In: *Edible Coatings and Films to Improve Food Quality* (J. M. Krochta, E. A. Baldwin and M. O. Nisperos-Carriedo, eds), pp. 105–120. Technomic Publishing, Lancaster, PA.

Ressouany, M., Vachon, C. and Lacroix, M. (1998). Irradiation dose and calcium effect on the mechanical properties of cross-linked caseinate films. *J. Agric. Food Chem.* **46**, 1618–1623.

Ressouany, M., Vachon, C. and Lacroix, M. (2000). Microbial resistance of caseinate films cross-linked by γ-irradiation. *J. Dairy Res.* **67**, 119–124.

Rinaudo, M. and Domard, A. (1989). Solution properties of chitosan. In: *Chitin and Chitosan* (G. Skjak-Braek, T. Anthonsen and P. Sandford, eds), pp. 71–83. Elsevier Science, New York, NY.

Robertson, G. L. (1993). *Food Packaging. Principles and Practice*. Marcel Dekker, New York, NY.

Rudrapatnam, N. T. and Farooqahmed, S. K. (2003). Chitin – the undisputed biomolecule of great potential. *Crit. Rev. Food Sci. Nutr.* **43**, 61–87.

Rutenberg, M. W. and Solarek, D. (1984). Starch derivatives: production and uses. In: *Starch: Chemistry and Technology*, 2nd edn (X. Chap, R. L. Whistler, J. N. BeMiller and E. F. Paschall, eds), pp. 311–387. Academic Press, New York, NY.

Sanderson, G. R. (1981). Polysaccharides in foods. *Food Technol.* **35**, 50–57, 83.

Sapers, G. M. and Dougglas, F. W. (1987). Measurement of enzymatic browning at cut surfaces and in juice of raw apple and pear fruits. *J. Food Sci.* **52**, 1258–1262.

Sashiwa, H., Kawasaki, N., Nakayama, A., Muraki, E., Yamamoto, N. and Aiba, S.-I. (2002). Chemical modification of chitosan derivatives by simple acetylation *Biomacromolecules* **3**, 1126–1128.

Schultz, T. H., Miers, J. C., Owens, H. S. and Maclay, W. D. (1949). Permeability of pectinate films to water vapour. *J. Phys. Colloid. Chem.* **53**, 1320–1330.

Seviour, R. J., Stasinopoulos, S. J., Auer, D. P. F. and Gibbs, P.A. (1992). Production of pullulan and other exopolysaccharides by filamentous fungi. *Crit. Rev. Biotechnol.* **12**, 279–298.

Sherwin, C. P., Smith, D. E. and Fulcher, R. G. (1998). Effect of fatty acid type on dispersed phase particle size distributions in emulsion edible films. *J. Agric. Food Chem.* **46**, 4534–4538.

Signini, R. and Campana-Filho, S. P. (1999). On the preparation and characterization of chitosan hydrochloride. *Polym. Bull.* **42,** 159–166.

Stollman, U., Hohansson, F. and Leufven, A. (1994). Packaging and food quality. In: *Shelf Life Evaluation of Foods* (C. M. D. Man and A. A. Jones, eds), pp. 52–71. Blackie Academic and Professional, New York, NY.

Stoloff, L. S., Puncochar, J. F. and Crowther, H. E. (1949). Curb mackerel fillet rancidity. *Food Ind.* **20,** 1130–1132.

Stubbs, C. A. and Cornforth, D. P. (1980). The effects of an edible pectinate film on beef carcass shrinkage and surface microbial growth. In: *Proceedings of the 26th European Meeting of Meat Research Workers, Kansas City, MO*, pp. 276–278. American Meat Science Association.

Thakur, B. R., Singh, R. K. and Handa, A. K. (1997). Chemistry and uses of pectin – a review. *Crit. Rev. Food Sci. Nutr.* **37,** 47–73.

Tharanathan, R. N. (2003). Biodegradable films and composite coatings: past, present and future. *Trends Food Sci. Technol.* **14,** 71–78.

Tharanathan, R. N. and Kittur, F. S. (2003). Chitin – the undisputed biomolecule of great potential. *Crit. Rev. Food Sci. Nutr.* **43(1),** 61–87.

Toulmin, H. A. Jr (1956a). US Patent 2,758,929.

Toulmin, H. A. Jr (1956b). US Patent 2,758,930.

Toulmin, H. A. Jr (1957). US Patent 2,790,721.

Wanstedt, K.G., Seideman, S. C., Donnelly, L. S. and Quenzer, N. M. (1981). Sensory attributes of precooked, calcium alginate-coated pork patties. *J. Food Protect.* **44,** 732–735.

Whistler, R. L. and BeMiller, J. N. (1973). *Industrial Gums, Polysaccharides and Their Derivatives,* 2nd edn., Academic Press, New York, NY.

Whistler, R. L. and Daniel, J. R. (1985). Functions of polysaccharides in foods. In: *Food Chemistry*, 2nd edn. (O. R. Fennema, ed.), pp. 69–137. Marcel Dekker, New York, NY.

Williams, S. K., Oblinger, J. L. and West, R. L. (1978). Evaluation of a calcium alginate film for use on beef cuts. *J. Food Sci.* **43,** 292–296.

Wong, D. W. S., Tillin, S. J., Hudson, J. S. and Pavlath, A. E. (1994). Gas exchange in cut apples with bilayer coatings. *J. Agric. Food Chem.* **42,** 2278–2285.

Wu, Y., Weller, C. L., Hamouz, F., Cuppett, S. and Schnepf, M. (2001). Moisture loss and lipids oxidation for precooked ground beef patties packaged in edible starch-alginate based composite films. *J. Food Sci.* **66,** 468–493.

Xu, J., McCarthy, S. P. and Gross, R. A. (1996). Chitosan film acylation and effects on biodegradability. *Macromolecules* **29,** 3436–3440.

Yalpani, M., Johnson, F. and Robinson, L. E. (1992). Antimicrobial activity of some chitosan derivatives. In: *Advances in Chitin and Chitosan* (C. J. Brine, P. A. Sandford and J. P. Zikakis, eds), p. 543. Elsevier Applied Science, London, UK.

Yan, X.-L., Khor, E. and Lim, L.-Y. (2000). PEC films prepared from chitosan–alginate coacervates. *Chem. Pharm. Bull.* **48,** 941–946.

Yan, X.-L., Khor, E. and Lim, L.-Y. (2001). *Chitosan–alginate Films Prepared with Chitosans of Different Molecular Weights*, pp. 358–365. John Wiley & Sons, New York, NY.

Yuen, S. (1974). Pullulan and its applications. *Process Biochem.* **9,** 7–9, 22.

21 Lipid-based edible films and coatings

Jong Whan Rhim and Thomas H. Shellhammer

Introduction

Protective edible films and coatings have long been used in the food industry to protect the quality and extend the shelf life of foods. Most commonly, edible lipid and shellac coatings have been used on fresh fruit and confections to improve appearance. Edible films and coatings can also function as barriers to moisture, oxygen, flavor, aroma, and oil, thus improving food quality and enhancing shelf life. An edible film or coating may also provide physical protection for a food, reducing bruising and breakage, and therefore improving food integrity. Sometimes the protective functions of edible films and coatings may be enhanced by the addition of antioxidants or antimicrobials. Edible coatings can also provide additional sensory attributes to foods, including gloss, color, and a non-greasy or non-sticky surface.

Edible films and coatings are generally prepared using biological polymers such as proteins, polysaccharides, lipids, and resins. Natural biopolymers have the advantage over synthetic polymers in that they are biodegradable and renewable, as well as edible. Each group of materials has certain advantages and disadvantages. Protein and polysaccharide films generally provide a good barrier against oxygen at low and intermediate relative humidity, and have good mechanical properties, but their barrier against water vapor is poor due to their hydrophilic nature (Guilbert, 1986; Kester and Fennema, 1986; Gennadios *et al.*, 1994). In contrast, films prepared with lipid materials have good water vapor barrier properties, but are usually opaque and relatively inflexible.

Innovations in Food Packaging
ISBN: 0-12-311632-5

The beneficial properties of some lipids, such as their reasonably good compatibility with other film-forming agents and their good barrier properties against water vapor and other gases, make them desirable candidates for increased use as edible films and coatings in foods (Greener and Fennema, 1992). Lipid compounds commonly used for the preparation of lipid-based edible films and coatings include neutral lipids, fatty acids, waxes, and resins (Kester and Fennema, 1986; Hernandez, 1994).

One way to achieve a better water vapor barrier is to produce a composite film by adding hydrophobic components such as lipid and wax materials. A composite hydrocolloid–lipid film or coating is particularly desirable, since it has acceptable structural integrity imparted by the hydrocolloid materials and good water vapor barrier properties contributed by the lipid materials (Greener and Fennema, 1989a).

The efficiency of the lipid materials in composite films and coatings depends on the nature of the lipid used, in particular on its structure, chemical arrangement, hydrophobicity, and physical state (e.g. solid or liquid), and on the lipid interactions with the other components of the film, such as proteins and polysaccharides.

In this chapter, the materials, preparation methods, characteristics, and applications of lipid-based edible films and coatings are discussed.

Materials used in lipid-based edible films and coatings

A very wide range of compounds is available for increasing the hydrophobicity of lipid-based edible films and coatings. Hydrophobic substances potentially used in this way include natural waxes (such as carnauba wax, candelilla wax, rice bran wax, and beeswax); petroleum-based waxes (such as paraffin and polyethylene wax); petroleum-based oils, mineral oils, and vegetable oils; and acetoglycerides and fatty acids. Resins are also used to impart gloss to the commodity (Hagenmaier and Baker, 1994, 1995; Morillon *et al.*, 2002).

Among the hydrophobic materials, wax has been the most widely used for the protective coating of fresh commodities. 'Wax' is the collective term for a series of natural or synthetically produced non-polar substances that normally possess properties such as being kneadable at room temperature, brittle to solid, coarse to finely crystalline, translucent to opaque, of relatively low viscosity even slightly above the melting point, not tending to springiness, having consistency and solubility depending on the temperature, and capable of being polished by slight pressure (Hamilton, 1995). Chemically, wax is an ester of a long-chain aliphatic acid with a long-chain aliphatic alcohol. Waxes are insoluble in bulk water, and do not spread to form a mono-layer on the surface. Their hydrophobicity is high, and this is proven by their solubility in typical organic solvents, such as hexane, chloroform, or benzene. These molecules either have no polar constituents or possess a hydrophilic part so small or so buried in the molecule that it cannot readily interact with water, thereby preventing the molecule from spreading. That explains why waxes are the most efficient barriers to water vapor transfer.

Table 21.1   Classification of waxes

Animal and insect waxes	Beeswax, spermaceti, wool grease, lanolin
Vegetable waxes	Carnauba wax, candelilla wax, ouricouri wax, sugar cane wax, jojoba oil, bayberry wax, Japan wax, rice bran oil
Mineral waxes	Ozocerite, montan wax, paraffin waxes, microcrystalline waxes
Synthetic waxes	Polyethylenes, Fischer-Tropsch waxes, synthetic esters, synthetic amides, carbowaxes

Table 21.2   Characteristics of commercial waxes

	AV	IN	SN	MP	SP	RI	EV	SG
Beeswax	7–36	7–16	90–149	62–65	60–63	1.4338–1.4527	60–84	0.9272–0.9676
Spermaceti	2–2.5	4.8–5.9	108–134	42–50	45.8			
Wool grease	7–15	5–30	100–110	35–42		1.4789–1.4822	85–100	0.9320–0.9450
Carnauba wax	2.9–9.7	7–14	79–95	78–85		1.4672–1.4720		
Candelilla wax	12.7–18.1	14.4–27	35–86.5	67–79		1.4545–1.4620		0.8850
Ouricouri wax	9–20	6–8	70–100	81.5–84		1.4478		0.9700–1.0500
Sugar cane wax	23–28	17	65–77	79–81		1.5100		0.9830
Jojoba oil	2	82	92	6.8–7.0		1.4650		
Bayberry wax	3.52	2.93	208	44.8				
Japan wax	6–20	4.5–12.5	206–237	45–53				0.8750–0.8770
Rice bran wax	2.1–7.3	11.2–19.4	56.9–104.4	75.3–79.9				
Ozocerite	31–38	14–18	87–104	83–89		1.4670		1.0200–1.0300
Paraffin wax				63.6		1.4497		
Microcrystalline wax						1.4450–1.4460		0.8900–0.9000

AV, acid value; IN, iodine number; SN, saponification number; MP, melting point in °C; SP, solidifying point in °C; RI, refractive index; EV, ester value; SG, specific gravity.

Generally, waxes can be divided into naturally occurring waxes and synthetic waxes, and the former group conventionally subdivided into animal, insect, vegetable, and mineral waxes according to their origin, as shown in Table 21.1 (Hamilton, 1995).

Commercial waxes are characterized by a number of properties, including organoleptic (color, odor, and taste), thermal (melting and setting points, flash temperature), physical (specific gravity, penetration, shrinkage), and chemical (acid value, ester value, iodine number, acetyl number) properties. The characteristics of some commercial waxes are summarized in Table 21.2 (Hamilton, 1995).

Triglycerides or neutral lipids are esters of fatty acids with glycerol. They have increased polarity relative to waxes, are insoluble in water, but will spread at the interface to form a stable mono-layer. The hydrophobicity of triglycerides depends on their structure. Long-chain triglycerides are insoluble in water, whereas short-chain molecules are partially water-soluble. Above a certain concentration, they form aggregates similar to micelles. Fatty acids are also considered to be polar lipids, and are used primarily as emulsifiers and dispersing agents. Table 21.3 lists the name of the most important fatty acids used for preparation of edible films and coatings, along with

Table 21.3	Characteristics of various fatty acids				
Common name	Systematic name	Carbon atoms	Double bonds	m.p. (°C)	Major occurring natural oils and fats
Capric	Decanoic	10	0	31.3	Palmae seed fat, milk fat
Lauric	Dodecanoic	12	0	43.9	Coconut oil
Myristic	Tetradecanoic	14	0	54.4	Butter, coconut oil, palm oil
Palmitic	Hexadecanoic	16	0	62.9	Palm oil, butter, lard, tallow
Stearic	Octadecanoic	18	0	69.6	Tallow, cocoa butter, lard, butter
Oleic	9-Octadecanoic	18	1	16.3	Olive, peanut, lard, palm, tallow, corn, rapeseed, canola
Linoleic	9,12-Octadecadienoic	18	2	−5	Soybean, safflower, sunflower, corn, cottonseed
Linolenic	9,12,15-Octadecatrienoic	18	3	−11	Soybean, canola
Arachidonic	5,8,11,14-Eicosatetraenoic	20	4	−49.5	Lard, tallow
Behenic	Docosanoic	22	0	80	Peanut, rapeseed

their chain lengths, the numbers of double bonds, and their melting points. Most fatty acids derived from vegetable oils are considered GRAS substances and are commonly used in the preparation of edible films and coatings (Hernandez, 1994; Baldwin *et al.*, 1997). The properties of fatty acids and of lipids derived from them are markedly dependent on their physical state, chain length, and degree of saturation. Unsaturated fatty acids have a significantly lower melting point than saturated fatty acids of the same chain length. For example, the melting point of stearic acid is 69.6°C, whereas those of oleic, linoleic, and linolenic acids are 16.3, −5, and −11°C, respectively. Chain length also affects the melting point of fatty acids, as shown in Table 21.3. Generally, the melting points of fatty acids increase with chain length and decrease with the number of double bonds. The water vapor permeability of fatty acid films is dependent on the degree of saturation and the chain length of the fatty acids.

Resins are represented by shellac, wood rosin, and coumarone indene, and these are the main coating components used to impart gloss to products (Hagenmaier and Baker, 1994, 1995). Shellac resin, which is secreted by the insect *Laccifer lacca*, has been used as a varnish and as edible coatings for pharmaceuticals, confectionery, fruits, and vegetables. Shellac is composed of a complex mixture of aliphatic alicyclic hydroxy acid polymers, such as aleuritic and shelloic acids (Griffin, 1979). It is soluble in alcohols and in alkaline solutions, and it is also compatible with most waxes, resulting in improved moisture-barrier properties and increased gloss for coated products. Its melting point ranges from 115 to 120°C (Martin, 1982). Shellac is not a GRAS substance, and is therefore only permitted as an indirect food additive in food coatings and adhesives.

Most natural waxes, such as beeswax, carnauba wax, and candelilla wax, also have emulsifying properties, as they are long-chain alcohols and esters (Baldwin *et al.*, 1997). Other compounds added to formulations of the coating materials include plasticizers, emulsifiers, lubricants, binders, de-foaming agents, and formulation aids. The common lipid compounds and additives permitted for use as components in the preparation of edible films and coatings have been listed by Hernandez (1994) and Baldwin *et al.* (1997). The choice of the material is mainly dependent on the target application.

When lipid-based edible films and coatings are a part of the food product and are consumed with its contents, it is not only important that the films or coatings be compatible with the product, but also that the edible films or coatings are benign from a sensory standpoint.

Two forces are driving the interest in natural lipid sources for coating foods; a consumer interest in natural products, and regulatory restrictions regarding the use of petroleum-based products. For instance, petroleum-based waxes, such as polyethylene and paraffin, are restricted or banned in some countries, including Norway, the United Kingdom, and Japan (Baldwin, 1994). Most likely, the naturally derived waxes, such as beeswax, carnauba wax etc., are considered more acceptable in the food industry. The same reasoning applies to oils, with mineral oil being replaced by vegetable-based oils. Resins are also under inspection, and have been banned in some countries (Hernandez, 1994).

Preparation

Generally, edible films and coatings are made from a solution or a dispersion of the film-forming agents, followed by a film-forming application method such as casting, spraying, dipping, extrusion, falling film enrobing, etc. Edible films contain at least one component with a high molecular weight polymer such as protein or polysaccharide, particularly if a self-supporting film is desired. Long-chain polymeric structures are required to yield film matrices with appropriate cohesive and tensile strength when deposited from a suitable solvent (Banker, 1966). The solvent system used for film-forming is also important in determining the properties of the finished films, and for edible films and coatings it is limited primarily to water, ethanol, or a combination of the two.

Environmental conditions during film formation can markedly influence the final film properties. Application of warm film-forming solutions to a warm receiving surface yields the most cohesive films. However, excessive temperatures resulting in an excessive rate of solvent evaporation during film drying may prematurely immobilize polymer molecules, before they have an opportunity to coalesce into a continuous, coherent film, possibly resulting in a brittle film (Banker, 1966). This may also result in film defects such as pinholes or non-uniform film thickness, both of which increase water vapor or gas permeability. Since solvents evaporate rapidly from atomized coating suspensions, the potential for premature immobilization of polymer chains is greater in spray-formed films than in films formed by casting or dipping (Kester and Fennema, 1986).

Another important factor is the relative humidity. Low relative humidity during film drying may increase the rate of solvent removal, probably resulting in brittle films, whereas high relative humidity may delay the drying time.

In contrast to the self-supporting films, edible coatings have been generally used as protective coatings for fresh fruits and vegetables to prevent the decrease in turgor and in weight and to maintain high quality during commercialization (Avena-Bustillos *et al.*, 1994). The most well known and oldest coating method was the application of natural waxes and lipid coatings on specific fruits and vegetables to reduce dehydration and abrasion during processing and handling, and to improve appearance by adding gloss (Baldwin, 1994; Hagenmaier and Baker, 1994; Baldwin *et al.*, 1997). Limitations to their use include their poor mechanical properties and oily appearance in some products. Composite films and coatings have been developed to combine the advantages of both lipid and hydrocolloid components (Baldwin *et al.*, 1997; Krochta and De Mulder-Johnston, 1997). The lipid component in the formulation can serve as a good barrier to water vapor, while the hydrocolloid component can provide a selective barrier to oxygen and carbon dioxide and the necessary supporting matrix (Guilbert, 1986; Kester and Fennema, 1986; Baldwin, 1994; Wong *et al.*, 1994; Baldwin *et al.*, 1997).

Generally, two kinds of composite films are known, according to their preparation method – lamination or emulsion. One is a bi-layer film in which a hydrophobic lipid layer is laminated over a preformed hydrophilic film, resulting in the lipid being a distinct layer within or atop the hydrophilic film (Park *et al.*, 1994; Gontard *et al.*, 1995; Weller *et al.*, 1998). The other is an emulsion film in which the lipid material is uniformly dispersed throughout the hydrophilic film (McHugh and Krochta, 1994c; Shellhammer and Krochta, 1997; Rhim *et al.*, 1999). Figures 21.1 and 21.2 show the procedures for the preparation of typical bi-layer and emulsion composite films, respectively.

Both bi-layer and emulsion films offer advantages. The laminate films are easier to apply with regard to the temperature, due to the distinct nature of the support matrix and lipid (Koelsch, 1994). During the casting of the lipid onto protein or polysaccharide film, the temperature of the film and lipid can easily be controlled separately.

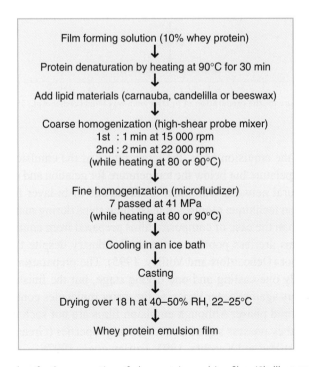

Figure 21.1 Procedure for the preparation of whey protein emulsion films (Shellhammer and Krochta, 1997).

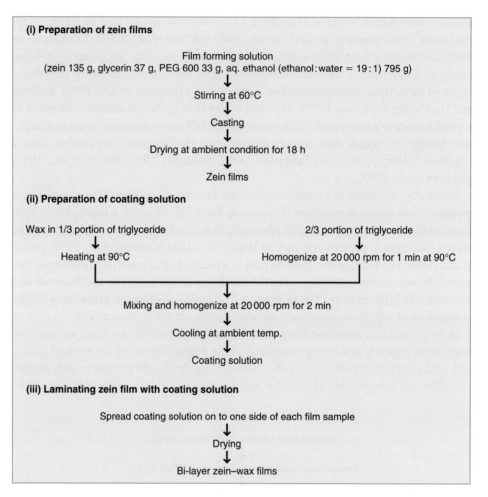

Figure 21.2 Procedure for the preparation of bi-layer zein–wax films (Weller *et al.*, 1998).

When producing the emulsion films, the temperature of the emulsion must be above the lipid-melt temperature but below the temperature for gelation and solvent volatilization of the structural network. The main disadvantage of bi-layer films, however, is that the preparation technique requires at least three steps during manufacture instead of only one or two in the case of composite films prepared from emulsion. This is why the laminated films are less popular in the food industry despite their good barrier against water vapor (Debeaufort and Voilley, 1995). The preparation of the emulsion films requires only one casting and one drying stage, but the finished films are still rather poor barriers against water vapor, as the water molecules continue to permeate through the non-lipid phase. Although emulsion films are not such effective barriers as bi-layer films, they possess superior mechanical properties (Greener and Fennema, 1989a; Avena-Bustillos and Krochta, 1993; Fairley *et al.*, 1997).

The technique of lipid-based edible film preparation method should be adapted to the application of the edible coatings. Generally, the use of an aqueous solution, colloidal dispersion or emulsion of the film-coating materials with a relatively high concentration is preferred. The application and distribution of the film-coating material in a liquid form can be achieved by (Guilbert, 1986):

- Hand spreading with a paint brush
- Spraying
- Falling film enrobing
- Dipping and subsequent dripping
- Distribution in a revolving pan (pan coating)
- Bed fluidizing or air brushing.

When considering possible edible coating applications, attention must be paid to the requirement that edible coating formulations have to be wet when spread on the food surface and upon drying form a film coating that has adequate adhesion, cohesion, and durability to function properly. These properties are influenced by both edible film formulation, and methods of coating and drying. Lack of attention to these facts has resulted in inconsistent and unsatisfactory results in many studies. In addition, edible coatings must provide a satisfactory appearance, aroma, flavor, and mouthfeel. Selecting the best application for the appropriate foods and good control of environmental conditions is necessary to ensure microbial stability.

Physical properties

The intended use of edible films and coatings requires a clear understanding of their moisture barrier and mechanical properties. These properties depend strongly on the film composition, its formation, and the methods of application onto the food products (Cuq *et al.*, 1995; Debeaufort *et al.*, 1998).

Many researchers have examined the water vapor permeability (WVP) and the mechanical properties of composite films made from proteins or polysaccharides with added lipids. For example, composite protein–lipid films had lower WVP values than control protein films from caseinates (Avena-Bustillos and Krochta, 1993), whey protein (McHugh and Krochta, 1994a, 1994b; Bannerjee and Chen, 1995; Pérez-Gago and Krochta, 1999), zein (Weller *et al.*, 1998), and wheat gluten (Gennadios *et al.*, 1993; Gontard *et al.*, 1994). Accordingly, the barrier against water vapor was improved with composite polysaccharide-lipid films made from methylcellulose (MC) (Greener and Fennema, 1989a, 1989b; Kester and Fennema, 1989a, 1989b; Koelsch and Labuza, 1992; Debeaufort and Voilley, 1995), and hydroxypropyl methylcellulose (HPMC) (Kamper and Fennema, 1984a, 1984b; Hagenmaier and Shaw, 1990).

Morillon *et al.* (2002) listed the WVP of lipid-based edible films as a function of their composition and structure, and part of their list is shown in Table 21.4. While some of the lipid and food-grade hydrophobic substances (principally waxes) have WVP values close to those of synthetic plastic films, such as low-density polyethylene (LDPE) or polyvinyl chloride (PVC), they are nonetheless more permeable to moisture transport by at least an order of magnitude. For the sake of comparison, the WVP of LDPE (25 μm thick, 38°C and 90/0% RH gradient) is 0.0722×10^{-12} to 0.0972×10^{-12} g m/m^2 s Pa and that of PVC (25 μm thick, 38°C and 90/0% RH gradient) is 0.0007 to 0.0024×10^{-12} g m/m^2 s Pa (Morillon *et al.*, 2002). Each

Table 21.4 Water vapor permeability (WVP) of lipid-based edible films

Films	Temp (°C)	RH gradient (%)	WVP ($\times 10^{-11}$ g m/m^2 s Pa)
Monolayer films			
Paraffin wax	25	0/100	0.02
Candelilla wax	25	0/100	0.02
Carnauba wax + glycerol monostearate	25	22/100	35
Microcrystalline wax	25	0/100	0.03
Beeswax	25	0/100	0.06
Capric acid	23	12/56	0.38
Myristic acid	23	12/56	3.47
Palmitic acid	23	12/56	0.65
Stearic acid	23	12/56	0.22
Shellac	30	0/100	0.42–1.03
Hydrogenated cottonseed oil	26.7	0/100	0.13
Hydrogenated palm oil	25	0/85	227
Hydrogenated peanut oil	25	0/100	390
Native peanut oil	25	22/44	13.8
Bilayer films			
MC/paraffin wax	25	22/84	0.2–0.4
MC/paraffin oil	25	22/84	2.4
MC/beeswax	25	0/100	0.058
MC/carnauba wax	25	0/100	0.033
MC/candelilla wax	25	0/100	0.018
MC/triolein	25	22/84	7.6
MC/hydrogenated palm oil	25	22/84	4.9
HPMC/stearic acid	27	0/97	0.12
Emulsion films			
MC + PEG400 + behenic caid	23	12/56	7.7
MC + triolein	25	22/84	14.4
MC + hydrogenated palm oil	25	22/84	13.3
MC + PEG400 + myristic acid	23	12/56	3.5
Wheat gluten + oleic acid	30	0/100	7.9
Wheat gluten + soy lecithin	30	0/100	10.5
Wheat gluten + paraffin wax	25	22/84	1.7
Wheat gluten + paraffin oil	25	22/84	5.1
Wheat gluten + triolein	25	22/84	9.7
Wheat gluten + hydrogenated palm oil	25	22/84	7.4
Na-caseinate + acetylated monoglyceride	25	0/100	18.3–42.5
Na-caseinate + lauric acid	25	0/92	11
Na-caseinate + beeswax	25	0/100	11.1–42.5
Whey protein isolate + palmitic acid	25	0/90	22.2
Whey protein isolate + stearyl alcohol	25	0/86	53.6
Whey protein isolate + beeswax	25	0/90	23.9–47.8

From Morillon *et al.* (2002).

hydrophobic substance has its own physicochemical properties, and thus edible films based on lipids have variable behaviors against moisture transfer. Generally, the WVP decreases with increasing hydrophobicity of the added lipid materials. Waxes are the most efficient substances to decrease the WVP because of their high hydrophobicity due to their high content of long-chain fatty alcohols and alkanes. The hydrophilic groups of lipid molecules normally promote water vapor sorption, which may bring about water vapor migration through the film. In addition, many other factors inside or outside the lipid-based edible films affect its barrier properties.

Influence of film preparation methods on WVP

The structure and barrier properties of lipid-based films strongly depend on the film preparation method (Kamper and Fennema, 1984a; Martin-Polo *et al.*, 1992; Debeaufort *et al.*, 1993). The barrier efficiency of bi-layer films is in the same order of magnitude as that of pure lipid or synthetic plastic films, and much lower than that of emulsion films (Table 21.4). The structure and the state of the surface of the film mainly depends on the film preparation technique. Debeaufort *et al.* (1993) used scanning electron microscopy (SEM) to characterize the surfaces of both emulsified and bi-layer films composed of MC and paraffin wax. The emulsified films had an irregular surface and a heterogeneous structure. In this type of film, paraffin wax was dispersed in the form of globules in the MC matrix. In contrast, the lipid component of the bi-layer or laminated film had a smooth surface and homogeneous structure. They determined the WVP of both types of films, and found that the more homogeneously the paraffin wax was distributed, the more the barrier efficiency increased. Nevertheless, Kamper and Fennema (1984a) reported opposite results with both bi-layer and emulsion films composed of HPMC and stearic and palmitic acids. Their emulsion films had 40 times lower permeability than the bi-layer films. These contradictory results could be explained by considering film preparation conditions. Although the initial emulsion prepared by Kamper and Fennema was homogeneous, a phase separation was observed during drying that led to an apparent bi-layer structure in the emulsified films. Such reorientation of lipid molecules during drying and storage of the films greatly influences the WVP of lipid-based composite films. In some cases, the WVP of lipid-based edible films varies greatly with the orientation of the lipid molecules within a film (Avena-Bustillos and Krochta, 1993; Kamper and Fennema, 1984a). In emulsion films, water migrates preferentially through the continuous hydrophilic matrix and the dispersed lipid phase only modifies the apparent tortuosity. McHugh and Krochta (1994b) reported that lipid particle size and distribution significantly affected the WVP of whey protein/beeswax emulsion films. They found that as mean emulsion particle diameters decreased, the WVP of the film decreased linearly.

Influence of fatty acids

Most fatty acids derived from vegetable oils are considered to be GRAS substances, and they are commonly used in the preparation of lipid-based edible films and coatings.

The properties of fatty acids are markedly dependent on their chain length and on their degree of saturation. As stated earlier, unsaturated fatty acids have a lower melting point than do saturated fatty acids of the same length (Table 21.3).

Similarly, the water vapor permeability of fatty acid films is dependent on the degree of saturation and the chain length of the fatty acids. As chain length increases, chain mobility decreases, making fatty acids with long chains good barriers to water vapor transmission (McHugh and Krochta, 1994a; Rhim *et al.*, 1999). On the contrary, fatty acids with shorter chain lengths tend to have greater chain mobility within the structure of composite films, resulting in higher water vapor permeabilities (McHugh and Krochta, 1994c). Although fatty acids such as stearic and palmitic acid lack the structural integrity to form continuous coatings, they have been used successfully in formulations for composite films. Furthermore, composite films containing unsaturated fatty acids have been shown to be more permeable to water vapor than those containing saturated fatty acids (Kamper and Fennema, 1984a). The poor barrier properties of unsaturated fatty acids are a result of the expansion in the molecular volume that occurs with the introduction of a double bond, and of the lower melting point with greater chain mobility (Greener and Fennema, 1992).

Reportedly, fatty acids showed a pronounced effect on the WVP of emulsified composite films with fatty acids and MC support matrix (Figure 21.3). Stearic acid was the most effective in decreasing the WVP of the film. The low permeability of the composite films containing stearic acid was explained by the fact that stearic acid forms an interlocking network; the effect of this network is significant enough to retard moisture transfer more effectively than those based on, say, myristic acid, which has a shorter chain length, or behenic acid, which has such long chains that they hinder

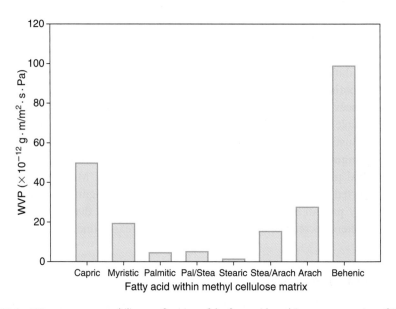

Figure 21.3 Water vapor permeability as a function of the fatty acid used (at a concentration of 30%) in emulsion barriers with a methyl cellulose support matrix (from Koelsch, 1994).

the formation of a tightly interlocking network. In essence, stearic acid provides the optimum chain length without hindering the formation of an interlocking network. However, this is not always the case. Tanaka *et al.* (2000) reported that among the lipid materials tested for preparation of composite films with fish water-soluble proteins, the incorporation of oleic acid was the most effective from the standpoint of the overall properties (tensile strength, elongation at break, and WVP) of the films. Rhim *et al.* (1999) showed that palmitic acid was more effective in decreasing the WVP of emulsified soy protein isolate films with fatty acids. The effect of fatty acids on the permeability properties of fatty acid-based composite films depended on the fatty acid characteristics, and on the interactions between the fatty acid and the hydrocolloid structural matrix.

Influence of the concentration of lipids

Lipid concentration is important in controlling the WVP of lipid-based edible films. Hagenmaier and Shaw (1990) tested the effect of stearic acid concentration on the WVP of HPMC composite films (Figure 21.4). The WVP of the composite films decreased about 300 times with the addition of 40–50% of stearic acid. With less stearic acid, the permeability was higher. The permeability was sharply dependent on the stearic acid content and decreased slowly with stearic acid concentration increases. The phenomenon of a WVP decrease in non-linear fashion was also observed with whey protein–lipid emulsion films (Shellhammer and Krochta, 1997). These results indicate that a critical concentration exists beyond which a sharp decrease in film WVP occurs. Excessive levels of lipid materials result in the film becoming brittle.

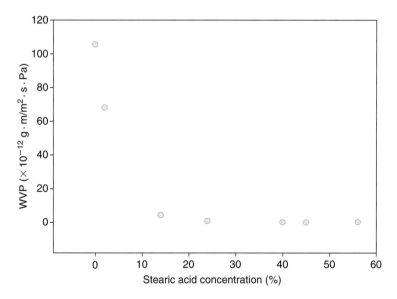

Figure 21.4   Effect of fatty acid concentration on WVP of hydroxypropyl methylcellulose composite films (from Hagenmaier and Shaw, 1990).

Therefore, the optimum concentration of lipid material in the preparation of lipid-based edible films should be determined by considering both the effect of decreasing the WVP and the physical strength of the films. As shown in Figure 21.4, stearic acid concentration of about 20% seems to be the optimum for modifying the water vapor barrier properties. Rhim *et al.* (1999) also found that the incorporation of 20% fatty acids was most effective from the standpoint of the overall properties, including the WVP.

Influence of film thickness

In contrast to hydrophobic, synthetic, polymeric films, the WVP of films prepared from biopolymer materials depends on the thickness of the film (McHugh *et al.*, 1993). The same is true with lipid-based edible films, but to a much lesser extent. The "thickness effect" is explained by the hydrophilic nature of most biopolymers. The diffusivity of moisture in these materials is moisture dependent; therefore at high relative humidities the material has a much higher moisture content and hence greater diffusivity. Thus as the film thickness increases, the equilibrium relative humidity of the film on the moisture side increases and the WVP increases.

With multicomponent films, such as lipid–biopolymer emulsions, the same issues exist. In addition, the state of the lipid phase will affect the WVP. In emulsion films the lipid phase is rarely completely stable, and thus during drying of the film there exists some degree of phase separation. The phase-separated layer contributes substantially to the decrease in WVP. Therefore, as the film thickness increases the amount of potential phase-separated material increases, thereby improving the resistance of the overall film. Hagenmaier and Shaw (1990) observed this phenomenon with HPMC composite films containing 42 ± 3% of stearic acid (Figure 21.5). The

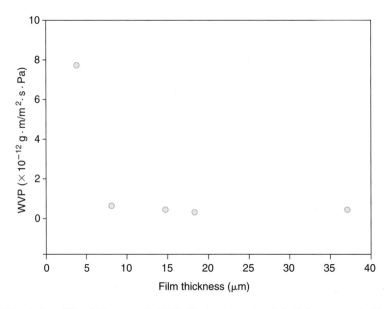

Figure 21.5 Effect of film thickness on the WVP of hydroxypropyl methylcellulose composite films containing 42% stearic acid (from Hagenmaier and Shaw, 1990).

WVP of the thinnest film (3.8 μm) was about 20 times its value for films of 14.7–37.1 μm thickness. These researchers explained that very thin films stretch more when peeled from the drying surfaces. They also observed that the HPMC–stearic acid films have much higher permeability when the films have been stretched slightly. However, films thicker than 15 μm seemed not to depend on the thickness of the film, and the WVP values were lower than LDPE of 0.0529×10^{-12} g m/m^2 s Pa. They did not test the film thickness above 37.1 μm. Martin-Polo *et al.* (1992) also noted that the water vapor transfer rate of cellophane films coated with paraffin wax or oil decreased when thickness increased from 30 to 60 μm, but remained almost constant up to 120 μm. Debeaufort and Voilley (1995) showed that the WVP decreased exponentially when the triglyceride layer thickness increased from 0 to 60 μm, and remained constant for higher thickness, although the critical value is about 300 μm in the case of paraffin-based films.

Influence of the RH differentials

The driving force of the transfer of water vapor is the vapor pressure difference between the two sides of the film. However, it is well known that the WVP of hydrophilic films depends both on the relative humidity (RH) difference and on the absolute humidity values. In hydrophilic films, the WVP is moisture-dependent – that is, at high relative humidity the moisture content of the film increases and the film becomes plasticized. Polymer chain mobility increases, thereby increasing moisture diffusivity. Lipid–biopolymer composite films display the same behavior, but the effect of the RH gradient is less pronounced until extremely high RH conditions are reached. Figure 21.6 shows the effect of RH differentials on the WVP of HPMC–stearic acid composite films. The film exhibited greater permeability as the RH differentials increased. At 97/0% RH, the WVP of the films increased to about four times its value at 94/0% RH, and at 99/0% RH it again doubled.

Kester and Fennema (1989a) also observed that the permeability to water vapor of stearic/palmitic–cellulose ether films increased approximately five-fold as the RH on the low water vapor-pressure side of the film was elevated from 0 to 65% (97/0% vs 97/65%). As the RH on the low humidity side of the film was increased, the equilibrium RH within the film matrix was elevated. The elevation of the absorbed moisture had a plasticizing effect, which increased the diffusion of the moisture through the film.

Influence of temperature

Temperature affects all thermodynamic and kinetic phenomena, and thus the transfer of water vapor through films, tremendously. Permeability is defined as the product of diffusivity and solubility coefficients, and therefore depends both on diffusion (kinetic factor) and sorption (thermodynamic factor). Generally, an increase in temperature causes an increase in diffusion due to an increase in molecular thermal motion. In contrast, water sorption is favored by a decrease in temperature. Consequently, the

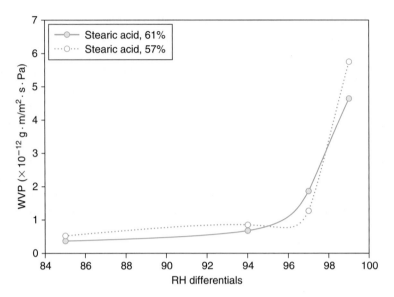

Figure 21.6 Effect of RH differentials on the WVP of hydroxypropyl methylcellulose composite films containing 57% or 61% of stearic acid (from Hagenmaier and Shaw, 1990).

permeability can either increase or decrease when temperature increases. Kamper and Fennema (1984b) observed approximately a six-fold increase in WVP through a bi-layer film of stearic/palmitic acids–HPMC upon decrease of the temperature from 25°C to 5°C, probably due to shrinking and breaking of the film as the components became more rigid. This also may be due to an increase in film hydration at lower temperatures, thus facilitating moisture transfer through the film. Kester and Fennema (1989a) also showed a decrease of the WVP when the temperature decreased from 40°C to 15°C with the activation energy of 59 kJ/mol, but the WVP increased at 4°C, which deviated from linearity obtained from higher temperatures in the Arrhenius plot (Figure 21.7). They explained that the deviation at 4°C was due to the lipid contraction at lower temperatures resulting in minor defects or flaws in the film, thereby lowering its resistance to water vapor resistance. Generally, wax-laminated films are not used at low temperatures because of the tendency for waxes to become brittle under these conditions. However, Kester and Fennema (1989a) asserted that the beeswax-coated lipid–cellulose ether film they developed was remarkably stable against fracture at low temperatures, indicating a potential utility in frozen foods.

Applications

The use of lipids in coatings has been practiced for centuries. Coating oranges and lemons with wax dates back to the twelfth century (Hardenburg, 1967), whereas coating foods with fats to prevent moisture loss (a practice known as "larding") was used in England during the sixteenth century (Labuza and Contreras-Medellin, 1981). Waxes (e.g. carnauba wax, beeswax, paraffin wax) and oils (mineral oil, vegetable oil)

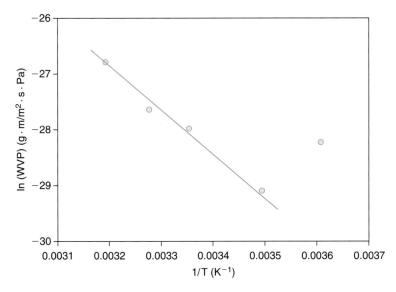

Figure 21.7 Effect of temperature on the water vapor permeability of a lipid-based edible film. (Reproduced from Kester and Fennema, 1989a, with permission from the Institute of Food Technologists.)

have been used commercially since the 1930s as protective coatings for fresh fruits and vegetables (Baldwin, 1994).

Potential applications of the lipid-based edible films and coatings include the following.

Fresh and processed food products

Currently, lipids are used as edible films and coatings for meat, poultry, seafood, fruits, vegetables, grains, candies, heterogeneous and complex foods, or fresh, cured, frozen, and processed foods.

In the case of fruits and vegetables, coatings are used to prevent weight loss (retain moisture), to slow down aerobic respiration, and to improve the appearance by providing gloss. Shellac and waxes are the most common coating agents used to improve fruit gloss. However, fruits coated with shellac have been reported to develop a whitening of the skin due to the condensation that develops when they are brought from cold storage to ambient temperatures (Hagenmaier and Baker, 1994). Hagenmaier and Baker (1996) also pointed that in some applications of fruit coatings, like citrus fruit, candelilla and other waxes are more advantageous than shellac or wood rosin because the former are more permeable to oxygen and carbon dioxide than the latter (Hagenmaier and Shaw, 1992). Coatings with too high a barrier to permanent gases can lead to fruit senescence and off-flavor production. Formulating shellac coatings with a lipid component, such as candelilla or carnauba wax, maintains the glossy appearance of a shellac coating but increases oxygen permeability, thereby reducing the production of ethanol that can occur in shellac-coated apples (Alleyne and Hagenmaier, 2000; Bai *et al.*, 2003).

Lipids and resins are useful in reducing surface abrasion during handling operations, by sealing tiny scratches that may occur on the surface of fruits and vegetables during harvest, handling and processing (Hernandez, 1994). Fruits and cleaned vegetables are often coated with waxes in cases where the peel will not be consumed – e.g. lemons, oranges, tangerines, and melons – or when the fruit must be transported over long distances. The colorless, odorless, and tasteless macrocrystalline slab waxes are especially valuable. Small quantities of hard waxes such as carnauba, ouricouri or montan waxes are incorporated to impart a gloss.

Wax and water spray emulsions have been used for apples, pears, citrus fruits, and tomatoes. The emulsion must have low surface tension, good wetting power, and dry rapidly after the coating process. A typical emulsion formulation is sodium hydroxide (6 parts), triethanolamine (20 parts), stearic acid (42 parts), paraffin wax (1–5 parts), carnauba (55 parts), shellac (100 parts) and water (2000 parts).

Paraffin wax is used in the poultry-processing industry for wax picking of ducks and geese. The microcrystalline slab waxes are too brittle for this task, and blends of macro- and microcrystalline waxes plus an additive of polyethylene waxes are used. Cheese is also coated with a paraffin wax to prevent its desiccation and the loss of flavor substances, and to protect the surface from undesired mold formation.

Stuchell and Krochta (1995) displayed efficient protection of frozen king salmon against lipid oxidation and water loss by using a coating composed of a mixture of whey proteins and acetylated monoglycerides. They reported that moisture loss was reduced by 42–65% during 3 weeks of storage at $-23°C$.

One of the main uses for shellac coatings is as a confectioner's glaze on candies and panned confections. Shellac is used primarily to impart high gloss to these items, but the hydrophobic character also serves as a barrier to recrystallization and to agglomeration of loose confections in high humidity.

Lipid coatings have been effective at reducing moisture uptake or loss, and minimizing or delaying oxidative rancidity and moisture uptake in raisins and nuts. For example, mixtures of various antioxidants (TBHQ, BHA, BHT) and citric acid in hydrogenated vegetable oil or in acetylated monoglycerides have been effective at minimizing oxidative rancidity in peanuts. Mineral oil has traditionally been used to prevent the clumping of raisins, as well as to improve their appearance (Kochar and Rossell, 1982). Furthermore, raisins have been stabilized against moisture loss by coating with beeswax (Watters and Brekke, 1961).

Minimally processed fruits and vegetables

The growing demand for convenience has influenced the product form in which fruits and vegetables are consumed. Pre-packaged salads, pre-cut fruits, broccoli and cauliflower florets, stir-fry mixes, and packaged carrot and celery sticks are examples of value-added produce that appeal to consumers seeking health and convenience.

Minimally processed fruits and vegetables are often exposed to substantial mechanical injury and wounding, inducing increased respiration, accumulation of secondary metabolites, increased ethylene synthesis, and cellular disruption (Rolle and Chism, 1987). Maintaining the cell membrane integrity and reducing the chemical and enzymatic reactions are major concerns in extending the shelf life of minimally processed fruits and vegetables. The application of semi-permeable edible coatings is promising in this field (Baldwin *et al.*, 1997). Avena-Bustillos *et al.* (1994) showed that using an edible coating of sodium caseinate–stearic acid emulsion on peeled carrots reduced surface dehydration and white blush formation. This is important, since white blush on the surface is a major factor in reducing consumer acceptance of minimally processed carrots.

Active packaging

Active packaging is an innovative packaging system in which the package, the product, and the environment interact to prolong the shelf life or to enhance the safety or sensory properties, while maintaining the quality of the product. This is particularly important in the area of fresh and extended shelf-life foods. Floros *et al.* (1997) reviewed active packaging and recognized antimicrobial packaging as one of the most promising versions of an active packaging system. Recently, Han (2000) and Vermeiren *et al.* (2002) reviewed antimicrobial packaging, and Suppakul *et al.* (2003) published a review article on active packaging with an emphasis on antimicrobial packaging.

The incorporation of preservatives into edible films and coatings to control surface microbial growth in foods is being explored. Film composition is one of the primary factors affecting the diffusivity of preservatives in edible films (Ozdemir and Floros, 2003). In particular, lipid materials are known to have pronounced effects on such diffusion processes. Redl *et al.* (1996) found that films containing lipids had lower diffusion coefficients than those without a lipid component. Similarly, Vojdani and Torres (1990) showed that the type and concentration of fatty acids used in the formulation of polysaccharide films affected the permeability of potassium sorbate through these films. Potassium sorbate permeation decreased. Palmitic, stearic, and arachidonic acids were effective in decreasing the potassium sorbate permeation through MC and HPMC films. The application of a lipid material such as beeswax in the form of a second layer on polysaccharide films containing potassium sorbate resulted in films with extremely low potassium sorbate permeability values (Vojdani and Torres, 1989).

Edible packaging

Edible packaging refers to the use of edible films, pouches, bags, and other containers for packaging of food products (McHugh and Krochta, 1994c). Potential uses for edible packages include the pre-measuring of ingredients, and controlling the atmosphere around fresh fruits and vegetables. When the consumption of the final package is not desirable, such as for short-term storage of foods using wraps or bags, edible packages could quickly be composted using domestic or municipal facilities. Edible packages offer environmental advantages, as well as cost and convenience advantages, over conventional synthetic packaging systems. Lipid-based edible films can be used where traditional plastic packaging materials cannot be applied – that is, they can separate and reduce migration of moisture in multi-component foods. Kamper and Fennema (1985) used a bi-layer film consisting of lipid and a layer of HPMC between tomato-paste ground crackers. The film substantially slowed the transfer of water from the salted tomato paste to the crackers.

Another major potential use of lipid-based edible films composed with hydrocolloids (e.g. starch) is as bags for fresh fruits, vegetables, and bread. Though the size of this market is still very limited as a result of the price difference in comparison with traditionally used synthetic plastics, the advantages of these films (selective permeability, biodegradability, and maintenance of the high quality of foods) may ultimately drive their commercial acceptance.

Conclusions

Edible films and coatings have shown potential for controlling the transfer of moisture, oxygen, lipids, aroma and flavor compounds in food systems, with a resulting increase in food quality and shelf life, while reducing the use of synthetic plastic packaging materials. Many materials have been used for film and coating formulations, such as carbohydrates, proteins, lipids, and mixtures of these. Due to the hydrophilic nature of most biopolymers, achieving adequate moisture barriers requires the formation of composite films that contain hydrophobic materials such as edible fatty acids or waxes. Two primary preparation systems exist for lipid-based edible films; emulsion films and bi-layer films. Pure lipid coatings are used in most cases, while freestanding lipid films cannot be manufactured due to their low mechanical strength. Film properties such as barrier, mechanical, thermal, and optical properties all vary depending on the film formulation and preparation methods, and on other external conditions such as RH differentials across the film, the temperature, and the packaging methods used. The specific functional properties of lipid-based edible films and coatings can be exploited for many foods, agricultural, pharmaceutical, and medical applications. Indeed, lipid-based edible films and coatings will continue to play an important role in the food industry by extending the shelf life or improving the quality of many products, and in some cases alleviating environmental packaging-related concerns.

References

Alleyne, V. and Hagenmaier, R. D. (2000). Candelilla-shellac: an alternative formulation for coating apples. *HortSci.* **35(4)**, 691–693.

Avena-Bustillos, R. J. and Krochta, J. M. (1993). Water vapor permeability of caseinate-based edible films as affected by pH, calcium crosslinking and lipid content. *J. Food Sci.* **58**, 904–907.

Avena-Bustillos, R. J., Krochta, J. M., Saltveit, M. E., Rojas-Villegas, R. J. and Sauceda-Pérez, J. A. (1994). Optimization of edible coating formulations on zucchini to reduce water loss. *J. Food Eng.* **21**, 197–214.

Bai, J., Hagenmaier, R. D. and Baldwin, E. A. (2003). Coating selection for 'Delicious' and other apples. *Postharvest Biol. Technol.* **28(3)**, 381–390.

Baldwin, E. A. (1994). Edible coatings for fruits and vegetables, past, present and future. In: *Edible Coatings and Films to Improve Food Quality* (J. M. Krochta, E. A. Baldwin and M. Nisperos-Carriedo, eds), pp. 25–64. Technomic Publishing Company, Inc., Lancaster, PA.

Baldwin, E. A., Nisperos-Carriedo, M. O., Hagenmaier, R. D. and Baker, R. A. (1997). Use of lipids in coatings for food products. *Food Technol.* **51(6)**, 56–64.

Banerjee, R. and Chen, H. (1997). Functional properties of edible films using whey protein concentrate. *J. Dairy Sci.* **78,** 1673–1683.

Banker, G. S. (1966). Film coating theory and practice. *J. Pharm. Sci.* **55,** 81–89.

Cuq, B., Gontard, N. and Guilbert, S. (1995). Edible films and coatings as active layers. In: *Active Food Packaging* (M. L. Rooney, ed.), pp. 112–142. Chapman and Hall, London, UK.

Debeaufort, F. and Voilley, A. (1995). Effect of surfactants and drying rate on barrier properties of emulsified edible films. *Int. J. Food Sci. Technol.* **30,** 183–190.

Debeaufort, F., Martin-Polo, M. and Voilley, A. (1993). Polarity homogeneity and structure affect water vapor permeability of model edible films. *J. Food Sci.* **58,** 426–429, 434.

Debeaufort, F., Quezada-Gallo, J. A. and Voilley, A. (1998). Edible films and coatings: Tomorrow's packagings: a review. *Crit. Rev. Food Sci.* **38,** 299–313.

Fairley, P., Krochta, J. M. and German, J. B. (1997). Interfacial interactions in edible emulsion films from whey protein isolate. *Food Hydrocolloids* **11,** 245–252.

Floros, J. D., Dock, L. L. and Han, J. H. (1997). Active packaging technologies and applications. *Food Cosmet. Drug Packag.* **20(1),** 10–17.

Gennadios, A., Weller, C. L. and Testin, R. F. (1993). Modification of physical and barrier properties of edible wheat gluten-based films. *Cereal Chem.* **70(4),** 426–429.

Gennadios, A., McHugh, T. H., Weller, C. L. and Krochta, J. M. (1994). Edible coatings and films based on proteins. In: *Edible Coatings and Films to Improve Food Quality* (J. M. Krochta, E. A. Baldwin and M. Nisperos-Carriedo, eds), pp. 201–277. Technomic Publishing Co., Inc., Lancaster, PA.

Gontard, N., Duchez, C., Cuq, J. L. and Guilbert, S. (1994). Edible composite films of wheat gluten and lipids: Water vapor permeability and other physical properties. *Intl. J. Food Sci. Technol.* **29,** 39–50.

Gontard, N., Marchesseau, S., Cuq, J. L. and Guilbert, S. (1995) Water vapor permeability of edible bilayer films of gluten and lipids. *Intl. J. Food Sci. Technol.* **30,** 49–56.

Greener, I. K. and Fennema, O. (1989a). Barrier properties and surface characteristics of edible bilayer films. *J. Food Sci.* **54,** 1393–1399.

Greener, I. K. and Fennema, O. (1989b). Evaluation of edible, bilayer films for use as moisture barriers for food. *J. Food Sci.* **54,** 1400–1406.

Greener, I. K. and Fennema, O. (1992). Lipid-based edible films and coatings. *Lipid Tech.* **4,** 34–38.

Griffin, W. C. (1995). Emulsions. In: *Kirk-Othmer Encyclopedia of Chemical Technology,* 3rd edn. Vol. 8, pp. 913–916. Wiley Interscience, New York, NY.

Guilbert, S. (1986). Technology and application of edible protective films. In: *Food Packaging and Preservation: Theory and Practice.* (M. Mathlouthi, ed.), pp. 371–394. Elsevier Applied Science Publishing Co., London, UK.

Guilbert, S., Gontard, N. and Gorris, L. G. M. (1996). Prolongation of the shelf-life of perishable food products using biodegradable films and coatings. *Lebensm.-Wiss. Technol.* **29,** 10–17.

Hagenmaier, R. D. and Shaw, P. E. (1990). Moisture permeability of edible films made with fatty acid and (hydroxypropyl) methylcellulose. *J. Agric. Food Chem.* **38,** 1799–1803.

Hagenmaier, R. D. and Shaw, P. E. (1992). Gas permeability of fruit coating waxes. *J. Am. Soc. Hort. Sci.* **117,** 105–109.

Hagenmaier, R. D. and Baker, R. A. (1994). Wax microemulsions and emulsions as citrus coatings. *J. Agric. Food Chem.* **42,** 899–902.

Hagenmaier, R. D. and Baker, R. A. (1995). Layered coatings to control weight loss and preserve gloss of citrus fruit. *HortSci.* **30,** 296–298.

Hagenmaier, R. D. and Baker, R. A. (1996). Edible coatings from candelilla wax microemulsions. *J. Food Sci.* **61,** 562–565.

Hamilton, R. J. (1995). Commercial waxes: Their composition and applications. In: *Waxes: Chemistry, Molecular Biology and Functions* (R. J. Hamilton, ed.), pp. 257–309. The Oily Press, Dundee, UK.

Han, J. H. (2000). Antimicrobial food packaging. *Food Technol.* **54(3),** 56–65.

Hardenburg, R. E. (1967). Wax and related coatings for horticultural products. A bibliography. *Agric. Res. Service Bull.* pp. 51–55. USDA, Washington, DC.

Hernandez, E. (1994). Edible Coatings from Lipids and Resins. In: *Edible Coatings and Films to Improve Food Quality* (J.M. Krochta, E.A. Baldwin and M. Nisperos-Carriedo, eds), pp. 279–303. Technomic Publishing Company, Inc., Lancaster, PA.

Kamper, S. L. and Fennema, O. (1984a). Water vapor permeability of edible bilayer films. *J. Food Sci.* **49,** 1478–1481, 1485.

Kamper, S. L. and Fennema, O. (1984b). Water vapor permeability of an edible, fatty acid bilayer film. *J. Food Sci.* **49,** 1482–1485.

Kamper, S. L. and Fennema, O. (1985). Use of an edible film to maintain water vapor gradient in foods. *J. Food Sci.* **50,** 382–284.

Kester, J. J. and Fennema, O. (1986). Edible films and coatings: A review. *Food Technol.* **40(12),** 47–59.

Kester, J. J. and Fennema, O. (1989a). An edible film of lipids and cellulose ethers: Barrier properties to moisture vapor transmission and structural evaluation. *J. Food Sci.* **54,** 1383–1389.

Kester, J. J. and Fennema, O. (1989b). An edible film of lipids and cellulose ethers: Performance in a model frozen-food system. *J. Food Sci.* **54,** 1390–1192, 1406.

Kochhar, S. P. and Rossell, J. B. (1982). A vegetable oiling agent for dried fruits. *J. Food Technol.* **17,** 661.

Koelsch, C. M. (1994). Edible water vapor barriers: properties and promise. *Trends Food Sci. Technol.* **5,** 76–81.

Koelsch, C. M. and Labuza, T. P. (1992). Functional, physical and morphological properties of methyl cellulose and fatty acid-based edible barriers. *Lebensm. Wiss. Technol.* **25,** 404–411.

Krochta, J. M. and De Mulder-Johnston, C. (1997). Edible and biodegradable polymer films: Challenges and opportunities. *Food Technol.* **51(2),** 61–74.

Labuza, T. P. and Contreras-Madellin, R. (1981). Prediction of moisture protection requirements for foods. *Cereal Foods World* **26,** 335–343.

Martin, J. (1982). Shellac. In: *Kirk-Othmer Encyclopedia of Chemical Technology,* pp. 737–747. Wiley Interscience, New York, NY.

Martin-Polo, M., Mauguin, C. and Voilley, A. (1992). Hydrophbic films and their efficiency against moisture transfer. 1. Influence of the film preparation technique. *J. Agric. Food Chem.* **40,** 407–412.

McHugh, T. H., Avena-Bustillos, R. and Krochta, J. M. (1993). Hydrophilic edible films: Modified procedure for water vapor permcability and explanation of thickness effects. *J. Food Sci.* **58,** 899–903.

McHugh, T. H. and Krochta, J. M. (1994a). Water vapor permeability properties of edible protein-lipid emulsion films. *J. Am. Oil. Chem. Soc.* **71,** 307–312.

McHugh, T. H. and Krochta, J. M. (1994b). Dispersed phase particle size effects on water vapor permeability of whey protein-beeswax edible emulsion films. *J. Food Process. Preserv.* **18,** 173–188.

McHugh, T. H. and Krochta, J. M. (1994c). Milk protein based edible films and coatings. *Food Technol.* **48(1),** 97–103.

Morillon, V., Debeaufort, F., Blond, G., Capelle, M. and Voilley, A. (2002). Factors affecting the moisture permeability of lipid-based edible films: A review. *Crit. Rev. Food Sci. Nutr.* **42,** 67–89.

Ozdemir, M. and Floros, J. D. (2003). Film composition effects on diffusion of potassium sorbate through whey protein films. *J. Food Sci.* **68,** 511–516.

Park, J. W., Testin, R. F., Park, H. J., Vergano, P. J. and Weller, C. L. (1994). Fatty acid concentration effect on tensile strength, elongation and water vapor permeability of laminated edible films. *J. Food Sci.* **59,** 916–919.

Pérez-Gago, M. B. and Kcrochta, J. M. (1999). Water vapor permeability of whey protein emulsion films as affected by pH. *J. Food Sci.* **64,** 695–698.

Redl, A., Gontard, N. and Guilbert, S. (1996). Determination of sorbic acid diffusivity in edible wheat gluten and lipid based films. *J. Food Sci.* **61,** 116–120.

Rhim, J. W., Wu, Y., Weller, C. L. and Schnepf, M. (1999). Physical characteristics of emulsified soy protein-fatty acid composite films. *Sci. Alimentas* **19,** 57–71.

Rolle, R. S. and Chism, G. W. III. (1987) Physiological consequences of minimally processed fruits and vegetables. *J. Food Quality* **10,** 157–177.

Shellhammer, T. H. and Krochta, J. M. (1997). Whey protein emulsion film performance as affected by lipid type and amount. *J. Food Sci.* **62,** 390–394.

Stuchell, Y. M. and Krochta, J. M. (1995). Edible coatings on frozen king salmon: effect of whey protein isolate and acetylated monoglycerides on moisture loss and lipid oxidation. *J Food Sci.* **60,** 28–31.

Suppakul, P., Miltz, J., Sonneveld, K. and Bigger, S. W. (2003). *J. Food Sci.* **68,** 408–420.

Tanaka, M., Ishizaki, S., Suzuki, T. and Takai, R. (2001). Water vapor permeability of edible films prepared from fish water soluble proteins as affected by lipid type. *J. Tokyo Univ. Fisher.* **87,** 31–37.

Vermeiren, L., Devlieghere, F. and Debvere, J. (2002). Effectiveness of some recent antimicrobial packaging concepts. *Food Add. Contam.* **19(suppl.),** 163–171.

Vojdani, F. and Torres, J. A. (1989). Potassium sorbate permeability of edible cellulose ether films. *J. Food Process. Preserv.* **13,** 417–430.

Vojdani, F. and Torres, J. A. (1990). Potassium sorbates permeability of methylcellulose and hydroxypropyl methylcellulose coatings: Effect of fatty acids. *J. Food Sci.* **55,** 841–846.

Watters, G. G. and Brekke, J. E. (1961). Stabilized raisins for dry cereal products. *Food Technol.* **5,** 236–238.

Weller, C. L., Gennadios, A. and Saravia, R. A. (1998). Edible bilayer films from zein and grain sorghum wax or carnauba wax. *Lebensm. -Wiss. u. -Technol.* **31,** 279–285.

Wong, D. W. S., Tillin, S. J., Hudson, J. S. and Pavlath, A. E. (1994). Gas exchange in cut apples with bilayer coatings. *J. Agric. Food Chem.* **42,** 2278–2285.

22 Emulsion and bi-layer edible films

María B. Pérez-Gago and John M. Krochta

Introduction

An edible film is defined as a thin layer of edible material, either formed on a food as a coating, or preformed and then wrapped around a food or placed between food components (Krochta, 1997a). The ability to provide a barrier to moisture, oxygen, carbon dioxide, oil, and flavor/aroma migration between adjacent food components, and/or between the food and the environment, is the reason for the great interest in edible films in recent years (Guilbert, 1986; Kester and Fennema, 1986; Krochta, 1992). The barrier characteristics of edible films can also enable a reduction and simplification of the packaging material required for a food product. Additional benefits are the improvement of the mechanical integrity or handling characteristics of the food, and functionality by incorporating food ingredients such as antioxidants, antimicrobials, or flavors.

Development of edible films has been mainly focused upon barriers containing proteins, polysaccharides, or lipids. The functional properties of the resulting film depend on the nature of the film-forming material. In general, polysaccharides and proteins, which are polymeric and hydrophilic by nature, are good film-formers and excellent oxygen, aroma, and lipid barriers at low relative humidity. However, they are poor moisture barriers compared to synthetic moisture-barrier films such as low-density polyethylene (LDPE) (Table 22.1). Lipids, which are hydrophobic, are better moisture barriers than polysaccharides and proteins, and are comparable to synthetic films (Table 22.1). However, their non-polymeric nature limits their ability to form cohesive films. In addition, some lipids require solvents or high temperatures for casting

Innovations in Food Packaging
ISBN: 0-12-311632-5

Table 22.1 Water vapor permeabilities of edible polymer films compared to lipids and synthetic polymer films

	Film[a]	Test conditions[b]	Permeability (g mm/m² d kPa)
Schultz et al. (1949)	Pectinate	25°C, 85/31%RH	73
Rankin et al. (1958)	Amylose	25°C, 100/0%RH	32
Hagenmaier and Shaw (1990)	HPMC	27°C, 0/85%RH	9.1
Kamper and Fennema (1984a)	HPMC : PEG (9 : 1)	25°C, 85/0%RH	6.5
Hanlon (1992)	MC	35°C, 90/0%RH	4.8
McHugh et al. (1994)	WPI : Gly (4 : 1)	25°C, 0/77%RH	70
Park and Chinnan (1995)	WG : Gly (3.1 : 1)	21°C, 85/0%RH	53
Stuchell and Krochta (1994)	SPI : Gly (4 : 1)	28°C, 0/78%RH	39
Avena-Bustillos and Krochta (1993)	Sodium caseinate	25°C, 0/81%RH	37
Park and Chinnan (1995)	CZ : Gly (4.9 : 1)	21°C, 85/0%RH	9.6
Gontard et al. (1992)	WG : Gly (5 : 1)	30°C, 100/0%RH	5.1
Lovegren and Feuge (1954)	Hydrogenated peanut oil	25°C, 100/0%RH	3.3
Lovegren and Feuge (1954)	AMG	25°C, 100/0%RH	1.9–13.0
Shellhammer and Krochta (1997b)	Tripalmitin	28°C, 0/100%RH	0.19
Shellhammer and Krochta (1997b)	CarW	28°C, 0/100%RH	0.098
Shellhammer and Krochta (1997a)	BW	26°C, 0/100%RH	0.089
Lovegren and Feuge (1954)	PW	25°C, 100/0%RH	0.019
Shellhammer and Krochta (1997a)	CanW	25°C, 0/100%RH	0.012
Shellhammer and Krochta (1997b)	PVC	28°C, 0/100%RH	0.62
Shellhammer and Krochta (1997b)	PET	25°C, 0/100%RH	0.17
Shellhammer and Krochta (1997b)	LDPE	28°C, 0/100%RH	0.031

[a] Compositions rounded to nearest whole number.
[b] RHs on top and bottom sides of film (top/bottom).
MC, methyl cellulose; HPMC, hydroxypropylmethyl cellulose; WPI, whey protein isolate; WG, wheat gluten; SPI, soy protein isolate; CZ, corn zein; PEG, polyethylene glycol; Gly, glycerol; AMG, acetylated monoglycerides; CarW, carnauba wax; BW, beeswax; PW, paraffin wax; CanW, candelilla wax; PVC, polyvinylchloride; PET, polyethylene terephthalate; LDPE, low-density polyethylene.

if they are solids at room temperature, which makes application more difficult (Krochta, 1997a).

An approach to improving film functionality is to combine lipid materials with polysaccharides or proteins to form composite films. This way, the polysaccharide or protein provides the film integrity while acting as an entrapping or supporting matrix for the lipid component, and the lipid provides the moisture barrier (Krochta, 1997a).

This chapter reviews the existing literature on polysaccharide–lipid and protein–lipid edible composite films. The specific objectives are to summarize the information on edible composite film formation and properties, as well as factors affecting the performance of edible composite films.

Composite film formation

A composite film can be produced as either a bi-layer or a stable emulsion. In bi-layer composite films, the lipid forms a second layer over the polysaccharide or protein layer. In emulsion composite films, the lipid is dispersed and entrapped in the supporting matrix of protein or polysaccharide.

Bi-layer film: Lipid on hydrophilic film

Coating technique: Two-step technique

Emulsion technique: One-step technique

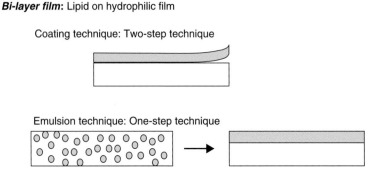

Emulsion film: Lipid droplets dispersed within the hydrophilic phase

Figure 22.1   Edible composite film formation.

Bi-layer film systems can be formed by two different techniques; the "coating technique" or the "emulsion technique". The coating technique is a two-step technique that involves casting a lipid layer, either molten or from solvent, onto a previously formed polysaccharide or protein film. The "emulsion technique" is a one-step coating technique that involves dispersing the lipid into the film-forming solution prior to film casting, with the bi-layer film resulting from the emulsion when the continuous phase cannot stabilize the emulsion and phase separation occurs during drying (Figure 22.1).

Emulsion film systems, on the other hand, can only be formed by the emulsion technique, which involves the dispersion of the lipid into either the polysaccharide or the protein film-formation solution to make a stable emulsion. In general, the emulsifying character of proteins makes them appropriate for this technique. However, polysaccharides are not as effective as emulsifiers, and emulsion film formation generally requires the addition of an emulsifier to improve emulsion stability (Figure 22.1).

Barrier properties that are commonly studied when determining the ability of edible films to protect foods from the environment and from adjacent ingredients are film water vapor permeability (WVP) and oxygen permeability (OP) (Pérez-Gago and Krochta, 2002). In addition, mechanical properties are evaluated to assess the ability of edible films to protect foods against mechanical abuse (Krochta, 1997b). Both barrier and mechanical properties of composite films are affected by the film composition and the distribution (uniform or non-uniform) of the hydrophobic substances in the composite films (Kamper and Fennema, 1985; Debeaufort *et al.*, 1993), which in turn are affected by the film-forming technique. In addition, for films of similar composition and structure, changes in test conditions (specifically temperature and relative humidity) affect their barrier properties. These effects make comparisons among films difficult; nonetheless, such comparisons are important and are made wherever possible in the following sections.

Properties of bi-layer films

Many studies have investigated the properties of bi-layer films prepared either by the coating technique or the emulsion technique. In general, investigations have shown that bi-layer films are more effective barriers against water vapor transfer than are emulsion films, due to the existence of a continuous hydrophobic phase in the film. However, the main disadvantages of the coating technique are that it requires several steps, and the use of solvents or handling of molten waxes. Below there is an overview of some of the most relevant works found in the literature with a description of the film-formation technique and the advantages and disadvantages of the method.

The first work published in the literature relating to edible bi-layer films dates from 1949. Schultz *et al.* (1949) prepared glycerol (Gly)-plasticized low methoxyl pectinate films with stearic acid (SA), lauric acid (LA), beeswax (BW) or paraffin wax (PW) by using the coating technique and the emulsion technique. In all the films formed, WVP decreased as the fatty acid chain length and the degree of saturation of the waxes increased. The coating technique produced bi-layer films by evaporation of the ether solvent from the wax coating on the surface of the base pectinate film. Among the hydrophobic components, the application of the PW layer to form the bi-layer film reduced the WVP of the base pectinate film by 99.8%. The same amount of PW dispersed in the pectinate solution before drying at room temperature to form an emulsion film produced a moisture barrier that was less effective than the bi-layer film (Table 22.2). Similarly, better moisture-barrier films were obtained when LA or BW was added as a surface coating to the preformed pectinate film, compared to suspending the lipids in the pectinate solution before casting the film. However, the disadvantages of forming the bi-layer film vs the emulsion film are the two-step nature of the technique, and the use of organic solvent.

Three papers describing the production of bi-layer films composed of a base of polyethylene glycol (PEG)-plasticized hydroxypropylmethyl cellulose (HPMC) and an upper layer of lipid were published by Kamper and Fennema (1984a, 1984b, 1985). Films were prepared by the coating technique and the emulsion technique, both of which produced bi-layer films. The coating technique involved painting the lipid material onto HPMC:PEG films which had been previously formed by drying aqueous-ethanol solutions at 90°C. The emulsion technique involved adding the lipid directly to the HPMC:PEG aqueous-ethanol film-forming solution at a temperature higher than the melting point of the lipid, then casting it hot, and drying at 90°C. As the solvent evaporated, and since the HPMC is not an effective emulsifier, the lipid phase separated from the mixture, forming two continuous layers with the lipid layer over the HPMC:PEG layer. In this work, as well as in the Schultz *et al.* (1949) work, it was found that composite films displayed a greater moisture barrier with an increase in the degree of saturation of the lipid used and of the chain length of fatty acids used (Kamper and Fennema, 1984a). The moisture barrier was also increased when a blend of stearic–palmitic acid (SA–PA) was incorporated into the film-forming system prior to film formation (emulsion technique) rather than being coated onto the surface of the HPMC:PEG films (coating technique) (Table 22.3). The WVP values obtained

Table 22.2 Water vapor permeabilities of bi-layer and emulsion pectinate–lipid edible films

Film[a]	Film preparation	Test conditions[b]	Permeability (g mm/m² d kPa)
Pectinate		25°C, 81/31%RH	73.0
Pectinate : Gly (3 : 1)		25°C, 81/31%RH	91.0
SA/pectinate : Gly (1/3 : 1)	Bi-layer film, two-step technique	25°C, 81/31%RH	51.0
LA/pectinate : Gly (1/3 : 1)	Bi-layer film, two-step technique	25°C, 81/31%RH	13.0
BW/pectinate : Gly (1/3 : 1)	Bi-layer film, two-step technique	25°C, 81/31%RH	7.8
PW/pectinate : Gly (1/3 : 1)	Bi-layer film, two-step technique	25°C, 81/31%RH	0.18
LA/pectinate : Gly (1/3 : 1)	Emulsion film	25°C, 81/31%RH	44.2
BW/pectinate : Gly (1/3 : 1)	Emulsion film	25°C, 81/31%RH	23.4
PW/pectinate : Gly (1/3 : 1)	Emulsion film	25°C, 81/31%RH	5.4
PW coating		25°C, 81/31%RH	0.035

Adapted from Schultz et al. (1949).
[a] Compositions rounded to nearest whole number.
[b] RHs on top and bottom sides of film (top/bottom).
BW, beeswax; PW, paraffin wax; LA, lauric acid; SA, stearic acid; Gly, glycerol.

Table 22.3 Water vapor permeabilities of bi-layer and emulsion HPMC–lipid edible films

	Film[a]	Preparation technique	Test conditions[b]	Permeability (g mm/m² d kPa)
Hagenmaier and Shaw (1990)	HPMC		27°C, 0/85%RH	9.12
Kamper and Fennema (1984a)	HPMC : PEG (9 : 1)		25°C, 85/0%RH	6.5
	SA–PA/HPMC : PEG (37/9 : 1)	Bi-layer film, two-step technique	25°C, 85/0%RH	2.0
	BW/HPMC : PEG (37/9 : 1)	Bi-layer film, two-step technique	25°C, 85/0%RH	0.064
	PW/HPMC : PEG (37/9 : 1)	Bi-layer film, two-step technique	25°C, 85/0%RH	0.034
	SA–PA/HPMC : PEG (3/9 : 1)	Bi-layer film, one-step technique	25°C, 85/0%RH	0.052
Kester and Fennema (1989a)	SA–PA/MC : HPMC : PEG (3.2 : 1 : 2)	Bi-layer film, one-step technique	25°C, 97/0%RH	0.27
	SA–PA/MC : HPMC : PEG (3/2 : 1 : 2)	Bi-layer film, one-step technique	25°C, 97/65%RH	1.4
	BW/SA–PA : MC : HPMC : PEG (2/1)	Bi-layer film, two-step technique	25°C, 97/0%RH	0.057
	BW/SA–PA : MC : HPMC : PEG (2/1)	Bi-layer film, two-step technique	25°C, 97/65%RH	0.15
Hagenmaier and Shaw (1990)	MA/HPMC (1/3)	Bi-layer film, one-step technique	27°C, 0/85%RH	3.42
	PA/HPMC (1/3)	Bi-layer film, one-step technique	27°C, 0/85%RH	2.28
	SA/HPMC (1/3)	Bi-layer film, one-step technique	27°C, 0/85%RH	0.095
	SA/HPMC (1/1)	Bi-layer film, one-step technique	27°C, 0/85%RH	0.026

[a] Compositions rounded to nearest whole number.
[b] RHs on top and bottom sides of film (top/bottom).
HPMC, hydroxypropylmethyl cellulose; PEG, polyethylene glycol; PA, palmitic acid; SA, stearic acid; BW, beeswax; PW, paraffin wax.

can be compared with WVPs of non-edible synthetic films, such as polyethylene terephthalate (PET) and low-density polyethylene (LDPE) (Table 22.1). The better performance of the emulsion technique was attributed to the orientation of the fatty acid molecules on the surface of the film. In the coating technique the fatty acids are deposited on the surface of the HPMC : EG film; whereas in the emulsion technique the fatty acid is allowed to orient at the air–water interface before the lipid solidifies at the HPMC : PEG surface. In addition, the bi-layer film made using the one-step technique was found to be very flexible and quite resistant to mechanical damage compared to the two-step bi-layer film, which was extremely brittle.

With the objective of enhancing the moisture barrier, Kester and Fennema (1989a, 1989b) studied the application of molten beeswax by using the coating technique over a preformed bi-layer film. This preformed bi-layer film was formed by drying an aqueous-ethanol solution of SA–PA and methyl cellulose (MC), HPMC, and PEG at 100°C for 15 minutes. Lamination of the SA–PA : MC : HPMC : PEG film with BW reduced WVP of the films by 79% and 89% when exposed to 97–0% RH and 97–65% RH differences, respectively (Table 22.3). In order to explain the permeability results, scanning electron microscopy was performed. A surprising result was that the degree of fatty acid surface crystallinity was much less in the SA–PA : MC : HPMC : PEG film than it was in the SA–PA : HPMC : PEG film developed by Kamper and Fennema (1984a, 1984b); however, on the contrary, the WVP was lower in the former. This was attributed to the greater thermal gelation potential of MC compared to HPMC, which might cause a greater percentage of the fatty acids to become entrapped in the bulk of the cellulose matrix during drying. In addition, the SA and PA functioned to increase the physical bonding of the BW layer to the cellulose matrix, resulting in bi-layer films with good stability against lipid fracture at low temperature.

Greener and Fennema (1989a, 1989b) studied bi-layer films made by either applying beeswax in a solvent (BW–S) or molten beeswax (BW–M) to a PEG–MC base film made from aqueous-ethanol solvent. MC appeared to provide adequate adhesion for the beeswax layer, probably due to the relatively hydrophobic nature of MC polymer. WVP of the BW–M/MC : PEG film was lower than that for the BW–S/MC : PEG film at a 100–0% RH difference, but similar at a 97–65% RH difference (Table 22.4). Scanning electron micrographs of the films showed that the BW–S based films did not have a uniform surface. This was probably due to the partial solubility of the BW in the ethanol solution, which might have translated into a greater number of small fissures and/or changes in the composition of the wax components. At the 97–65% RH difference, the moisture barrier of both films was inferior to the moisture barrier of BW–M/SA–PA : MC : HPMC : PEG films produced by Kester and Fennema (1989a, 1989b) (Table 22.3). This was attributed to the thicker BW layer and the fatty acid content of the base film used for the BW–M/SA–PA : MC : HPMC : PEG films. However, neither the BW–S nor the BW–M based films showed delamination after testing, whereas delamination of the BW–M/SA–PA : MC : HPMC : PEG film was extensive during testing. Physical abuse tests revealed that the BW–M/MC : PEG film was more resistant to surface impingement than the BW–S/MC : PEG film, with only a 20–40% increase in WVP after abuse compared to a 6- to 9-fold increase. On the other hand, a bi-layer film produced by the emulsion method suffered substantial

Table 22.4 Water vapor permeabilities of bi-layer and emulsion methyl cellulose–lipid edible films

	Film[a]	Preparation technique	Test conditions[b]	Permeability (g mm/m^2 d kPa)
Martin-Polo et al. (1992a)	MC : PEG (3 : 1)		25°C, 84/23%RH	9.3
Greener and Fennema (1989a)	BW–S/MC : PEG (4.3 : 1)	Bi-layer film, two-step technique	25°C, 100/0%RH	0.30
	BW–S/MC : PEG (4/3 : 1)	Bi-layer film, two-step technique	25°C, 97/65%RH	1.39
	BW–M/MC : PEG (6/3 : 1)	Bi-layer film, two-step technique	25°C, 100/0%RH	0.1
	BW–M/MC : PEG (6/3 : 1)	Bi-layer film, two-step technique	25°C, 97/65%RH	1.5
Martin-Polo et al. (1992a)	PW–M/MC : PW (?/40 : 1)	Bi-layer film, two-step technique	25°C, 84/23%RH	0.2
	PW–M/MC : PEG (3/3 : 1)	Emulsion film	25°C, 84/23%RH	8.2
Debeaufort and Voilley (1995)	PW–MC : PEG (5/5 : 1)	Bi-layer film, two-step technique	25°C, 22/84%RH	0.02
	PW : MC : PW (5 : 5 : 1)	Emulsion film	25°C, 22/84%RH	1.1
Park et al. (1994)	CZ : SA–PA : Gly : PEG/MC : PEG	Bi-layer film, two-step technique	25°C, 50/100%RH	1.5
Koelsch and Labuza (1992)	SA : MC : PEG (40 : 10 : 1)	Emulsion film	25°C, 33/56%RH	0.17
Sapru and Labuza (1992)	SA : MC : PEG (2 : 10 : 1)	Emulsion film	25°C, 75/0%RH	1.7

[a] Compositions rounded to nearest whole number.
[b] RHs on top and bottom sides of film (top/bottom).
MC, methyl cellulose; CZ, corn zein; PEG, polyethylene glycol; SA, stearic acid; PA, palmitic acid; BW, beeswax; PW, paraffin wax; –M, molten state; –S, in a solvent.

damage from the impingement test, reflected in a 77-fold increase in WVP. Both the BW–M/MC : PEG film and the BW–S/MC : PEG film showed 3- to 4-fold increase in WVP after weighted deformation, but they maintained their integrity.

Hagenmaier and Shaw (1990) prepared bi-layer edible films made from HPMC and lauric acid, myristic acid, palmitic acid, and stearic acid, by using the one-step emulsion technique, and found that film WVP decreased as fatty acid chain length and concentration increased. In this case, SA/HPMC bi-layer film, made from 95.6% ethanol by casting a hot solution and then drying at 75°C for 1 hour, had a low WVP quite similar to that obtained earlier by Kamper and Fennema (1984a) (Table 22.3). However, it was observed that on day 1 after preparation the films shrank by 7% in length, and the HPMC–SA films became covered with stearic acid crystals. On day 6 the surface powder could be wiped off, resulting in films that were fairly transparent, and a 10-fold increase in the permeability was observed. This result indicates the need for further research to ensure the integrity of bi-layer films.

Martin-Polo et al. (1992a) studied the influence of film preparation techniques on moisture barrier. The films were composed of a hydrophobic substance, either paraffin wax (PW) or oil, with MC as the supporting matrix. Films were prepared by emulsifying the PW with MC : PEG and drying them at room temperature for 24 hours, or by coating the MC : PEG films with molten PW. The best results in reducing WVP were obtained using the two-step coating technique (Table 22.4), which could be explained by the deposition of the PW in a continuous layer. Scanning electron microscopy of the emulsified film showed a non-uniform surface, which allowed moisture transmission to occur through the non-crystalline areas free of PW. These results indicated that a bi-layer film was not formed using the emulsion technique, and showed that the ability of a hydrophobic substance to retard moisture transfer depends on its homogeneity and distribution in the system.

Results similar to those of Martin-Polo *et al.* (1992a) were obtained by Debeaufort *et al.* (1993). MC and PW films were also prepared using both the emulsion technique and the coating technique. The results showed that a film of laminated PW over the MC support was 10 times more efficient as a barrier to moisture transfer than an emulsion film, due to the inability to form a bi-layer film using the emulsion technique (Table 22.4). These results contrast with the ones obtained by Kamper and Fennema (1984a), and could indicate that bi-layer films using the emulsion technique may only be obtainable using fatty acids and/or by drying at elevated temperatures (Krochta, 1997a).

Park *et al.* (1994) prepared laminated films by casting corn zein–fatty acid solutions onto MC : PEG films. Results showed that WVP decreased as chain length and concentration of fatty acid increased. The best WVP obtained corresponded to films containing 40% SA–PA blends (Table 22.4). The results were similar to those reported by Kamper and Fennema (1984a) using the one-step emulsion technique. Such differences could be due to differences in RH gradients and fatty acid concentrations. Since the corn zein–fatty acid layer is insoluble in water, such films may be useful in high moisture foods. However, the use of organic solvents is a significant disadvantage.

Debeaufort *et al.* (2000) prepared two bi-layer films, composed of a MC base layer coated with a triglyceride or alkane layer, and studied the effect of solid fat content, thickness, and melting point of the lipid layer on the mechanical and barrier properties of the films. The MC–PEG film was prepared in a water–ethyl alcohol solution and dried at $50 \pm 2°C$ and $7 \pm 1\%$ RH, and the lipid mixtures were applied in the molten state. The triglyceride lipid layer consisted of a mixture of hydrogenated palm oil (HPO) and triolein (Trio), whereas the alkane layer consisted of a mixture of paraffin oil (PO) and PW. The mechanical properties were mainly attributed to the methyl cellulose matrix. However, liquid lipids, both triolein and PO, had an antiplasticizing effect on the hydrocolloid network. The moisture barrier decreased 5- to 20-fold as a function of the lipid nature (alkanes or triglycerides) when the solid fat content varied from 0 to 80%. When emulsion films were prepared with a similar composition, Quezada-Gallo *et al.* (2000) observed that the nature of the lipid phase had little influence on the mechanical properties of emulsified films, but had a great effect on the water vapor barrier efficiency, with alkanes providing better moisture barriers than triglycerides. The solid–liquid ratio of the lipid phase had little effect on the moisture barrier, since water vapor could pass through the hydrophilic MC.

Weller *et al.* (1998) determined the mechanical properties, WVP, and colorimetric values of single-layer and bi-layer zein–lipid films. Zein films plasticized with glycerin and polyethylene glycol were cast from heated (60°C) aqueous-ethanol solutions. Bi-layer films were prepared by coating dried zein films with either medium chain length triglyceride oil (MCTO), sorghum wax (SW)–MCTO, or carnauba wax

Table 22.5   Water vapor permeabilities of bi-layer and emulsion WG–lipid edible films

Film[a]	Preparation technique	Test conditions[b]	Permeability (g mm/m^2 d kPa)
WG : Gly (5 : 1)		30°C, 100/0%RH	8.3
BW–M/WG : Gly (7/2 : 1)	Bi-layer film, two-step technique	30°C, 100/0%RH	0.059
BW–M/WG : DTM : Gly (7/5 : 2 : 1)	Bi-layer film, two-step technique	30°C, 100/0%RH	0.036
BW : WG : Gly (2 : 5 : 1)	Emulsion film	30°C, 100/0%RH	3.0

Adapted from Gontard et al. (1994, 1995).
[a] Compositions rounded to nearest whole number.
[b] RHs on top and bottom sides of film (top/bottom).
WG, wheat gluten; Gly, glycerol; BW, beeswax; DTM, diacetyl tartaric ester of monoglyceride; –M, molten state.

(CarW)–MCTO. Application of the lipid layer reduced WVP by up to 98.7% and produced films with greater elongation. However, tensile strength was not modified by lipid application. Bi-layer films were less yellow and more opaque than single-layer films.

Gontard et al. (1995) investigated the moisture barrier properties of bi-layer films consisting of wheat gluten (WG) as the structural layer and a thin layer of lipid as a moisture barrier. Films were prepared by depositing the lipid layer, either in solvent or molten state, over the surface of dried Gly–plasticizer WG film. Among the lipids studied, BW was the most effective in reducing WVP, followed by PW and CW. In the solvent method, an ethanol suspension of BW was cast over the base film at 70°C and then dried at 30°C. In the molten lipid method, the BW was melted at 100°C and spread over the previously formed base film, which had also been pre-heated to 100°C. The molten lipid method gave a bi-layer film with lower WVP than the solvent method. This was attributed to the fact that BW is only partially soluble in hot ethanol, which led to the formation of a much less continuous BW layer on the film surface. The problem observed in these films was that the lipid layer became easily detached from the WG-based film. Lipid adhesion was improved by the incorporation of a diacetyl tartaric ester of monoglyceride (DTM) in the WG : Gly base film, which also helped to reduce WVP (Table 22.5). The quantity of lipid cast over the WG was also studied in this work, and a sharp decrease in film WVP was found as the lipid quantity increased.

Hutchinson and Krochta (2002) prepared three types of two-step whey protein isolate (WPI)–BW composite films, with an aqueous emulsion (AE), ethanolic dispersion (ED), or hot melt (M–BW) of BW as the second layer. A one-step bi-layer film was also attempted, using a large lipid particle-size WPI–BW emulsion. The two-step films were found to be highly variable in quality, making it difficult to draw any conclusions about their effectiveness as moisture barriers. The M–BW bi-layer film had the lowest WVP, reducing the WVP by two orders of magnitude over the base film alone. The one-step bi-layer film was not successfully formed. However, the authors showed that heating the one-step partial bi-layer film produced a reduction of WVP of approximately 50%, giving WVP equivalent to the AE and ED bi-layers, but containing approximately 50% less BW (Table 22.6).

Table 22.6 Water vapor permeabilities of bi-layer and emulsion WPI–lipid edible films

	Film[a]	Preparation technique	Test conditions[b] (g mm/m² d kPa)	Permeability
Hutchinson and Krochta (2002)	WPI:BW:Gly (6:3:1)	WPI:Gly (3:1)	25°C, 0/77%RH	88.1
	WPI:BW:Gly (6:3:1)	Emulsion film, dried at 25°C	25°C, 0/97%RH	43.9
	AE–BW/WPI:BW: Gly (4/6:3:1)	Emulsion film, dried at 70°C	25°C, 0/92%RH	35.0
	ED–BW/WPI:BW: Gly (4/6:3:1)	Bi-layer film, two-step technique	25°C, 0/97%RH	21.4
	M–BW/WPI:BW: Gly (4/6:3:1)	Bi-layer film, two-step technique	25°C, 0/99%RH	11.0
		Bi-layer film, two-step technique	25°C, 0/100%RH	0.96
Shellhammer and Krochta (1997a)	CarW:WPI:Gly (11:15:1)	WPI:Gly (15:1)	25°C, 0/65%RH	45.6
	CanW:WPI:Gly (11:15:1)	Emulsion film	25°C, 0/90%RH	33.6
	HAMFF:WPI:Gly (11:15:1)	Emulsion film	25°C, 0/90%RH	31.2
	HAMFF:WPI:Gly (24:15:1)	Emulsion film	25°C, 0/92%RH	21.8
	BW:WPI:Gly (4:15:1)	Emulsion film	25°C, 0/99%RH	4.8
	BW:WPI:Gly (11:15:1)	Emulsion film	25°C, 0/92%RH	34.1
	BW:WPI:Gly (24:15:1)	Emulsion film	25°C, 0/94%RH	10.8
		Emulsion film	25°C, 0/98%RH	7.7

[a] Compositions rounded to nearest whole number.
[b] RHs on top and bottom sides of film (top/bottom).
WPI, whey protein isolate; Gly, glycerol; BW, beeswax; CarW, carnauba wax; CanW, candelilla wax; HAMFF, hard anhydrous milk fat fraction.
AE, aqueous emulsion; ED, ethanolic dispersion; M, molten.

Properties of emulsion films

The nature of the interactions between proteins and lipids, or between polysaccharides and lipids, determines the characteristics of emulsion formulations. In protein–lipid emulsions, proteins play a major role in stabilizing the system. Due to the amphiphilic character of proteins, they orient at the protein–lipid interface so that the non-polar groups align toward the oil and the polar groups align towards the aqueous phase. Therefore, the stabilization of the emulsions results from a balance between forces of a different nature, mainly electrostatic and hydrophobic. While protein–lipid emulsions are mainly stabilized by electrostatic forces, polysaccharides stabilize emulsions by stearic effects. To be good stabilizers, polysaccharides should be strongly attached to the surface of the lipid and also protrude significantly into the continuous phase to form a polymeric layer or a network of appreciable thickness (Callegarin *et al.*, 1997). However, in many cases polysaccharides possess limited amphiphilic character, so the addition of emulsifiers is required to improve emulsion stability.

Several studies have been conducted in an attempt to produce effective water vapor barriers by dispersing lipids into either a polysaccharide or a protein film-formation solution to form an emulsion film (Krochta, 1997a). Issues of lipid type, location, volume fraction, polymorphic phase, and drying conditions in emulsion composite films have been studied as affecting the barrier properties of protein and polysaccharide-based emulsion films.

As in bi-layer films, the water vapor resistance of emulsion films depends on the lipid type. Many works have reported that the barrier properties depend on the polarity and the degree of saturation of lipids (Martin-Polo *et al.*, 1992a, 1992b; Gontard *et al.*, 1994). Thus, high melting-point fatty acids, monoglycerides, hydrogenated fats, and waxes are useful edible lipid barriers (Shellhammer and Krochta, 1997a; Morillon *et al.*, 2002).

McHugh and Krochta (1994a) showed improved moisture-barrier properties in whey protein–lipid emulsion films when the hydrocarbon chain length for fatty alcohols and monoglycerides increased from 14 to 18 carbon atoms (Table 22.7). In these protein–lipid emulsion films, BW and fatty acids were more effective at reducing WVP of WPI-based emulsion films than fatty acid alcohols, which is consistent with the magnitude of lipid polarity. WVP measurements also revealed some emulsion separation during drying, since different WVP values were obtained depending on whether the lipid-enriched side was oriented towards the high or the low RH environment.

Koelsch and Labuza (1992) also showed an improvement in moisture-barrier properties of MC and fatty acids-based edible films as chain length increased from 12 to 18 atoms. However, the behavior changed as chain length increased from 18 to 22 carbon atoms. These results were correlated to the physical state and morphological arrangement of the fatty acid within the film. Fluorescence analysis showed that the

Table 22.7 Effect of lipid chain length and degree of saturation on the water vapor permeability of dispersed lipid (emulsion) edible films

	Film[a]	Test conditions[b]	Permeability (g mm/m² d kPa)
McHugh and Krochta (1994b)	WPI : Sor (4 : 1)	23°C, 0/80%RH	51.8
McHugh and Krochta (1994a)	TD : WPI : Sor (2 : 4 : 1)	23°C, 0/88%RH	50.9
McHugh and Krochta (1994a)	HD : WPI : Sor (2 : 4 : 1)	23°C, 0/87%RH	48.5
McHugh and Krochta (1994a)	STAL : WPI : Sor (2 : 4 : 1)	23°C, 0/86%RH	46.3
McHugh and Krochta (1994a)	MA : WPI : Sor (2 : 4 : 1)	23°C, 0/93%RH	23.7
McHugh and Krochta (1994a)	PA : WPI : Sor (2 : 4 : 1)	23°C, 0/93%RH	19.2
McHugh and Krochta (1994b)	BW : WPI : Sor (2 : 4 : 1)	23°C, 0/98%RH	5.3
Koelsch and Labuza (1992)	LA : MC : PEG (4 : 10 : 1)	23°C, 33/56%	2.9
Koelsch and Labuza (1992)	MA : MC : PEG (4 : 10 : 1)	23°C, 33/56%	0.91
Koelsch and Labuza (1992)	PA : MC : PEG (4 : 10 : 1)	23°C, 33/56%	0.18
Koelsch and Labuza (1992)	SA : MC : PEG (4 : 10 : 1)	23°C, 33/56%	0.17
Koelsch and Labuza (1992)	AA : MC : PEG (4 : 10 : 1)	23°C, 33/56%	1.1
Koelsch and Labuza (1992)	BA : MC : PEG (4 : 10 : 1)	23°C, 33/56%	2.6
Rhim *et al.* (1999)	SPI : Gly (2 : 1)	25°C, 0/100%	5.0
Rhim *et al.* (1999)	SPI : OA : Gly (2 : 2 : 1)	25°C, 0/100%	2.4
Rhim *et al.* (1999)	SPI : LA : Gly (2 : 2 : 1)	25°C, 0/100%	1.5
Rhim *et al.* (1999)	SPI : SA : Gly (2 : 2 : 1)	25°C, 0/100%	0.5

[a] Compositions rounded to nearest whole number.
[b] RHs on top and bottom sides of film (top/bottom).
MC, methyl cellulose; WPI, whey protein isolate; SPI, soy protein isolate; BW, beeswax; TD, tetradecanol; HD, hexadecanol; STAL, stearyl alcohol; LA, lauric acid; MA, myristic acid; PA, palmitic acid; SA, stearic acid; AA, arachidic acid; BA, behenic acid; OA, oleic acid; Sor, sorbitol; PEG, polyethylene glycol.

films containing stearic acid exhibit more micro-dispersed acid globules and a more complex interlocking network than do the films formulated with other fatty acids. This complexity probably resulted in an increased tortuosity and path distance for a water molecule to travel through the system, giving lower WVP (Krochta, 1997a).

The degree of saturation of fatty acids also affects the moisture-barrier properties of emulsion films, since they are more polar than saturated lipids. Rhim *et al.* (1999) observed an increase in WVP for soy protein isolate (SPI)–fatty acid emulsion films as the degree of saturation decreased. Emulsion films prepared with oleic acid (OA) had higher WVP than saturated SA. Similarly, their results showed a decrease in WVP as chain length and fatty acid content increased, with a reduction of about 260% for films with 40% SA (Table 22.7). The mechanical properties of emulsified SPI–fatty acid films were also affected by the amount of fatty acid incorporated into the film, the chain length, and the degree of saturation. Tensile strength decreased when adding fatty acids up to 20%, then decreased with the addition of fatty acids to more than 20%; whereas elongation decreased as fatty acid concentration increased. Unsaturated OA showed the lowest tensile strength and the highest elongation.

Shellhammer and Krochta (1997a) observed that WVPs of dispersed-lipid films made with WPI and Gly plasticizer did not correlate with the WVPs of the pure lipid components but with the viscoelastic properties of the lipids. Emulsion films were prepared with candelilla wax (CanW), carnauba wax (CarW), beeswax (BW), and hard anhydrous milk fat fraction (HAMFF). CanW and CarW, the materials with the lowest WVPs, gave emulsion films with the highest WVPs; BW and HAMFF, the materials with the highest WVPs, allowed formation of emulsion films with the highest lipid content and the lowest WVPs (Table 22.6). Interestingly, these results correspond with the viscoelastic properties of the lipids and suggest that the BW and HAMFF may have yielded more easily to the internal forces related to shrinkage of the drying protein structure, thus preventing film breakage with a high lipid content. The large drop in film WVP at 40–50% BW or HAMFF content could be explained by formation of an interconnecting lipid network within the film, which could be made possible by deformation of BW and HAMFF due to film internal stress as water evaporates from the film-forming solution.

In spite of a continuous reduction in WVP in emulsion films as the lipid content increases, some studies have found a critical volume fraction beyond which the barrier properties of the emulsion films did not improve or even got worse. For instance, SA : MC : PEG emulsion films showed a minimum WVP with about 14% SA content, beyond which WVP increased (Sapru and Labuza, 1994). Wheat gluten (WG)-based emulsion films also showed a decrease in WVP with increasing lipid content. However, the effect depended on the lipid hydrophobicity, the melting temperature and degree of unsaturation, and the interactions and organization of WG with the lipids in the film (Gontard *et al.*, 1994).

McHugh and Krochta (1994b) observed that, given a constant volume fraction of beeswax, decreasing the lipid particle size significantly correlated with a linear decrease in film WVP of WPI-based emulsion films (Table 22.8). This result could be explained by the presence of a large number of lipid particles uniformly dispersed in the system, which (1) increased the tortuous migration path length of water molecules diffusing through the composite film, and/or (2) immobilized proteins at the lipid interfaces, with a resulting more ordered and tightly cross-linked structure with lower permeability. However, these conclusions were somewhat tentative due to the changing lipid distribution within the film as particle size changed (i.e. large lipid particle sizes induced unstable emulsions versus small lipid particle sizes that induced stable emulsions). In this work, the mean lipid particle diameter ranged between approximately 0.9 and 2.0 μm. Measurement of WVP revealed some lipid phase separation during drying, because different WVP values were obtained depending on whether the film upper (lipid-enriched) side was oriented toward the high or low RH environment. In order to further elucidate the effect of particle size in determining the performance of whey protein–lipid emulsion films, Pérez-Gago and Krochta (2001) studied the effect of lipid particle size on WVP and mechanical properties for emulsion films with a low and high lipid content, and at different drying temperatures. In this work, it was shown that the effect of lipid particle size on film WVP and mechanical properties was influenced by lipid content and film orientation during WVP measurements (Table 22.8). As lipid content increased, a decrease in lipid particle size reduced the WVP of the WPI–BW emulsion films, probably due to an increase in protein immobilization at the lipid–protein interface as lipid content became more important in the film. This effect seemed to be supported by the increase in film tensile strength. For those films that showed lipid phase separation during drying, WVP was not affected by lipid particle size when the enriched lipid phase was facing the high RH side. This could be attributed to the greater barrier that the water vapor experienced when the enriched lipid phase was exposed to the high RH side during the WVP measurement overwhelming any effect due to lipid particle size.

Sherwin et al. (1998) showed an effect of fatty acid type on the film microstructure of microfluidized whey protein–fatty acid emulsion films, which could explain some of the functional property differences of WPI–fatty acid films. Emulsions of saturated

Table 22.8 Water vapor permeabilities of emulsion WPI–lipid edible films

	Film[a]	Lipid particle size (μm)	Test conditions[b]	Permeability (g mm/m² d kPa)
McHugh and Krochta (1994b)	WPI : Sor (4 : 1)		23°C, 0/80%RH	51.8
	BW : WPI : Sor (2 : 4 : 1)	0.83	23°C, 0/98%RH	8.4
	BW : WPI : Sor (2 : 4 : 1)	1.40	23°C, 0/94%RH	22.8
	BW : WPI : Sor (2 : 4 : 1)	1.95	23°C, 0/93%RH	38.4
Pérez-Gago and Krochta (2001)	BW : WPI : Gly (6 : 3 : 1)	0.54	25°C, 0/93%RH–Up	26.1
	BW : WPI : Gly (6 : 3 : 1)	1.01	25°C, 0/94%RH–Up	23.6
	BW : WPI : Gly (6 : 3 : 1)	1.82	25°C, 0/92%RH–Up	34.0
	BW : WPI : Gly (6 : 3 : 1)	2.28	25°C, 0/91%RH–Up	39.3
	BW : WPI : Gly (6 : 3 : 1)	0.54	25°C, 0/93%RH–Down	25.1
	BW : WPI : Gly (6 : 3 : 1)	1.01	25°C, 0/94%RH–Down	26.1
	BW : WPI : Gly (6 : 3 : 1)	1.82	25°C, 0/93%RH–Down	21.4
	BW : WPI : Gly (6 : 3 : 1)	2.28	25°C, 0/93%RH–Down	27.9

[a] Compositions rounded to nearest whole number.
[b] RHs on top and bottom sides of film (top/bottom).
WPI, whey protein isolate; Sor, sorbitol; Gly, glycerol; BW, beeswax.

fatty acids from C_{14} to C_{22} and WPI were prepared by microfluidization, and were analyzed by polarized light microscopy coupled with digital image analysis. The results showed that particle size increased with increasing fatty acid chain length, and at least two populations of fatty acid particle size were observed in each film, which could reflect differences in the manner in which crystals are formed during drying.

In addition to the effect of lipid particle size, Debeaufort and Voilley (1995) determined that the more homogeneous the distribution of the lipid particles, the better the water vapor barrier of PW : MC : PEG emulsion films. Films were prepared by adding PW to the MC : PEG in water-ethanol solution at 75°C in the presence of different emulsifiers, and then drying at several temperatures (25–45°C), relative humidities (15–40% RH), and air speeds (0–2 m/s). Whatever the drying conditions, a strong destabilization of the emulsion was observed during drying, with resulting increased mean globule diameter. In the slowest film-drying conditions, the average PW particle diameter doubled in size from the original emulsion; with the fastest drying, the particle diameter increased by a factor of 55. The slowest drying (low temperature, high RH, stagnant air) gave the highest tensile strength and elongation, and the lowest WVP. The relatively low WVP for the slowest-drying films was attributed to the small PW particle diameter, and equal PW particle distribution throughout the film. In spite of the quite large average PW particle diameter, the fastest-drying (high temperature, low RH, circulating air) film WVP was only 45% greater than the WVP for the slowest-drying film. The effectiveness of films dried quickly was attributed to a high proportion of large liquid paraffin globules, and enhanced creaming and coalescence due to rapid solvent evaporation. The resulting proposed film structure was an imperfect bi-layer, with a concentration of large coalesced PW at the film surface exposed to the high RH conditions of the WVP test procedure. However, drying at a rapid rate produced some films with many bubbles and holes.

Péroval et al. (2002) studied the effect of lipid type on the barrier and mechanical properties of arabinoxylan-based films containing palmitic acid, stearic acid, triolein, and hydrogenated palm oil. In comparison with other hydrocolloid-based films, the WVP of emulsified films was suggested to depend mostly on the nature of the continuous arabinoxylan phase, and secondarily on the lipid type. In this work, the relation between lipid particle size and WVP could not be established because lipid particle size and distribution changed with lipid type. To better understand the influence of the structure of arabinoxylan-hydrogenated palm kernel oil (HPKO) films on their functional properties, Phan The et al. (2002a) studied the effect of sucro-esters with different sterification levels, used as emulsifiers, on the stabilization of the emulsified film structure. The results showed that the addition of emulsifier had a great effect on emulsion stability, and the effectiveness of the sucro-esters depended not only on the esterification degree, but also, and mainly, on its concentration. From their results the WVP and lipid mean particle of the different films could not be correlated, because both the hydrophobicity of emulsifiers and the film structure (lipid distribution across the film) changed. However, emulsified films with small-sized HPKO globules homogeneously distributed within the matrix and the less stable systems corresponding to the film prepared without emulsifier both produced low permeability values. Therefore, the authors concluded that an improvement in the moisture-barrier property of composite films could be achieved by either great stability of the emulsion during drying and the homogeneous distribution of very small lipid globules within the film cross section, or destabilization of the emulsion by creaming, aggregation, and/or coalescence of the HPKO at the evaporation surface – which leads to an apparent bi-layer film structure.

Pérez-Gago and Krochta (1999) studied the role of emulsion stability, as affected by pH, on the final morphology and WVP of whey protein–lipid emulsion films. The films were cast from aqueous solutions of WPI (5% w/w), BW, and glycerol. Phase separation was observed at pH values different from the isoelectric point (pI $\approx$ 4–5) due to large lipid particle size distributions, in spite of the electrostatic repulsion between lipid particles charged by the adsorbed protein layer. However, no significant differences among WVPs of emulsion films prepared at values of pH $\neq$ pI were observed. Furthermore, rather than enhanced-phase separation due to lipid particle coalescence at the pI, phase separation was inhibited, and the film WVP was significantly higher that at pH different from the pI. This was attributed to an increase in emulsion viscosity at the pI, and the formation of a weak gel due to protein–protein aggregation, which lowered lipid mobility and inhibited any phase separation.

Since lipid distribution in the matrix has been shown to affect the moisture-barrier properties of emulsion films, some works have focused on the effect of the principal processes that can affect emulsion stability and, as a consequence, the morphology of the resulting film. Phan The *et al.* (2002b) studied the influence of the structure of films obtained from emulsions based on arabinoxylans, hydrogenated palm kernel oil, and emulsifiers on their functional properties during drying. Films were prepared with sucrose esters (emulsifier) having different HLB values, and films were dried at 30, 40 or 80°C. It was found that the emulsifier had a great effect on the stabilization of the emulsified film structure. In addition, an increase in the drying temperature affected emulsion film stability (higher creaming and coalescence), which gave films a bi-layer-like structure and thus reduced film WVP.

Pérez and Krochta (2001) studied the effect of the drying temperature on the properties of whey protein–lipid emulsion films. Emulsion films were prepared by emulsifying Gly-plasticized WPI films with CanW, BW or AMFF. The results confirmed that a decrease in WVP of emulsion films occurred as drying temperature increased from 40°C to 80°C. A large drop was observed in WVP of WPI–BW emulsion films at 20% BW content, reaching values as low as those found by Shellhammer and Krochta (1997a), even though the plasticizer level was five times greater than that used by these authors. In addition, the large drop in WVP of the WPI–BW emulsion films shifted to lower BW contents, compared to the 40% BW content effect shown by Shellhammer and Krochta (1997a). This could be an indication of a change in lipid distribution in the emulsion films, creating regions of higher BW content and, as a consequence, lower film permeability.

Conclusions

Incorporation of lipid materials into polysaccharide and protein films to form edible composite films has the potential to improve film moisture-barrier properties. The hydrophobic lipid component can form a continuous layer over the hydrophilic polysaccharide and protein phase, or it can be dispersed in the hydrophilic matrix to form dispersed-lipid films. In both bi-layer and emulsion films, the WVP depends mainly on the polarity and the degree of saturation of the lipid type. This way, high melting-point fatty acids, monoglycerides, hydrogenated fats, and waxes are useful edible lipid barriers.

Bi-layer films can be formed in one or two steps. Bi-layer films that are formed in two steps provided the best moisture barriers. This technique requires two different casting and drying stages, and the application of the hydrophobic layer requires the use of high temperatures or solvents. However, bi-layer films tend to delaminate, and exhibit poor mechanical properties compared to emulsion films. In the one-step method, the bi-layer film is formed due to the destabilization of the emulsion. This has only been achieved using a blend of PA–SA in an aqueous-ethanolic HPMC solution, and drying at temperatures above the fatty acid melting point.

Emulsion composite films require only a single casting and drying stage for film formation. However, they give poorer moisture-barrier properties than bi-layer films. Issues of lipid type, location, volume fraction, and drying conditions in emulsion composite films have been studied, as affecting the barrier properties of protein- and polysaccharide-based emulsion films. The results show that the moisture-barrier properties of emulsion films can be improved by using viscoelastic lipids, increasing lipid content, reducing lipid particle size, and improving film-drying conditions.

References

Avena-Bustillos, R. J. and Krochta, J. M. (1993). Water vapor permeablility of caseinate-based edible films as affected by pH, calcium crosslinking and lipid content. *J. Food Sci.* **58,** 904–907.

Callegarin, F., Quezada-Gallo, J. A., Debeaufort, F. and Voilley, A. (1997). Lipids and biopackaging. *J. Am. Oil Chem. Soc.* **74,** 1183–1192.

Debeaufort, F. and Voilley, A. (1995). Effect of surfactant and drying rate on barrier properties of emulsified edible films. *Intl J. Food Sci. Technol.* **30,** 183–190.

Debeaufort, F., Martin-Polo, M. and Voilley, A. (1993). Polarity homogeneity and structure affect water vapor permeability of model edible films. *J. Food Sci.* **58,** 426–434.

Debeaufort, F., Quezada-Gallo, J. A., Delporte, B. and Voilley, A. (2000). Lipid hydrophobicity and physical state effects on the properties of bilayer edible films. *J. Membrane Sci.* **180,** 47–55.

Gontard, N., Guilbert, S. and Cuq, J. L. (1992). Edible wheat gluten films: influence of the main process variables on film properties using response surface methodology. *J. Food Sci.* **57,** 190–195, 199.

Gontard, N., Duchez, C., Cuq, J. L. and Guilbert, S. (1994). Edible composite films of wheat gluten and lipids: Water vapour permeability and other physical properties. *Intl J. Food Sci. Technol.* **29,** 39–50.

Gontard, N., Marchesseau, S., Cuq, J. L. and Guilbert, S. (1995). Water vapour permeability of edible bilayer films of wheat gluten and lipids. *Intl J. Food Sci. Technol.* **30,** 49–56.

Greener, I. K. and Fennema, O. (1989a). Barrier properties and surface characteristics of edible, bilayer films. *J. Food Sci.* **54,** 1393–1399.

Greener, I. K. and Fennema, O. (1989b). Evaluation of edible, bilayer films for use as moisture barriers for food. *J. Food Sci.* **54,** 1400–1406.

Guilbert, S. (1986). Technology and application of edible protective films. In: *Food Packaging and Preservation – Theory and Practice* (M. Mathlouthi, ed.), pp. 371–394. Elsevier Applied Science, New York, NY.

Hagenmaier, R. D. and Shaw, P. E. (1990). Moisture permeability of edible films made with fatty acid and (hydroxypropyl)methylcellulose. *J. Agric. Food Chem.* **38,** 1799–1803.

Hanlon, J. F. (1992). *Handbook of Package Engineering*, 2nd edn. Technomic Publishing, Lancaster, PA.

Hutchinson, F. M. and Krochta, J. M. (2002). Moisture barrier properties of whey protein–beeswax bilayer composite films. In: *Moisture Barrier Properties of Whey Protein Isolate–Lipid Composite Films and Coatings*, pp. 19–48. Master Thesis Dissertation, University of California, Davis, CA.

Kamper, S. L. and Fennema, O. (1984a). Water vapor permeability of edible bilayer films. *J. Food Sci.* **49,** 1478–1481, 1485.

Kamper, S. L. and Fennema, O. (1984b). Water vapor permeability of an edible, fatty acid, bilayer film. *J. Food Sci.* **49,** 1482–1485.

Kamper, S. L. and Fennema, O. (1985). Use of an edible film to maintain water vapor gradients in foods. *J. Food Sci.* **50,** 382–384.

Kester, J. J. and Fennema, O. (1986). Edible films and coatings: a review. *Food Technol.* **40,** 47–59.

Kester, J. J. and Fennema, O. (1989a). An edible film of lipids and cellulose ethers: barrier properties to moisture vapor transmission and structural evaluation. *J. Food Sci.* **54,** 1383–1389.

Kester, J. J. and Fennema, O. (1989b). An edible film of lipids and cellulose ethers: performance in a model frozen-food system. *J. Food Sci.* **54,** 1390–1392, 1406.

Koelsch, C. M. and Labuza, T. P. (1992). Functional, physical and morphological properties of methyl cellulose and fatty acid-based edible barriers. *Lebensm. Wiss. u. Technol.* **25,** 404–411.

Krochta, J. M. (1992). Control of mass transfer in foods with edible coatings and films. In: *Advances in Food Engineering* (R. P. Singh and M. A. Wirakartakasumah, eds), pp. 517–538. CRC Press, Boca Raton, FL.

Krochta, J. M. (1997a). Edible composite moisture-barrier films. In: *Packaging Yearbook: 1997* (B. Blakistone, ed.), pp. 38–51. National Food Processors Association, Washington, DC.

Krochta, J. M. (1997b). Edible protein films and coatings. In: *Food Proteins and Their Applications* (S. Damodaran and A. Paraf, eds), pp. 529–549. Marcel Dekker, New York, NY.

Lovegren, N. V. and Feuge, R. O. (1954). Permeability of acetostearin products to water vapor. *J. Agric. Food Chem.* **2,** 558–563.

Martin-Polo, M., Mauguin, C. and Voilley, A. (1992a). Hydrophobic films and their efficiency against moisture transfer: 1. Influence of the film preparation technique. *J. Agric. Food Chem.* **40,** 407–412.

Martin-Polo, M., Voilley, A., Blond, G., Colas, B., Mesnier, M. and Floquet, N. (1992b). Hydrophobic films and their efficiency against moisture transfer: 2. Influence of the physical state. *J. Agric. Food Chem.* **40,** 413–418.

McHugh, T. H. and Krochta, J. M. (1994a). Water vapor permeability properties of edible whey protein–lipid emulsion films. *J. Am. Oil Chem Soc.* **71,** 307–312.

McHugh, T. H. and Krochta, J. M. (1994b). Dispersed phase particle size effects on water vapor permeability of whey protein-beeswax edible emulsion films. *J. Food Process. Preserv.* **18,** 173–188.

McHugh, T. H., Aujard, J. F. and Krochta, J. M. (1994). Plasticized whey protein edible films: water vapor permeability properties. *J. Food Sci.* **59,** 416–419, 423.

Morillon, V., Debeaufort, F., Blond, G., Capelle, M. and Voilley, A. (2002). Factors affecting the moisture permeability of lipid-based edible films: a review. *Crit. Rev. Food Sci. Nutr.* **42,** 67–89.

Park, H. J. and Chinnan, M. S. (1995). Gas and water vapor barrier properties of edible films from protein and cellulosic materials. *J. Food Eng.* **25,** 497–507.

Park, J. W., Testin, R. F., Park, H. J., Vergano, P. J. and Weller, C. L. (1994). Fatty acid concentration effect on tensile strength, elongation, and water vapor permeability of laminated edible films. *J. Food Sci.* **59,** 916–919.

Pérez-Gago, M. B. and Krochta, J. M. (1999). Water vapor permeability of whey protein emulsion films as affected by pH. *J. Food Sci.* **64,** 695–698.

Pérez-Gago, M. B. and Krochta, J. M. (2001). Lipid particle size effect on water vapor permeability and mechanical properties of whey protein-beeswax emulsion films. *J. Agric. Food Chem.* **49,** 996–1002.

Pérez-Gago, M. B. and Krochta, J. M. (2002). Formation and properties of whey protein films and coatings. In: *Protein-based Films and Coatings* (A. Gennadios, ed.), pp. 159–180. CRC Press, Boca Raton, FL.

Péroval, C., Debeaufort, F., Despré, D. and Voilley, A. (2002). Edible arabinoxylan-based films: 1. Effects of lipid type on water vapor permeability, film structure, and other physical characteristics. *J. Agric. Food Chem.* **50,** 3977–3983.

Phan The, D., Péroval, C., Debeaufort, F., Despré, D., Courthaudon, J. L. and Voilley, A. (2002a). Arabinoxylan-lipids-based edible films and coatings. 2. Influence of sucroester nature on the emulsion structure and film properties. *J. Agric. Food Chem.* **50,** 266–272.

Phan The, D., Debeaufort, F., Péroval, C., Despré, D., Courthaudon, J. L. and Voilley, A. (2002b). Arabinoxylan-lipid-based edible films and coatings. 3. Influence of drying temperature on film structure and functional properties. *J. Agric. Food Chem.* **50,** 2423–2428.

Quezada-Gallo, J. A., Debeaufort, F., Callegarin, F. and Voilley, A. (2000). Lipid hydrophobicity, physical state and distribution effects on the properties of emulsion-based edible films. *J Membrane Sci.* **180,** 37–46.

Rankin, J. C., Wolff, I. A., Davis, H. A. and Rist, C. E. (1958). Permeability of amylose film to moisture vapor, selected organic vapors, and the common gases. *I&EC* **3,** 120–123.

Rhim, J. W., Wu, Y., Weller, C. L. and Schnepf, M. (1999). Physical characteristics of emulsified soy protein-fatty acid composite films. *Sci. Aliments* **19,** 57–71.

Sapru, V. and Labuza, T. P. (1994). Dispersed phase concentration effect on water vapor permeability in composite methyl cellulose-stearic acid edible films. *J. Food Process. Preserv.* **18,** 359–368.

Schultz, T. H., Miers, J. C., Owens, H. S. and Maclay, W. D. (1949). Permeability of pectinate films to water vapor. *J. Phys. Colloid Chem.* **53,** 1320–1330.

Shellhammer, T. H. and Krochta, J. M. (1997a). Whey protein emulsion film performance as affected by properties of the dispersed phase. *J. Food Sci.* **62,** 390–394.

Shellhammer, T. H. and Krochta, J. M. (1997b). Water vapor barrier and rheological properties of simulated and industrial milkfat fractions. *Trans. ASAE* **40,** 1119–1127.

Sherwin, C. P., Smith, D. E. and Fulcher, R. G. (1998). Effect of fatty acid type on dispersed phase particle size distribution in emulsion edible films. *J. Agric. Food Chem.* **46,** 4534–4538.

Stuchell, Y. M. and Krochta, J. M. (1994). Enzymatic treatments and thermal effects on edible soy protein films. *J. Food Sci.* **59,** 1332–1337.

Weller, C. L., Gennadios, A. and Saraiva, R. A. (1998). Edible bilayer films from zein and grain sorghum wax or carnauba wax. *Lebensm. Wiss. u. Technol.* **31,** 279–285.

Plasticizers in edible films and coatings

Rungsinee Sothornvit and John M. Krochta

Introduction

Edible films and coatings are intended to help maintain the quality and shelf life of food products by controlling the transfer of moisture, oxygen, carbon dioxide, lipids, aromas, flavors, and food additives (Krochta, 1997a, 1997b; Krochta and De Mulder-Johnston, 1997). Film formation as a coating is dependent on two types of interaction; cohesion (attractive forces between film polymer molecules) and adhesion (attractive forces between film and substrate). Polymer properties such as molecular weight, polarity, and chain structure are relevant to cohesion and adhesion. Cohesive forces in polymer films can result in the undesirable property of brittleness. To overcome this limitation, food-grade plasticizers are added to the film formulation to decrease intermolecular forces resulting from chain-to-chain interaction. By reducing intermolecular forces and thus increasing the mobility of the polymer chains, plasticizers lower the glass transition temperature of films and improve film flexibility, elongation and toughness (Banker, 1966).

Commonly used plasticizers in film systems are monosaccharides, disaccharides or oligosaccharides (e.g. glucose, fructose-glucose syrups, sucrose, and honey), polyols (e.g. glycerol, sorbitol, glyceryl derivatives, and polyethylene glycols), and lipids and derivatives (e.g. phospholipids, fatty acids, and surfactants). Plasticizers are generally

Innovations in Food Packaging
ISBN: 0-12-311632-5

required at approximately 10–60% on dry basis, depending on the stiffness of the polymer (Guilbert, 1986; Daniels, 1989). The main drawback of plasticizers is the increase in gas, water vapor, and solute permeability that results from decreased film cohesion (Gontard *et al.*, 1993).

The objectives of this chapter are:

1. To define the term "plasticizer", classify plasticizer types, and present plasticizer theories
2. To summarize the advantages and disadvantages of plasticizers, plasticizers used, and plasticized-film properties of both polysaccharide- and protein-based films and coatings
3. To discuss the challenges and opportunities in food application and research.

Definition and purpose of plasticizers

Generally, plasticizers are defined by two purposes, which are to aid processing and modify the properties of the final product. As a processing aid, plasticizers lower the processing temperature, reduce sticking in molds, and enhance wetting. As a modified final product, plasticizers increase the temperature range of usage; increase flexibility, elongation, and toughness; and lower the glass transition temperature (Sears and Darby, 1982). For films or coatings, there are different definitions of plasticizers depending on the purpose of the polymer–plasticizer system. Plasticizers might be defined as small low molecular weight, non-volatile compounds added to polymers to reduce brittleness, impart flexibility, and enhance toughness for films. As a specific definition for coatings, plasticizers reduce flaking and cracking by improving coating flexibility and toughness. Generally speaking, plasticizers reduce intermolecular forces along the polymer chains, thus increasing free volume and chain movements (Immergut and Mark, 1965; Daniels, 1989).

Types of plasticizing

Internal and external plasticization

In polymer science, two types of plasticizers are defined: internal and external (Immergut and Mark, 1965). Internal plasticizers are part of the polymer molecules, which are either co-polymerized into the polymer structure, or reacted with the original polymer. Internal plasticizers generally have bulky structures that provide polymers with more space to move around and prevent polymers from coming close together. Therefore, they soften polymers by lowering the glass transition temperature (T_g) and thus reducing the elastic modulus. On the other hand, external plasticizers are low-volatility substances that are added to polymers. They do not chemically react, but they interact with polymers and produce swelling. The benefit of using external plasticizers over internal plasticizers is the opportunity to select from a variety of plasticizers, depending on the film properties desired (Banker, 1966; Wilson, 1995).

Common plasticizers for edible films and coatings are monosaccharides, oligosaccharides, polyols, lipids, and derivatives (Guilbert, 1986; Baldwin *et al.*, 1997). Water is also an important plasticizer for edible films and coatings. Water content is dependent on the polymer and external plasticizer selected.

Some plasticizer types and chemical structures are shown in Table 23.1. Most plasticizers contain hydroxyl groups which will form hydrogen bonds with biopolymers, and thus increase the free volume and flexibility of the film matrix. Various plasticizers consist of different numbers of hydroxyl groups and have different physical states (solid or liquid); thus they show differences in degree of softening film stiffness. Water is the smallest molecular weight ($M_w = 18$), and it is well known as an excellent plasticizer. However, most films still need an additional plasticizer to obtain desired film flexibility. Moisture sorption of various plasticizers plays an important role in affecting different film properties, as will be discussed later.

Other plasticizers besides those shown in Table 23.1 can be used; for example, a variety of M_w of polyethylene glycol (PEG). These include PEG 300 (Gioia *et al.*, 1998), PEG 600 (Lieberman, 1973; Guo, 1993; Gioia *et al.*, 1998), PEG 1450 (Donhowe and Fennema, 1993; Turhan *et al.*, 2001), PEG 1500 (Heinamaki *et al.*, 1994), PEG 4000 (Guo, 1993; Heinamaki *et al.*, 1994; Turhan *et al.*, 2001), PEG 8000 (Donhowe and Fennema, 1993; Guo, 1993; Turhan *et al.*, 2001), and PEG 20000 (Donhowe and Fennema, 1993), depending on the biopolymer materials used. Combinations of a variety of plasticizers have also been used to obtain averaged properties (Cherian *et al.*, 1995; Cao and Chang, 2002).

Lipid plasticizers such as fatty acids and derivatives, lecithin, oils, and waxes are also commonly used in edible coatings. Generally, the purpose of adding lipid is to reduce film water vapor permeability, since lipids are non-polar nature or hydrophobic, and thus provide a good barrier against moisture migration. Moreover, lipids can provide gloss and enhance the visual appearance of food products. However, lipids exhibit poor mechanical properties because of their lack of cohesive structural integrity (Gontard *et al.*, 1995). Therefore, incorporating lipid in protein- or polysaccharide-based films may produce a plasticizing effect, including reduction of film strength and increase of film flexibility, as seen in milk protein (Shellhammer and Krochta, 1997) and wheat gluten films (Gontard *et al.*, 1994). Common fatty acids used are shown in Table 23.2.

Oleic acid has been used as a plasticizer in zein films to provide modified atmosphere packaging for fresh broccoli (Rakotonirainy *et al.*, 2001). Palmitic acid-plasticized methylcellulose films (Kester and Fennema, 1989; Rico-Pena and Torres, 1991; J. W. Park *et al.*, 1994), lauric acid- and stearic–palmitic acid-plasticized laminated methyl cellulose/corn zein films (J. W. Park *et al.*, 1994) have been studied. In addition, fatty acids from C_{14} to C_{22} (myristic, palmitic, stearic, arachidic, and behenic acids) have been studied as plasticizers in whey protein isolate (WPI) films (Sherwin *et al.*, 1998). Stearic acid, palmitic acid, myristic acid, lauric acid, stearyl alcohol, hexadecanol, tetradecanol, and beeswax have been used as plasticizers in WPI–lipid emulsion films (McHugh and Krochta, 1994a). Carnauba wax, candelilla wax, beeswax, and a hard milkfat fraction have been found to plasticize WPI–glycerol films (Shellhammer and Krochta, 1997). Generally, increasing the chain length of fatty

Table 23.1 Plasticizer types, chemical structure and applications for edible films and coatings

Plasticizer type	M_w	Chemical structure*	Reference**
Xylitol	$C_5H_{12}O_5$ 152	HO–CH₂–CH–CH–CH–CH₂–OH (with OH OH OH)	4, 5
Mannitol	$C_6H_{14}O_6$ 182		5
Propylene glycol	$C_3H_8O_2$ 76		1,8,14,15
Glycerol	$C_3H_8O_3$ 92		1,2,3,4, 5,6,8,9,10, 13,14,15,17, 18,19,20
Sorbitol	$C_6H_{14}O_6$ 182		1,2,3,4,5,6, 10,12,19
Polyethylene glycol 200	$H(OCH_2\text{-}CH_2)_4OH \sim 200$		1,2,3
Sucrose	$C_{12}H_{22}O_{11}$ 342		1,2,3,8,18
Polyethylene glycol 400	$H(OCH_2\text{-}CH_2)_8OH \sim 400$		1,7,8,11, 14,15,16
Triethylene glycol	$C_6H_{14}O_4$ 150	HO–CH₂–CH₂–O–CH₂–CH₂–O–CH₂–CH₂–OH	9
Ethylene glycol	$C_2H_6O_2$ 62	HO–CH₂–CH₂–OH	12

(Continued)

Table 23.1 (*Continued*)

Plasticizer type	M_w	Chemical structure*	Reference**
1.4 butanediol	$C_4H_{10}O_2$ 90	$HO-(CH_2)_4-OH$	12
1,6 hexanediol	$C_8H_{14}O_2 \sim 200$	$HO-(CH_2)_6-OH$	12
Triacetin	$C_9H_{14}O_6$ 218	$AcO-CH_2-\overset{\overset{OAc}{\vert}}{CH}-CH_2-OAc$	16
Water	H_2O 18	H_2O	17
Glucose	$C_6H_{12}O_6$ 180		17
Urea	CH_4N_2O 60	$H_2N-\overset{\overset{O}{\Vert}}{C}-NH_2$	17
Diethanolamine	$C_4H_{11}NO_2$ 105	$HO-CH_2-CH_2-NH-CH_2-CH_2-OH$	17
Dibutyl phthalate	$C_{16}H_{22}O_4$ 278		20
Glycerol tributyrate	$C_{15}H_{26}O_6$ 302		21
Tributyl citrate	$C_{18}H_{32}O_7$ 360		21
Diethyl tartrate	$C_8H_{14}O_6$ 206		21

* Some obtained from the Sci Finder.
** Application of plasticizer used in edible films and coatings as seen in these references:
1, Sothornvit and Krochta, 2001; 2, Lin and Krochta, 2003; 3, Hong and Krochta, 2003; 4, Shaw *et al.*, 2002; 5, Kim *et al.*, 2002; 6, Kim and Ustunol, 2001a, 2001b; 7, Turhan *et al.*, 2001; 8, Lee *et al.*, 2002; 9, Cao and Chang, 2002; 10, Chick and Ustunol, 1998; 11, Heinamaki *et al.*, 1994; 12, Lieberman, 1973; 13, Park and Chinnan, 1995; 14, Donhowe and Fennema, 1993; 15, Park *et al.*, 1993; 16, Johnson *et al.*, 1991; 17, Gioia *et al.*, 1998; 18, Cherian *et al.*, 1995; 19, McHugh and Krochta, 1994b; 20, Benita *et al.*, 1986; 21, Beck and Tomka, 1996.

Table 23.2 Fatty acids used in edible films and coatings

Type of fatty acid	M_w	Chemical structure	Reference[*]
Lauric acid	$C_{12}H_{24}O_2$ 200	COOH	1
Stearic acid	$C_{18}H_{36}O_2$ 284	COOH	1, 2, 6
Palmitic acid	$C_{16}H_{32}O_2$ 256	COOH	1, 3, 4, 6
Linoleic acid	$C_{18}H_{32}O_2$ 280	COOH	5
Linolenic acid	$C_{18}H_{30}O_2$ 278	COOH	5
Myristic acid	$C_{14}H_{28}O_2$ 228	COOH	6
Behenic acid	$C_{22}H_{44}O_2$ 340	COOH	6
Oleic acid	$C_{18}H_{34}O_2$ 282	COOH	6
Arachidic acid	$C_{20}H_{40}O_2$ 312	COOH	6

[*]Application of fatty acids used in edible films and coatings as seen in these references:
1, J. W. Park et al., 1994; 2, Sherwin et al., 1998; 3, Rico-Pena and Torres, 1991; 4, Kester and Fennema, 1989; 5, Tanaka et al., 2001; 6, Rakotonirainy et al., 2001.

acids decreases the WVP of laminated MC/corn zein–fatty acid films (J. W. Park et al., 1994) and WPI–lipid emulsion films (McHugh and Krochta, 1994a).

Polymer molecular weight and plasticization

Another potential approach to increasing film flexibility is to reduce the polymer molecular weight, thus reducing intermolecular forces along polymer chains and increasing polymer chain end groups and polymer free volume (Sears and Darby, 1982). Therefore, polymer molecular weight (M_w) can play a plasticization role for edible films and coatings. Park et al. (1993) found that film oxygen permeability (OP), water vapor permeability (WVP), and tensile strength (TS) increased as methylcellulose (MC) and hydroxypropyl cellulose (HPC) M_w increased. However, film WVP decreased with increasing hydroxypropyl methyl cellulose (HPMC) and MC M_w (Ayranci et al., 1997). The differences may be attributed to the more hydrophobic behavior of the methyl group in HPMC compared to the HPC (Ayranci *et al.*, 1997). Reducing whey protein isolate (WPI) M_w also showed no significant effect on film WVP and OP (Sothornvit and Krochta, 2000), but the hydrolyzed WPI showed a significant plasticization effect by reducing film EM and TS compared to unhydrolyzed WPI (Sothornvit and Krochta, 2000). Chitosan M_w effect showed the same trend as film TS and WVP as hydrolyzed WPI (S. Y. Park *et al.*, 2002). Generally, studies have found that lowering the polymer M_w increased film flexibility without changing film WVP.

External plasticizer requirements

Besides cost, the selection of a plasticizer requires consideration of three basic criteria: compatibility, efficiency, and permanence.

Compatibility

It is necessary to use a plasticizer that is compatible with the intended polymer. Compatibility depends on polarity, structural configuration (shape), and size (M_w) of plasticizer. Good compatibility results from the plasticizer and polymer having a similar chemical structure (Immergut and Mark, 1965; Banker, 1966; Wilson, 1995). Therefore, different polymers require different plasticizers (Sears and Darby, 1982). In addition, plasticizers should have low volatility, as well as being non-toxic and aroma free.

Efficiency

Generally, good plasticizers provide high plasticization at low concentration and exhibit rapid polymer diffusion and interaction. The "plasticizer efficiency" is defined as the quantity of plasticizer required to produce the desired film mechanical properties. One method to define the efficiency is the lowering of T_g at a given amount or volume fraction of plasticizer. There is no exact number to indicate the efficiency of each plasticizer, because it depends on the polymer properties. Not only the size (M_w) but also the rate of plasticizer diffusion into the polymer matrix is another important factor in defining plasticizer efficiency (Immergut and Mark, 1965). A higher diffusion rate results in higher plasticizer efficiency. Small molecules have high diffusion rates, but possess higher volatility and therefore it is easier to lose the plasticizer in the process.

Permanence

The permanence of plasticizers in polymers depends on the size of the plasticizer molecule and on the rate of diffusion in polymers. Larger plasticizer molecules possess lower volatility, resulting in greater permanence. Moreover, polarity and hydrogen bond capability will influence the volatility of plasticizers. Greater plasticizer efficiency, defined by rapid diffusion into the polymer matrix, may also result in lower plasticizer permanence due to diffusion out of the polymer matrix. Thus it is often necessary to compromise in selecting which type of plasticizer is more suitable for each polymer material (Immergut and Mark, 1965; Banker, 1966).

The most effective plasticizers show the following properties (Sears and Darby, 1982):

1. A smaller plasticizer (lower M_w) is more effective in lowering film T_g
2. Plasticizer efficiency is proportional to the T_g of plasticized films; this means that small amounts of plasticizer are the most effective and larger amounts have less effect
3. Lesser affinity between plasticizer and polymer compared to the affinity between polymer and polymer gives more efficient plasticization – in other words, a good plasticizer is a bad solvent
4. Lesser affinity between plasticizer and plasticizer compared to the affinity between polymer and polymer gives more efficient plasticization, thus low-viscosity plasticizers are more efficient.

Theories of plasticization

There are four theories to explain plasticizing effects. The earliest are the free volume theory, which involves the intermolecular spaces in polymer, and the coiled spring theory, which deals with tangled macromolecules (Sears and Darby, 1982). The gel theory and the lubricity theory were later theories introduced to explain plasticization. The gel theory explores the rigidity in an unplasticized gel formed by weak polymer–polymer interactions along the polymer chains (Doolittle, 1965; Sears and Darby, 1982). The lubricity theory considers the plasticizer to act as a lubricant to facilitate the movement of polymer chains over each other, consequently lowering resistance to deformation (Doolittle, 1965; Sears and Darby, 1982). These two theories account for different portions of the total phenomenon of plasticization and the deformation of plasticized polymers. For edible films, the most useful concepts are the gel theory and the free volume theory. Thus, these two theories will be considered in greater detail.

Gel theory

The gel theory in food hydrocolloids pertains to an internal, three-dimensional rigid structure of polymer. Plasticizer molecules attach along the polymer chains, replacing polymer–polymer attachments at places and hindering the forces holding polymer chains together (Van der Waals, London, Debye, hydrogen bonding, crystal, or primary valence forces). This reduces the rigidity of the gel structure, resulting in increased gel flexibility. In addition, plasticizer molecules that are not attached to polymer form aggregated plasticizer domains that facilitate the movement of polymer molecules. This also enhances the gel flexibility.

Free volume theory

The free volume of amorphous materials or polymers is the volume unoccupied by the material's molecules (Sears and Darby, 1982). Another definition is the volume difference between the temperature at absolute zero and the temperature of interest. Free volume can be expressed as $V_f = V_t - V_o$, where V_f is the free volume, V_t is the specific volume at temperature T, and V_o is the specific volume at reference temperature (e.g. absolute zero).

Approaches to increasing the free volume of polymer system could include:

1. Using low M_w polymer, or reducing the M_w of polymer, to increase the number of polymer end groups
2. Increasing the length of side chains, thus increasing steric hindrance and lowering chain intermolecular forces to increase polymer chain motion, related to internal plasticization
3. Using low M_w plasticizers that are compatible with the polymer molecules to increase the motion of chain ends, side chains, and main chain, related to external plasticization
4. Increasing temperature.

Free volume theory is used to describe many things, such as plasticizing action, T_g, viscosity, cross-linking, diffusion, film drying and film properties (Wicks, 1986; Duda and Zielinski, 1996).

Plasticization models

Plasticization models are used to predict the effect of components on the T_g of food systems, in order to control or estimate the properties of food materials (Roos, 1995).

Gordon and Taylor equation

The Gordon and Taylor equation can be applied to predict plasticization of food components such as carbohydrates and proteins, as well as of pharmaceutical materials. This equation requires an empirical constant (k) from experimental data for T_g at different plasticizer contents. For water acting as a plasticizer, the equation covers the range of 0–50% water content.

$$T_g = \frac{w_1 T_g + k w_2 T_{g2}}{w_1 + k w_2} \qquad (23.1)$$

Thus, the constant k can be obtained from the following equation:

$$k = \frac{w_1 T_{g1} - w_1 T_g}{w_2 T_g - w_2 T_{g2}} \qquad (23.2)$$

where w_1 and w_2 are the weight fractions of solid component and plasticizer, respectively. T_g, T_{g1} and T_{g2} are the glass transition temperatures of the system, solid and plasticizer, respectively.

If the k and T_g of the system are known, the weight fraction of solid (w_1) can be obtained with the following equation:

$$w_1 = \frac{k(T_g - T_{g2})}{k(T_{g1} - T_{g2}) + T_{g1} - T_g} \qquad (23.3)$$

Application of the Gordon and Taylor equation is only possible for a binary system. A related empirical equation is the Kwei equation, which has another term, qw_1w_2:

$$T_g = \frac{w_1 T_{g1} + kw_2 T_{g2}}{w_1 + kw_2} + qw_1w_2 \tag{23.4}$$

where q is another constant.

Couchman and Karasz equation

The Couchman and Karasz equation is based on the heat capacity changes at the T_g of a binary system. This equation is equivalent to the Gordon and Taylor equation if $k = \Delta C_{p2}/\Delta C_{p1}$, where C_{p1} is the heat capacity of polymer and C_{p2} is the heat capacity of the plasticizer.

$$T_g = \frac{w_1 \Delta C_{p1} T_{g1} + w_2 \Delta C_{p2} T_{g2}}{w_1 \Delta C_{p1} + w_2 \Delta C_{p2}} \tag{23.5}$$

This equation requires knowing ΔC_{p1} and ΔC_{p2} at T_{g1} and T_{g2}, respectively, from DSC. In biological material such as proteins and starch, it is difficult to determine ΔC_p of the anhydrous polymer, due to thermal decomposition. Furthermore, the wide range of T_g is another problem in determining ΔC_p. ΔC_p trends to decrease when water content decreases. The usefulness of this equation depends upon obtaining the T_g for many biopolymers.

Couchman and Karasz equation – exact form

The equation below is the exact form of the Couchman and Karasz equation:

$$\ln T_g = \frac{w_1 \Delta C_{p1} \ln T_{g1} + w_2 \Delta C_{p2} \ln T_{g2}}{w_1 \Delta C_{p1} + w_2 \Delta C_{p2}} \tag{23.6}$$

$$\ln \left(\frac{T_g}{T_{g1}} \right) = \frac{w_2 \Delta C_{p2} \ln \left(\frac{T_{g2}}{T_{g1}} \right)}{w_1 \Delta C_{p1} + w_2 \Delta C_{p2}} \tag{23.7}$$

This equation is the same as Couchman and Karasz equation when $\ln (1 + y) = y$.

It is important to be aware that sources of error come from the determination of ΔC_p at T_g. The main error in using the simplified form of the Couchman and Karasz equation is the assumption that $\ln (1 + y) = y$.

Other equations

Other equations that are related to T_g but do not apply for water plasticization are:

1. The Fox equation, which is used in predicting T_g of binary blends of certain materials:

$$\frac{1}{T_g} = \frac{w_1}{T_{g1}} + \frac{w_2}{T_{g2}} \tag{23.8}$$

2. The Pochan–Beatty–Hinman equation, which is used in predicting the composition dependence of T_g:

$$\ln T_g = w_1 \ln T_{g1} + w_2 \ln T_{g2} \qquad (23.9)$$

3. The linear equation, which is the simplification of the Pochan–Beatty–Hinman equation:

$$T_g = w_1 T_{g1} + w_2 T_{g2} \qquad (23.10)$$

4. The Huang equation, which is based on the Couchman and Karasz equation and gives somewhat higher prediction of T_g than the Gordon and Taylor equation at intermediate water content:

$$T_g = \frac{[w_1 \Delta C_{p1}(T_{g1} + T_{g2}) + 2w_2 \Delta C_{p2} T_{g2}] T_{g1}}{[w_1 \Delta C_{p1}(T_{g1} + T_{g2}) + 2w_2 \Delta C_{p2} T_{g1}]} \qquad (23.11)$$

Plasticization and T_g

The effect of plasticizer on biopolymer film T_g is difficult to detect, because biopolymer chain mobility is much less than that of synthetic polymers (Mitchell, 1998). Furthermore, T_g determination is complicated by crystallization and enthalpic relaxation (Sichina, 2000), as well as chain-to-chain interactions such as hydrogen bonding, hydrophobic interactions, ionic interactions, and disulfide bonds (Mitchell, 1998). Nevertheless, it is suggested that T_g be determined from a distinct transition in the motion of the backbone, rather than smaller transitions (Seyler, 1994).

Water is an excellent plasticizer for biopolymers, shown by its effective lowering of T_g with increasing water content. Plasticization occurs in the higher molecular-mobility amorphous region, and the ability of plasticizers to interrupt hydrogen bonding along the protein chains depends on the plasticizer type and amount. Generally, increasing the amount of plasticizer decreases the film T_g.

However, the plasticizer type, but not the plasticizer amount, significantly affected the film T_g in both β-lactoglobulin (Sothornvit et al., 2002) and methylcellulose (Debeaufort and Voilley, 1997). Sothornvit and Krochta (2001) explained the lack of change in film T_g with increasing plasticizer content or water content by a mechanism of external plasticization involving an excess of plasticizer that produced domains of plasticizer aggregates. Another explanation is that these plasticizers acted as a lubricant between molecules in these systems (Debeaufort and Voilley, 1997).

An effective plasticizer needs to be compatible with and remain permanent when combined with the polymer. Thermal transitions for PEG 400-plasticized β-lactoglobulin (β-Lg) films could not be observed, because crystallizing and melting peaks of water occurred during heating (Sothornvit et al., 2002), similar to the PEG 400-plasticized methylcellulose (MC) films, which showed two water melting peaks (Debeaufort and Voilley, 1997). Using a combination of immiscible plasticizers showed two transitions in wheat gluten films (Cherian et al., 1995).

The physical state of the plasticizer is another important factor that affects film T_g. For example, sorbitol (Sor)- and sucrose (Suc)-plasticized β-Lg films possessed higher film T_g than glycerol (Gly)-, PEG 200- and propylene glycol (PG)-plasticized films, presumably because the solid state of Sor and Suc reduced their efficiency in disrupting hydrogen bonding along protein chains compared to the liquid state plasticizers (Gly, PEG 200 and PG) (Sothornvit *et al.*, 2002). Small molecular size (M_w) plasticizers also appear to be more effective in reducing film T_g. Furthermore, the diverse hygroscopicity of plasticizers affects film T_g differently. Sor is less hygroscopic and exhibits less efficient plasticization compared to Gly. The last factor is plasticizer shape, such as straight vs ring structure. Presumably, the bulky ring structure of Suc resulted in higher T_g than PEG 200 in β-Lg films (Sothornvit *et al.*, 2002).

In summary, the effect of various plasticizers on film T_g is related to film flexibility and other properties. Thus, film T_g is quite important for applications of edible films and coatings.

Advantages and disadvantages of edible films and coatings

Most edible films and coatings are quite brittle due to extensive intermolecular forces such as hydrogen bonding, electrostatic forces, hydrophobic bonding, and disulfide bonding. Plasticizers are required to interrupt intermolecular forces and lower the T_g, resulting in greater flexibility and toughness. Unfortunately, plasticizers increase film permeability to moisture, oxygen, aroma, and oils. Therefore, there are advantages and disadvantages of using plasticizers in edible films and coatings.

Plasticizers do not only produce plasticization. They can also produce antiplasticization. Plasticization is known to involve the reduction of intermolecular forces between polymer chains, thus improving film and coating flexibility. On the other hand, antiplasticization can occur at lower plasticizer concentrations and temperatures. Antiplasticization in films has been attributed to several mechanisms, such as reduction of polymer free volume, interaction between the polymer and plasticizer, and film stiffness due to the presence of rigid plasticizer molecules adjacent to the polar groups of the polymer (Guo, 1993; Seow *et al.*, 1999). In addition, adding a low concentration of plasticizer can lead to an increase in polymer crystallinity, due to lowering the energy barrier for a change of polymer state and thus enhancing the formation of ordered structures within amorphous regions (Wilson, 1995). Antiplasticization decreases edible film and coating WVP, and increases the elastic modulus and brittleness, which is the opposite behavior to plasticization. Water has been shown to have both plasticizer and antiplasticizer effects on tapioca starch film properties over the low to intermediate moisture range (Chang *et al.*, 2000). Moreover, triacetin, PEG 600, PEG 4000, and PEG 8000 have shown plasticization and antiplasticization effects on WVP of cellulose acetate films (Guo, 1993). Therefore, plasticizers must be used in the correct amount to obtain the advantage of enhancing the film and coating properties.

The main advantage of using plasticizers is to increase film toughness or resiliency by increasing flexibility, decreasing tensile strength, and increasing elongation, as seen in most edible films and coatings (Donhowe and Fennema, 1993; Guo, 1993; Park *et al.*, 1993; Gontard *et al.*, 1994; Galietta *et al.*, 1998; Anker *et al.*, 1999; Sothornvit and Krochta, 2000a; Sothornvit and Krochta, 2001). However, plasticizers also appear to have the disadvantage of increasing film permeability, depending on the type of plasticizer. For example, hydrophilic plasticizers are low barriers to moisture and tend to have a more pronounced effect on increasing film WVP compared to OP. On the other hand, hydrophobic plasticizers have a greater effect on increasing permeability to oxygen, aroma, and oil compared to WVP. Other concerns are that plasticizers are subject to aging, extraction, migration, volatility, and exudation (Daniels, 1989). These phenomena limit effectiveness, and leave levels of plasticizer that can be detrimental to film properties.

Migration of plasticizer from the film matrix to the surface can result from weak interactions between polymer molecules and plasticizer, allowing excess plasticizer to migrate easily through the film. An indicator of plasticizer migration is the appearance of greasiness and cloudiness on the film surface, as found with PEG 400-plasticized β-Lg films (Sothornvit *et al.*, 2002). To slow down the migration rate of plasticizers, a mixture of different plasticizers such as Gly and PEG in zein films can be used (H. J. Park *et al.*, 1994). This will help to maintain the mechanical properties of films during storage.

Properties of edible films and coatings

The main properties of edible films and coatings are their barrier properties (permeability to moisture, oxygen, aroma, and oil), mechanical properties, and moisture sorption.

Barrier properties are important to separate food from the environment, which causes food deterioration. The barrier properties of polymer films are generally related to the physical and chemical nature of the polymers. The chemical structure of the polymer backbone, the degree of crystallinity and orientation of molecular chains, and the nature of the plasticizer added all affect barrier properties (Salame, 1986). No polymer films, including edible films, are perfect barriers. However, each edible film is a barrier to the degree that it limits the permeability of water, oxygen, aroma, and oil. Therefore, the barrier is measured in terms of permeability. Briefly, the permeant molecules (water, oxygen, etc.) collide with the polymer film, adsorb to the polymer surface, diffuse through the polymer network, then desorb from the other side of the film. The most common barriers of interest in edible films and coatings include:

1. *Water barriers.* Water can enhance the rate of several reactions (browning, lipid oxidation, vitamin degradation, enzyme activity), increase the rate of micro-organism growth, and cause texture change, all of which are related to food shelf life and quality. Thus, controlling water permeability is of substantial importance in the development of an edible film and its application. Different edible polymer films have their own strengths. While polysaccharide- and protein-based films provide good barriers against oxygen, aromas, and oil (non-polar molecules), they are poor barriers against water (polar molecules). Much effort has gone into decreasing film

permeability to water vapor, such as combining polysaccharide or protein with lipids, cross-linking with transgluminase, irradiation, and heat curing. While adding lipid decreases polysaccharide- and protein-film WVP, modification of these polymers mainly affects their mechanical properties. The type and amount of plasticizer also affects the WVP. The potential applications of edible films include coatings on fruits (e.g. apples, citrus fruits, and pears) to prevent moisture loss and improve gloss (Grant and Burns, 1994).

2. *Oxygen barriers*. Oxygen transfer from the environment to food affects the deterioration of food components (lipids, vitamins, colors, and flavors) and micro-organism growth (especially molds), leading to sensory and nutrient changes. Depending on the formation materials used, the properties of edible films and coatings can be adjusted to achieve each food application (Cuq *et al.*, 1998). Most protein-based films are excellent oxygen barriers. Nuts provide an example of the use of protein films to avoid oxygen contact. Fruits and vegetables need a balance of oxygen and carbon dioxide flow to control respiration and ripening. Generally speaking, films that are better water barriers are poorer oxygen barriers. Increasing the plasticizer amount always increases oxygen permeability through edible films, because of the higher free volume in the film network.

3. *Grease/oil barriers*. Grease/oil barrier packaging is essential for foods containing fats or oils. Typically, synthetic polymers such as polyethylene (PE) are used to provide grease resistance for coated paperboard. Biopolymer-coated paperboard is an alternative that can be used in fast-food packaging and for other forms of grease/oil barrier packaging (Lin and Krochta, 2003).

Mechanical properties are important for edible films and film coatings, as they reflect the durability of films and the ability of a coating to enhance the mechanical integrity of foods. As discussed earlier, addition of plasticizers to biopolymers has a large effect on film mechanical properties, such as increasing film flexibility and resilience.

Finally, moisture sorption is an important characteristic of hydrophilic films, since the sorbed water acts as a plasticizer. Hydrophilic plasticizers increase the water-holding capacity of the plasticized films, such as Gly-plasticized corn zein (Padua and Wang, 2002). The plasticizing effect of moisture sorption on edible films indicates that the best applications for long-term food protection are for low water-activity foods.

Polysaccharide-based films and coatings

Polysaccharides used to form films and coatings are cellulose derivatives (e.g. methylcellulose (MC), hydroxypropyl methylcellulose (HPMC), hydroxypropylcellulose (HPC), starches and starch derivatives (e.g. amylose, amylopectin, hydroxypropyl amylose), and dextrins, pullulan, konjar, carrageenan, pectin, alginate, and other gums. Polysaccharide-based films and coatings are hydrophilic, and thus exhibit poor water-barrier properties. Nonetheless, polysaccharide films are good barriers against oxygen, aromas, and oil at low to intermediate RH. They generally require plasticizers to achieve the desirable mechanical properties.

As polysaccharide-based films are generally water-based, the most effective plasticizers are similar to the polysaccharide structure; therefore, hydrophilic plasticizers containing hydroxyl groups are best suited to this use. The plasticizers commonly used for polysaccharide-based films are glycerol, sorbitol, xylitol (X), mannitol (M), PEG (with M_w from 400 to 8000), ethylene glycol (EG), and PG.

Properties

Polysaccharide-based films are poor barriers against water vapor and other polar substances, especially at high RH. However, at low to intermediate RH they are good barriers against oxygen and other non-polar substances, such as aromas and oils. Polysaccharide-based films are relatively stiff, and therefore plasticizers are needed to facilitate handling. The effects of various plasticizers have been explored for films made from MC, HPC (Donhowe and Fennema, 1993; Park *et al.*, 1993; Ayranci and Tunc, 2001), HPMC (Ayranci *et al.*, 1997), locust bean gum (Aydinli and Tutas, 2000), gellan (Yang and Paulson, 2000), highly carboxyl methylated starch (Kim *et al.*, 2002), pullulan (Kim *et al.*, 2002), and high-amylose cornstarch (Ryu *et al.*, 2002).

Permeability properties

Table 23.3 shows the water vapor permeability (WVP) and oxygen permeability (OP) of different polysaccharide-based films with various plasticizers. Comparisons are difficult because data were collected in different studies using different film compositions, at different test conditions (RH, temperature), and with different methods of measurement and numbers of replications. It can be concluded from Table 23.3 that Gly is a commonly selected and effective plasticizer, as seen with MC films and high carboxy methylated starch (HCMS) films. At similar levels (wt/wt), the smaller M_w hydrophilic plasticizers such as PG and Gly produced higher WVP than larger M_w plasticizer such as PEG 400 in MC films (Donhowe and Fennema, 1993; Park *et al.*, 1993) and HPC films (Park *et al.*, 1993). Also, higher film OP was found with PEG 400 compared to higher M_w PEG 2000 (Donhowe and Fennema, 1993). However, the same plasticizer content but with higher M_w gave slightly higher WVP in locust bean gum (LBG) films (Aydinli and Tutas, 2000). Therefore, the selection of plasticizer M_w can tailor a plasticized film, including lowering its permeability. Increasing the plasticizer content generally increases permeability. Besides the plasticizer type, the polysaccharide type also affects film properties. For example, pullulan films possess better water barrier properties at mechanical properties similar to HCMS films. To improve water barrier, lipids such as fatty acids are applied to films and can be very effective, as seen with plasticized MC–fatty acid films (Ayranci and Tunc, 2001) and laminated MC–corn zein–fatty acid films (J. W. Park *et al.*, 1994).

Mechanical properties

Table 23.3 shows that, generally, polysaccharide film tensile strength (TS, stress at maximum force before break), and elastic modulus (EM or Young's modulus, film stiffness as calculated from the slope of force–deformation curve) decreases with

Table 23.3 Water vapor permeability (WVP), oxygen permeability (OP), and mechanical properties (tensile strength (TS), elastic modulus (EM), and % elongation (%E)) of various plasticizers used in polysaccharide-based films

Polysaccharide/plasticizer (wt/wt)	WVP (g mm/ m² d kPa)	OP (cm³ µm/ m² d kPa)	Mechanical properties			Reference
			TS (MPa)	EM (MPa)	%E	
MC¹ : Gly = 1.52 : 1–5.88 : 1	9.07–10.80	101.03	20.83–37.5	–	50–100	1
HPC² : Gly = 1.52 : 1–5.88 : 1	7.00	201.46	8.33–12.5	–	100–125	1
MC¹ : PEG 400 = 1.52 : 1–5.88 : 1	7.78	242	16.67–41.67	–	78–100	1
HPC² : PEG 400 = 1.52 : 1–5.88 : 1	4.75	242	9–16.67	–	100–198	1
MC¹ : PG = 1.52 : 1–5.88 : 1	8.64–30.24	242–1512.5	40–50	–	25–50	1
HPC² : PG = 1.52 : 1–5.88 : 1	12.96–20.74	605–1028.5	20–24	–	63–67	1
LBG : PEG 200 = 0.58 : 1–2.3 : 1	1.51–1.83	–	–	–	–	2
LBG : PEG 400 = 0.58 : 1–2.3 : 1	1.51–2.19	–	–	–	–	2
LBG : PEG 600 = 0.58 : 1–2.3 : 1	1.67–2.78	–	–	–	–	2
LBG : PEG 1000 = 0.58 : 1–2.3 : 1	1.89–2.76	–	–	–	–	2
Gellan : Gly = 0.5 : 1	36	–	30	25	30	3
Gellan : PEG 400 = 0.5 : 1	–	–	27	44	8	3
HCMS : Sor = 3.3 : 1–10 : 1	103–129	–	4.7–10.2	–	2.0–2.8	4
HCMS : M = 3.3 : 1–10 : 1	155–181	–	4.0–6.7	–	2.0–2.4	4
HCMS : X = 3.3 : 1–10 : 1	112–155	–	6.8–13.0	–	2.2–5.6	4
HCMS : Gly = 3.3 : 1–10 :1	103–190	–	9.7–15.3	–	2.6–7.7	4
Pullulan : Sor = 3.3 : 1	2.88	–	29.2	–	2.6	4
Pullulan : M = 3.3 : 1	3.6	–	15.7	–	9.5	4
MC : PEG 400 = 2.3 : 1	12.1	623	41.3	–	33.0	5
MC : PEG 1450 = 2.3 : 1	11.6	472	50.3	–	41.2	5
MC : PEG 8000 = 2.3 : 1	11.2	460	43.7	–	17.8	5
MC : PEG 2000 = 2.3 : 1	10.3	351	45.0	–	13.3	5
MC : Gly = 2.3 : 1	13.8	242	48.6	–	36.7	5
MC : PG = 2.3 : 1	8.62	200	70.9	–	11.6	5
MC³-PEG 400 : LA = 2.5 : 1–20 : 1	2.1–2.9	–	–	–	–	6
MC³-PEG 400 : PA = 2.5 : 1–20 : 1	1.3–2.8	–	–	–	–	6
MC³-PEG 400 : SA = 2.5 : 1–20 : 1	1.1–1.8	–	–	–	–	6
MC : PEG 400 = 2.5 : 1	3	–	–	–	–	6
Laminated film⁴	35.30	–	33.00	–	28.41	7
Laminated film⁴ : LA = 9 : 1–1.5 : 1	6.0–32.7	–	16.5–20.96	–	57.40–68.38	7
Laminated film⁴ : PA = 9 : 1–1.5 : 1	1.6–12.3	–	17.76–24.92	–	37.35–65.21	7
Laminated film⁴ : SP = 9 : 1–1.5 : 1	1.9–8.64	–	18.55–19.81	–	60.89–69.03	7
HPMC⁵ : PEG 200 or PEG 10 000 = 1.67 : 1	1.13–1.63	–	–	–	–	8
HACS⁶ : Gly = 1 : 1–5 : 1	1011–1270	–	2–32	–	6–22	9
HACS⁶ : Sor = 1 : 1–5 : 1	1011–1270	–	7–47	–	6–38	9

MC¹, methylcellulose (M$_w$ = 20 000); HPC², hydroxypropyl cellulose (M$_w$ = 1 000 000); LBG, locust bean gum; HCMS, highly carboxy methylated starch; MC³, methylcellulose (M$_w$ = 41 000); laminated⁴, laminated MC/corn zein plasticized with PEG 400 and Gly; HPMC⁵, hydroxypropyl methylcellulose; HACS⁶, high-amylose cornstarch; Gly, glycerol; PG, propylene glycol; Sor, sorbitol; Suc, sucrose; PEG, polyethylene glycol; PPG, polypropylene glycol; M, mannitol; X, xylitol; LA, lauric acid; PA, palmitic acid; SA, stearic acid; SP, stearic–palmitic acid mixture; OA, oleic acid.

References:

1, Park *et al.*, 1993; 2, Aydinli and Tutas, 2000; 3, Yang and Paulson, 2000; 4, Kim *et al.*, 2002; 5, Donhowe and Fennema, 1993; 6, Ayranci and Tunc, 2001; 7, J. W. Park *et al.*, 1994; 8, Ayranci *et al.*, 1997; 9, Ryu *et al.*, 2002.

increasing plasticizer content. Film elongation (%E, maximum percentage change in length of film before breaking) normally increases with increase of certain plasticizer content. Table 23.3 indicates that flexible and high-permeability films sometimes possess good film strength, as seen with MC and HPC films (Park *et al.*, 1993). Thus, flexible films are not necessarily weak films. Different plasticizer types also result in similar film tensile properties. Incorporation of fatty acid has been seen to lower film TS in laminated MC–corn zein–fatty acid films (J. W. Park *et al.*, 1994).

Moisture sorption

Moisture sorption must be considered for hydrophilic films that are sensitive to moisture change. Furthermore, plasticizers with a greater degree of hygroscopicity cause higher moisture sorption, with a resulting increase in plasticization. The equilibrium moisture contents for both amylopectin and amylose films were lower at low humidity (>50% equilibrium relative humidity, ERH), with or without Gly added (Myllarinen *et al.*, 2002). Films with a large amount of Gly had high water content above 50% ERH. Above 70% ERH, amylose sorbed more water than amylopectin; therefore, amylose films were much more flexible than amylopectin films. Polymer crystallization in hydrophilic films during storage is known to decrease film moisture sorption.

Protein-based films and coatings

Proteins used to form films can be placed into one of two categories; animal or plant proteins. Animal proteins include whey protein, casein, gelatin, collagen, fish myofibrillar protein, egg-white protein, and keratin. Plant proteins include wheat gluten, corn zein, soy protein, peanut protein, and cottonseed protein. Similar to polysaccharide-based films, protein-based films have high affinity with polar substances such as water, but low affinity with non-polar substances such as oxygen, aromas, and oils. Nevertheless, both polar and non-polar permeability depend on plasticizer concentration and relative humidity (RH).

The plasticizers commonly used for protein-based films include Gly, PG, Sor, PEG, Suc, and (X). Plasticizers used for corn zein films have included liquid organic compounds such as polyols, and solids compounds such as monosaccharides, oligosaccharides, lipids, and lipid derivatives (Padua and Wang, 2002). Plasticizers used for wheat gluten films have included Sor, mannitol (M), diglycerol, PG, triethylene glycol, polyvinyl alcohol, and polyethylene glycol; however, Gly and ethanolamine are widely used (Guilbert *et al.*, 2002). Plasticizers that have been used with soy protein include Gly, Sor, PEG, sugars, and lipids (S. K. Park *et al.*, 2002). Other protein sources, such as cottonseed protein, whey protein, casein, fish myofibrillar protein, egg-white protein, collagen, and gelatin, have been most commonly plasticized with Gly, Sor, and PEG.

Properties

Table 23.4 shows the effect of plasticizer on WVP, OP, and mechanical properties of protein films. All these properties are important to food applications and food stability.

Table 23.4 Water vapor permeability (WVP), oxygen permeability (OP), and mechanical properties (tensile strength (TS), elastic modulus (EM) and % elongation (%E)) from various plasticizers used in protein–based films

Protein/plasticizer (wt/wt)	WVP (g mm/ m²d kPa)	OP (cm³ μm/ m²d kPa)	Mechanical properties			Reference
			TS (MPa)	EM (MPa)	%E	
SC : Gly = 0.89 : 1–1.67 : 1	5.4–11.4	–	–	–	–	1
SC : PEG 400 = 0.81 : 1–1.32 : 1	14.8–22.6	–	–	–	–	1
SC : Gly = 2 : 1	–	–	10.9–11.7	73.7–84.2	–	1
SC : PEG 400 = 1.9 : 1	–	–	10.9–13.9	25.4	–	1
β-Lg : Gly = 1.70 : 1–3.20 : 1	–	20–43	4.98–16.01	150.1–705.6	11.36–76.46	2, 3
β-Lg : PG = 2.06 : 1–3.87 : 1	–	17–27	13.2–21.8	1476.4–1922.3	–	2
β-Lg : Sor = 0.86 : 1–1.62 : 1	–	3–8	2.71–10.06	99.6–383.8	24.75–65.85	2, 3
β-Lg : Suc = 0.46 : 1–0.86 : 1	–	<0.05	1.74–9.71	64.1–340.8	30.33–89.41	2, 3
β-Lg : PEG 200 = 0.78 : 1–1.47 : 1	–	110–700	1.80–6.46	67.4–255.2	41.67–77.09	2, 3
β-Lg : PEG 400 = 0.39 : 1–0.73 : 1	–	1050–2220	0.72–2.88	28.7–117.2	25.53–32.31	2, 3
WPI : Gly = 1 : 1–2 : 1	116–144	–	1–3.5	20–110	35–48	4
WPI : X = 1 : 1–2 : 1	84–89	–	0.5–8.5	80–275	2–15	4
WPI : Sor = 1 : 1–2 : 1	84–112	–	2.5–9	75–325	12–22	4
WPI : Gly = 1 : 1–1.6 : 1	119.8–154.56	–	–	–	–	5
WPI : Sor = 1 : 1–1.6 : 1	61.92–84.72	–	–	–	–	5
WPI : PEG 200 = 1 : 1	134.64	–	–	–	–	5
WPI : PEG 400 = 1 : 1	129.6	–	–	–	–	5
WPI : Gly = 2.3 : 1	–	76.1	13.9	–	30.8	6
WPI : Sor = 1 : 1	–	8.3	14.7	–	8.7	6
EA : Gly = 2.0 : 1–3.3 : 1	8.77–10.68	–	1.26–4.12	–	12.4–32.2	7
EA : Sor = 1.7 : 1–2.0 : 1	4.90–5.69	–	2.22–3.71	–	15.0–18.6	7
EA : PEG 400 = 1.7 : 1–2.0 : 1	6.21–6.22	–	3.37–3.84	–	59.7–88.1	7
WG : Gly : Suc = 15 : 6 : 0	121.55	–	4.2	–	89	8
WG : Gly : Suc = 15 : 4 : 2	99.14	–	5.6	–	39	8
WG : Gly : Suc = 15 : 3 : 3	81.90	–	6.0	–	11.0	8
WG : Gly : Suc = 15 : 0 : 6	15.52	–	3.8	–	3.2	8
WG : Gly : Sor = 15 : 3 : 3	105.17	–	5.7	–	57.2	8
LAC : Gly = 0.6 : 1–1.4 : 1	54.7–59.3	0.73–2.18	0.42–2.51	–	121.4–253.6	9
LAC : Sor = 0.6 : 1–1.4 : 1	34.0–45.0	0.65–0.81	2.43–11.65	–	50.6–170.7	9
RC : Gly = 0.6 : 1–1.4 : 1	45.2–58.2	1.84–7.06	0.83–4.5	–	123.2–223.5	9
RC : Sor = 0.6 : 1–1.4 : 1	39.6–49.6	0.71–1.02	3.83–15.12	–	4.9–17.9	9
Zein : Gly = 2.3 : 1	35.52	–	7.0	498	2.6	10
Zein : Gly = 5.67 : 1	24.24	–	–	–	–	10
Zein : PEG 400 = 2.3 : 1	–	–	5.7	199	44.4	10
Zein : PPG 400 = 2.3 : 1	16.8	–	9.4	608	2.8	10
Zein : Gly : PPG 400 = 9.3 : 1 : 3	25.68	–	5.1	135	117.8	10
Zein : Gly : PEG 400 = 9.3 : 1 : 3	–	–	4.9	180	26.6	10
Zein : Gly : PPG 400 = 9.3 : 3 : 1	34.8	–	5.6	275	5	10
Zein : Gly : PPG 400 = 4.67 : 1 : 1	31.2	–	3	100	112	10
WGI : EG = 2.7 : 1	–	–	4.1	–	243	11
WGI : DEG = 2.7 : 1	–	–	2.5	–	479	11
WGI : TEG = 2.7 : 1	–	–	2.8	–	504	11
WGI : TEEG = 2.7 : 1	–	–	1.6	–	557	11
WGI : DEGMET = 2.7 : 1	–	–	11.3	–	3.8	11
FMP : Gly = 2.5 : 1	6.53	–	–	–	–	12
FMP : Sor = 2.5 : 1	5.15	–	–	–	–	12
FMP : Suc = 2.7 : 1	3.9	–	–	–	–	12
FMP[1] : LA = 1.5 : 1–4 : 1	3.80–4.75	–	5.6–6.6	–	55–73	13

(Continued)

Table 23.4 (*Continued*)						
Protein/plasticizer (wt/wt)	WVP (g mm/ m² d kPa)	OP (cm³ μm/ m² d kPa)	Mechanical properties			Reference
			TS (MPa)	EM (MPa)	%E	
FMP[1] : PA = 1.5 : 1–4 : 1	5.27–7.17	–	4.1	–	5–16	13
FMP[1] : SA = 1.5 : 1–4 : 1	4.49–5.79	–	2.9–3.7	–	3–13	13
FMP[1] : Beewax = 1.5 : 1–4 : 1	3.80–7.52	–	3.2–4.6	–	82–120	13
FMP[1] : OA = 1.5 : 1–4 : 1	2.25–3.46	–	5.7	–	145–200	13
FMP[1] : LL = 1.5 : 1–4 : 1	3.80–4.41	–	3.8–4.1	–	115–135	13
FMP[1] : LN = 1.5 : 1–4 : 1	3.80–5.01	–	4.1–4.4	–	98–130	13
FMP[1] : PO = 1.5 : 1–4 : 1	8.38–10.37	–	3	–	90–130	13
FMP[1] : CO = 1.5 : 1–4 : 1	5.44–6.13	–	3.5–4.2	–	100–165	13
FMP[1] : SO = 1.5 : 1–4 : 1	6.13–7.34	–	4.1–4.4	–	130–145	13
FMP[1] : CL = 1.5 : 1–4 : 1	7.17–8.64	–	4.1–5	–	130–145	13
EA : PEG 400 : MMM = 6 : 4 : 1–3 : 2 : 1	196.8–204	–	4.36–4.75	–	47.2–54.9	14
EA : PEG 400 : HMM = 6 : 4 : 1–3 : 2 : 1	194.4–213.6	–	4.06–4.54	–	43.1–44.1	14
EA : PEG 400 : OA = 6 : 4 : 1	199.2	–	5.77	–	89.2	14
EA : PEG 400 : LPL = 6 : 4 : 1–3 : 2 : 1	211.2–220.8	–	4.56–4.80	–	77.7–84.5	14
Soybean : Gly = 0.25 : 1–0.67 : 1	–	–	1.55–2.08	–	250–275	15
Soybean : TEG = 0.25 : 1–0.67 : 1	–	–	1.98–2.38	–	240–267	15
Soybean : Gly : TEG = 0.5 : 1 : 1–1.33 : 1 : 1	–	–	1.75–2.3	–	240–255	15
SC : S : Gly : water = 9.5 : 9.5 : 0 : 1	0.22	35.4	–	–	–	16
SC : S : Gly : water = 9.0 : 9.0 : 1 : 1	1.81	80.4	38.2	–	8.2	16
SC : S : Gly : water = 8 : 8 : 3 : 1	4.50	46.6	32.0	–	18.5	16
SC : S : Gly : water = 6.5 : 6.5 : 6 : 1	13.05	30.2	22.5	–	26.8	16
SC : S : Sor : water = 9 : 9 : 1 : 1	2.85	12.1	36.3	–	9.5	16
SC : S : Sor : water = 8 : 8 : 3 : 1	5.53	82.9	29.4	–	20.9	16
SC : S : Sor : water = 6.5 : 6.5 : 6 : 1	18.40	63.1	20.2	–	29.7	16
SC : S : xylose : water = 9 : 9 : 1 : 1	3.37	4.3	33.1	–	2.4	16
SC : S : xylose : water = 8 : 8 : 3 : 1	6.57	67.4	27.2	–	15.2	16
SC : S : xylose : water = 6.5 : 6.5 : 6 : 1	21.25	45.8	19.0	–	18.0	16
Zein : PEG+Gly = 4 : 1–10 : 1	345.6–985.0	–	19.5–22.8	–	1.3–4.3	17
Zein : OA = 4 : 1–10 : 1	259.2–388.8	–	17.5–21.7	–	1.8–5	17

SC, sodium caseinate, β-Lg, β-lactoglobulin, WPI; whey protein isolate; EA, egg albumen; WG, wheat gluten; LAC, lactic acid casein; RC, rennet casein; WGl, wheat gliadin; FMP, fish myofibrillar protein; FMP[1], fish myofibrilla plasticized with Gly at 1 : 1; S, soluble starch; Gly, glycerol; PG, propylene glycol; Sor, sorbitol; Suc, sucrose; PEG, polyethylene glycol; X, xylitol; PPG, polypropylene glycol; EG, ethylene glycol; DEG, diethylene glycol; TEG, triethylene glycol; TEEG, tetraethylene glycol; DEGMET, diethylene glycol monomethyl ether; LL, linoleic acid; LN, linolenic acid; OA, oleic acid; PO, peanut oil; CO, corn oil; SO, salad oil; CL, cod liver; MMM, middle-melting milkfat; HMM, high-melting milkfat; LPL, lysophospholipid.

References:

1, Siew *et al.*, 1999; 2, Sothornvit and Krochta, 2000; 3, Sothornvit and Krochta, 2001; 4, Shaw *et al.*, 2002; 5, McHugh *et al.*, 1994; 6, McHugh and Krochta, 1994a; 7, Gennadios *et al.*, 1996; 8, Cherian *et al.*, 1995; 9, Chick and Ustunol, 1998; 10, Parris and Coffin, 1997; 11, Sanchez *et al.*, 1998; 12, Cuq *et al.*, 1997; 13, Tanaka *et al.*, 2001; 14, Handa *et al.*, 1999; 15, Cao and Chang, 2002; 16, Arvanitoyannis and Biliaderis, 1998; 17, Ryu *et al.*, 2002.

Mechanical properties

Similar to polysaccharide-based films, protein-based films are brittle without plasticizer. Comparing Tables 23.3 and 23.4, it can be seen that protein-based films generally have lower TS than polysaccharide-based films. As with polysaccharide films, increasing the plasticizer content generally has a large impact on film mechanical properties by reducing TS and EM and increasing %E. However, PG-plasticized β-Lg films were very brittle and did not experience any change in mechanical properties with increasing PG content (Sothornvit and Krochta, 2001). This may have been a result of less interaction with β-Lg due to the lower polarity of PG compared to the other plasticizers in the study. This was in agreement with PG-plasticized MC films (Sothornvit and Krochta, 2001). The overall effect of plasticizers on mechanical properties of β-Lg films can be ranged, from the greatest to the least effect, as Gly, Sor, Suc, PEG 200, and PEG 400, respectively (Sothornvit and Krochta, 2001). TS and EM with each of the five plasticizers exhibited negative exponential dependence on plasticizer content (EM $= 1500 \, e^{-kEM\,X}$ or TS $= 37.28 \, e^{-kTS\,X}$, X $=$ plasticizer content). Plasticizer efficiency was defined by Sothornvit and Krochta (2001) for the first time in terms of k_{EM}, k_{TS} and k_E determined from the fitted EM, TS and %E data, respectively. Moreover, the effect of plasticizer related to the size (M_w), shape, and number of plasticizer oxygen atoms was elucidated. Smaller M_w plasticizers and straight-chain plasticizers were most effective – i.e., Gly and PEG 200 were the most efficient plasticizers. Related work has shown an exponential relationship between the mechanical properties and OP of β-Lg films that included a ratio of the mechanical and OP plasticizer efficiencies defined as k_{MO} (Sothornvit and Krochta, 2000a). Gly has also been found to be the most efficient plasticizer in WPI films (McHugh and Krochta, 1994b; Shaw et al., 2002) and peanut protein films (Jangchud and Chinnan, 1999). PEG 400-plasticized egg-white (EA) films possess greater TS and %E than Sor- and Gly-plasticized films (Gennadios et al., 1996). Sor is required in higher amounts than Gly to achieve similar mechanical properties, due to the larger size (i.e. lower efficiency) of Sor (Gennadios et al., 1996), which is in agreement with plasticized-β-Lg films (Sothornvit and Krochta, 2001).

Other research has studied Gly–Suc and Gly–Sor mixtures of plasticizers to improve wheat gluten (WG) film properties (Cherian et al., 1995). Here, Gly alone with WG gave the largest film WVP and %E, but combining Suc or Sor with Gly decreased WVP with little effect on film TS, as shown in Table 23.4. Addition of Sor had less effect on reducing WVP than Suc, but gave less reduction in %E. Results with lactic acid casein (LAC) and rennet casein (RC) confirmed that Sor-plasticized films had better water-barrier properties, while Gly-plasticized films had better elongation (Chick and Ustunol, 1998). PEG 400 plasticized-zein films had greater flexibility than Gly-plasticized films, because Gly tended to migrate to the zein film surface within a few hours of preparation (Parris and Coffin, 1997). If plasticized with the more hydrophobic plasticizer, polypropylene glycol (PPG), zein films remained brittle. However, a mixture of PPG–Gly produced flexible zein films (Parris and Coffin, 1997). A synergy between PPG and Gly, perhaps related to the methyl side-chain of PPG that reduced chain-to-chain interaction to enhance film elongation, did not occur with PEG–Gly. Wheat gliadin (WGI) film plasticized with the same level of glycol of increasing M_w showed increase of film elongation (Sanchez et al., 1998). The most effective plasticizers for WGI films in terms of increase in %E were Gly and tetraethylene glycol. However, the study concluded that the plasticizer effect depends on the nature of the plasticizer and the polymer, and the same trend in mechanical properties might not be found for different polymer-plasticizer pairs.

Incorporating fatty acids and other lipids was found to improve the tensile strength of fish myofibrillar protein (FMP) films (Tanaka *et al.*, 2001), egg albumin (EA) films (Handa *et al.*, 1999), and soybean films (Cao and Chang, 2002). Unsaturated fatty acids (linoleic acid, LL; linolenic acid, LN), except for oleic acid (OA), decreased the TS of FMP films; however, fatty acid content did not exhibit any change in film TS (Tanaka *et al.*, 2001). Beeswax and saturated fatty acids (palmitic acid, PA; stearic acid, SA), except lauric acid (LA), decreased the TS of FMP films (Tanaka *et al.*, 2001). It was observed that increasing the M_w of saturated fatty acids dramatically reduced the film flexibility. Edible oils (peanut oil, PO; corn oil, CO; salad oil, SO; and cod liver oil, CL) decreased TS as well. It can be concluded from this study that fatty acids and other lipids produce flexible films with a certain level of lipids. Generally, the main purpose of using fatty acids and other lipids is for a water barrier.

Permeability properties

Protein-based films have polar hydrophilic characteristics; consequently, they have poor water vapor barrier properties but good oxygen-, aroma-, and oil-barrier properties. Increasing the plasticizer content of protein films increases permeability. At similar plasticizer contents, sodium caseinate (SC) films plasticized with PEG 400 had higher WVP than films plasticized with Gly (Siew *et al.*, 1999). The effects of various plasticizers on the OP of β-Lg films using different bases of comparison have been studied. The bases of comparison used were mole of plasticizer multiplied by the number of oxygen atoms per mole of β-Lg (Figure 23.1a); mole of plasticizer per mole of β-Lg (Figure 23.1b); and mass of plasticizer per mass of β-Lg (Figure 23.1c) (Sothornvit and Krochta, 2000b). Plasticizer size, shape, and number of plasticizer oxygen atoms were shown to have a significant effect on plasticizer efficiency (k_{OP}). The relationship between OP data and plasticizer amount was linear on all three bases (Figure 23.1). Plasticizer efficiency for OP (k_{OP}) of β-Lg films, from low to high efficiency values, is in the following order for all bases: Suc < Sor < PG, Gly < PEG 200 < PEG 400. PG and Gly did not give any significant difference in OP, likely due to their similarity of plasticizer structure and hygroscopicity. Sor possesses a larger size and lower hygroscopicity than Gly, resulting in lower OP values and correspondingly higher tensile properties (TS and EM). PEG 400-plasticized β-Lg films exhibited higher OP than PEG 200-plasticized films, reflecting their larger and less efficient size. The relationship between OP and the mechanical properties of β-Lg films was expressed in terms of plasticizer efficiency ratios ($k_{MO} = k_{EM}/k_{OP}$ or k_{TS}/k_{OP}), which were used in a negative exponential correlation (EM = 1500 $e^{-(k_{EM}/k_{OP})}$OP or TS = 37.28 $e^{-(k_{TS}/k_{OP})}$OP). Suc- and Sor-plasticized β-Lg films were the best oxygen barriers. High plasticizer efficiency ratios are desirable because they reflect a plasticizer's ability to improve film flexibility while having little effect on OP (Sothornvit and Krochta, 2000b).

In other studies comparing plasticizers at similar content in films, Gly gave higher WVP in WPI films (McHugh *et al.*, 1994; Shaw *et al.*, 2002), LAC and RC films (Chick and Ustunol, 1998), and FMP films (Cuq *et al.*, 1997). Gly also gave higher

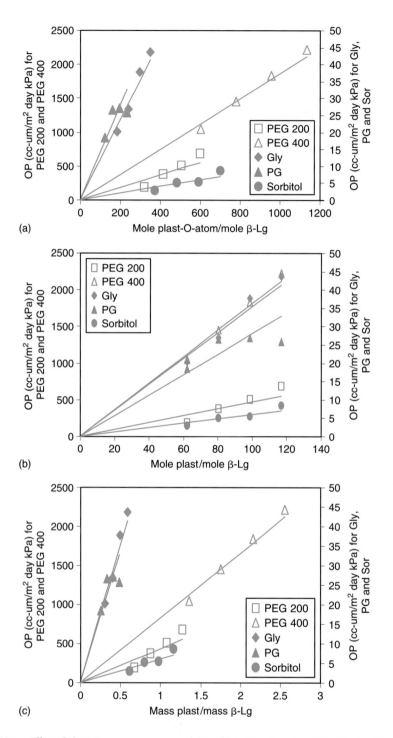

Figure 23.1 Effect of plasticizer on oxygen permeability of β-Lg films based on (a) mole plasticizer-oxygen-atom/mole β-Lg, (b) mole plasticizer/mole β-Lg, and (c) mass plasticizer/mass β-Lg. The symbols represent the data sets and the lines represent the empirical fitting results with zero intercept. Reprinted with permission from Sothornvit and Krochta, 2000b. Copyright 2000 American Chemical Society.

WVP than PEG 400 in EA films (Gennadios *et al.*, 1996) and WPI films (McHugh and Krochta, 1994b). Furthermore, Gly gave higher OP than Sor in WPI films (McHugh and Krochta, 1994b), LAC, and RC films (Chick and Ustunol, 1998). Thus, Gly generally appears to be the most effective plasticizer in terms of tensile properties, but also gives the highest permeability. To reduce the effect on permeability increase, using different plasticizer mixtures such as Gly–Suc or Gly–Sor in WG films (Cherian *et al.*, 1995) and Gly–PPG 400 and Gly–PEG 400 in zein films (Parris and Coffin, 1997) helped to lower film WVP. Incorporation of fatty acids and other lipids as plasticizers reduced the permeability in FMP films (Tanaka *et al.*, 2001).

Moisture sorption

Similar to polysaccharide films, moisture sorption is important in protein films because moisture acts as a plasticizer. Gly-plasticized WPI films at higher Gly content and higher RH had greater flexibility, and the Guggenheim–Anderson–de Boer (GAB) model gave the best fit of moisture sorption data (Coupland *et al.*, 2000). Transglutaminase cross-linked EA films plasticized with higher Gly content possessed higher film hydrophilicity (Lim *et al.*, 1998). Film moisture sorption isotherms were well defined by the GAB equation. With increasing moisture content, plasticization and swelling of the polymer matrix provided additional binding sites and resulting enhanced water sorption at elevated a_w (Lim *et al.*, 1998). The sorption isotherm and thickness of peanut protein films also reflected film hydrophilicity (Jangchud and Chinnan, 1999). Gly content was found to have greater effect on film tensile properties than film permeability. Data showing the effect of various plasticizers (Gly, Sor and mixture of Gly–Sor) on the moisture sorption isotherm of soy protein isolate (SPI) films were fitted to the GAB model (Cho and Rhee, 2002). Gly-plasticized SPI films absorbed more water than Sor-plasticized films at the same a_w, due to the higher hygroscopicity of Gly. The high water affinity of Gly contributes to the excellent plasticizing effect of Gly at even low Gly content and low RH. RH, along with plasticizer type and content, affects hydrophilic film flexibility. The minimum plasticizer content required decreases with increase in RH (Cho and Rhee, 2002).

Challenges and opportunities

Although plasticizers generally comprise a relatively small part of edible film formulations, they are needed to enhance the ability of edible polymers to form usable films and coatings. Understanding of plasticizers can enable tailoring of film permeability and mechanical properties to make them more suitable for food applications. The amount and type of plasticizer, and the effect of RH, are important issues to be considered for each polymer being investigated with regard to edible films and coatings.

While basic research on plasticizer effects on edible film properties is still needed to improve understanding, enough information is available to pursue food applications. To achieve the implementation of plasticized edible films and coatings for commercial use, it is important to maximize the desirable mechanical properties while minimizing the effect on barrier properties for each polymer film in order to explore and optimize food applications. Mathematical modeling of plasticizer efficiency and/or plasticizer efficiency ratios needs to be continued to confirm the validity of existing equations or adjust the equations for greater reliability. The incorporation of antimicrobials or antioxidants into plasticized films or coatings is a recent trend worthy of investigation. Furthermore, study of the possible migration of plasticizer out of the polymer film matrix is essential for long-term shelf-life applications. Using a smaller M_w polymer is another approach to providing the desirable film flexibility with lower plasticizer content, as shown with hydrolyzed WPI film (Sothornvit and Krochta, 2000a; Sothornvit and Krochta, 2000c). Sensory evaluation of foods with edible films and coatings as a new product line is also necessary.

Food system applications

The objective of introducing edible films and coatings is to improve food quality and shelf life while reducing the use of expensive and non-recyclable synthetic polymer films. In addition to protecting foods from moisture change, oxidation, and aroma loss, edible coatings can protect food from micro-organisms and mechanical damage, as well as sometimes providing an additionally impressive appearance for customers. Many food applications of edible films and coatings have been commercialized, and many more explored, as summarized in several review articles (Kester and Fennema, 1986; Conca and Yang, 1993; Grant and Burns, 1994; Nisperos-Carriedo, 1994; Guilbert and Gontard, 1995; Krochta and De Mulder-Johnston, 1997; Debeaufort et al., 1998; Arvanitoyannis and Gorris, 1999; Petersen et al., 1999; Krochta, 2002).

Selection of a suitable plasticizer for individual food applications will benefit food stability. For example, Suc-plasticized WPI coating provides the highest and most stable gloss on chocolate compared to Gly-, PEG 400-, and PG-WPI coatings (Lee et al., 2002). Suc-WPI films also have the lowest OP, and highest EM and TS. Suc-plasticized WPI coating on polypropylene films also gives the best oxygen barrier compared to Gly-, Sor-, PG- and PEG 200-WPI coatings, although OP increased exponentially with RH (Hong and Krochta, 2003). Furthermore, Suc–whey protein concentrate (WPC)-plasticized coating on paperboard have been found to provide a good grease barrier and minimum plasticizer migration (Lin and Krochta, 2003). Suc appears to be a more compatible and effective plasticizer in certain food coatings in certain aspects over Gly, because it has a high k_{MO}.

To use edible films for food pouches, it is necessary to know the heat-sealing temperature that will give an effective heat seal with good integrity. The optimum heat-sealing temperature for Sor-plasticized WPI–lipid emulsion edible films appears to be 130°C, while the temperature for Gly-plasticized films is 110°C (Kim and Ustunol, 2001). Different plasticizers revealed different thermal transition onset temperatures. This is also an important plasticizer issue, essential to continuous film formation by the extrusion process.

Finally, as mentioned earlier, the sensory attributes of edible films and coatings must be investigated to ensure consumer satisfaction. WPI coatings on chocolate compared favorably with shellac coatings in a consumer test (Lee et al, 2002). WPI-emulsion films plasticized with either Sor or Gly had no characteristic milk aroma, but they were slightly sweet (Kim and Ustunol, 2001). In addition, Sor-plasticized films were more adhesive than Gly-plasticized films, based on results from trained sensory panelists. This might be the advantage of using polyhydric alcohols plasticizers such as Sor, Gly, M, X, and Suc.

Conclusions

Most polysaccharide- and protein-based films and coatings are brittle in nature. To solve this problem, plasticizers are incorporated to enhance the film flexibility and resilience. However, film permeability always increases with increasing plasticizer content. Proper selection of a plasticizer for a given polymer will allow optimization of the film mechanical properties with a minimum increase in film permeability. Common plasticizers for edible films and coatings include monosaccharides, oligosaccharides, polyols, lipids, and derivatives. Different moisture sorption by various plasticizers contributes to differences in film permeability and mechanical properties. Selection of plasticizers needs consideration of the issues of plasticizer compatibility, efficiency, permanence, and economics.

Plasticization occurs in the higher molecular mobility (amorphous) region, and the ability of plasticizers to interrupt hydrogen bonding along the protein chains depends on the amount and type of plasticizer. Generally, increasing the amount of plasticizer decreases the film T_g.

Plasticizers generally act to produce plasticization, but can also provide antiplasticization. Therefore, plasticizers should only be used at the minimum amount required to obtain the advantage of enhancing the film and coating properties.

Another potential approach to increase film flexibility is reducing polymer M_w, thus increasing the polymer-chain end groups, reducing the intermolecular forces along polymer chains and increasing the polymer free volume.

Polysaccharide- and protein-based films are poor barriers against water and polar substances, especially at high RH, but good barriers against oxygen and other non-polar substances. Nevertheless, both polar and non-polar compound permeabilities depend on the plasticizer concentration and the relative humidity (RH). Incorporating fatty acids or other lipids can help improve water-barrier properties, whilst providing a certain degree of plasticization. Combining plasticizer types can utilize the advantages of each plasticizer and reduce the disadvantages. Addition of a polysaccharide substance to a protein film can also improve film tensile properties. Thus, film permeability and tensile properties depend on both polymer and plasticizer. Migration of plasticizer during film formation and storage must be studied to guarantee the feasibility of applying edible films and coatings to food products.

The effect of the amount and type of plasticizer on achieving desirable mechanical properties with optimal permeability is of continuing interest to researchers. The mathematical models of plasticizer efficiencies should be explored further to verify the equations and allow application to food systems. Selection of the appropriate plasticizer for an edible film or coating applied to a food system should, ideally, be based on basic understanding. Thus, study of the interesting topic of plasticizer effects on edible film and coating properties will continue as an adventure in food science, technology, and engineering.

References

Anker, M., Stading, M. and Hermansson, A.-M. (1999). Effects of pH and the gel state on the mechanical properties, moisture contents, and glass transition temperature of whey protein films. *J. Agric. Food Chem.* **47(5)**, 1878–1886.

Arvanitoyannis, I. and Gorris, L. G. M. (1999). Edible and biodegradable polymeric materials for food packaging or coating. In: *Processing Foods* (F. A. R. Oliveira, J. C. Oliveira, M. E. Hendrickx, D. Korr and L. G. M. Gorris, eds), pp. 357–371. CRC Press, New York, NY.

Aydinli, M. and Tutas, M. (2000). Water sorption and water vapour permeability properties of polysaccharide (locust bean gum) based edible films. *Lebensm. Wiss. u. Technol.* **33(1)**, 63–67.

Ayranci, E. and Tunc, S. (2001). The effect of fatty acid content on water vapour and carbon dioxide transmissions of cellulose-based edible films. *Food Chem.* **72**, 231–236.

Ayranci, E., Buyuktas, B. S. and Cetin, E. E. (1997). The effect of molecular weight of constituents on properties of cellulose-based edible films. *Lebensm. Wiss. u. Technol.* **30(1)**, 101–104.

Baldwin, E. A., Nisperos, M. O., Hagenmaier, R. D. and Baker, R. A. (1997). Use of lipids in coatings for food products. *Food Technol. Chicago* **51(6)**, 56–62, 64.

Banker, G. S. (1966). Film coating theory and practice. *J. Pharm. Sci.* **55(1)**, 81–89.

Beck, M. I. and Tomka, I. (1996). *Macromolecules* **29(27)**, 8759–8769.

Benita, S., Dor, Ph., Aronhime, M. and Marom, G. (1986). *Intl J. Pharm.* **33(1–3)**, 71–80.

Cao, Y. M. and Chang, K. C. (2002). Edible films prepared from water extract of soybeans. *J. Food Sci.* **67(4)**, 1449–1454.

Chang, Y. P., Cheah, P. B. and Seow, C. C. (2000). Plasticizing-antiplasticizing effects of water on physical properties of tapioca starch films in the glassy state. *J. Food Sci.* **65(3)**, 445–451.

Cherian, G., Gennadios, A., Weller, C. and Chinachoti, P. (1995). Thermomechanical behavior of wheat gluten films: effect of sucrose, glycerin and sorbitol. *Cereal Chem.* **72(1)**, 1–6.

Chick, J. and Ustunol, Z. (1998). Mechanical and barrier properties of lactic acid and rennet precipitated casein-based edible films. *J. Food Sci.* **63(6)**, 1024–1027.

Cho, S. Y. and Rhee, C. (2002). Sorption characteristics of soy protein films and their relation to mechanical properties. *Lebensm. Wiss. u. Technol.* **35(2)**, 151–157.

Conca, K. R. and Yang, T. C. S. (1993). Edible food barrier coatings. In: *Biodegradable Polymers and Packaging* (C. Ching, D. L. Kaplan and E. L. Thomas, eds), pp. 357–369. Technomic Publishing, Lancaster, PA.

Coupland, J. N., Shaw, N. B., Monahan, F. J., O'Riordan, E. D. and O'Sullivan, M. (2000). Modeling the effect of glycerol on the moisture sorption behavior of whey protein edible films. *J. Food Eng.* **43**, 25–30.

Cuq, B., Gontard, N., Cuq, J.-L. and Guilbert, S. (1997). Selected functional properties of fish myofibrillar protein-based films as affected by hydrophilic plasticizers. *J. Agric. Food Chem.* **45(3)**, 622–626.

Cuq, B., Gontard, N. and Guilbert, S. (1998). Proteins as agricultural polymers for packaging production. *Cereal Chem.* **75(1)**, 1–9.

Daniels, C. A. (1989). *Polymers: Structure and Properties*. Technomic Publishing, Lancaster, PA.

Debeaufort, F. and Voilley, A. (1997). Methylcellulose-based edible films and coatings: 2. Mechanical and thermal properties as a function of plasticizer content. *J. Agric. Food Chem.* **45(3),** 685–689.

Debeaufort, F., Quezada-Gallo, J.-A. and Voilley, A. (1998). Edible films and coatings: tomorrow's packagings: a review. *Crit. Rev. Food Sci. Nutr.* **38(4),** 299–313.

Donhowe, I. G. and Fennema, O. (1993). The effects of plasticizers on crystallinity permeability, and mechanical properties of methylcellulose films. *J. Food Process. Preserv.* **17(4),** 247–257.

Doolittle, A. K. (1965). The effects of plasticizers on crystallinity permeability, and mechanical properties of methylcellulose films. In: *Plasticizer Technology* (P. F. Bruins, ed.), pp. 1–20. Chapman and Hall, London.

Duda, J. L. and Zielinski, J. M. (1996). Free-volume theory. In: *Diffusion in Polymers* (P. Neogi, ed.), pp. 143–171. Marcel Dekker, New York, NY.

Galietta, G., Gioia, L. D., Guilbert, S. and Cuq, B. (1998). Mechanical and thermomechanical properties of films based on whey proteins as affected by plasticizer and crosslinking agents. *J. Dairy Sci.* **81(12),** 3123–3130.

Gennadios, A., Weller, C. L., Hanna, M. A. and Froning, G. W. (1996). Mechanical and barrier properties of egg albumen films. *Cereal Chem.* **70(4),** 426–429.

Gioia, L. D., Cuq, B. and Guilbert, S. (1998). Effect of hydrophilic plasticizers on thermomechanical properties of corn gluten meal. *Cereal Chem.* **75(4),** 514–519.

Gontard, N., Guilbert, S. and Cuq, J.-L. (1993). Water and glycerol as plasticizers affect mechanical and water vapor barrier properties of and edible wheat gluten film. *J. Food Sci.* **58(1),** 206–211.

Gontard, N., Duchez, C., Cuq, J.-L. and Guilbert, S. (1994). Edible composite films of wheat gluten and lipids: water vapor permeability and other physical properties. *Intl J. Food Sci. Technol.* **29(1),** 39–50.

Gontard, N., Marchesseau, S., Cuq, J.-L. and Guilbert, S. (1995). Water vapour permeability of edible bilayer films of wheat gluten and lipids. *Intl J. Food Sci. Technol.* **30(1),** 49–56.

Grant, L. A. and Burns, J. (1994). Application of coatings. In: *Edible Coatings and Films to Improve Food Quality* (J. M. Krochta, E. A. Baldwin and M. O. Nisperos-Carriedo, eds), pp. 189–200. Technomic Publishing, Lancaster, PA.

Guilbert, S. (1986). Technology and application of edible protective films. In: *Food Packaging and Preservation Theory and Practice* (M. Mathlouthi, ed.), pp. 371–394. Elsevier Applied Science, New York, NY.

Guilbert, S. and Gontard, N. (1995). Edible and biodegradable food packaging. In: *Foods and Packaging Materials – Chemical Interactions* (P. Ackermann, M. Jagerstad and T. Ohlsson, eds), pp. 159–174. The Royal Society of Chemistry, Cambridge, UK.

Guilbert, S., Gontard, N., Morel, M. H., Chalier, P., Micard, V. and Redl, A. (2002). Edible and biodegradable food packaging. In: *Protein-based Films and Coatings* (A. Gennadios, ed.), pp. 69–122. CRC Press, New York, NY.

Guo, J.-H. (1993). Effects of plasticizers on water permeation and mechanical properties of cellulose acetate: antiplasticization in slightly plasticized polymer. *Drug Devel. Ind. Pharm.* **19(13),** 1541–1555.

Handa, A., Gennadios, A., Froning, G. W., Kuroda, N. and Hanna, M. A. (1999). Tensile, solubility, and electrophoretic properties of egg white films as affected by surface sulfhydryl groups. *J. Food Sci.* **64(1),** 82–85.

Heinamaki, J. T., Lehtola, V.-M., Nikupaavo, P. and Yliruusi, J. K. (1994). The mechanical and moisture permeability properties of aqueous based hydroxypropyl methylcellulose coating systems plasticized with polyethylene glycol. *Intl J. Pharm.* **112(2),** 191–196.

Hong, S.-I. and Krochta, J. M. (2003). Oxygen barrier properties of whey protein isolate coatings on polypropylene films. *J. Food Sci.* **68(1),** 224–228.

Immergut, E. H. and Mark, H. F. (1965). Principles of plasticization. In: *Plasticization and Plasticizer Processes* (R. F. Gould, ed.), pp. 1–26. American Chemical Society, Washington, DC.

Jangchud, A. and Chinnan, M. S. (1999). Properties of peanut protein film: sorption isotherm and plasticizer effect. *Lebensm. Wiss. u. Technol.* **32,** 89–94.

Johnson, K., Hathaway, R., Leung, P. and Franz, R. (1991). Effect of triacetin and polyethylene glycol 400 on some physical properties of hydroxypropyl methylcellulose free films. *Intl J. Pharm.* **73(3),** 197–208.

Kester, J. J. and Fennema, O. R. (1986). Edible films and coatings: a review. *Food Technol. Chicago* **December,** 47–59.

Kester, J. J. and Fennema, O. (1989). An edible film of lipids and cellulose ethers: barrier properties to moisture vapor transmission and structural evaluation. *J. Food Sci.* **54(6),** 1383–1389.

Kim, S.-J. and Ustunol, Z. (2001a). Sensory attributes of whey protein isolate and candelilla wax emulsion edible films. *J. Food Sci.* **66(6),** 909–911.

Kim, S.-J. and Ustunol, Z. (2001b). Thermal properties, heat sealability and seal attributes of whey protein isolate/lipid emulsion edible films. *J. Food Sci.* **66(7),** 985–990.

Kim, K. W., Ko, C. J. and Park, H. J. (2002). Mechanical properties, water vapor permeabilities and solubilities of highly carboxymethylated starch-based edible films. *J. Food Sci.* **67(1),** 218–222.

Krochta, J. M. (1997a). Edible protein films and coatings. In: *Food Proteins and Their Applications* (S. Damodaran and A. Paraf, eds), pp. 529–549. Marcel Dekker, New York, NY.

Krochta, J. M. (1997b). Film, edible. In: *The Wiley Encyclopedia of Packaging Technology* (A. L. Brody and K. S. March, eds), pp. 397–401. John Wiley & Sons, New York, NY.

Krochta, J. M. (2002). Proteins as raw materials for films and coatings: definitions, current status, and opportunities. In: *Protein-based Films and Coatings* (A. Gennadios, ed.), pp. 1–41. CRC Press, New York, NY.

Krochta, J. M. and De Mulder-Johnston, C. L. C. (1997). *Edible and Biodegradable Polymer Films: Challenges and Opportunities*. IFT.

Lee, S.-Y., Dangaran, K. L. and Krochta, J. M. (2002). Gloss stability of whey-protein/plasticizer coating formulations on chocolate surface. *J. Food Sci.* **67(3),** 1121–1125.

Lieberman, E.R. (1973). Gas permeation of collagen films as affected by cross-linkage, moisture, and plasticizer content. *J Polym. Sci. Symp.* **41,** 33–43.

Lim, L.-T., Mine, Y. and Tung, M. A. (1998). Transglutaminase cross-linked egg white protein films: tensile properties and oxygen permeability. *J. Agric. Food Chem.* **46(10),** 4002–4029.

Lin, S.-Y. and Krochta, J. M. (2003). Plasticizer effect on grease barrier and color properties of whey-protein coating on paperboard. *J. Food Sci.* **68(1),** 229–233.

McHugh, T. H. and Krochta, J. M. (1994a). Water vapor permeability properties of edible whey protein-lipid emulsion films. *J. Am. Oil Chem. Soc.* **71(3),** 307–311.

McHugh, T. H. and Krochta, J. M. (1994b). Sorbitol-vs glycerol-plasticized whey protein edible films: integrated oxygen permeability and tensile property evaluation. *J. Agric. Food Chem.* **42(4),** 841–845.

McHugh, T. H., Aujard, J. F. and Krochta, J. M. (1994). Plasticized whey protein edible films: water vapor permeability properties. *J. Food Sci.* **59(2),** 416–419, 423.

Mitchell, J. R. (1998). *Water and Food Macromolecules.* Aspen Publishers, Gaithersburg, MD.

Myllarinen, P., Partanen, R., Seppala, J. and Forssell, P. (2002). Effect of glycerol on behaviour of amylose and amylopectin films. *Carbohydr. Polym.* **50,** 355–361.

Nisperos-Carriedo, M. O. (1994). Edible coatings and films based on polysaccharides. In: *Edible Coatings and Films to Improve Food Quality* (J. M. Krochta, E. A. Baldwin and M. O. Nisperos-Carriedo, eds), pp. 305–336. Technomic Publishing, Lancaster, PA.

Padua, G. W. and Wang, Q. (2002). Formation and properties of corn zein films and coatings. In: *Protein-based Films and Coatings* (A. Gennadios, ed.), pp. 43–67, CRC Press, New York, NY.

Park, H. J. and Chinnan, M. S. (1995). Gas and water vapor barrier properties of edible films from protein and cellulosic materials. *J. Food Eng.* **25(4),** 497–507.

Park, H. J., Weller, C. L., Vergano, P. J. and Testin, R. F. (1993). Permeability and mechanical properties of cellulose-based edible films. *J. Food Sci.* **58(6),** 1361–1364, 1370.

Park, H. J., Bunn, J. M., Weller, C. L., Vergano, P. J. and Testin, R. F. (1994). Water vapor permeability and mechanical properties of grain protein-based films as affected by mixtures of polyethylene glycol and glycerin plasticizers. *Trans. ASAE* **37(4),** 1281–1285.

Park, J. W., Testin, R. F., Park, H. J., Vergano, P. J. and Weller, C. L. (1994). Fatty acid concentration effect on tensile strength, elongation, and water vapor permeability of laminated edible films. *J. Food Sci.* **59(4),** 916–919.

Park, S. K., Hettiarachchy, N. S., Ju, Z. Y. and Gennadios, A. (2002). Formation and properties of soy protein films and coatings. In: *Protein-based Films and Coatings* (A. Gennadios, ed.), pp. 123–137. CRC Press, New York, NY.

Park, S. Y., Marsh, K. S. and Rhim, J. W. (2002). Characteristics of different molecular weight chitosan films affected by the type of organic solvents. *J. Food Sci.* **67(1),** 194–197.

Parris, N. and Coffin, D.R. (1997). Composition factors affecting the water vapor permeability and tensile properties of hydrophilic zein films. *J. Agric. Food Chem.* **45(5),** 1596–1599.

Petersen, K., Nielsen, P. V., Bertelsen, G. *et al.* (1999). Potential of biobased materials for food packaging. *Trends Food Sci. Technol.* **10,** 52–68.

Rakotonirainy, A. M., Wang, Q. and Padua, G. W. (2001). Evaluation of zein films as modified atmosphere packaging for fresh broccoli. *J. Food Sci.* **66(8)**, 1108–1111.

Rico-Pena, D. C. and Torres, J. A. (1991). Sorbic acid and potassium sorbate permeability of an edible methylcellulose-palmitic acid film: water activity and pH effects. *J. Food Sci.* **56(2)**, 497–499.

Roos, Y. H. (1995). *Phase Transitions in Foods*. Academic Press, New York, NY.

Ryu, S. Y., Rhim, J. W., Roh, H. J. and Kim, S. S. (2002). Preparation and physical properties of zein-coated high-amylose corn starch film. *Lebensm. Wiss. u. Technol.* **35(8)**, 680–686.

Salame, M. (1986). Barrier polymers. In: *The Wiley Encyclopedia of Packaging Technology* (M. Bakker, ed.), pp. 48–54. John Wiley & Sons, New York, NY.

Sanchez, A. C., Popineau, Y., Mangavel, C., Larre, C. and Gueguen, J. (1998). Effect of different plasticizers on the mechanical and surface properties of wheat gliadin films. *J. Agric. Food Chem.* **46(11)**, 4539–4544.

Sears, J. K. and Darby, J. R. (1982). *Mechanism of Plasticizer Action*. John Wiley & Sons, New York, NY.

Seow, C. C., Cheah, P. B. and Chang, Y. P. (1999). Antiplasticization by water in reduced-moisture food systems. *J. Food Sci.* **64(4)**, 576–581.

Seyler, R. J. (1994). Opening discussion. In: *Assignment of the Glass Transition* (R. J. Seyler, ed.), pp. 13–16. ASTM, Philadelphia, PA.

Shaw, N. B., Monahan, F. J., O'Riordan, E. D. and O'Sullivan, M. (2002). Physical properties of WPI films plasticized with glycerol, xylitol, or sorbitol. *J. Food Sci.* **67(1)**, 164–167.

Shellhammer, T. H. and Krochta, J. M. (1997). Whey protein emulsion film performance as affected by lipid type and amount. *J. Food Sci.* **62(2)**, 390–394.

Sherwin, C. P., Smith, D. E. and Fulcher, R. G. (1998). Effect of fatty acid type on dispersed phase particle size distributions in emulsion edible films. *J. Agric. Food Chem.* **46(11)**, 4534–4538.

Sichina, W. J. (2000). *Enhanced Characterization of the Glass Transition Event using Stepscan DSC*. PerkinElmer Instruments, Norwalk, CT.

Siew, D. C. W., Heilmann, C., Easteal, A. J. and Cooney, R. P. (1999). Solution and film properties of sodium caseinate/glycerol and sodium caseinate/polyethylene glycol edible coating systems. *J. Agric. Food Chem.* **47(8)**, 3432–3440.

Sothornvit, R. and Krochta, J. M. (2000a). Oxygen permeability and mechanical properties of films from hydrolyzed whey protein. *J. Agric. Food Chem.* **48(9)**, 3913–3916.

Sothornvit, R. and Krochta, J. M. (2000b). Plasticizer effect on oxygen permeability of beta-lactoglobulin (β-Lg) films. *J. Agric. Food Chem.* **48(12)**, 6298–6302.

Sothornvit, R. and Krochta, J. M. (2000c). Water vapor permeability and solubility of films from hydrolyzed whey protein. *J. Food Sci.* **65(4)**, 700–703.

Sothornvit, R. and Krochta, J. M. (2001). Plasticizer effect on mechanical properties of beta-lactoglobulin (β-Lg) films. *J. Food Eng.* **50(3)**, 149–155.

Sothornvit, R., Reid, D. S. and Krochta, J. M. (2002). Plasticizer effect on the glass transition temperature of beta-lactoglobulin (β-Lg) films. *Trans. ASAE* **45(5)**, 1479–1484.

Tanaka, M., Ishizaki, S., Suzuki, T. and Takai, R. (2001). Water vapor permeability of edible films prepared from fish water soluble proteins as affected by lipid type. *J. Tokyo Univ. Fisheries* **87**, 31–37.

Turhan, K. N., Sahbaz, F. and Guner, A. (2001). A spectrophotometric study of hydrogen bonding in methylcellulose-based edible films plasticized by polyethylene glycol. *J. Food Sci.* **66(1),** 59–62.

Wicks, Z. W. (1986). Free volume and the coatings formulator. *J. Coating Technol.* **58(743),** 23–33.

Wilson, A. S. (1995). *Plasticisers: Principles and Practice.* The Institute of Materials, London, UK.

Yang, L. and Paulson, A. T. (2000). Mechanical and water vapour barrier properties of edible gellan films. *Food Res. Intl* **33,** 563–570.

24 Sensory quality of foods associated with edible film and coating systems and shelf-life extension

Yanyun Zhao and Mina McDaniel

Introduction

Edible films and coatings have been studied extensively as a means of extending the shelf life of foods, especially fresh and minimally processed produce. Generally, an edible coating is defined as a thin layer of edible material formed on a food, whereas an edible film is a preformed thin layer of edible material placed on or between food components (Krochta and De Mulder-Johnston, 1997). In other words, edible coatings are applied in liquid form on food by dipping, spraying, or panning. Edible films, on the other hand, are preformed into solid sheets and then applied on or between food components. Edible films and coatings are natural polymers obtained from agricultural products, such as animal and vegetable proteins, celluloses, gums, and lipids, and are biodegradable (Debeaufort *et al.*, 1998).

By regulating the transfer of moisture, oxygen, carbon dioxide, lipids, aroma, and flavor compounds in food systems, edible films and coatings can increase food-product

Innovations in Food Packaging
ISBN: 0-12-311632-5

shelf life and improve food quality. Edible films and coatings can also carry functional ingredients, such as antioxidants, antimicrobials, nutrients, and flavors, to further enhance food stability, quality, functionality, and safety (Krochta *et al.*, 1994; Krochta and De Mulder-Johnston, 1997; Debeaufort *et al.*, 1998) (Table 24.1). The concept of using edible films or coatings to extend the shelf life of foods and protect them from harmful environmental effects has been emphasized in recent years. Increased consciousness of environmental conservation and protection, the need for higher quality foods, the demand for new food processing and storage technologies, and the scientific discovery of the functionality of new materials have renewed researchers' and industrial interests in edible films and coatings.

The success of an edible film or coating in extending the shelf life and enhancing the quality of food strongly depends on its barrier properties to moisture, oxygen, and carbon dioxide, which in turn depends on the chemical composition and structure of the film-forming polymer and the conditions of storage. For example, due to the hydrophilic nature of most film- and coating-forming materials, some edible films and coatings are not satisfactory for controlling the moisture loss of high-moisture foods. Also, one principle disadvantage of using edible coating on fresh produce is the potential development of off-flavors if the inhibition of O_2 and CO_2 exchange results in anaerobic respiration. In addition, undesirable sensory qualities may develop on some coated products, where non-uniform, white, and sticky surfaces become very unattractive to consumers. This chapter focuses on the application of edible films and coatings on fresh and minimally processed produce, illustrates the major quality attributes of edible films and coatings, discusses the applications of edible coatings for improving quality and extending the shelf life of fresh fruits and vegetables through case studies, gives

Table 24.1 Applications and functions of edible coatings and films

Foods	Film materials	Functionality
Poultry, meat, fish, seafood	Gelatin; carraghenan and alginate coatings; whey protein; collagen, casein, cellulose derivatives	Mold prevention; O_2, lipid and moisture barrier; antioxidant carrier; abuse protection; texture improvement
Fruits, nuts, grains, vegetables	Wheat gluten, whey protein, corn zein, waxes, cellulose derivatives, pectins	O_2, lipid, and moisture barrier; antioxidant carrier; stickiness prevention; salt binding
Cheese, icecream, yogurt	Gelatin, seaweed, pectinate	Fruit separation; elimination of dripping; moisture barrier
Confections	Corn zein, milk and whey proteins, wax, MC	O_2, lipid and moisture barrier; antioxidant carrier
Elaborate and heterogeneous foods (e.g. paste, puree, cake, icecream cones)	Mixture of stearic-palmitic acids and HPMC; MC and palmitic acid coating	Moisture barrier

Source: summarized information from Kester and Fennema (1986), Gennadios and Weller (1990), Krochta *et al.* (1994), McHugh and Krochta (1994), Krochta and De Mulder-Johnston (1997), and Debeaufort *et al.* (1998).

recommendations of sensory test methods for coated products, and finally provides insights regarding future research directions.

Sensory quality attributes associated with edible films and coatings

Theoretically, the use of edible films as protective coatings for food can help to retain or improve product quality by:

1. Forming an efficient barrier to prevent moisture loss, thus controlling dehydration
2. Delaying the ripening process through their selective permeability to gases
3. Controlling the migration of water-soluble solutes to retain the natural color, pigments and nutrients
4. Incorporating coloring, flavors, or other functional ingredients to enhance quality.

Hence the sensory qualities of foods are strongly associated with the applications of edible films and coatings. Meanwhile, many active compounds used in the manufacture of edible films and coatings, including edible polymers, plasticizers, and other active agents, may impact the sensory attributes of wrapped or coated products, since most active agents have their own characteristic flavor and color, and the interactions among the compounds may generate unique flavors. Since functional edible films and coatings are edible portions of the packaged or coated foods, it is expected that all components of the edible films and coatings should not interfere with the organoleptic characteristics of the food product. Generally, tasteless edible films and coatings are desirable to minimize any taste interference. Fortunately, the concentration of most active compounds used in edible films and coatings is usually very low, and hence their taste effects may be negligible. When high concentrations of natural active agents are added to edible films and coatings, the film and coating layers may possess a strong flavor of the incorporated active agents. This phenomenon becomes more significant when plant and herb essential oils/extracts or phenolic flavors are added to edible film layers. Unfortunately, few studies have examined the sensory qualities of edible films and coating layers as well as warped or coated food products. The following sections will illustrate some sensory attributes of foods associated with the applications of edible films and coatings.

Appearance

Many quality attributes contribute to the appearance of foods, including surface dehydration, whitening, waxiness, shininess/glossiness, and discoloration (e.g. enzymatic browning). The use of edible coatings may significantly impact the surface appearance of products. For example, by reducing moisture loss, edible coatings can control surface dehydration and discoloration. By selecting the right material, a coating may enhance the surface shininess of food. The O_2 barrier function of edible coatings can also prevent enzymatic browning discoloration in some fresh-cut fruits and vegetables. Hence an appropriately designed coating has the potential to retain or enhance surface appearance of food during storage.

The delay of whitening on the surface of fresh carrots and the control of darkness on fresh strawberries by the use of edible coatings has been demonstrated in several studies. White discoloration on the surface of mini-peeled carrots is the result of reversible surface dehydration and irreversible formation of lignin, which affect their storage quality and shelf life (Bolin and Huxsoll, 1991; Howard and Griffin, 1993; Cisneros-Zevallos et al., 1995). Edible coatings controlled white surface discoloration and enhanced orange color intensity by acting as a moisture barrier or surface moisturizer (Avena-Bustillos et al., 1993a, 1993b; Howard and Dewi, 1995, 1996; Cisneros-Zevallos et al., 1995; Li and Barth, 1998). The same results were demonstrated with sensory evaluation, where baby carrots coated with xanthan gum received less white surface discoloration and greater orange color intensity ratings than those of uncoated ones, hence the coating resulted in products with the best color attributes (Mei et al., 2001) (Table 24.2).

Enzymatic browning is a major factor contributing to quality deterioration in some fruits and vegetables, especially in fresh-cut produce. Using edible coatings to reduce O_2 transfer into the surface of the product and/or to carry anti-browning agents into the coating formulation can prevent enzymatic browning. In addition, some coating materials, such as whey protein, can act as O_2 scavengers (McHugh and Senesi, 2000; Lee et al., 2003). The benefit of using edible coatings to prevent enzymatic browning has become more attractive and important along with increased market demands for fresh and minimally processed fruits and vegetables.

Table 24.2   Effects of edible coatings on sensory attributes of baby carrots during storage

Time	Treatment*	White discoloration	Orange intensity	Crispness	Sweetness	Bitterness	Fresh aroma	Fresh flavor	Slipperiness
1st week	1	3.83[a]	4.54[c]	5.88[a]	5.13[a]	1.92[a]	5.76[a]	5.7[a]	1.91[c]
	2	1.68[bc]	6.45[ab]	5.92[a]	4.52[ab]	1.99[a]	4.83[ab]	5.25[a]	5.14[b]
	3	2.48[b]	6.03[b]	6.06[a]	4.48[ab]	2.23[a]	4.43[b]	4.89[ab]	6.58[a]
	4	0.92[c]	7.27[a]	6.13[a]	4.03[b]	2.98[a]	3.83[b]	3.89[b]	7.57[a]
2nd week	1	6.11[d]	3.82[d]	5.67[a]	6.16[c]	1.71[b]	4.78[ab]	4.82[ab]	1.12[d]
	2	2.36[b]	6.49[ab]	5.42[a]	5.31[a]	1.93[ab]	5.68[a]	5.71[a]	3.10[c]
	3	2.29[b]	6.19[b]	5.61[a]	4.48[ab]	1.95[ab]	5.41[ab]	4.66[ab]	5.27[b]
	4	1.40[c]	7.44[a]	6.01[a]	4.41[ab]	2.83[a]	4.39[b]	3.91[b]	6.76[a]
3rd week	1	6.19[d]	3.52[c]	5.69[a]	5.40[ac]	2.20[a]	4.72[ab]	4.33[ab]	1.22[d]
	2	2.41[b]	6.44[ab]	6.06[a]	5.53[ac]	2.17[a]	4.51[b]	6.39[a]	3.23[c]
	3	2.39[b]	6.40[ab]	6.02[a]	5.24[a]	2.07[a]	5.72[a]	5.08[bc]	5.10[b]
	4	0.90[c]	7.39[a]	5.59[a]	4.90[ab]	2.93[a]	5.52[ab]	5.54[a]	6.62[a]

Source: adapted from Mei et al. (2002).

* Treatment: 1, control; 2, xanthan gum alone; 3, xanthan gum/vitamin E; 4, xanthan gum/calcium.

[a-d] Mean values with different superscript letters in the same column differ ($p < 0.05$). Sensory scale: 0 = none, 10 = intense.

The color, shininess, and transparency of edible films and coating layers vary significantly depending on the chemical composition and structure of the polymer used, which in turn impact the surface appearance of wrapped or coated foods. For example, shellac-based coatings are commonly used in the fresh citrus fruit industry to increase shine and marketability. Although imparting shine, the lack of durability of shellac-based coatings reduces their use when commodities must be stored and shipped for prolonged periods of time. If shellac-coated fruits experience too many changes in temperature and humidity, causing them to "sweat", the shellac coatings may whiten, causing the fruit to become unmarketable. This is especially true of fruits moved from cold storage to a humid environment, and exposed to warm and dry conditions such as during store displays. The color stability of whey protein isolate (WPI) coatings is also an issue, as milk proteins can undergo Millard reactions during storage, causing yellowing (McHugh and Krochta, 1994; Miller *et al.*, 1998). The rate of brown pigment formation in whey powders has been shown to increase as storage temperature and water activity increase from 25 to 45°C and 0.33 to 0.6, respectively (Labuza and Saltmarch, 1981). Trezz and Krochta (2000) compared the yellowing rates of commercial coatings to those of whey protein coatings during storage at 23°, 40°, and 55°C at 75% RH, and found that WPI coatings had lower yellowing rates than whey protein concentrate (WPC) and the same rates as shellac; hydroxypropyl methylcellulose (HPMC) coatings had the lowest yellowing rates, and zein coatings became less yellow during storage. It was concluded that WPI coatings could be used as an alternate coating to shellac or HPMC when low color change is desired, and WPC coatings have potential applications when color development is desired.

Texture

Firmness and crispness are very important textural quality criteria of fresh produce, and may be lost quickly during postharvest storage due to moisture loss and postharvest ripening. While an edible coating reduces moisture loss and delays the ripening process of fresh produce, it is also expected to help maintain product texture quality. In addition, an edible coating may improve the mechanical integrity or handling characteristics of products. It has been well documented in many studies that a coating has the potential to maintain the firmness of fresh fruits and vegetables in comparison with uncoated samples (Sumnu and Bayindirli, 1995; Mei *et al.*, 2002; Han *et al.*, 2004).

Edible coatings can improve the integrity of foods, especially frozen products. Chitosan-based coatings result in at least a 24% reduction in the drip loss of frozen–thawed raspberries (Han *et al.*, 2004). Thus, edible coatings formed on the surface of the fruit can survive the freezing process and storage, help to hold the liquid, and prevent migration of moisture from the fruit to the environment during freezing and thawing. Reduced drip loss, in turn, improves the texture quality of frozen–thawed products. In the case of frozen–thawed raspberries, a chitosan-coating containing calcium increased firmness by about 25% in comparison with uncoated fruits. Adding calcium to a coating formulation provided additional benefit in maintaining the integrity and texture quality of frozen–thawed berries (Han *et al.*, 2004).

Flavor, odor, taste, and other sensory attributes

Edible coatings can retard the production of ethylene and delay the ripening process due to their selective permeability to O_2 and CO_2, thus preventing development of off-flavor and off-odor during postharvest storage of fresh produce. However, inappropriately designed coatings may create an anaerobic condition, leading to the development of off-flavors and off-odors.

Plasticizers are the major component of edible films and coatings, and could affect the flavor and taste of the films and coatings. Common edible plasticizers are glycerol, sorbitol, and polyethylene glycol. Polyethylene glycol is tasteless. Glycerol and sorbitol taste sweet, but the sweetness of glycerol in protein films is negligible, while the sweetness of sorbitol is noticeable. Hence, a plasticizer should be chosen carefully to avoid an undesirable flavor and taste.

Some coating materials may also generate undesirable odors, flavors, or mouthfeel due to the interactions among the components used for making films and coatings. Citrus fruits and some apple varieties develop "off" flavors when coated with shellac. Astringency, an undesirable mouthfeel associated with chitosan-based coatings, has been a concern. When preparing chitosan-based edible coatings, acids (usually acetic or lactic acid) are required to assist the solubility of chitosan in water-based solutions. The pH of the solution is usually in the range of 3.9 to 4.2. In this pH range, the sensation of bitterness and astringency develops, which has been a major limitation that has made chitosan-coated foods less attractive on the market. The chemical reactions promoting astringency in acids are still not well understood. One common hypothesis is that acids precipitate the proteins in saliva or cause sufficient conformational changes so that lubrication is lost when acids form complexes with salivary proteins or mucopolysaccharides (Bate-Smith, 1954). The astringency associated with chitosan could be the result of a rise in amine protonated groups when dissolved in an acid solution. The pKa of the amino group of glucosamine residue is about 6.3 (Muzzarelli,

Table 24.3 Consumer liking mean ratings of chitosan-coated strawberries after 1-day storage at 2°C and 89% RH

Coating treatments*	CON	AA	LA	LAE
Appearance liking***	7.27[a] (1.29)	7.40[a] (1.33)	7.41[a] (1.28)	6.85[b] (1.46)
Overall liking**	6.08[ab] (1.93)	6.36[a] (1.73)	6.08[ab] (1.77)	5.70[b] (1.71)
Flavor liking (NS)	5.55 (2.26)	6.07 (2.09)	5.74 (2.11)	5.47 (1.98)
Sweetness liking (NS)	5.33 (2.29)	5.75 (2.13)	5.34 (2.01)	5.00 (1.95)
Firmness liking (NS)	6.17 (1.90)	6.10 (2.07)	6.14 (1.75)	5.61 (1.96)

Source: adapted from Han *et al.* (2004), using means of 92 consumer panelists.
[a-b] Different letter superscripts indicate significant differences at $p < 0.05$ for each row separated by Tukey's HSD.
Nine-point intensity scale: 1 = dislike extremely; 9 = like extremely.
*Treatments: CON, control, uncoated samples; AA, acetic acid dissolved chitosan coating solution; LA, lactic acid dissolved chitosan coating solution; LAE, lactic acid dissolved chitosan coating solution containing 0.2% vitamin E.
, * indicates significance at $p < 0.05$, and $p < 0.001$, respectively.

1985); hence chitosan is polycationic at acidic pH value (Hwang and Damodaran, 1995; Fernander and Fox, 1997). Chitosan can selectively bind desired materials, such as proteins (Kubota and Kihuchi, 1998). When binding with salivary proteins, the affinity of chitosan in acidic solutions might be increased. Rodriguez *et al.* (2003) reported that at a higher pH of 4.6 to 6.3, less astringency is perceived.

A recent study in the authors' laboratory has focused on improving the sensory quality of chitosan-based coatings, especially on overcoming astringency. It was found that adjusting the pH of the coating solutions to above 4.0 can minimize astringency. Table 24.3 shows the results regarding consumer liking of five sensory attributes of strawberries subjected to different chitosan-coating treatments 1 day after coating. No differences ($p > .05$) were identified in flavor, sweetness, and firmness liking among coated and uncoated samples, or in samples coated with different coating formulations.

Edible coatings to improve the quality and extend the shelf life of foods – case studies

In general, an ideal coating should be safe, invisible, have no off-flavor and taste, be a desirable moisture and gas barrier, and have a certain mechanical strength. Many factors determine the success of edible coatings in improving the quality and extending the shelf life of foods. Among these factors, the chemical composition, structure, methods used for forming films or coatings, storage conditions, and the properties of the food itself (maturity stage, moisture and lipid content, etc.) are important factors. When applying edible films and coatings, these factors have to be carefully considered. Applications of selected edible coatings, based on coating material characteristics, for improving the quality and extending the shelf life of fresh and minimally processed fruits and vegetables are discussed in the following sections.

Whey protein

Whey protein has excellent nutritional and functional properties, and the ability to form films. Whey protein has been shown to produce transparent, bland, flexible, water-based edible films with excellent oxygen and oil barriers, and is flavorless and odorless (McHugh and Krochta, 1994). WPI coatings give a gloss comparable to those of shellac and zinc coatings, and a higher gloss than HPMC coatings. Whey-protein coatings have been utilized in coating peanuts and walnuts to increase their shelf life, and in coating fresh-cut fruits and vegetables to control enzymatic browning.

Four different WPI/plasticizer formulations were compared to determine which provided the most gloss and which was most stable with time when applied on chocolates (Lee *et al.*, 2002a). The four plasticizers studied were glycerol, polyethylene glycol 400 (PEG 400), propylene glycol (PG), and sucrose, all in a 1 : 1 ratio with WPI. WPI/sucrose coatings provided the highest and most stable gloss. It was found that with optimization, water-based WPI/sucrose coatings could be an alternative source of glaze to alcohol-based shellac coatings in the confectionery industry. In addition, consumer acceptance of WPI-coated chocolate-covered almonds was compared with that for shellac-coated chocolate (Lee *et al.*, 2002b). Four WPI coating formulations were tested: two formulations were without lipid, and two were with lipid. The shellac formulation consisted of 30% solids, of which 90% was shellac and 10% was propylene glycol. A central location consumer test was carried out for attributes such as overall degree of liking. Results indicate that water-based WPI–lipid coatings can be used as an alternative glaze, with higher consumer acceptance than alcohol-based shellac.

Lipid oxidation is one of the leading causes of deterioration in peanuts. Oxygen concentration plays an important role in oxidation. Oxygen uptake can be impeded by specialized packaging systems or an edible coating, which in turn will decrease the rate of lipid oxidation. WPI-based films were shown to be excellent O_2 barriers at low to intermediate relatively humidity (Mate and Krochta, 1996). WPI coatings were shown significantly to reduce the oxidative rancidity of coated dry-roasted peanuts (Lee *et al.*, 2002c), thus extending their shelf life (Lee and Krochta, 2002). Descriptive sensory analysis results revealed that the rancidity was significantly lower for whey protein-coated peanuts than for uncoated ones (Lee *et al.*, 2002c). This finding was also confirmed by static headspace gas chromatography (GC) analysis.

WPI coatings have also been used on fresh-cut fruits and vegetables to prevent enzymatic browning (Le Tien *et al.*, 2001; Lee *et al.*, 2003; Pérez-Gago *et al.*, 2003a). Color analysis of apple and potato slices coated with whey-protein solutions showed that the coatings efficiently delayed browning by acting as oxygen barriers and/or reactive oxidative species scavengers (Le Tien *et al.*, 2001). Lactose present in commercial whey can also account for the increased antioxidative activity. Moreover, the addition of a polysaccharide-like carboxymethyl cellulose (CMC) to the formulations can further improve the antioxidative potential of whey-protein films. Combinations of WPC coating with antibrowning agents (ABA) were developed for extending the shelf life of minimally processed (MP) Fuji apple slices packaged with Cryovac bags and stored at 3°C (Lee *et al.*, 2003). The coatings and ABA extended the shelf life of MP apple slices by 2 weeks, maintaining their color and reducing the counts of mesophilic and psychrotrophic micro-organisms. WPC reduced initial respiration rates by 20%. Coated apple slices maintained acceptable sensory scores for color, firmness, flavor, and overall preference, even after 14 days; WPC containing 1 g/100 ml each of ascorbic acid and calcium chloride was the optimal treatment. In another study, edible coatings from WPI–beeswax (BW) with various total solids (8%, 12%, 16%, 20%) and BW content (0%, 20%, 40%, 60%, dry bases) were applied on apple slices (Pérez-Gago *et al.*, 2003a). Coated apples had higher lightness and lower Brown Index (BI) than uncoated apples, indicating that whey proteins exert an antibrowning effect. The BI decreased as the solid content of the coating emulsion increased. Increasing BW content decreased enzymatic browning. Coating application did not reduce weight loss in fresh-cut apples, probably due to the product's high relative humidity.

Semperfresh™

Semperfresh™, manufactured by Agricoat Industries Ltd (7B Northfield Farm, Great Shefford, Berkshire, England) is a commercial coating formulated with sucrose esters of fatty acids, mono- and di-glycerides, and the sodium salt of carboxymethylcellulose. It forms an invisible coating which is odorless and tasteless, and creates a barrier that is differentially permeable to oxygen and carbon dioxide. In this way, a modified atmosphere of reduced oxygen with slightly raised carbon dioxide levels results. Semperfresh™ coating has been applied successfully to retard ripening on bananas (Banks, 1984), apples (Drake et al., 1987; Bauchot et al., 1995; Sumnu and Bayindirli, 1995), and mangoes (Dhalla and Hanson, 1988) in aqueous solution containing 1.5, 1.0, and 0.75% (w/v), respectively. Effectiveness of coatings for increasing the shelf life and post storage life of Amasya apples was studied (Sumnu and Bayindirli, 1995). Semperfresh™ applied after harvesting maintained apple firmness for 25 days. Increasing the coating concentration caused increased firmness values and extended the shelf life by 10, 25, 30 and 35% at treatment levels of 5, 10, 15 and 20 g/l, respectively. Semperfresh™ coating reduced weight loss and loss of ascorbic acid from apples. Application of Semperfresh™ was effective in the retention of soluble solids, in reducing ascorbic acid loss, and in the retention of color change. It is concluded that >10 g/l of Semperfresh™ is needed for coating apples.

Application of Semperfresh™ for extending the shelf life and improving the quality of fresh sweet cherries and tomatoes has also been studied (Yaman and Bayondrl, 2001, 2002). It was demonstrated that Semperfresh™ effectively reduced the weight loss and increased the firmness, ascorbic acid content, titratable acidity, and skin color of cherries during storage. However, soluble solid content and sugar content were not affected by coating. Semperfresh™ increased the shelf life of cherries by 21% at $30 \pm 3°C$ and by 26% at 0°C without perceptible losses in quality. By dipping fully matured tomato fruits (cv. Pusa Early Dwarf) in a 0.2–0.4% Semperfresh™ solution for 10 minutes, the shelf life of the tomatoes at 24–31.5°C, 77–90% RH was extended by delaying ripening and decreased physiological weight loss.

Chitosan

Chitosan, a high molecular weight cationic polysaccharide and tasteless fiber obtained by the deacetylation of chitin, might be an ideal preservative coating material for fresh fruits and other food products because of its excellent film-forming and biochemical properties (Ghaouth et al., 1991). Chitosan, a by-product of the seafood industry, was approved by the United States Food and Drug Administration (USFDA) as a feed additive in 1983 (Knorr, 1986). Chitosan has been shown to inhibit the growth of several fungi (Ghaouth et al., 1991; Jiang and Li, 2001) and to induce defense enzymes, such as chitinase and β-1,3-glucanase (Zhang and Quantick, 1998). Although the FDA has not approved the application of chitosan in food, significant research has been conducted to understand its safety and potential application in foods (Arai et al., 1968; Hirano et al., 1990). Chitosan-based coatings have been studied for their wide application in food, including fresh berries (Ghaouth et al., 1991; Garcia et al., 1998; Zhang and Quantick, 1998), raw eggs (Bhale et al., 2003), fresh-cut Chinese water chestnuts (Pen and Jiang, 2003), and peaches (Li and Yu, 2001).

Chitosan coatings have improved the quality and storability of strawberries and raspberries (Ghaouth et al., 1991; Garcia et al., 1998; Zhang and Quantick, 1998).

A chitosan coating significantly reduced the decay of strawberries and raspberries stored at 13°C and induced a significant increase of chitinase and β-1,3-glucanase activities of the berries as compared to uncoated berries (Zhang and Quantick, 1998). Chitosan coating proved almost as effective as the fungicide TBZ (thiabendazole) in controlling the decay of berries stored at 13°C caused by *Botrytis cinerea* and *Rhizopus* sp. Chitosan coating had the beneficial effects of retaining firmness, titratable acidity, vitamin C content, and anthocyanin content of strawberries and raspberries stored at 4°C. Control of decay by chitosan coating was also demonstrated with peaches (Li and Yu, 2001). The 5–10 mg chitosan/ml treatment showed the potential to protect peaches from brown rot by prolonging the incubation period of spores, reducing the rot area, and reducing the incidence of lesions. This protective effect appeared to be related to chitosan induction of the defense response, as well as an antifungal activity. Chitosan semipermeable coatings decelerated aging by reducing respiration rate and ethylene production, reducing malondialdehyde formation, stimulating superoxide dismutase activity, and maintaining membrane integrity.

In a study of chitosan coating applied to fresh-cut Chinese water chestnuts (cv. Guilin) stored at a low temperature, it was found that the coatings retarded the development of browning in a dose-dependent fashion by reducing the activities of PAL, PPO, and POD, and reducing phenol levels (Pen and Jiang, 2003). Sensory quality was maintained longer in coated slices, and levels of total soluble solids, acidity, and ascorbic acid were better retained, also in a dose-dependent manner. Overgrowth by spoilage micro-organisms was delayed by the coatings.

A recent study evaluated the internal and sensory quality of eggs coated with chitosan during 5 weeks of storage at 25°C (Bhale *et al.*, 2003). Three chitosans with high (HMW, 1100 KDa), medium (MMW, 746 KDa), and low (LMW, 470 KDa) molecular weights were used to prepare coating solutions. Coating with LMW chitosan was more effective in preventing weight loss than with MMW and HMW chitosans (Pen and Jiang, 2003). The Haugh unit and yolk index values indicated that the albumen and yolk quality of coated eggs can be preserved for up to 5 weeks at 25°C, which is at least 3 weeks longer than was observed for non-coated eggs. Consumers could not differentiate between the external quality of coated eggs and that of non-coated eggs. Overall acceptability of the coated eggs was not different from that of the control and commercial eggs.

Lipid–HPMC composite coatings

Polysaccharides and proteins are good film-forming materials, but make poor moisture barriers. Lipids, on the other hand, provide a better moisture barrier, but present low mechanical integrity and require solvents or high temperatures for casting. Each group of coating materials has certain advantages and disadvantages. For this reason, many formulations are composite coatings of both groups. For example, there are some studies regarding the development and application of lipid–HPMC composite coatings for fresh fruits and vegetables (Pérez-Gago *et al.*, 2002, 2003a, 2003b, 2004c).

HPMC–lipid composite coatings were developed and applied on mandarin oranges and plums (Pérez-Gago *et al.*, 2003b, 2003c). The coatings consisted of beeswax, carnauba wax, or shellac, at two lipid contents (20% and 60%). It was found that the weight loss of coated fruits decreased significantly as the lipid content increased. Beeswax (60%) emulsions were associated with the lowest weight loss. Coated fruits had higher internal CO_2, lower internal O_2, and higher ethanol contents than uncoated fruit. At 20% lipid content, internal O_2 was lower and ethanol content was higher than at 60% lipid content, which could be related to the low oxygen permeability of hydroxypropyl methylcellulose and the higher viscosity of these emulsions, which could have affected the final coating thickness. For plums, however, no differences in weight loss were observed between uncoated and 20% lipid-coated plums, indicating that the natural waxes of plums are as effective as coatings having 20% lipid. Fruit texture was not affected by coating after short-term storage at 20°C. However, for prolonged storage at 20°C, the coatings significantly reduced texture loss and internal breakdown compared to uncoated and water-dipped plums. In order to improve the moisture barrier of "Autumn Giant" plums, a coating containing more than 20% lipids needed to be applied.

Sensory evaluation of edible films, coatings and coated products

Sensory evaluation involves the use of human subjects, and has been conducted for as long as there has been human life on earth. As it is practiced now in relation to food products, sensory evaluation is of incomparable value to food companies producing products for an ever-changing and expanding market. Sensory science is not an old science; however, it has its landmarks, as discussed by Lawless and Heymann (1998). Most significant was the change from individual "experts" to a panel of subjects to discriminate among samples, describe what they perceive, or relate how they feel about the products presented.

Through the years sensory methods have been modified, largely based on what worked and what didn't work, by food scientists, psychologists, and statisticians. There has been little testing of one method against another. Critical to the choice of any method is a clear statement of objectives. When working with edible films and coatings, the major questions are, "What are the sensory properties of the product before and after coating?", "What changes occur during storage and or subsequent processing?", and "Does the consumer notice and care about any difference that is perceivable?" It is impossible in this brief chapter to discuss how all sensory methods might be applied to coated products. Rather, we will direct you to excellent references such as Lawless and Heymann's (1998) ambitious and thorough covering of sensory principles and practices. For those new to sensory evaluation, Meilgaard *et al.* (1999) is a user-friendly guide to sensory common sense, laden with excellent case studies. For assistance in implementing sensory evaluation in quality control, Munoz *et al.* (1992) is helpful. Carpenter *et al.* (2000) is a good source for both quality control and product development sensory guidelines.

The American Society for Testing and Materials (ASTM) Committee E18 has a long history of standardizing sensory methodology and providing manuals on special topics written by industry members and academics. An example is STP 433, *Basic Principles of Sensory Evaluation*, written in 1968. STP 545, *Sensory Evaluation of Appearance of Materials*, was published in 1973, and provides good discussions of the relationship between sensory and instrumental measures of color and appearance. Chambers and Wolf (1996) revised the original ASTM methods manual, STP 434, resulting in *Sensory Testing Methods*, 2nd edition, MNL 26. As the final judgement of any new process or product is by the consumer, the ASTM published two volumes on consumer testing edited by Wu (1989) and Wu and Gelinas (1992).

Whether in research, product development or quality control, understanding a product's sensory characteristics is paramount. With the application of edible films or coatings to a product, it is important first to understand the initial product. The goal of descriptive analysis, using a trained panel, is to describe fully the appearance, aroma, taste, texture, and aftertaste or residual mouthfeel of a product in sensory terms. With some descriptive methods the lexicon of terms is taught to the panel, as is the case with the Spectrum Method developed by Gail Vance Civille (Meilgaard *et al.*, 1999). QDA, developed by Herb Stone and Joel Sidel at Tragon, has the panel develop the lexicon during training (Stone and Sidel, 1993). Both techniques provide valuable information. The Spectrum Method results in terms more directly useful to research and product development, while QDA results in terms more closely related to the consumer. The results of a typical descriptive study are described later in this chapter, with examples using carrots (Mei *et al.*, 2002) and strawberries (Han *et al.*, 2005).

Understanding a product's sensory characteristics and how they change with the application of edible films or coatings is the first step in determining the product's quality. The consumer must be consulted to determine if the treatment efficacy satisfies the project goals. The success of consumer testing is determined by selecting the appropriate consumers (users of the product), by testing enough consumers to get the desired power from the test (Lawless and Heymann, 1998), by using methodology sensitive to the need, and by presenting representative samples. A general rule of thumb is to test at least 100 consumers, but in some cases many more than this are required. In recent years Preference Mapping has become very popular both to study the effect of individual consumers and to give a multidimensional representation of products. Macfie and Thomson (1988) provide a good introduction to this topic.

Statistical analysis of test results, appropriate for the design used, is usually conducted once the data have been collected. While sensory methods remain basically the same, their statistical analysis options have expanded greatly over the past few years. O'Mahony (1986) provides an introduction to the basic statistical techniques used in sensory evaluation. Many sensory data sets are multivariate in nature, and require selection of an appropriate multivariate technique whose statistical assumptions fit with the goals of the research (Lawless and Heymann, 1998). Lawless and Heymann (1998) provide a good overview of multivariate statistical techniques.

Edible films and coatings incorporating functional ingredients

A unique feature of edible films and coatings is their capacity to carry many functional ingredients, including antimicrobial agents, antioxidants, flavorings, and colorants. Integration of these minor ingredients can enhance food stability, quality, functionality, and safety. For example, the antimicrobial agents nisin (Janes *et al.*, 1999), and p-aminobenzoic and sorbic acids (Cagri *et al.*, 1999), were added directly to edible coatings to suppress surface microbial growth in cooked chicken and model food systems; and antioxidants, such as tocopherol (Wu *et al.*, 1999) and butylated hydroxytoluene (BHT) (Huang and Weng, 1998), were integrated into edible coatings to inhibit lipid oxidation in precooked beef patties and fish muscle, respectively.

Edible coatings could also serve as excellent carriers of nutraceutical ingredients. There has been an explosion of consumer interest in the health-enhancing role of specific foods or physiologically active food components – so-called nutraceuticals or functional foods (Hasler, 1998). Edible coating could provide an excellent vehicle to enhance the nutritional value of fruits and vegetables by carrying nutrients or nutraceuticals that are lacking or present in low quantities. This is a natural extension of their primary function as moisture and gas barriers.

Efforts to integrate high concentrations of minerals and vitamins into edible films and coatings have been made in the authors' laboratory. Mei *et al.* (2002) developed xanthan gum coatings containing a high concentration of calcium and vitamin E, and applied such coatings to fresh baby carrots. Applying such coatings, per serving of carrots (85 g), provides ~6.7% DRI (Dietary Reference Index) calcium and ~70% DRI vitamin E, respectively, without degrading the primary barrier function of coatings. Mei and Zhao (2003) also examined the functionality of calcium casienate- (CC) and WPI-based films containing 5% or 10% Gluconal Cal$^{®}$ (GC), a mixture of calcium lactate and gluconate, or 0.1% or 0.2% α-tocopheryl acetate (VE), respectively, and reported that CC and WPI films have the capabilities to carry a high concentration of calcium or vitamin E, although some of the film functionality may be compromised. CC and WPI films containing VE exhibited increased elongation at break; addition of 0.2% VE decreased WVP of CC films and the tensile strength of both CC and WPI films. Incorporation of GC reduced the tensile strength of CC films ($p < 0.05$), with 10% GC decreasing both elongation at break and WVP ($p < 0.05$).

Park and Zhao (2003) further developed chitosan-based films containing high concentrations of calcium, zinc or vitamin E, and studied their functionality. GC incorporation significantly increased the pH and decreased the viscosity of film-forming solutions, but the addition of ZL or VE did not. The water-barrier property of the films was improved by increasing the concentration of mineral or vitamin E in the film matrix. However, the tension strength of the films may be reduced by incorporation of high concentrations of GC or VE, although film elongation, puncture strength, and puncture deformation are not impacted. The study demonstrated the capability of a chitosan-based film matrix to carry high concentrations of mineral or vitamin E.

Han *et al.* (2004) then demonstrated the application of these coatings on fresh and frozen strawberries. Adding high concentrations of calcium or vitamin E to chitosan-based coatings did not alter their antifungal and moisture-barrier functions. The coatings significantly decreased the decay incidence and weight loss, and delayed the change in color, pH, and titratable acidity of strawberries and red raspberries during cold storage (Table 24.4). Coatings also reduced the drip loss and helped maintain the texture quality of frozen strawberries after thawing. In addition, chitosan-based coatings containing calcium or vitamin E significantly increased the content of these nutrients in both fresh and frozen fruits. A weight of 100 g of coated fruits contained about 34–59 mg of calcium or 1.7–7.7 mg of vitamin E, depending on the types of fruits and the time of storage, while uncoated fruits only contain 19–21 mg of calcium or 0.25–1.15 mg of vitamin E.

In addition to incorporating antimicrobial agents, antioxidants, and nutraceuticals, flavor and coloring agents may also be added to an edible film/coating matrix to improve or enhance the sensory quality of wrapped or coated products. Unfortunately, little study has been reported regarding this area.

Table 24.4 Trained panel mean ratings of descriptors showing changes of chitosan-coated strawberries during storage at 2°C and 89% RH

Descriptors	Strawberry sample conditions*							
	FRESH	CON1W	LA1W	LA2W	LA3W	LAE1W	LAE2W	LAE3W
Appearance								
Glossiness	9.4^e	7.8c^{de}	9.4^e	8.8^{de}	6.7^{bc}	5.8^b	2.8^a	7.2^{bcd}
Shrived	2.6^a	3.9^{ab}	3.6^{ab}	4.4^{bc}	5.7^{cd}	5.1^{bc}	9.1^e	5.1^{bc}
Dryness	2.8^a	4.5^b	5.6^{bc}	7.6^d	9.2^e	6.6^{cd}	9.2^{ef}	10.8^f
Damaged	2.3^{ab}	2.4^{ab}	1.6^a	3.6^{bc}	3.2^{ab}	2.1^{ab}	5.1^c	3.2^{ab}
White/waxy	0.1^a	0.2^a	0.1^a	0.8^a	3.9^b	4.1^b	5.7^b	4.9^b
Texture								
Crispness (NS⁻)	6.3^a	6.5^a	6.2^a	6.8^a	7.5^a	5.8^a	6.4^a	6^a
Firmness	6.6^{abc}	6.3^{abc}	5.7^{ab}	6.9^{abc}	7.7^c	5.6^a	6.2^{abc}	7.4^{bc}
Juiciness	6.7^{ab}	6.7^{ab}	7.8^b	5.9^a	6.0^a	8.0^b	6.3^a	6.4^a
Flavor/basic taste								
Overall strawberry	8.0^c	7.9^c	7.4^{bc}	6.9^{abc}	5.7^a	7.9^c	8.0^c	6.3^{ab}
Fresh strawberry	7.3^c	7.2^{bc}	6.0^{bc}	4.0^a	5.5^{ab}	6.8^{bc}	4.2^a	6.3^{bc}
Sour (NS)	5.4^a	5.9^a	4.8^a	5.4^a	5.5^a	4.9^a	4.7^a	4.8^a
Sweet (NS)	5.0^a	5.3^a	5.4^a	4.4^a	5.1^a	4.9^a	5.3^a	5.8^a
Astringency (NS)	3.6^a	3.4^a	3.7^a	3.3^a	3.1^a	3.5^a	3.3^a	3.8^a

Source: adapted from Han *et al.* (2005). Means of nine trained panelists by use of free-choice profile method.

* FRESH, fresh, uncoated samples; CON1W, fresh, uncoated samples stored for 1 week; LA, lactic acid dissolved chitosan coating solution; LAE, lactic acid dissolved chitosan coating solution containing 0.2% vitamin E; 1W, 2W, 3W, Samples stored for 1, 2, or 3 weeks.

a-f Different letter superscripts indicate significant differences at $p < 0.05$ for each row separated by Tukey's HSD.

Fifteen-point structured intensity scales (0 = none, 7 = moderate, and 15 = extreme).

(NS⁻, not significant at $p < 0.001$).

Future research

Edible films and coatings, due to their unique features, provide a promising technology to food companies for enhancing the quality and extending the shelf life of their products. However, commercial applications on fresh and minimally processed fruits and vegetables are still very limited. Part of the reason is the lack of film and coating materials that have desirable functionality. More efforts are required to develop new materials and understand their functionality and interactions among the components used in the edible films and coatings. Meanwhile, studies to improve the functionality of existing materials are important. This may be done by adjusting the formulation of film-forming solutions, incorporating functional ingredients, and modifying film-forming conditions. In addition, it is important to conduct more sensory studies to understand consumer acceptance of edible films and coatings, as this information would help to determine quality criteria for the measurement of shelf life and quality of wrapped or coated foods. These research needs are further illustrated in the following sections.

Development of new materials

Shellac (wax), milk proteins, and polysaccharides are the most common coating materials used and studied for fresh produce. Each of these materials has its advantages and disadvantages. When applying coating on fresh-cut produce, one major challenge is the wet surface. If the coating material is highly hydrophilic, such as milk proteins or polysaccharides, it would be difficult to dry and to form a uniform layer on the surface. Hence novel materials that are invisible, less water-soluble, and that have certain moisture- and gas-barrier properties are essential.

Fruit puree-based films and coatings have been patented with claims of desirable color, nutrients, and barrier properties (McHugh and Senesi, 2000). Composite films and coatings made of two or more materials may also meet some of the desired functionality that a single material cannot achieve. Efforts have been made to combine lipids and polysaccharides, or lipids and milk proteins, to complement each other. More efforts in developing composite materials should continue, and new ideas for using other agricultural by-products are imperative. In addition, the functionality of edible films and coatings may be improved by simply redesigning their formulation and modifying processing procedures/conditions. More attempts should be made to understand fully the chemical structure and interactions of each component in film/coating matrices.

Incorporating functional ingredients

The use of edible films and coatings as carriers of functional ingredients is a natural extension of their barrier properties. By taking advantage of the carrying/incorporating feature of edible films and coatings, functional ingredients such as antimicrobial agents, antioxidants, flavor, and coloring agents may be added to film/coating formulations to further extend the shelf life and enhance the sensory quality of wrapped or coated food. Exposure in this area is still very limited, and more attempts should be made. One critical issue is the possibility of altering the basic functional properties of edible films and coatings, such as their resistance to vapor, gas, or solute transport, when incorporating other functional ingredients. The influence of a given ingredient on coating functionality depends on its concentration in the coating, its chemical structure, its degree of dispersion in the coating, and the extent of its interaction with the polymer (Kester and Fennema, 1986). For example, the solubility and stability of added ingredients and their impact on functional properties have to be investigated and characterized because these ingredients may induce static cross-linking and affect the film structure and interactions, and in turn change the barrier properties of the films and coatings. More research on novel ingredients and their impacts on film functionality should be conducted.

Sensory studies for determining quality criteria in the measurement of shelf life

When applying coatings on food, quality criteria must be determined carefully. Color change, firmness loss, ethanol fermentation, decay ratio, and weight loss are among the most important quality attributors. Sensory studies conducted by both trained panels and consumer panels are essential. The information would help in the development of edible films and coatings that will meet market and consumer needs.

References

Arai, K., Kinumaki, T. and Fujita, T. (1968). Toxicity of chitosan. *Bull. Tokai Reg. Fish. Res. Lab.* **56,** 89.

Avena-Bustillos, R., Cisneros-Zevallos, L., Krochta, J. M. and Saltveit, M. E. (1993a). Application of casein–lipid edible film emulsions to reduce white blush on minimally processed carrots. *Postharv. Biol. Biotechnol.* **4,** 319–329.

Avena-Bustillos, R., Cisneros-Zevallos, L., Krochta, J. M. and Saltveit, M. E. (1993b). Optimization of edible coatings on minimally processed carrots using response surface methodology. *Trans. ASAE* **36,** 801–805.

Banks, N. H. (1984). Studies of the banana fruit surface in relation to the effects of TAL prolong coating on gaseous exchange. *Scientia Hort.* **24(3/4),** 279–286.

Bate-Smith, E. C. (1954). Astringency in foods. *Food* **23,** 124–135.

Bauchot, A. D., John, P., Soria, Y. and Recasens, I. (1995). Sucrose ester-based coatings formulated with food-compatible antioxidants in the prevention of superficial scald in stored apples. *J. Am. Soc. Hort. Sci.* **120(3),** 491–496.

Bhale, S., No, H. K., Prinyawiwatkul, W., Farr, A. J., Nadarajah, K. and Meyers, S. P. (2003). Chitosan coating improves shelf life of eggs. *J. Food Sci.* **68(7),** 2378–2383.

Bolin, H. R. and Huxsoll, C. C. (1991). Control of minimally processed carrot (*Daucus carota*) surface discoloration caused by abrasion peeling. *J. Food Sci.* **56,** 416–418.

Cagri, A., Ustunol, Z. and Ryser, E. (1999). Mechanical, barrier and antimicrobial properties of low-pH whey protein isolate edible films containing paminobenoic acid or sorbic acid. Paper No. 16-6. 1999 IFT Annual Conference, July 23–26.

Carpenter, R. P., Lyon, D. H. and Hasdell, T. A. (2000). *Guidelines for Sensory Analysis in Food Product Development and Quality Control.* Aspen, Gaithersburg, MD.

Chambers, E. and Wolf, M. (eds) (1996). Sensory testing methods. In: *American Society for Testing and Materials*, 2nd edn. American Society for Testing and Materials, West Consohocken, PA.

Cisneros-Zevallos, L., Saltveit, M. E. and Krochta, J. M. (1995). Mechanism of surface discoloration of peeled (minimally processed) carrots during storage. *J. Food Sci.* **60,** 320–323.

Debeaufort, F., Quezada-Gallo, J. A. and Voilley, A. (1998). Edible films and coatings: tomorrow's packagings: a review. *Crit. Rev. Food Sci. Nutr.* **38(4),** 299–313.

Dhalla, R. and Hanson, S. W. (1988). Effect of permeable coatings on the storage life of fruits. II. Pro-long treatment of mangoes (*Mangifera indica* L. cv. Julie). *Intl J. Food Sci. Technol.* **23,** 107–112.

Drake, S. R., Fellman, J. K. and Nelson, J. W. (1987). Postharvest use of sucrose polyesters for extending the shelf-life of stored "Golden Delicious" apples. *J. Food Sci.* **52(5),** 1283–1285.

Fernandez, M. and Fox, P. F. (1997). Fractionation of cheese nitrogen using chitosan. *Food Chem.* **58(4),** 319–322.

Garcia, M. A., Martino, M. N. and Zaritzky, N. E. (1998). Plasticized starch-based coatings to improve strawberry (*Fragaria ananassa*) quality and stability. *J. Agric. Food Chem.* **46,** 3758–3767.

Gennadios, A. and Weller, C. L. (1990). Edible films and coatings from wheat and corn proteins. *Food Technol.* **44(10),** 63–69.

Ghaouth, A. E., Arul, J., Ponnampalam, R. and Boulet, M. (1991). Chitosan coating effect on storability and quality of fresh strawberries. *J. Food Sci.* **56,** 1618–1620.

Han, C., Zhao, Y., Leonard, S. W. and Traber, M. G. (2004). Edible coatings to improve storability and enhance nutritional value of fresh and frozen strawberries (*Fragaria ananassa*) and raspberries (*Rubus ideaus*). *Postharv. Biol. Technol.* **33(1),** 67–78.

Han, C., Lederer, C., McDaniel, M. and Zhao, Y. (2005). Sensory quality of strawberries coated by chitosan-based edible coatings. *J. Food Sci.* **70(3),** S172–178.

Hasler, C. M. (1998). Functional foods: their role in disease prevention and health promotion. *Food Technol.* **52(11),** 63–70.

Hirano, S., Itakura, C. and Seino, H. *et al.* (1990). Chitosan as an ingredient for domestic animal feeds. *J. Agric. Food Chem.* **38,** 1214–1217.

Howard, L. R. and Dewi, T. (1995). Sensory, microbiological and chemical quality of mini-peeled carrots as affected by edible coating treatments. *J. Food Sci.* **60,** 142–144.

Howard, L. R. and Dewi, T. (1996). Minimal processing and edible coating effects on composition and sensory quality of mini-peeled carrots. *J. Food Sci.* **61(3),** 643–645.

Howard, L. R. and Griffin, L. E. (1993). Lignin formation and surface discoloration of minimally processed carrot sticks. *J. Food Sci.* **58,** 1065–1067.

Huang, C. H. and Weng, Y. M. (1998). Inhibition of lipid oxidation in fish muscle by antioxidant incorporated polyethylene film. *J. Food Process. Preserv.* **22,** 199–209.

Hwang, D. and Damodaran, S. (1995). Selective precipitation and removal of lipids from cheese whey using chitosan. *J. Agric. Food Chem.* **43,** 33–37.

Janes, M. E., Nannapaneni, R. and Johnson, M. G. (1999). Control of *Listeria monocythognes* on the surface of refrigerated, ready-to-eat chicken coated with edible zein films with nisin. Paper No. 37D-21, 1999 Annual IFT Conference, Chicago, 23–26 July.

Jiang, Y. and Li, Y. (2001). Effects of chitosan coating on postharvest life and quality of longan fruit. *Food Chem.* **73,** 139–143.

Kester, J. J. and Fennema, O. R. (1986). Edible films and coatings: a review. *Food Technol.* **40(12),** 47–59.

Knorr, D. (1986). Nutritional quality, food processing, and biotechnology aspects of chitin and chitosan: a review. *Proc. Biochem.* **6,** 90–92.

Krochta, J. M. and De Mulder-Johnston, C. (1997). Edible and biodegradable polymer films: challenges and opportunities. *Food Technol.* **51(2),** 61–74.

Krochta, J. M., Baldwin, E. A. and Nisperos-Carriedo, M. O. (1994). *Edible Coatings and Films to Improve Food Quality.* Technomic Publishing, Lancaster, PA.

Kubota, N. and Kihuchi, Y. (1998). Macromolecular complexes of chitosan. In: *Polysacchrides: Structural, Diversity and Functional Versatility* (S. Dumitrin, ed.), pp. 602–603. Marcel Dekker, New York, NY.

Labuza, T. P. and Saltmarch, M. (1981). The nonenzymatic browning reaction as affected by water in foods. In: Water Activity: Influences on Food Quality, pp. 605–650. Academic Press, New York, NY.

Lawless, H. T. and Heymann, H. (1998). *Sensory Evaluation of Food: Principles and Practices.* Chapman and Hall, New York, NY.

Lee, S. Y. and Krochta, J. M. (2002) Accelerated shelf-life testing of whey-protein-coated peanuts analyzed by static headspace gas chromatography. *J. Agric. Food Chem.* **50,** 2022–2028.

Lee, S. Y., Dangaran, K. L. and Krochta, J. M. (2002a). Gloss stability of whey-protein/plasticizer coating formulations for chocolate. *J. Food Sci.* **67,** 1121–1125.

Lee, S. Y, Dangaran, K.-L, Guinard, J. X. and Krochta, J. M. (2002b). Consumer acceptance of whey-protein-coated as compared with shellac-coated chocolate. *J. Food Sci.* **67(7),** 2764–2769.

Lee, S. Y., Trezza, T. A., Guinard, J. X. and Krochta, J. M. (2002c) Whey-protein-isolate (WPI) coated peanuts assessed by sensory evaluation and head-space gas chromatography. *J. Food Sci.* **67,** 1212–1218.

Lee, J. Y., Park, H. J., Lee, C. Y. and Choi, W. Y. (2003). Extending shelf-life of minimally processed apples with edible coatings and antibrowning agents. *Food Sci. Technol.* **36(3),** 323–330.

Le Tien, C., Vachon, C., Mateescu, M. A. and Lacroix, M. (2001). Milk protein coatings prevent oxidative browning of apples and potatoes. *J. Food Sci.* **66(4),** 512–516.

Li, P. and Barth, M. M. (1998). Impact of edible coatings on nutritional and physiological changes in lightly-processed carrots. *Postharvest Biol. Biotechnol.* **14,** 51–60.

Li, H. and Yu, T. (2001). Effect of chitosan on incidence of brown rot, quality and physiological attributes of postharvest peach fruit. *J. Sci. Food Agric.* **81(2),** 269–274.

MacFie, H. J. H. and Thomson, D. M. H. (1988). Preference mapping and multidimensional scaling. In: *Sensory Analysis of Foods*, 2nd edn. (J. R. Piggott, ed.). Elsevier Science, London, UK.

Mate, J. I. and Krochta, J. M. (1996). Comparison of oxygen and water vapor permeability of whey protein isolate and b-lactoglobulin edible films. *J. Agric. Food Chem.* **44(10),** 3001–3004.

McHugh, T. H. and Krochta, J. M. (1994). Water vapor permeability properties of edible whey protein–lipid emulsion films. *J. AOCS* **71**, 307–312.

McHugh, T. H. and Senesi, E. (2000). Apple wraps: a novel method to improve the quality and extend the shelf-life of fresh-cut apples. *J. Food Sci.* **65(1)**, 480–485.

Mei, Y., Zhao, Y. and Farr, H. (2002). Enhancement of nutritional and sensory qualities of fresh baby carrots by edible coatings. *J. Food Sci.* **67(5)**, 1964–1968.

Mei, Y. and Zhao, Y. (2003). Barrier and mechanical properties of milk protein-based edible films incorporated with nutraceuticals. *J. Agric. Food Chem.* **51(7)**, 1914–1918.

Meilgaard, M., Civille, G. V. and Carr, B. T. (1999). *Sensory Evaluation Techniques*, 3rd edn. CRC Press, New York, NY.

Miller, K. S., Upadhyaya, S. K. and Krochta, J. M. (1998). Permeability of d-limonene in whey protein films. *J. Food Sci.* **63(2)**, 244–247.

Munoz, A. M., Civille, G. V. and Carr, B. T. (1992). *Sensory Evaluation in Quality Control.* Van Rostrand Reinhold, New York, NY.

Muzzarelli, R. A. A. (1985). Chitin. In: *The Polysaccharides* (G. O. Aspinall, ed.), pp. 417–450. Academic Press, New York, NY.

O'Mahony, M. (1986). *Sensory Evaluation of Food.* Marcel Dekker, New York, NY.

Park, H. J., Chinnan, M. S. and Shewfelt, R. L. (1994). Edible coating effects on storage life and quality of tomatoes. *J. Food Sci.* **59(3)**, 568–570.

Park, S. I. and Zhao, Y. (2004). Incorporation of high concentrations of minerals or vitamins into chitosan-based films. *J. Agric. Food Chem.* **52**, 1933–1939.

Pen, L. T. and Jiang, Y. M. (2003). Effects of chitosan coating on shelf life and quality of fresh-cut Chinese water chestnut. *Lebensm. Wiss. u. Technol.* **36(3)**, 359–364.

Pérez-Gago, M. B., Rojas, C. and del Rio, M. A. (2002). Effect of lipid type and amount of edible hydroxypropyl methylcellulose-lipid composite coatings used to protect postharvest quality of mandarins cv. *Fortune. J. Food Sci.* **67(8)**, 2903–2910.

Pérez-Gago, M. B., Serra, M., Alonso, M., Mateos, M. and del Río, M. A. (2003a). Effect of solid content and lipid content of whey protein isolate-beeswax edible coatings on color change of fresh-cut apples. *J. Food Sci.* **68(7)**, 2186–2191.

Pérez-Gago, M. B., Rojas, C. and del Rio, M. A. (2003b). Edible coating effect on postharvest quality of mandarins cv. 'Clemenules'. *Acta Hort.* **600**, 91–94.

Pérez-Gago, M. B., Rojas, C. and del Rio, M. A. (2003c). Effect of hydroxypropyl methylcellulose-lipid edible composite coatings on plum (cv. Autumn giant) quality during storage. *J. Food Sci.* **68(3)**, 879–885.

Rodriguez, M. S., Albertengo, L. A., Vitale, I. and Agullo, E. (2003). Relationship between astringency and chitsoan-saliva solutions turbidity at different pH. *J. Food Sci.* **68(2)**, 665–667.

Shewfelt, R. L. and Bruckner, B. (2000). *Fruit and Vegetable Quality: An Integrated View.* Technomic Publishing, Lancaster, PA.

Stone, H. and Sidel, J. (1993). *Sensory Evaluation Practices*, 2nd edn. Academic Press, San Diego, CA.

Sumnu, G. and Bayindirli, L. (1995). Effects of coatings on fruit quality of Amasya apples. *Lebensm. Wiss. u. Technol.* **28(5)**, 501–505.

Trezza, T. A. and Krochta, J. M. (2000). Color stability of edible coatings during prolonged storage. *J. Food Sci.* **65(7),** 1166–1169.

Wu, L. S. (1989). *Product Testing with Consumers for Research Guidance.* ASTM STP 1035, American Society for Testing and Materials, Philadelphia, PA.

Wu, L. S. and Gelinas, A. D. (1992). *Product Testing with Consumers for Research Guidance: Special Consumer Groups*, 2nd volume, ASTM STP 1155. American Society for Testing and Materials, Philadelphia, PA.

Wu, Y., Weller, C. L., Hamouz, F., Cuppett, S. and Schnepf, M. (1999). Moisture loss and lipid oxidation for precooked ground beef patties packaged in edible starch-alginate based composite films. Paper No. 11B-21, 1999 Annual IFT Conference, 23–26 July, Chicago, IL.

Yaman, O. and Baymdirh, L. (2001). Effects of an edible coating, fungicide and cold storage on microbial spoilage of cherries. *Eur. Food Res. Technol.* **213,** 53–55.

Yaman, Ö. and Bayondrl, L. (2002). Effects of an edible coating and cold storage on shelf-life and quality of cherries. *Lebensm. Wiss. u. Technol.* **35(2),** 146–150(5).

Zhang, D. L. and Quantick, P. C. (1998). Antifungal effects of chitosan coating on fresh strawberries and raspberries during storage. *J. Hort. Sci. Biotechnol.* **73(6),** 763–767.

Commercial aspects of new packaging technologies

PART

5

Commercial uses of active food packaging and modified atmosphere packaging systems

25

Aaron L. Brody

Introduction

Addressing the complex subject of active packaging from a commercial perspective is a challenge. The topic of modified atmosphere packaging is confounded by controlled atmosphere packaging, including largely a storage and transportation technology, and vacuum packaging, which is sometimes not regarded by purists as a part of the modified atmosphere packaging family. Active packaging's commercial applications are complicated by intelligent packaging, which is not included because mostly it falls under the realm of distribution logistics rather than of food science and technology.

This chapter is divided essentially into two parts, one section is a brief on modified atmosphere packaging and the other is a 2004 update on active packaging. The paucity of commercial applications for edible coatings suggests not addressing this family of technologies in this context. On the other hand, this chapter includes modified atmosphere packaging, which was effectively born during the 1950s and long ago took a major commercial position, now exceeding canning, freezing or aseptic packaging in volume in the United States and Canada.

Innovations in Food Packaging
ISBN: 0-12-311632-5

Active packaging

Although some active packaging technologies have been commercialized for more than 20 years, commercial applications in the United States to date have been relatively disappointing. Today, active packaging technologies are largely developmental and subject to both incremental and breakthrough change, with only about 4000–4500 million packages (less than 1% of all packages) using an active packaging technology in the United States in 2003. Developments appear to project a 20% compound annual growth rate for the 2004–2009 period.

As indicated above, active packaging refers to packaging materials that sense environmental change and respond by changing their properties, on their own, to better retain product safety, quality, shelf life, and functionality, beyond the protection of the base package structure against the varying distribution environments. Intelligent packaging refers to structures that sense and signal a variation of the conditions of the package or the contained product, but do not change the properties of the package.

Active packaging encompasses about ten broadly defined functions, each applying a different technology. Within most of the technologies there are multiple methods to obtain the desired result. Active packaging functions and technologies include:

1. Moisture control
   - Water/purge or fluid squeeze or drip from the product
   - Water vapor in equilibrium with the water within the product
   - Relative humidity control
2. Oxygen-permeable films to obviate respiratory anaerobiosis
3. Oxygen scavengers/absorbers
   - Ferrous iron
   - Ascorbates
   - Nylon MXD6
   - Unsaturated ethylenics such as butadiene
   - Benzoacrylates
   - Antioxidants
4. Oxygen generators
5. Carbon dioxide controllers
   - Removers
   - Generators
6. Odor controllers
7. Flavor enhancement
8. Ethylene removal
9. Antimicrobials
10. Microwave susceptors – usually controlled by a user rather than by sensing a product, and thus not included in this chapter.

Depending on the technology, active packaging can sense changes in the internal package, in external temperatures, or time; or can sense changes in the gaseous, physical, chemical, or biological environments. The package material and/or the active packaging device responds to these changes by altering its properties to effect the change.

As shown in Table 25.1, three active packaging technologies appear to account for almost all of the active packaging unit usage – oxygen scavengers, in several forms (50%); desiccants, also in several forms (34%); and purge absorbers (14%). By 2006, these three technologies will decline slightly relative to others, accounting for about 95% of the active packaging units used. Oxygen scavengers, the largest active packaging technology in terms of volume, will have the lowest average annual growth rate at about 12%. Relative humidity controllers will have the highest growth rate of more than 58%, albeit from a very small base of only 10 million packages in 2003.

More than 50 end-use product applications for active packaging have been identified to create Tables 25.1 and 25.2. Although most end-use applications have focused on only one active packaging technology, several applications – including bakery products, processed meat, poultry, and military rations – currently apply more than one active packaging technology.

Oxygen scavenging packaging

Active packaging has enjoyed good growth over the past ten years, to 3500–4500 million units in 2003. Once regarded as a failure in the United States, the oxygen scavenger market grew 15-fold during the 1990s and currently appears to be on an exponential growth curve. Recognition and appreciation by food packaging technologists of the ability of oxygen scavengers to better retain the quality of packaged products has led to a multiplication of the application of basic inorganic scavenger-containing sachets (e.g. ferrous salts) and the incorporation of organic oxygen scavengers such as nylon MXD6 and benzoacrylates directly into package materials.

The concept of incorporating oxygen scavengers directly into plastic materials has challenged polymer and packaging technologists for more than 20 years. Early problems included regulatory concerns over the involved compounds and obvious secondary

Table 25.1 US active packaging applications, 2003 (units)

Application	Million units (2003)	Annual growth % 2003–2006	% of units 2003
Moisture controllers*			
Purge absorbers	700–750	38	15
RH controllers	20–25	59	1
Desiccants	1000–1500	15	30
Oxygen scavengers	1500–2000	12	44
Controlled gas permeability	40–45	27	1
Ethylene absorbers	20–25	15	<1
Antimicrobials	0		0
Odor controllers	0		0
Total	3400–4400	19	100

* Excludes pulp pads.
Source: Packaging/Brody, Inc.

Table 25.2 Commercial and commercial trials for food applications for active packaging technologies: United States (current or past)

End-use applications	Purge	High RH controller	Desiccants	Gas permeability	O₂ scavengers	CO₂ controllers	Add aroma	Remove odor	Flavor enhancer	Ethylene remover	Antimicrobial
Bakery goods					X	X					X
Bakery goods, bread					X						
Bakery goods, bread, gluten-free					X						
Bakery goods, soft pretzels					X						
Beer bottle, closure liners					X						
Beer bottles					X						
Beverage, fruit											
Cereal products											
Cheese, block											X
Cheese, shredded											X
Cheese, slices						X					X
Coffee, roasted and ground					X			X			
Condiments, fluid					X						
Dry foods, dry milk solids			X								
Dry foods, instant coffee			X								
Fish, fresh – trays	X			X							
Fish, MAP	X			X		X					
Fruit, bananas				X						X	
Fruit, berries		X								X	
Fruit, climactic produce, berries				X						X	
Fruit, fresh				X							
Fruit, fresh-cut – melon/watermelon	X			X						X	
Fruit, fresh-cut – trays/pouches	X										
Fruit, strawberries	X	X		X							X
Meat, beef cuts	X										X

Meat, dry, bacon
Meat, dry, jerky
Meat, ground beef patties
Meat, modified atmosphere packaged
Meat, sausage collagen casings
Meats, processed/cured
Military rations
Military rations, bakery, long shelf life
Military rations, nut products
Milk, dry
Nuts
Oils, edible
Pasta, fresh
Peanut products
Pet food, soft, moist
Poultry, fresh – pouches
Poultry, fresh – trays
Poultry, modified atmosphere packaged
Prepared foods, reduced oxygen packed
Salty snacks
Tomato sauces
Vegetable, fresh-cut – pkgs, tomatoes
Vegetables, fresh
Vegetables, high respiration, asparagus
Vegetables, high respiration, broccoli
Vegetables, high respiration, cauliflower

Source: Packaging/Brody, Inc.

Table 25.3 Oxygen-absorbing technologies, 2003 (units)		
Technology	Million units (2003)	Projected annual growth rate 2003–2006
Oxygen absorber in bottle	840	13%
Cap/liner/lidding	612	15%
Oxygen absorber film	105	54%
Sachet	283	−9%
Total	1840	12%

Source: Packaging/Brody, Inc.

effects. Previous package structures were able to approach actually mimicking the zero oxygen permeation of sealed glass. When coupled with the later introduction of oxygen scavenger-containing oxygen-barrier closures, the resulting bottles and jars demonstrated no measurable oxygen movement. Such outcomes were heralded as major breakthroughs to the packaging of highly oxygen-sensitive products such as beer and fruit beverages in lightweight, non-breakable, semi-rigid plastic structures. A multi-billion-package market represents a highly inviting investment opportunity for polymer manufacturers, package converters, and others. Currently, depending on how oxygen scavenging is defined, more than half a dozen competitive systems are offered, with more certain to come.

Ferrous iron oxygen scavengers in in-package sachets are often dismissed as archaic, but retain a leadership position in a wider variety of packaged products. More reliable adherence to package interiors and fabrication in flat label form has helped to grow their position. Ferrous iron-based oxygen-scavenger sachets have proved to be functional, reliable and effective. Despite its dark color, ferrous iron is a preferred oxygen scavenger for incorporation into plastics, which suggests a continuing future.

As costs for oxygen scavengers decline, and when the hurdles have in large part been overcome, the use of oxygen controllers such as scavengers should increase for packaging oxidation-sensitive foods. Adverse biochemical oxidation reactions and aerobic microbiological spoilage growth can be retarded in a host of foods, including fruit beverages, tomato products, home meal replacements, shredded cheese, and processed meats, whose distribution life is now limited. Gradually, the use of passive oxygen barriers will be enhanced by active oxygen scavengers to enhance quality retention.

Table 25.3 shows the four major oxygen-absorbing technologies, and Table 25.4 lists the major end product users of oxygen-absorbing technologies in the US.

High gas-permeability package structures

Probably the best example of classical active packaging is Landec's Intellipac®. This material senses temperature increases and, unlike passive packaging, markedly increases its gas permeability as a result. In modified atmosphere packaging (MAP), increases in temperature lead to increases in product respiration rate, which can consume the oxygen and lead to anaerobic conditions that in turn can produce off-flavors. The high

Table 25.4 US oxygen scavengers, 2003 (units)	
Product	**Million units (2003)**
Beer	900–1000
Juice	300–350
Ketchup	200
Beef jerky	120
Case-ready meats	75
Pasta	65–75
Processed meats	60–70
Military rations	40
Gluten-free bread	15–20
Prepared foods	10
Cheese	20
Total	1800–2000

Source: Packaging/Brody, Inc.

gas permeability packaging concept might possibly be applicable to low-acid wet food products vulnerable to pathogenic anaerobic microbial growth, e.g. home meal replacements, thus obviating this deterrent to category growth.

High relative humidity control

The control of in-package moisture is practically a very useful commercial technique, because:

1. Moisture control, although long ago dismissed as a casual or matter-of-fact method, is probably the leading active packaging technology, and will remain so
2. Moisture control is being coupled to other active packaging technologies, such as oxygen scavengers and antimicrobials, to produce synergistic active packaging.

Always a source of spoilage and a poor, messy appearance, purge or liquid drip from fresh and minimally processed foods has been linked to the adverse effects of relative humidity in closed-package environments. Control of purge by gel-forming polymeric absorbers has been demonstrated in commercial testing for the purpose of measurable enhancement of quality retention. Among the consequences is that these active packaging technologies are moving from independent pulp pads or "diapers" in the tray base to functional incorporation into plastic package tray and pouch structures; further technology development will concentrate on hiding the purge and controlling internal package relative humidity.

By mixing moisture-activated antimicrobials into the gel-forming polymers, the preservation benefits of the active packaging may be multiplied. Beyond the dual effect, moisture-activated oxygen scavengers may become a third material to function in synergy – to provide controlled relative humidity, oxygen, and microbial growth within the package. Food products such as fresh poultry can be envisioned to benefit from this potential enhancement of hurdle technologies.

Low relative humidity control

At the other end of the moisture spectrum of humidity controlling technology are dry products sensitive to even minute quantities of environmental moisture. While effective water vapor barriers are used for the packaging system, their functionality is exponentially increased by the presence of desiccant sachets and/or cartridges. Such devices are limited by capacity, proximity, and lack of persistence.

Probably one of the most significant active packaging technologies is the concept of developing high surface-area channels on interior package surfaces on which desiccants are spread to capture moisture. This structure is so powerful that it can continue to remove water vapor even after the package has been opened and exposed to air, thereby providing low humidity conditions after reclosure.

Still in its infancy, the notion of exponentially increasing the internal package surface area and coupling this with active packaging technology – in this instance, desiccants – represents a strong argument for the growth of active packaging. Although confined today to desiccants, it is possible that this could be extrapolated to combine moisture-triggered oxygen scavengers with antimicrobials to form an extraordinary food preservation hurdle technology. As food scientists and technologists evolve from process to packaging technologies to enhance quality retention, hurdle or combination technologies such as active packaging on internal surfaces will emerge as effective tools to achieve this objective.

Antimicrobials

Headlined and apparently used widely in Japan, in-package antimicrobials have not been applied in the United States for a variety of reasons, including:

- Regulatory concerns
- The fear of adverse secondary effects
- A lack of evidence on the efficacy
- The fact that too many are contact functional only, limiting their potential
- Cost.

Experience in Japan and, to some degree, in Europe has not translated to the United States. Too many antimicrobials, such as silver salts and Microban®, function only in contact with micro-organisms present on smooth food surfaces. Antimicrobials that are claimed to function at a distance, such as chlorine dioxide or ethanol, have been limited by secondary effects and a narrow spectrum of activity. There is a great deal of excitement regarding the concept of building microbiostatic and/or microbiocidal agents into package materials to extend microbiological shelf life. To date, food packagers have been hesitant with proceeding to commercialization. Controlling visible fungal growth on fresh fruit or cheese would be a desired result for those organizations that become the early adopters and lead the category into active packaging.

Odor controllers

Odor control derived from active packaging has been the subject of considerable philosophical debate. If the odor of a packaged product is enhanced by a desirable aroma, the consumer may be deceived into the perception that the product is better than it is. When food and food-packaging scientists and technologists employ the flavor add-back in instant coffee or juice concentrate, there have been no questions about "synthetically enhanced" flavor. However, when the source of the aroma is the package material, questions may arise. The opposite end of the odor control spectrum is the overt removal of confinement or even minor spoilage odors. Some argue that minute sulfurous or amine odors are trivial and indigent to many protein foods, and are therefore harmless; thus their removal should be a benefit since some consumers might reject the product for no good reason. Lipid oxidation odors, such as aldehydes or ketones at low levels, are generally inconsequential, and their removal should be equally beneficial for consumers. On the other hand, some experts argue that food spoilage odors – including sulfide and lipid oxidation odors – are overt signals to consumers of an incipient hazard, and should not be artificially suppressed. Since the driving forces for odor-control systems have been from both users and technology suppliers, the question will not be answered until consumers are educated on the issues and make their own decisions.

Conclusions on active packaging

Active packaging is very much in its infancy, with both suppliers and potential users needing and wanting to spark a broad front of new applications. However, too many elements must be comprehended and remain to be refined. The result so far has been slow and uneven growth in the segments.

Active packaging today is largely supplier-driven. Furthermore, many of the technologies originated and are being nurtured in offshore locations, while North American organizations puzzle over the benefit versus cost. Unlike Europe, where independent central laboratories such as the Campden & Chorleywood Food Research Association Group (UK) or TNO (The Netherlands) evaluate active packaging technologies, no such agencies exist in the United States or Canada, and North American-based organizations such as universities have barely touched on active packaging technology.

The short-term potential for any single active packaging technology capturing more than a handful of niche applications today is small. Several large petrochemical companies, such BP Chemical and Chevron Phillips, are drivers in unsaturated hydrocarbon oxygen-scavenger technology. A long-time pioneer in oxygen scavengers, Mitsubishi Gas Chemical (Japan) still leads in ferrous iron sachets, with Multisorb Technologies (of Buffalo, NY) close behind. The Japanese company is also the supplier of the oxygen-scavenging nylon MXD6. Multisorb Technologies, whose sole business is active packaging, is by far the leader in desiccant sachets. A large specialty chemical company, Ciba, is also in this market.

The package material converters that are active in active packaging include Cryovac Sealed Air, and Cadillac, in flexible packaging; Continental PET Technologies, Amcor,

Pechiney, and Crown Cork & Seal, for bottles; and Silgan, for plastic cans. The numbers are too few for flexible packaging and cans, but there is the opportunity for strong competition in bottles if a market should emerge.

Mainstream moisture-absorbing pads are not regarded as active packaging. The new super-absorbent gels fall into the definition of active packaging and belong to entrepreneurial firms such as Maxwell-Chase, which are still seeking their market segments. Desiccant sachets/cartridges are produced by Multisorb Technologies, SÜD-Chemie, and United Desiccants, which are all relatively modest sized companies. The progenitor of the in-package desiccant is CSP Technologies (Capitol Specialty Plastics Technologies), which is a relatively small company whose core competency is in injection molding of small vials for M&M® milk chocolate candies and for pharmaceuticals.

Odor controllers have been pushed by two different operations; DuPont, and Cellresin Technologies (a development company).

Most of the proposed in-package antimicrobials come from relatively small entrepreneurial development companies, such as AgION, Japan's Certex, and Bernard Technologies, plus a small operation in the textile giant Milliken.

The variety of organizations involved in active packaging might be classified as heterogeneous. The active packaging technology market is dominated by start-up suppliers and developers, with a sprinkling of specialty operations in large chemical companies, apparently targeting the potentially lucrative plastic beer bottle market. There are too few converters actively studying. Although the history dates back more than a decade, few (if any) organizations have entered and subsequently dropped out of the development projects of active packaging technologies. The array of technologies continues to attract a range of participants.

Modified atmosphere packaging

Although modified atmosphere and vacuum-packaged foods are not highly visible in food marketing, they constitute a substantial and growing proportion of US food supplies, mostly in distribution packaging. Vacuum packaging, modified atmosphere packaging (MAP) and controlled atmosphere packaging (CAP) are all regarded by most observers as part of the same technology.

History of CAP and MAP

For decades, food and food-packaging technologists have been taught the principles of microbial growth and retardation, including reducing temperature to inhibit microbial activity rates and applying heat to destroy micro-organisms. Micro-organisms slow down their respiratory and growth rates when oxygen is reduced and the respiratory gases such as carbon dioxide are increased. Aerobic respiration is the basis for the degradation of plant materials after removal from the growing plant. Plant-product respiration is slowed by increasing carbon dioxide and reducing oxygen and ethylene.

In the early years of the twentieth century, fresh meat shipped from the Antipodes to England was sometimes chilled by solid carbon dioxide. Mutton, beef, and lamb in the shipment were noted to have shelf lives longer than carcass meat held under wet ice only, and this phenomenon was later attributed to the disruption of the gaseous atmosphere. During the 1930s, fresh apples and pears were placed in enclosed warehouses. The natural respiratory reaction of the fruit reduced the oxygen and increased the carbon dioxide within the storage areas sufficiently to slow the respiration rate markedly. The stored apples or pears could be consumed six months after the original harvest – an extension of about double the normal chilled storage shelf life. The use of natural respiratory-controlled storage for apples and pears was expanded rapidly during the 1950s in both New York and the Pacific Northwest.

Tectrol

During the 1950s and 1960s in the US, Whirlpool Corporation's food technologists developed methods to control the atmospheres surrounding meat, fruit, and vegetable products. The concept was adapted to the bulk and industrial distribution of fresh fruit and meat products. The name Tectrol (**T**otal **E**nvironmental **ConTROL**) was applied to total control of warehouses. By the mid- to late-1960s, hundreds of apple and pear warehouses throughout the United States and Europe were equipped with Tectrol systems to extend the shelf life of the fruits.

Transfresh

The Tectrol concept was spun off to the produce grower, Bruce Church, in California, and reorganized under the name Transfresh. The Tectrol/Transfresh concept has since been expanded to transport containers in order partially to control the internal gas content surrounding the contents. The Tectrol/Transfresh system is now used to deliver bulk and consumer-size packages of fresh-cut vegetables to HRI and retail outlets, and pallet loads of strawberries to subdistributors. In 1990, the process was refined into complete control of shipboard container loads. A further spin-off from Transfresh is Fresh Express, the largest producer and packager of retail and foodservice fresh-cut vegetables.

Cryovac

During the 1960s, barrier shrink-film vacuum packaging that was used to protect frozen turkeys in distribution was applied to fresh red meat by the Cryovac organization, which also gave its name to the process. Under reduced oxygen, the growth of micro-organisms responsible for meat spoilage are retarded. Simultaneously, however, the original purple myoglobin color of fresh red meat is retained. During the mid-1960s, Cryovac and Iowa Beef Packers (now IBP, Tyson) joined in the concept of slaughtering cattle at a central location and, rather than shipping in hanging carcass form, breaking the meat into primal and sub-primal cuts. These products were subsequently vacuum-skin packed into high gas-barrier multiple-layer plastic film bags. The reduced oxygen in the vacuum pack retarded microbial growth and oxidation of fat. Furthermore, the bag did not permit the passage of water vapor, and so weight loss due to evaporation was significantly reduced. These filled bags were then packed in water-resistant corrugated fiberboard shipping cases (hence the name "boxed" beef) for shipment to foodservice operations. By the mid-1970s, this process had been successfully introduced to retail supermarkets, thus reducing the number of butchers required to serve consumer-size cuts, and subsequent packaging.

Processed meats

During the early 1950s, vacuum packaging of cured meats such as frankfurters, hams, ham slices, bologna, etc. was developed using nylon-based thermoforming webs of flexible oxygen-barrier materials. The processed meat was placed within the formed web and heat-sealed with a second gas-barrier material while drawing a vacuum or, later, back-flushing with inert gas. This concept was later applied to employ carbon dioxide flushing to displace oxygen and to prolong the quality retention of aged cheese.

Bakery goods

In the mid-1960s, the British Flour Milling and Bakery Research Association investigated the use of elevated carbon dioxide to retard mold growth on the surfaces of perishable bakery products. After the late 1970s, when the West German Government instituted a regulation to declare all food additives on the package label, many commercial bread bakers opted for air-sealed barrier packages containing elevated carbon dioxide and reduced oxygen to permit the desired shelf life of preservative-free products. In the mid-1980s, the principles were applied to meat-filled sandwiches in both the United States and Canada. Those sandwiches containing cured meat fillings, e.g. sausage, retained their quality under modified atmospheres for many weeks (at chilled temperatures), stimulating a small, but solid and growing, product niche. Instructions to consumers are given to heat the product before eating, which is a simple process that partially overcomes the staling effects in the bakery product over long storage.

Retail red meat

During the 1960s, investigations by Kalle, a German plastic film converter, demonstrated that fresh red meat could be preserved under refrigeration with its desirable "cherry red" oxymyoglobin color if a high carbon dioxide/high oxygen environment were present. The presence of high oxygen levels was contrary to normal practice because common belief had been that high oxygen accelerates microbial growth, respiration, and fat oxidation. Elevated carbon dioxide was shown to be sufficiently effective in retarding microbial growth, and so the adverse oxidative effects of the high oxygen could be compensated. These findings were put into practice in West Germany, where thermoform/vacuum/gas flush-seal packaging systems with high oxygen-barrier materials were employed.

Pasta

During the 1980s, a small New York store initiated the integration of excellent sanitation of both ingredients and process plus modified atmosphere within sealed plastic packages to enable it to distribute 30% moisture pasta under chilled conditions. These measures permitted prolongation of safe quality up to 40 days. More importantly, this concept encouraged others to apply the principles for entrees, side dishes and meat/fish salads.

Red meat

Of the 39 000 million pounds of red meat produced in the US annually, approximately 63% is beef, 36% is pork, and the remainder is lamb, veal, etc. About two-thirds of American pork enters further processing (to form cured meats such as hams, bacon, and sausages). Only 12% of beef is processed further. Of the 21 000 million pounds of beef that is not processed further, approximately 40% (or 8500 million pounds) becomes ground beef. Therefore, about 5000–6000 million pounds of beef cuts are generated, and overwhelmingly cut and packaged at retail grocery level. Most of this beef is cut and air-packaged in the supermarket backroom, employing expanded polystyrene trays with plasticized PVC flexible-film overwrap plus paper labels. In addition, nearly 5000 million pounds of ground beef is marketed through retail stores, largely in expanded polystyrene trays with PVC wraps. Almost all of this ground beef is coarsely ground at federal government-regulated factory level. It is distributed under low-oxygen conditions, packaged in flexible barrier-liner chub casings, and sent under refrigeration to retail stores, where final grinding and retail packaging occurs. Almost 500 million pounds of beef is finely ground in factories into barrier-film chub packaging, also under low-oxygen conditions, for retail sale.

Table 25.5 depicts the current case-ready red meat situation – about 2000 million package units out of nearly 9000 million total fresh red meat packages, or slightly more than 20% market share in 2003. It incorporates the data of chub packages of ground beef, sub-primal cut pouches sold at retail, and expanded polystyrene trays overwrapped with PVC film that have been packaged at a remote location but are distributed with no special secondary packaging. To date, a relatively small proportion of intact muscle beef cuts have converted into case-ready, and these are found largely at Wal-Mart Super Center outlets.

Perhaps up to 100 million pounds annually from all the major pork packers, or about 3% of the total, is centrally packaged, including master packaged and vacuum packaged, some with added antioxidants.

Table 25.5 Case-ready meat – packaging units (million packages, 2003)

Case-ready	
Barrier trays – elevated oxygen	
Expanded polystyrene-barrier	185
Solid plastic	955
Overwrap trays – with masterpack	
Expanded polystyrene	630
Overwrap trays – no masterpack	700
New technology	60
Vacuum/reduced oxygen	
Chubs	300
Pouches for subprimals	50
Total	2880

Source: Brody and Bieler (2001). *Case Ready Meat Packaging*. Packaging Strategies. Reproduced with permission (see www.packstrat.com).

Poultry

It is remarkable that case-ready red meat technology has effectively transferred to the centralized packaging of poultry. Nearly one-third of the rapidly expanding US poultry market – which now exceeds beef – is centrally prepackaged and branded. Until the early 1980s, nearly all poultry products were dressed whole birds packaged in ice in wire-bound wood crates. In the retail store backroom, poultry were removed from the ice by the butcher, cut, and repackaged in expanded polystyrene trays with plasticized PVC film overwrap, before adding an adhesive store label. As a result of the introduction of chill pack followed by MAP in master packs, the concept of "shelf-ready" poultry emerged, where the retailer removes the factory prepackaged poultry from the master pack and places it on the shelf. Today, 2000 million pounds of poultry are master packed, about half of them are under modified atmosphere in the PVC film-wrapped expanded polystyrene trays.

Precooked and fresh-ground poultry represents a 200+ million pound market, and 20% of this is packaged on thermoform/vacuum/back-gas flush machines such as Multivac or Tiromat, using barrier expanded polystyrene/ethylene vinyl alcohol film-laminated trays with heat-sealed flexible lids. Most of the remainder is packaged in conventional expanded polystyrene trays overwrapped with co-extruded barrier film on horizontal flow wrapping equipment. The internal atmosphere for precooked poultry is high carbon dioxide, or high oxygen/carbon dioxide for fresh ground poultry, to retard microbial growth and retain red color. Partly as a result of employing centralized MAP coupled with controlled distribution and modern marketing, poultry has overtaken red meat as the animal protein source of choice.

Fish

Per capita consumption of seafood remains low at close to 14 pounds, with relatively limited prospects for major growth in the near future. Owing to concerns about the potential growth of pathogenic anaerobes such as type E *Clostridium botulinum*, which can grow and produce toxin at low temperatures, regulatory officials have discouraged reduced oxygen packaging of fish. Some seafood products are vacuum skin packaged and frozen for retail slackout, which gains some of the benefits of low-oxygen packaging and somewhat skirts the regulations.

Fruits and vegetables

The strongest growth segment for MAP from the 1990s through 2003 has been fresh-cut produce, also regarded as the fastest-growing food product category in the country. Starting in the mid-1980s with vacuum packaging of fresh-cut lettuce for use by quick foodservice chains in their penetration into the salad market, expansion has continued into the US retail market, with $4 billion retail value in 2003 – representing nearly 10% of fresh produce sales. The foodservice market is valued at about $10 US billion. The major product within the category, with about two-thirds of the volume, is fresh-cut lettuce valued at $2.4 US billion, with up to 15% of the produce in some form of packaging under MAP. Other important products include broccoli and cauliflower florets, diced onions, baby carrots, sliced peppers, and a series of fresh-cut vegetables packed with meat, pasta, croutons, salad dressing, etc. to create a complete salad-based meal.

From a technological perspective, most fresh-cut produce is cut, cleaned, and cooled using chlorinated water. It is packaged on vertical form–fill–seal machines for retail distribution, and on preformed bag fillers such as CVP Systems or M-Tek for food-service distribution. Relatively little is packaged with gas mixtures, but the bag systems evacuate the air so that the product contents consume the residual oxygen and produce carbon dioxide. Anaerobic respiration is minimized by employing high surface-to-volume ratio pouches in the case of bulk packages, and high gas permeability plastics for both bulk and retail size pouches. Gas permeabilities are around $3000 \, \mathrm{ml \, m}^{-2}$. It is axiomatic that the materials have low water vapor permeabilities and that generally, for the retail sizes, antifog agents are incorporated in the film to permit visibility.

One major category of MAP produce is pallet wrapping of prepackaged strawberries in both Florida and California. In the latter, almost all of the produce exported from the state, either to other states or to other countries, is shrouded in high-density ethylene vinyl acetate/polyethylene film, evacuated to give a vacuum, and the air replaced by a mixture of carbon dioxide and nitrogen to suppress mold growth. Probably as many as 100 000 pallets annually are treated to effect the modified atmosphere condition within the pallet wrap.

For bulk packs, mono-layer high-density ethylene vinyl acetate films are often employed, although CVP Systems uses co-extrusion to achieve its requirement of gas permeabilities. For consumer packs, laminations of styrene block copolymer and metallocene-based low-density polyethylene, or polypropylene plus metallocene-based polyethylene, are used. The California packer Fresh Western uses FreshHold mineral-filled polypropylene over large orifices in the polyethylene film substrate to achieve high oxygen permeation. Increasingly, co-extrusions incorporating metallocene polythylenes are being applied to gain some economies.

Cut pineapple is largely packaged with high carbon dioxide levels, vacuum packaging, or in air followed by passive equilibrium atmosphere. Cut melons (which are low acid), table grapes, and fresh-cut apples can be treated with chemical antioxidants such as ascorbic acid to retard browning. Modified atmosphere packaging has been tested with almost every fruit consumed as convenience snacks, such as strawberries, peaches, and pears. Almost $250 million worth of MAP fruits were retailed in 2003.

Soft bakery goods

Breads, rolls, cakes, and muffins are subject to rapid quality loss, with the most prevalent deteriorative vectors being moisture loss, staling, and mold growth. Staling is a partially reversible crystallization of starch leading to texture hardening. Microbiological deterioration in soft bakery goods is mostly mold growth on surface and crumb. Moisture loss can be retarded by packaging in low water-vapor permeability materials, such as low-density polyethylene film. Use of such materials retains water vapor within the package, but can create conditions optimal for mold growth. Extending the shelf life of soft bakery products can:

- Reduce the number and percentage of returned goods
- Extend the distribution range of bakery goods

- Reduce the frequency of delivery
- Permit a bakery to produce a broader range of products.

Reduction of oxygen and its replacement with an inert atmosphere is ineffective unless the oxygen content of the internal atmosphere is reduced to below 1%. The most common reduced-oxygen packaging system for bakery goods uses horizontal form–fill–seal with PVdC-coated oriented polypropylene film in a single web. This bakery packaging system requires extensive gas flushing while forming and sealing the flexible pouch. The system is used for relatively simple breads, such as German rye meal. In vacuum-packaging systems, twin-web thermoform-vacuum/gas flush–seal systems require more expensive packaging materials and equipment than horizontal fill–seal systems, but deliver better seals. Multivac and Tiromat thermoform/vacuum/gas flush-seal equipment is used for MAP of specialty goods such as crumpets. An interesting MAP product concept is brown-and-serve French bread. The commercial baker partially bakes the loaves and places them into a thermoformed plastic cavity on a thermoform/vacuum/gas flush–seal machine. While still hot, and expelling the carbon dioxide of baking, the packages are closed and cooled. The carbon dioxide-saturated packages a shelf life of over 3 months at ambient temperature. The consumer then places the loaf in a conventional oven, where a brown crust is developed and the crumb remoisturized. The result is comparable to a freshly baked French baguette.

In Canada, crumpets (similar to English muffins) are modified atmosphere packaged on thermoform/vacuum/gas flush–seal equipment, with carbon dioxide back-flushed, on a commercial scale. Sandwiches are packaged under vacuum using similar systems, but the sandwiches are distributed under refrigeration to preserve the fillings.

Several American companies are packaging sandwiches for refrigerated distribution on a regional basis. These products may appear in cured meat or special refrigerated displays, with or without expiration dates. Perhaps the most widely commercialized applications are the little-known MAP of sausage and biscuits, and of precooked hamburgers. About 90 days of actual refrigerated shelf life is achieved by MAP on Multivac thermoform/vacuum/gas flush–seal equipment, with a target expiration date of about 45 days. Since sausage is cured meat, problems of microbiological safety are minimized. The instruction to the consumer to heat the product drives moisture from the meat into the biscuit to refresh it. All such products are nationally distributed in the US.

Prepared foods

The arena of prepared foods under MAP is led by the very successful high-moisture pastas that are now packaged, with pasta sauces, under reduced oxygen by virtue of hot filling. Probably $1.5 US billion worth of such products are in widespread chilled distribution, but after significant growth over several years the category appears to have reached maturity. The three major technologies employed include oxygen replacement, in which nitrogen is the sole gas injected after evacuation. This system, used by Kraft for its DiGiorno brand pastas, also incorporates internal oxygen scavenger sachets affixed to the interior of the bottom tray to remove residual oxygen. This system claims 90 days of actual shelf life under refrigeration.

The second system, used by most producers, is displacement of the internal environmental oxygen with a mixture of at least 25% carbon dioxide plus nitrogen with no further active packaging. This system is claimed to be microbial-satisfactory for up to 90 days of refrigerated shelf life. All systems rely on the sanitation of the initial product, process, and packaging, with chilling. They further integrate product formulation to ensure a sufficiently low water activity to minimize the likelihood of the growth of pathogenic anaerobic micro-organisms. This product category is one of several in which natural antimicrobials and/or natural antioxidants are incorporated into the formulation to complement the basic food preservation actions.

The third system, applied to the sauces, is hot filling, in which the condensation of steam generates a vacuum that inhibits aerobic spoilage micro-organisms together with the heat and subsequent refrigerated distribution.

Most pasta is packaged on thermoform/vacuum/gas flush–seal machines such as Multivac or Tiromat, using semi-rigid PVC as the thermoformed web and a lamination of polyester/PVdC and polyethylene as the flat sealing web. The pasta sauces are usually in polypropylene tubs with flexible polypropylene heat-seal closures. Some pasta sauces, as well as many other soups and pumpable foods, are packaged in flexible barrier pouches fabricated from the structures of polyester and/or nylon plus linear low-density polyethylene. These pouches may be distributed in paperboard sleeves to help protect them from abuse and damage. Beyond pasta and pasta sauces, a conservative estimate would place the size of the prepared foods MAP market at well over 100 million units and growing at more than 9% annual rate.

Another category for prepared foods under reduced oxygen and chilled distribution is cook–chill bulk packaging for pumpable foods such as pasta sauces, soups, and chili. The product is packaged hot into nylon/polyethylene pouches capable of holding about one gallon. The hot packaged product is immediately chilled in cold water and distributed at temperatures of about 0°C to satellite foodservice outlets for reheating and serving. This 10+ million unit market is technologically very similar to the sous-vide. Most cook–chill is made and applied internally by mass feeding operations and central commissaries, and so is under a single control throughout its short distribution cycle. Retail consumers almost never see or are exposed to cook–chill pouches of food, and so specific rules governing this technology alone are not in effect.

Related to cook–chill technology, there are products that are inserted into high oxygen-barrier pouches, sealed, cooked to pasteurization temperatures, chilled, and subsequently distributed under chilled conditions to the many delicatessen operations. These products are effectively vacuum packaged, and must be chilled to ensure microbial safety and retail quality. This product category includes roast beef, turkey, and similar solid pack products in which cooking and pasteurization are effected by the same cook cycle.

More than 90% of the beef in the United States is shipped from meat packers to retailers and foodservice operations in the form of vacuum-packaged primal cuts. About half of all fresh poultry in North America is master packed in bulk under modified atmospheres for distribution to retail grocery and foodservice outlets. A growing retail category is precooked poultry packaged under modified atmosphere or vacuum, and marinated and precooked poultry packaged under vacuum. Virtually all cured or processed meat and cured cheese products in retail distribution are packaged under either vacuum or inert (i.e. nitrogen) atmospheres. Most cured and cooked meats for in-store delicatessens are vacuum or inert atmosphere packaged. Twenty percent of all lettuce and more than 80% of all California strawberries are distributed under modified atmosphere conditions. About one-third of all retail supermarkets display case-ready fresh beef, mostly ground beef. Sausage and biscuits, sandwiches, and lunch kits are atmosphere packaged. Fresh pasta is distributed under MAP. The spectrum of vacuum packaging, CAP, and MAP extends from fresh red meat through precooked meals and embraces most of the categories of perishable and minimally processed products currently in the food chain.

The magnitude of the MAP food market is approaching 10 000 million package units, containing nearly 100 000 million pounds of food contained. The growth is 5–10% annually. The most significant MAP items in North America are fresh-cut vegetables, 30% moisture pasta, sausage and biscuits, lunch kits, and precooked poultry.

Further reading

Anonymous (1988). *Modified Atmosphere Packaging: The Quiet Revolution Begins.* Packaging Strategies, West Chester, PA.

Anonymous (1990). *Modified Atmosphere Packaging.* Food Development Division, Agriculture Canada, Ottawa, Canada.

Brody, A. L. (1989). *Controlled/Modified Atmosphere/Vacuum Packaging of Foods.* Food & Nutrition Press, Trumbull, CA.

Brody, A. L. (1995). In: *Proceedings of the M A Pack Leading Edge Conference on Modified Atmosphere Packaging of Foods.* Institute of Packaging Professionals, Herndon, VA.

Brody, A. L. (1997). A Vision of Packaging for the 21st Century. Paper presented at the *Refrigerated Foods Association Annual Meeting,* Orlando, FL.

Brody, A. L. (1997). The impact of minimally processed – and often minimally packaged – foods on packaging. In: *Proceedings of Packaging Strategies Conference.* Atlanta, GA.

Brody, A. L. (2002). *Active Packaging Beyond Barriers.* BRG Townsend, Inc., Mt Olive, NJ.

Brody, A. L., and Bieler, A. (2001). *Case-Ready Meat Packaging.* Packaging Strategies, West Chester, PA.

Brody, A. L. and Shepherd, L. (1987). *Controlled/Modified Atmosphere Packaging: An Emergent Food Marketing Revolution.* Scotland Business Research, Princeton, NJ.

Blakistone, B. (1998). *Principles and Applications of MAP of Food*, 2nd edn. Blackie, Glasgow, UK.

Delaney, C. (2003). *Retail Fresh-Cut Produce Sales Approaching $4 Billion a Year.* Fresh Cut, November.

US Food and Drug Administration approach to regulating intelligent and active packaging components

26

Yoon S. Song and Mark A. Hepp

Introduction

Items manufactured for use in contact with food in the US, whether for home or commercial use, fall under the jurisdiction of the Federal Food, Drug, and Cosmetic Act (FFDCA, Title 21 United States Code 348) whenever they are intended to be marketed in interstate commerce. Section 201(s) of the FFDCA provides a definition for the term "food additive" that includes components of food-contact articles like packaging materials whenever they migrate (or are reasonably expected to migrate) to food as a result of their intended use, or if they otherwise affect the characteristics of food. Generally, food-packaging materials are not intended to have any technical effect in the food, and are considered food additives because their components can migrate to the food that contacts the packaging. In the case of active packaging, however, sometimes the package

Innovations in Food Packaging
ISBN: 0-12-311632-5

may be designed to deliver a chemical to the food for the purpose of exerting some effect in the food.

The food additive provisions of the FFDCA provide a framework to ensure the safe use of all chemicals that fall within the scope the food additive definition. These food additive provisions became law in 1958, and may be found mostly in section 409 of the FFDCA. The Food and Drug Administration (FDA) administers the main part of the FFDCA, including section 409. The FDA's administration of the food additive provisions of the FFDCA focuses mainly on reviewing the proposed uses and safety information relating to new food additives and, once a safety determination has been made, writing regulations, or allowing food contact notifications to become effective that stipulate conditions of safe use for the new food additive. The FFDCA also provides a standard of safety and a standard of review to which FDA must adhere.

In addition to the definition of a food additive in section 201(s), other important provisions of the FFDCA that relate to the regulation of food packaging materials include:

- Section 409(a), which provides that the use of a food additive is unsafe unless that use conforms to a food additive regulation, a food contact substance notification, or an exemption (consequently, all uses of food additives are deemed to be unsafe until FDA has completed a review on the safety of its use)
- Section 402, *Adulterated Food*, which provides that a food is adulterated if it is (or if it contains) any food additive that is unsafe
- Section 301, *Prohibited Acts*, which prohibits the introduction or delivery into interstate commerce of any food that is adulterated or misbranded.

Together, these sections of the FFDCA provide a mechanism for FDA to take enforcement action against unsafe food additive products in the marketplace.

Although the definition of the term "food additive" broadly includes any substance that is reasonably expected to result, directly or indirectly, in its becoming a component of (or otherwise affecting) the characteristics of any food, the definition also provides exceptions to that term. For example, substances that are *generally recognized as safe* (GRAS) do not fall within the scope of that definition. Substances that are GRAS are those that are generally recognized, among experts qualified in scientific training and experience to evaluate their safety, as having been shown through scientific procedures (or in the case of a substance used in food prior to 1 January 1958, through either scientific procedures or experience based on common use in food) to be safe under the conditions of its intended use. Section 201(s) also excludes pesticide chemicals, pesticide chemical residues, color additives, substances approved for use prior to 1958, new animal drugs, and dietary supplements from the food additive definition. Although such substances may become components of food, they are regulated under provisions of the FFDCA other than section 409.

The FDA generally classifies food additives by whether or not they have an intended technical effect in food. Federal food additive regulations are listed in Parts 170 through 189 of Title 21 Code of Federal Regulations (CFR) (Government Printing Office, Pittsburgh, 2004). Direct food additives are substances that are added directly to the food and have an intended technical effect in the food. They usually result in higher exposures and require more safety data to arrive at a safety determination. Examples of direct food additives include preservatives for food, high-intensity sweeteners, and flavoring agents, to name just a few. Direct food additive approvals are obtained through a food additive petition process, and result in a food additive regulation. Indirect food additives are those substances that are components of food contact substances, like food packaging, and may migrate to food, but they have no intended technical effect in the food. The indirect food additive regulations contain listings for substances that may be used in the manufacture of packaging, including Part 175 (*Adhesives and Components of Coatings*), Part 176 (*Paper and Paperboard Components*), Part 177 (*Polymers*), and Part 178 (*Adjuvants Production Aids and Sanitizers*). Part 179 contains regulations governing the safe use of radiation in the production, processing, and handling of food, including components of packaging that may be irradiated incidental to the irradiation of prepackaged food. Indirect food additives generally result in much lower exposures, and a safety determination is usually based upon much simpler data packages. The food contact substance approval is obtained through a food-contact notification process. Section 409 describes both the petition process and the food contact notification processes.

The food additive petition process

Section 409 of the FFDCA provides that any person may petition the FDA to issue a regulation prescribing the conditions under which a new food additive may be safely used. Such a petition should contain information relating to the safety of such use, including an unambiguous chemical identity of the proposed new additive, its impurities and contaminants, a clear description of the proposed conditions of use, and data relating to its technical effect. The statement of use should include enough detail to estimate a probable human exposure to the additive, and its impurities. The probable human exposure to chemicals migrating from food-contact substances is often estimated using the results of migration studies of the additive into food simulation solvents. Data relating to the technical effect are used to determine the minimum amount of the additive that is required to accomplish the intended technical effect either in the food, or in the food-contact substance in the case of an indirect food additive. These data are used to ensure that the human exposure to the food additive is not more than is necessary.

Food additive petitions also contain an analytical method for quantifying the additive (such methods can be used for enforcement purposes), and complete reports relating to the safety of the proposed use of the additive. Such reports generally include relevant toxicological testing studies on the additive and its impurities. Information on the manufacturing methods used to produce the additive is also provided so that the FDA may anticipate possible contaminants or impurities in the additive.

Section 409 of the FFDCA also has provisions relating to the FDA's required actions on the petition. The FDA is required to complete its review of a food additive petition within 180 days, and must at that time either establish a regulation stipulating conditions of safe use for the new food additive, or deny the petition. The FDA may only deny the petition if a fair evaluation of all data before the FDA fails to establish that the proposed use of the food additive, under the conditions of use specified in the regulation, will be safe, or if those data show that the proposed use of the additive would promote deception of the consumer, or otherwise result in adulteration or misbranding of food. There is also a provision that the petition must be denied if the additive has been found to induce cancer when ingested by man or animal.

Food contact substance notifications

Since 1958, the food additive petition process had been the major process by which food additive approvals were granted for both direct and indirect food additives. However, in 1997 Congress further amended the FFCDA to include a notification process for the approval of food-contact substances. A food-contact substance means "any substance that is intended for use as a component of materials used in manufacturing, packing, packaging, transporting, or holding food if such use is not intended to have any technical effect on the food" (FFDCA Section 409(h)). Under the notification process, a manufacturer may notify the FDA of its intent to market a new food-contact substance at least 120 days prior to introducing it, or delivering it for introduction, into interstate commerce. The notification must provide the basis for the determination by the manufacturer that the intended use of the food contact substance is safe under the general safety standard for all food additives. Thus, the food contact substance notification will contain essentially the same information relating to the identity, conditions of use, technical effect, manufacture, and safety, as a food additive petition contains.

There are, however, several important differences between the food additive petition process and the food contact notification process. First, the food contact notification becomes effective 120 days after the date of its receipt by the FDA, unless the FDA objects to the notification. If the FDA objects, the objection must be based on a determination that the data and information before the secretary do not establish that the proposed use of the food contact substance would be safe. If the FDA does not object, the food contact substance notification becomes effective, and from that time the notified material may be introduced, or delivered for introduction, into interstate commerce. Consequently, the FDA does not promulgate a food additive regulation as required for a petition. Instead, the notification serves to describe the specifications and conditions under which it may be safely used. Second, unlike food additive regulations, which are generic and apply equally to all manufacturers, an effective food-contact notification is effective only for the manufacturer or supplier identified in the notification. Third, while the petition process results in a food additive regulation printed in the Code of Federal Regulations (CFR), the notification process results in a listing on the FDA's website. A chronological list of effective food contact substance notifications may be found on the CFSAN website. The listings of effective notifications are often linked to conditions of use described in food additive regulations.

Special considerations for antimicrobial food additives

The general provisions applicable to all indirect food additives state, in part, "that the quantity of any food additive that may be added to food, as a result of its use in articles in that contact food, shall not exceed that which results from the use of the substance in an amount not more than reasonably required to accomplish the intended technical effect in the food contact article." As stated previously, the FDA requests data relating to the technical effect of a proposed new food additive in both food additive petitions and food contact substance notifications. These technical-effect data are not used to set performance standards for additive technical effects, nor to compare the effectiveness of one additive to another; rather they are used to ensure that the amount of the additive migrating to food is at the lowest possible level reasonably required to accomplish the intended technical effect. Because the FDA's authority to deny a food additive petition is limited to safety issues, and issues that may lead to deception of the consumer, adulteration of food or misbranding, the FDA views technical effect data only from these perspectives. Consequently, there are no distinctions in the approval processes between components of traditional or passive packaging materials (like polymeric plasticizers and stabilizers) and the "active" components of active or intelligent packaging materials. Importantly, no distinction needs to be made to arrive at a safety determination.

Although our current food additive approval processes readily accommodate food additive petitions and food contact substance notifications for components of active and intelligent packaging, certain types of packaging materials, whether they are active or passive, deserve additional consideration. Importantly, as stated above, the food contact substance notification process is intended only for the approval of materials that exert no technical effect on the food. Consequently, substances that are employed in active packaging materials and that have no intended technical effect on the food, such as freshness indicators, may be approved through the notification process. However, approvals for substances that exert a technical effect on the food, even though they may be incorporated in or delivered from the package, would be processed through the food additive petition program.

In particular, packaging materials that contain antimicrobial additives may function in either a passive or an active manner. If the antimicrobial additive is functioning as a material preservative in the package, then its function is passive, and there is no effect on the food in the package. Such a substance is a candidate for approval through the food contact substance notification process. However, if the antimicrobial additive is intended to be slowly delivered to the packaged food by migration, or any other type of time-release mechanism, then the intended technical effect is on the food, and a food additive approval would generally be sought through the food additive petition process. Marketing claims such as "improved shelf life" generally indicate a technical effect on the food.

There are other ways that antimicrobial food additives, whether passive or active, are treated differently from other food additives in the approval process. While many food additive products improve the quality, appearance or flavor of foods, antimicrobial food additive products, often called preservatives, have the potential to improve the safety of food. Since food safety is of paramount importance, the FDA has developed an expedited petition review process for antimicrobial food additives to ensure that safe antimicrobial food additives may enter the marketplace as quickly as possible. Also, when an antimicrobial additive is intended to have a technical effect on the food, whether it is added directly to the food or delivered to the food from an active package, that antimicrobial is a food preservative and is subject to the food labeling requirements for all food preservatives under section 403(k) of the FFDCA.

Antimicrobials, like all biocides, are generally intended to exhibit toxic effects, and are therefore expected to be more toxic than other food additives. For this reason, the FDA has established lower exposure thresholds for requesting additional toxicological data. For example, whereas the use of a polymeric stabilizer resulting in an exposure just above 50 parts per billion may require two additional subchronic studies to establish that the use is safe, an exposure in excess of 10 parts per billion to an antimicrobial additive may require the same additional studies.

Because the technical effect of any antimicrobial is to control the growth of micro-organisms, the technical effect data in food additive petitions, or food contact notifications, for antimicrobial additives are reviewed by the FDA's microbiological review scientists. Whether the antimicrobial effect is on the food or in the package, as in the case of a material's preservative, the microbiological review focuses on whether the petitioner (or notifier) has demonstrated that the intended technical effect is achieved, and whether it is achieved using the minimum amount of the additive reasonably required. Whenever the antimicrobial technical effect is intended to be in or on the food, some additional safety-related questions arise. The first relates to the spectrum of organisms against which the additive may be effective. In particular, the petitioner should address the potential for the proposed use of the antimicrobial additive to establish conditions that may control the growth of spoilage organisms, while allowing for the opportunistic growth of pathogens. Second, if the antimicrobial additive is an oxidizing agent, the petitioner should address the extent of oxidative degradation of the nutrients in the food being preserved.

Finally, antimicrobial additives that are intended to control micro-organisms in or on a food packaging material may also be subject to registration under the Federal Insecticide, Fungicide and Rodenticide Act (FIFRA, Title 7 US Code 136 *et seq.*) by the US Environmental Protection Agency (EPA) as pesticides. Antimicrobials that are intended to have a technical effect on processed foods, whether they are added directly to the foods or delivered from an active package, are not subject to FIFRA registration. Such substances are food preservatives by law, and not pesticides.

Other active or intelligent packaging materials

Many food packages are designed to contain diffusion barriers or oxygen scavengers, either to prevent gases from either entering the package or becoming a component of food, or to prevent gases from leaving the package. Other polymeric additives are intended to scavenge minor chemical constituents of the package material itself to prevent them from migrating to food and affecting its flavor or aroma. Such uses of polymeric additives may or may not be "active packaging," depending on the definition of active packaging to which one subscribes. Because the FDA's premarket review, under section 409 of the FFDCA, focuses upon safety as it relates to the migration and exposure to the package components, there is no need to distinguish between active and passive packaging materials. Therefore, the FDA has not attempted to define the term. Importantly, diffusion barriers and scavengers have no direct technical effect on the food, and are therefore food-contact substances that are subject to the notification approval process. Although the use of diffusion barriers and oxygen scavengers has the potential to create a modified atmosphere within the package that could promote the growth of anaerobic pathogens, the production of safe food is assured by good manufacturing practice regulations found mainly in 21 CFR Parts 110–129. The criteria and definitions in these parts are used, in part, to determine whether food is adulterated, or whether it has been prepared under conditions that would make it unfit for consumption.

Other types of functional food packaging include indicator and tamper-resistant packaging. Indicator packaging may be used to indicate the state of freshness of the packaged food. Tamper-resistant packaging generally indicates whether the package has been opened prior to purchase. Again, the FDA's premarket safety review focuses on the safety of the chemicals that may be expected to migration from the package to the food. Whether such packages are called "active" or "intelligent" does not impact the FDA's safety review. The FDA encourages innovation in these types of applications, but reminds developers that such materials are subject to the food additive approval process if the components of the package are reasonably expected to become components of food, or otherwise to affect the characteristics of food as a result of their intended use.

Conclusions

In summary, food additives are deemed to be unsafe until the FDA writes a regulation, allows a food contact substance notification to become effective, or writes an exemption. Regulations, effective notifications, and exemptions all stipulate an identity, specifications, and conditions of safe use for that food additive. Food contact substances and indirect food additives have no intended technical effect on food. Although food additive regulations are generic and apply to anyone, and notifications apply only to the manufacturer named in the effective notification, neither approval is a specific product approval. Importantly, because the FDA's premarket safety review focuses on the safety of the chemicals that may be expected to migrate from the package to the food and not on the relative effectiveness of the additives, it is irrelevant whether packaging materials are seen to be active, passive, or intelligent. Although the safe use of chemicals in the production of food-contact materials is assured by compliance with the food additive provisions in section 409 of the FFDCA, the production of safe food is assured by compliance with the provisions in section 402 of the FFDCA and, more specifically, the good manufacturing practice regulations in Parts 110–129 of Title 21 of the Code of Federal Regulations.

27 Packaging for non-thermal food processing

Seacheol Min and Q. Howard Zhang

Introduction

Pulsed electric fields (PEF), high-pressure processing (HPP), irradiation, and pulsed light have been developed as non-thermal food preservation methods to satisfy consumers, who like fresh foods. These non-thermal food preservation methods process foods at temperatures below those used for thermal pasteurization, and inactivate both spoilage and pathogenic micro-organisms without significant loss of flavor, color, taste, nutrients, and functionality of the foods.

The success of extending the shelf life of an initially high-quality food greatly depends on the packaging. Proper packaging materials and methods need to be selected to maintain the initial high quality of non-thermally processed food flavors, colors, and nutrients.

Plastic packaging materials have been widely used as moldings, thermoforms, films, label stocks, closures, and coatings (Brown, 1992). The selection of plastic packaging materials depends on the packaging requirements of mechanical, thermal, barrier, and optical properties, as well as the targeted shelf-life length and the cost (Brown, 1992). The exposure to different processing conditions may alter physical and/or chemical properties of the plastic packaging materials, and this alteration can influence the quality of the packaged food products (Ozen and Floros, 2001). HPP and irradiation may require the processing of foods inside packages. The interaction between packaging materials and these processing methods is important, and should be addressed.

Innovations in Food Packaging
ISBN: 0-12-311632-5

Current research in food packaging involves the development of active packaging. Active packaging contributes to the preservation of food products by providing antioxidant and/or antimicrobial properties. Such packaging materials interact directly with the food and the environment, to extend shelf life and improve food quality (Hotchkiss, 1995). This chapter presents an overview of the non-thermal food preservation methods of PEF, HPP, irradiation, and pulsed light, a summary of packaging materials used for non-thermal food preservation methods, and a direction for future research in food packaging for non-thermal foods.

Non-thermal food processing

The mechanisms and critical factors of non-thermal food processing methods such as pulsed electric fields (PEF), high-pressure processing (HPP), irradiation, and pulsed light are summarized in Table 27.1.

PEF

PEF treatment uses high-intensity electric field generated between two electrodes. A large flux of electrical current flows through the food product when a high-intensity electric field is generated. A non-thermal treatment is obtained by the use of a very short treatment time in pulses (i.e. microseconds).

Various types of PEF treatment chambers, which house electrodes and deliver PEF to foods, have been used (Zhang et al., 1995). A uniform distribution of electric field strength in the PEF treatment chamber is desirable to ensure that each microbial cell within a population receives the same PEF treatment, and thus to develop mathematical kinetic models for the prediction of microbial inactivation and quality control (Fiala et al., 2001). The high-voltage pulse generator converts low voltage into high voltage and provides the high voltage to PEF systems (Qin et al., 1995a). The square, exponential decay, and oscillatory waveforms are generally used for PEF treatments (Barbosa-Canovas et al., 1999).

Structural damage of cell membranes, which leads to ion leakage and metabolite losses, has been used to explain the microbial inactivation by PEF (Kinosita and Tsong, 1977; Benz and Zimmermann, 1980). Chang and Reese (1990) introduced the effects of PEF on microbial cells. Primary effects include structural fatigue due to induced membrane potential and mechanical stress. Secondary effects include material flow after the loss of the integrity of cellular membranes caused by the electric field, local heating, and membrane stress. Tertiary effects include cell swelling or shrinking, and disruption due to the unbalanced osmotic pressure between the cytosol and the external medium.

The cell membrane is regarded as an insulator due to its electrical conductivity, which is six to eight times weaker than that of the cytoplasm (Barbosa-Canovas et al., 1999). Electrical charges are accumulated in cell membranes when microbial cells are exposed to electric fields. The accumulation of negative and positive charges in cell

Table 27.1 Brief descriptions of pulsed electric fields, high-pressure processing, irradiation, and pulsed light

Process	Conditions	Mechanism for microbial inactivation	Critical factors
Pulsed electric fields (PEF)	PEF treatment occurs inside PEF treatment chambers, which houses electrodes and deliver PEF (at 20–80 kV/cm for <1 s) to foods. The square, exponential decay, and oscillatory waveforms are generally used	Membrane structural or functional damage. PEF temporarily increases the trans-membrane potential of cells by accumulating compounds of opposite changes in membrane surroundings. Continuous increase in trans-membrane potential can cause pore formation. Electroporation in protein channels and lipid domains results in osmotic swelling of the cell and membrane weakening until the cell bursts.	Electric field strength, PEF treatment time, pulse width, pulse shape, treatment temperature, electric conductivity, density, viscosity, pH, water activity, microbial characteristics
High-pressure processing (HPP)	HPP refers to the exposure of foods within vessels to high pressures (300–700 MPa) for a short time (few seconds–several minutes) at <0°C–>100°C)	Deprotonation of charged groups and disruptions of salt bridges and hydrophobic bonds. Changes in cell morphology involve the collapse of intercellular gas vacuoles, anomalous cell elongation, and cessation of movement of micro-organisms.	Pressure range, temperature, pH, protein structure, solvent composition, water activity, composition of foods, microbial characteristics
Irradiation	Food irradiation involves exposing prepackaged or bulk foods to γ-ray, X-ray, or electrons. The radioisotope used in most commercial γ-irradiation facilities is ^{60}Co. The mean energy of ^{60}Co γ-irradiation is 1.25 MeV. Electrons and X-ray are restricted to 10 and 5 MeV, respectively	Damaging genetic materials (DNA). This damage prevents multiplication and terminates most cell functions	Molecular weight of organic compounds, composition of foods, efficiency of repair mechanisms for DNA of micro-organisms
Pulsed light	Pulsed light is composed of about 25% ultraviolet, 45% visible, and 30% infrared radiation. Energy density ranges about 0.01 to 50 J/cm^2. Duration range: 1 μs–0.1 s. Flash: 1–20 flashes/s	The antimicrobial effects of ultraviolet wavelengths are primarily mediated through absorption by highly conjugated carbon-to-carbon double-bond system in proteins and nucleic acids. DNA mutations are induced by DNA absorption of ultraviolet light	Efficiency of penetration of the pulsed light (transmissivity), geometry of foods, the power wavelength, arrangement of light sources

membranes forms the transmembrane potential. A high transmembrane potential gives rise to pressure on the cell membranes. This increase in pressure decreases the thickness of cell membranes, and ultimately causes pore formation (Zimmermann, 1986). The critical transmembrane potential (i.e. that which causes pore formation) varies depending on the pulse duration time, number of pulses, and PEF treatment temperatures (Barbosa-Canovas *et al.*, 1999). Electroporation in protein channels and lipid domains by PEF results in osmotic swelling of the cell and membrane weakening until the cell bursts (Tsong, 1991).

Harrison *et al.* (1997) reported that transmission electron microscopy (TEM) micrographs of PEF treated *Saccharomyces cerevisiae* in apple juice exhibited disruption of organelles and lack of ribosomes. They proposed that the damaged organelles and lack of ribosomes after PEF treatment are an alternative inactivation mechanism to the electroporation theory.

The extent of microbial inactivation depends on the electric field strength, PEF treatment time, pulse width and shape, treatment temperature, electric conductivity, density, viscosity, pH, water activity, and microbial characteristics.

The potential for commercialization of PEF technology has drawn attention to the food industry. A commercial-scale PEF system with flow rates of 500–2000 l/h has been constructed, and has processed orange and tomato juice successfully (Min *et al.*, 2003a, 2003b).

HPP

In HPP, foods are exposed to high pressures (300–700 MPa) for a short period, typically ranging from a few seconds to several minutes. Foods are pressurized by direct and indirect methods using a pressure-transmitting medium such as water (Farr, 1990; Mertens and Deplace, 1993). HPP at 100–1000 MPa inactivates micro-organisms and enzymes (Knorr, 1995).

HPP damages microbial cell membranes (Farr, 1990). Changes in cell morphology involve the collapse of intercellular gas vacuoles, cell elongation, and cessation of movement in micro-organisms (Barbosa-Canovas and Rodriguez, 2002). High pressure causes deprotonation of charged groups and disruption of salt bridges and hydrophobic bonds, resulting in conformational changes and denaturation of proteins (Barbos-Canovas *et al.*, 1998). The effect of HPP in the reduction of microbial populations has been reviewed in detail by Smelt (1998).

The factors determining the efficiency of microbial inactivation by HPP include the type and number of micro-organisms, the magnitude and duration of HPP treatment, and the temperature, pH, and composition of the suspension media or foods (Hoover *et al.*, 1989; Barbosa-Canovas *et al.*, 1998). Generally, bacteria are more resistant to HPP than are yeasts and molds. Spores of bacteria are extremely resistant to HPP (Nakayama *et al.*, 1996). The inactivation of microbial spores by HPP is strongly influenced by temperature (Barbosa-Canovas *et al.*, 1998).

Enzymes are inactivated by HPP as a result of the conformational changes at enzyme active sites. Recovery in activity after decompression depends on the degree of distortion of the molecule. The chances of the recovery decrease with an increase in pressure beyond 3000 atm (Jaenicke, 1981).

DNA molecules are more stable at high pressures compared to proteins. However, DNA transcription and replication are disrupted by high pressures due to the inactivation of enzymes by HPP (Landau, 1967).

Irradiation

More than 40 countries permit the use of irradiation for over 60 types of foods (Ozen and Floros, 2001). Food irradiation involves exposing prepackaged or bulk foods to γ-rays, X-rays or electrons.

Radiation inactivates micro-organisms by damaging genetic materials (Grecz et al., 1983). This damage prevents multiplication, and terminates most cell functions. A photon of energy or an electron randomly strikes the genetic material of the cell, and causes a lesion in the DNA. The lesion breaks strands of DNA. Large numbers of single-strand lesions may result in the death of microbial cells (Dickson, 2001). Radiation also directly and indirectly damages other components of microbial cells, such as membranes, enzymes, and plasmids (Dickson, 2001).

The sensitivity of a micro-organism to irradiation is related to the efficiency of its DNA repair mechanisms. Micro-organisms that have a more efficient DNA repair mechanism are more resistant to irradiation (Dickson, 2001). Other factors determining the efficiency of microbial inactivation by irradiation include the molecular weight of organic compounds, and the composition of foods (Barbosa-Canovas et al., 1998). The inactivation of gram-positive bacteria that have significance to the public health significance by irradiation has been discussed in detail by Dickson (2001).

Vitamins A, C, and E are known to be sensitive to irradiation. Carbohydrates, lipids, and proteins will not be noticeably degraded by irradiation, but minor components may be disproportionately depleted by free radicals formed during irradiation (Bloomfield, 1993).

Irradiation can be used for frozen foods, and is the only preservation method available for the inactivation of pathogenic micro-organisms in frozen foods (Barbosa-Canovas et al., 1998).

Pulsed light

Pulsed light is a rapid, intense, and magnified flash of light or electrical energy (Dunn, 1996). Pulsed light is applicable in inactivating micro-organisms on the surfaces of foods, of packaging materials and of processing equipment (Barbosa-Canovas et al., 1998). The spectrum of light used for sterilization includes wavelengths in the ultraviolet to those in the near infrared region. Pulsed light consists of about 25% ultraviolet, 45% visible, and 30% infrared light (Dunn, 1996). The energy density of one pulse of light ranges from 0.01 to $50 \, J/cm^2$ (Dunn et al., 1991). The pulse width is in the range of $1 \, \mu s$ to $0.1 \, s$, and the flashes are typically applied at a rate of 1–20 flashes per second (Barbosa-Canovas et al., 1998).

Pulsed light is effective when it can penetrate food surfaces or packaging materials (Dunn, 1996). Pulsed light inactivates a wide range of micro-organisms, including bacterial and fungal spores. Dunn (1996) showed inactivation of micro-organisms in fresh juices by pulsed light without impairment of the sensory quality. Food applications of pulsed-light treatments have been reviewed in detail by Hoover (1997).

The antimicrobial effects of ultraviolet light are primarily due to its absorption by highly conjugated double-bond systems in proteins and nucleic acids of micro-organisms. Comparison of the antimicrobial effects obtained using pulsed light with those obtained using non-pulsed or continuous ultraviolet light indicated that pulsed light inactivated more micro-organisms than did continuous ultraviolet light (Barbosa-Canovas *et al.*, 1998).

The effectiveness of pulsed light on the shelf-life extension of fresh fruits and vegetables is limited due to their irregular surfaces. The microbial inactivation by pulsed light depends on the efficiency of the penetrating pulsed light (transmissivity), the geometry of the foods, the power wavelength, and the arrangement of the light sources (Dunn, 1996).

Plastic packaging materials

Food packaging is one of the major fields for the application of plastic materials (Baner and Piringer, 1999). Plastic packaging materials have been developed for the packaging of fresh quality products (Yam and Lee, 1995).

The shelf life of foods packaged in plastics depends on the permeation of gas and water vapor through the packages. This is because a significant amount of food deterioration results from oxidation and changes in the water content (Varsanyi, 1986).

Factors affecting the permeability include the chemical structure, crystallinity and molecular weight of polymers, the degree of polymerization, the molecular orientation, the presence of double bonds, the film thickness, the polarity, the additives (including plasticizers), the permeate size, and the density, temperature and pressure (Pascat, 1986).

Polyolefins, styrenics, polyesters, and vinyls are the principal families of thermoplastics in food packaging (Brown, 1992). Polyolefins include polyethylene (PE) and polypropylene (PP). Polystyrene (PS) is a type of styrenic. Polyethylene terephthalate (PET) and polycarbonates (PC) are examples of polyesters.

Various materials with specific functionality are available. Amorphous polyamides (AmPA) have recently appeared in food packages as gas-barrier materials. The barrier property of AmPA is intermediate between the high barrier ethylene vinyl alcohol (EVOH) and polyvinylidene chloride (PVDC), and the lower barrier PE. The barrier property increases with increasing relative humidity (RH). AmPAs have oxygen permeabilities around 2 cc-mil/100 in^2 day atm (800 cc μm/m^2 day atm) when the RH is low. The oxygen permeability decreases to about 1 cc ml/100 in^2 day atm at 80% RH (Brown, 1992). Brown (1992) presented a polycondensation product of adipic acid and metaxylylene diamine, MXD6, which is not sensitive to moisture compared to other polyamides (PA) in use. The oxygen permeability at 20°C of biaxially oriented MXD6 increases by a factor of ~3.3, whereas nylon 6, a PA, increases by a factor of ~7 at 70 and 100% RH. MXD6 is five times less permeable than nylon 6 when the RH is low, and twelve times less permeable when the RH is 75% (Brown, 1992). Liquid crystals of PET (LCP) have high barrier properties to gas transmission. They have recently been produced in pilot plant quantities to test their strength and high heat resistance (Brown, 1992). New materials have been incorporated with inorganic–organic nanocomposites, giving high toughness and hardness measurements, and these can tremendously decrease the gas and moisture transmission rates (Ruan, 2000). These materials need to be examined for their suitability for use as food packaging materials.

Packaging for non-thermal food processing

Aseptic food packaging is effective for producing shelf-stable food products that have many advantages over conventionally processed products (Barbosa-Canovas et al., 1999). Aseptic technology has been used for foods for more than 30 years (Singh and Nelson, 1992). Packaging materials for aseptic packaging have no limitations on container size, and do not need to withstand high temperatures as in conventional processing methods (Singh and Nelson, 1992). Plastic- and paper-laminated materials are widely used as packaging materials for aseptic food packaging. Aseptically packaged foods may be stored at refrigerated or ambient temperatures, depending on the type of foods. The effectiveness of aseptic packaging of several PEF processed foods has been demonstrated (Qin et al., 1995b; Zhang et al., 1997; Yeom et al., 2000).

Plastic packaging materials have been used for non-thermal processing. Those materials used for foods processed by PEF, high-pressure, and irradiation are listed in Table 27.2.

PEF

Ayhan et al. (2001) investigated the effects of packaging materials on the quality of orange juice treated by a pilot plant-scale PEF system at 35 kV/cm for 59 µs. The PEF-treated orange juice was put into four different packaging materials; sanitized glass, PET, high-density polyethylene (HDPE) bottles, and low-density polyethylene (LDPE) bottles, inside a sanitized glove box. They reported that glass and PET bottles were effective in maintaining flavor compounds and vitamin C, as well as the color of PEF-treated orange juice during storage at 4°C for 112 days. They suggested that this might be due to the higher oxygen-barrier properties of glass and PET compared to PE (Ayhan et al., 2001).

Min et al. (2003a) observed that the losses of lycopene of thermally processed, PEF-processed and untreated tomato juices in PP tubes were most significant during the first 7 days of storage at 4°C, regardless of the processing method. The stability of carotenoids in foods depends on the packaging conditions and the oxygen availability, since the main cause of carotenoid degradation is oxidation (Rodriguez-Amaya, 1993; Thakur et al., 1996). The significant reduction of lycopene during early storage may be due to the high

Table 27.2 Packaging materials for foods processed by pulsed electric fields, high pressure, and irradiation

Processing method	Food	Packaging material	Packaging method and storage temperature	Source
PEF	Orange juice, a protein fortified fruit beverage	200 ml thermoformed plastic container, glass bottle	Thermo-forming and -sealing using an aseptic packaging machine, sterilized by heat and H_2O_2. Storage at 4°C	Orange juice, Qiu et al. (1998); a protein fortified fruit beverage, Sharma et al. (1998)
	Cranberry juice	Glass vials sealed with small headspace	Storage at 4 and 20–24°C	Jin and Zhang (1999)
	Apple juice and apple cider, orange juice, cranberry juice, chocolate milk	Materials: base material, HIPS/PVDC/LDPE (Allista Plastic Packaging Co., Muncie, IN); lid material, Nylon/Al/LDPE (Rollprint, Addison, IL). Thermoforming. Size of thermoformed plastic container: apple juice and cider, 180 ml; orange juice, cranberry juice, chocolate milk, 200 ml	Thermoforming and thermosealing using an aseptic packaging machine, sterilized by heat and H_2O_2. Storage temperature: apple juice and cider, cranberry juice, chocolate milk, 4, 22 and 37°C; orange juice, cranberry juice, chocolate milk, 4 and 22°C	Apple juice and cider, Evrendilek et al. (2000); orange juice, Yeom et al. (2000); cranberry juice, chocolate milk, Evrendilek et al. (2001)
	Orange juice	500 ml glass, PET, LDPE, HDPE bottles with 28 mm PP caps (glass, PET, HDPE – General Bottles Supply Co., Los Angeles, CA; LDPE – Consolidated Plastic Co., Twinsburg, OH)	Packed in a glove box, sanitized by H_2O_2 and UV. Storage at 4 and 22°C	Ayhan et al. (2001)
	Tomato juice, orange juice	50 ml pre-sterilized PP tubes (Corning, Acton, MA)	Packed in a glove box, sanitized by H_2O_2 and UV. Storage at 4°C	Tomato juice, Min et al. (2003a); orange juice, Min et al. (2003b)
HPP	Kimchi (Korean fermented vegetable product)	PE bag	Heat sealing without entrapping any air bubbles. Storage in ice water	Sohn and Lee (1998)
	Orange juice	Plastic bag with EVOH	Storage at 0 and 10°C	Takahashi et al. (1998)
	Sliced cooked ham	PE/Nylon pouches	Vacuum sealing. Storage at 3 and 9°C	Carpi et al. (1999)
	Orange juice	500 ml PET bottle	Storage at 4°C	Goodner et al. (1999)

(Continued)

Table 27.2 (*Continued*)

Processing method	Food	Packaging material	Packaging method and storage temperature	Source
	Pork sausage	Nylon/PE bag (0.75 mil nylon, 2.27 mil PE) (Koch supplies, Inc., Kansas City, MO)	Vacuum sealing. Storage at 4°C	Murano et al. (1999)
	Low-fat yogurt	250 ml PE bottle	Chilled storage	Ancos et al. (2000)
	Guacamole	Plastic whirl-pak sampling bags (Nasco, Fort Atkinson, WI)	Heat sealing. Storage at 5, 15, and 25 (±0.5)°C	Palou et al. (2000)
	Salsa	PE bag	Storage at 4 and 21–23°C	Raghubeer et al. (2000)
	Orange juice, orange–lemon–carrot juice	Glass, PP, teflon, and Barex 210 (modified acetonitrile-methyl acrylate copolymers) flasks or PE pouches	Flasks were closed by screw lids. PE pouches were heat sealed. Storage at 4°C	Garcia et al. (2001)
	Orange juice	HPP: Nylon-EVA pouches (Winpak Ltd,, Winnipeg, Canada) Shelf life study: 8 oz PET bottle with screw-cap closure (Novapak Corp., Hazelton, PA)	Storage at 4, 15, 26, and 37°C	Nienaber and Shellhammer (2001)
	Orange juice	250 ml PE bottle	Storage at 4°C	Ancos et al. (2002)
	Fatty duck liver	PA/PE/EVOH/PE, PA/PP/EVOH/PP, PE/PA/PE, PET/Al/PE films (Soplaril, Elf-Atochem, Dax, France)	Storage at 4°C	Cruz et al. (2003)
	Turkey meat	Laminated PE–PA foil (Multiseven-80)	Storage at 4°C	Tuboly et al. (2003)
	Potato	Steel tray, covered with a plastic film	Controlled atmosphere storage (0.03, 0.5, 5, and 15% (v/v) CO_2 in air)	Ziegler et al. (1968)
Irradiation	Precooked lobster	3 mil mylar/saran/PE bags, wrapped in PE bags	Storage at 3°C	Dagbjartsson and Solberg (1973)

Chicken meat	Air packaging – LDPE. Vacuum packaging – unplasticized bags of PA and PE (UPA/PE 15-60, Sudpack Verpackungen, Germany)	Storage at 5°C	Calenberg et al. (1999)
Pork patties, meats, turkey breast meat	Air packaging – PE bag (2 mil, Associated Bag Company, Milwaukee, WI). Vacuum packaging – Nylon/PE (Koch, Kansas City, MO)	Storage at 4°C	Pork patties, Ahn et al. (1998); meats, Kim et al. (2002); turkey breast meat, Nam and Ahn (2002)
Cooked pork sausages	Nylon/PE bag (Koch, Kansas City, MO)	Storage at 4°C	Jo et al. (2000)
Ground beef patties	(1) Nylon/PE bags (Koch Supplies, Inc., Kansas City, MO), consisting of 0.75 mil nylon and 2.25 mil PE, with a moisture transmission rate of 0.73 g/100 in² 24 h atm and oxygen permeability of 3.9 cc/100 in² 24 h atm. (2) Saran/polyester/PE bags (Koch Supplies, Inc., Kansas City, MO), consisting of a top layer of 0.48 mil Saran + 2 ml PE, and a bottom layer of 2 mil PE + 0.48 metallized polyester, with a moisture transmission rate of 0.22 g/100 in² 24 h atm and oxygen permeability of 0.49 cc/100 in² 24 h atm. (3) Saran film overwrap (Dow Brands, Indianapolis, IN) plus a Styrofoam tray (Albertson's College Station, TX) on the bottom, with Saran having a moisture transmission rate of 0.45 g/100 in² 24 h atm and an oxygen permeability of 1.0 cc/100 in² 24 h atm. Bags were sealed in air using a model CE95 modified atmosphere packaging machine (Koch Supplies, Inc., Kansas City, MO)	Storage at 5°C	Lopez-Gonzalez et al. (2000)
Chicken meat	Air packaging – PE bag. Vacuum packaging – PVDC film (Cryovac, Duncan, SC)	Storage at 4°C	Lee et al. (2001)

PEF, pulsed electric fields; HPP, high-pressure processing; PE, polyethylene; EVOH, ethylene vinyl alcohol; PA, polyamide; LDPE, low-density polyethylene; HIPS, high-impact polystyrene; PVDC, polyvinylidene chloride; Al, aluminum; PET, polyethylene terephthalate; HDPE, high-density polyethylene; PP, polypropylene; EVA, ethylene vinyl acetate.

oxygen availability in the headspace of the PP tubes. Modified atmosphere packaging (MAP), which limits oxygen in the headspace, may be applied as a complement to PEF, to reduce oxidation and maintain high freshness in PEF processed foods.

HPP

Foods that are treated by HPP are generally first packaged and then placed in a pressure chamber. Therefore, it is important to study the effect of HPP on packaging materials.

Masuda *et al.* (1992) reported that water vapor and oxygen permeabilities of several laminated plastic films, (PP/EVOH/PP, orientated polypropylene (OPP)/ PVOH/PE, KOP/cast polypropylene (CPP), and PET/Al/CPP) were not affected by high pressures at 400–600 MPa. It was demonstrated that packaging materials constituted with EVOH and polyvinyl alcohol (PVOH) are compatible with high-pressure treatments.

Caner *et al.* (2000) reported that metallized PET was the only film with a significant increase in permeabilities to oxygen, carbon dioxide, and water vapor among the eight high-barrier laminated films tested (PET/SiO$_x$/polyurethane (PU) adhesive/LDPE, PET/Al$_2$O$_3$/PU adhesive/LDPE, PET/PVDC/Nylon/HDPE/PP, PE/Nylon/EVOH/PE, PE/Nylon/PE, PET/ethylene vinyl acetate (EVA), PP, and metallized PET) after a high-pressure process at 600–800 MPa.

Kubel *et al.* (1996) investigated the effect of HPP on the absorption of aroma compounds, p-cymene and acetophenone, into plastic films. They found that the absorption of the aroma compounds was lower in films exposed to 500 MPa pressure compared to non-pressurized films. The transition of the plastic films to the glassy state at higher pressures was suggested as the reason for the decrease in absorption of the aroma compounds. Masuda *et al.* (1992) also reported a decrease in the absorption of d-limonene into LDPE and EVA films as a result of HPP at 400 MPa for 10 minutes.

Lambert *et al.* (2000) presented important aspects of packaging materials for HPP in terms of package properties (barriers and flexibility) and package integrity. HPP was carried out at 200, 350 and 500 MPa for 30 minutes at ambient temperatures. They compared the performance of pressurized multi-layer packaging materials with that of untreated materials. A cast-co-extruded 100 μm PA/glue/free radical linear PE (40 μm/20 μm/40 μm) was found to be incompatible with high-pressure treatments, mainly due to delamination. The other packaging materials, a tubular-co-extruded 100 μm PA/glue/medium density PE (20 μm/10 μm/70 μm), a cast-co-extruded 100 μm PA/glue/free radical linear PE (30 μm/10 μm/60 μm), a 65 μm PET/glue/PVDC/glue/ free radical linear PE (12 μm/1 μm/1 μm/1 μm/50 μm), a 110 μm PA/glue/PE surlyn (40 μm/15 μm/55 μm) and a 110 μm PA/PP/glue/free radical linear PE (30 μm/ 10 μm/60 μm), did not show any significant changes in tensile strength, heat-seal strength and laminations after HPP. Barrier properties (oxygen and moisture) of tested materials were not significantly changed after HPP.

Garcia *et al.* (2001) evaluated the antioxidative capacities, the nutrient contents and the sensory properties of orange juice and orange–lemon–carrot (OLC) juice in different packaging materials after HPP and during storage. Juices were packaged in glass, PP, Teflon, and Barex 210 (modified acetonitrile-methyl acrylate copolymers) flasks (0.51 in volume) or in PE pouches covered by aluminum foil. The flasks were closed by screw lids, and the PE pouches were heat-sealed. HPP and storage for 21 days at 4°C did not cause any significant differences in the antioxidative capacity and in the vitamin C, sugar, and carotene contents. They studied the effects of HPP and the effects of packaging materials on the sensory properties using the triangular test technique. Results from the sensory evaluation indicated that the odor and flavor of pressurized orange juice were occasionally altered by HPP and cultivars, but not by the packaging materials.

Cruz *et al.* (2003) packaged fatty duck liver in four different multi-layer films, PA/PE/EVOH/PE, PA/PP/EVOH/PP, PE/PA/PE, and PET/Al/PE films (Soplaril, Elf-Atochem, Dax, France), and applied HPP to the packaged fatty duck liver. The PET/Al/PE films showed mechanical damages after HPP. The PA/PE/EVOH/PE and PA/PP/EVOH/PP films were durable after HPP. The barrier properties to oxygen transmission were not affected by the HPP treatment.

Irradiation

Foods are generally pre-packaged before irradiation to prevent recontamination (Sahasrabudhe, 1990).

The impact of irradiation on the packaging materials must be considered, and any packaging materials must be accepted by the FDA before use in food irradiation, since gases and low molecular weight polymers formed during irradiation have the potential to migrate into the product and to influence the product quality (Lee *et al.*, 1996; Olson, 1998). Irradiation of packaging materials may generate gases, such as hydrogen, and may produce low molecular weight hydrocarbons and halogenated polymers (Kilcast, 1990). Packaging materials for irradiated foods should not transmit toxic substances and undesirable odors or flavors to foods (Barbosa-Canovas *et al.*, 1998). Materials for irradiation are listed in 21 CFR 179.45.

Some chemical and physical properties of plastic materials can be changed as a result of irradiation (Ozen and Floros, 2001). Irradiation can lead to cross-linking and chain scission of polymers. The changes in the polymers by irradiation depend on the composition of the plastic materials and the conditions of irradiation. Cross-linking is the predominant reaction during irradiation in most plastic packaging materials, including PE, PP, and PS. Cross-linking can decrease elongation, crystallinity, and solubility, and increase the mechanical strength of the plastic materials. Chain scission decreases the chain length of plastic materials, and thus provides free volume in the materials (Ozen and Floros, 2001). For example, cellulose undergoes chain scission under irradiation, resulting in the loss of mechanical properties (Buchalla *et al.*, 1993). Free radicals formed during irradiation can be trapped in the polymers in the crystalline regions. This may also change the mechanical properties of plastic materials (Buchalla *et al.*, 1993).

The degree of crystallinity of LDPE, HDPE, PP, PET, PVC and PVDC was not changed after irradiation at 8 kGy (Varsanyi, 1975). The structures of LDPE and OPP were significantly changed at 10–50 kGy. Carbon dioxide and the number of double bonds increased in the polymer structures at 100 kGy. Oxygen permeabilities of LDPE and OPP films were not significantly changed after irradiation at up to 25 kGy (Rojas De Gante and Pascat, 1990). Pilette (1990) reported that oxygen and water vapor permeabilities of PE pouches were not changed after irradiation with γ-photons (^{60}Co) and accelerated electrons. Kim and Gilbert (1991) reported that oxygen permeability of glycol modified PET/PVDC/PE laminate decreased significantly after ^{60}Co treatment. Pentimalli *et al.* (2000) studied the effects of γ-irradiation on the structures and the mechanical properties of PS, poly-butadiene, styrene-acrylonitrile, high-impact polystyrene, and acrylonitrile-butadiene-styrene using NMR. They found that PS samples did not show any detectable differences after 100 kGy irradiation, and concluded that PS was the best material, among the tested plastics, for packaging for irradiation.

The effects of irradiation on the characteristics of the diffusion and the absorption of octan, ethyl hexanoate, and d-limonene were investigated by Matsui *et al.* (1991). With increasing radiation dose, diffusion coefficients in electron-beam irradiated EVA films increased, and the solubility coefficients decreased. The chain scission in EVA films caused by irradiation possibly resulted in the increase of the diffusion coefficient.

Marque *et al.* (1995) detected alkyl radicals after ionization treatment of PP at 40 kGy. Rojas De Gante and Pascat (1990) reported that hydroperoxides and carbonyl compounds such as ketones and aldehydes were formed after irradiation of LDPE and OPP (<25 kGy). El Makhzoumi (1994) reported that irradiation formed 63 different volatile compounds from PET, PE, and OPP films.

Ground beef patties were packaged in air with nylon/polyethylene bags, Saran/polyester/polyethylene bags, and Saran film overwrap plus a styrofoam tray. Samples were irradiated at 2 kGy by either gamma rays or an electron beam, and evaluated by a trained sensory panel for seven flavors, three mouthfeels, and seven taste attributes. No differences were detected among the packaging materials used (Lopez-Gonzalez *et al.*, 2000).

Pulsed light

Plastic packaging materials that cannot tolerate autoclave temperatures are prohibited from use for pulsed-light sterilization. The resistance to heat of PVC make it the preferred packaging material for pulsed light (Katz, 1999).

The use of pulsed light as a secondary treatment to sterilize packaging materials is disclosed in the US Patent 5,925,885, assigned to PurePulse Technologies, Inc. Pulsed light inactivates micro-organisms on packaging materials used for aseptic packaging processing methods (Katz, 1999).

Future research

The degradation of flavor compounds can occur not only by oxidation but also by the permeation of the flavor compounds through packaging materials, and the absorption of the compounds into the packaging materials. A greater absorption rate is found if the flavor compound has a similar chemical structure or a similar polarity to the functional groups in the packaging materials (Landois-Garza and Hotchkiss, 1987). As an example, more d-limonene and α-pinene (volatile flavor compounds of orange juice) are absorbed more easily in LDPE than in PET, PVDC or EVOH. LDPE absorbs more d-limonene and α-pinene than the other materials because the non-polar hydrocarbon of LDPE has a strong affinity to the non-polar terpene hydrocarbons of the flavor compounds (Sheung, 1995). Many researchers have investigated the absorption and diffusion phenomena of flavor compounds of foods packaged in various food packaging materials (Baner *et al.*, 1991; Ikegami *et al.*, 1991; Nielsen *et al.*, 1992; Van Willige *et al.*, 2000). To reduce the absorption of the fresh flavor compounds of foods, a packaging material with low diffusivity and solubility for the fresh flavor compounds should be used.

As consumers demand more non-thermally processed foods, the opportunities for active packaging may increase. The use of active packaging would be effective in retarding the degradation of the "fresh-like quality" of non-thermally processed foods during storage. Modified atmosphere packaging (MAP) can be applied to control oxygen in the headspace by reducing the oxidation degradation of foods. MAP limits the oxygen in the headspace by reducing the oxygen or replacing the air with a controlled mixture of gases such as nitrogen and carbon dioxide (Robertson, 1993). MAP integrated with aseptic packaging has been used in minimally processed refrigerated foods (Brody, 1996). MAP reduces the growth rate of aerobic micro-organisms, and extends the product's shelf life. However, MAP can selectively change the microbiology of foods. The aerobic spoilage micro-organisms' growth will decrease, which will then provide sufficient time for the growth of pathogenic micro-organisms, especially if the MAP foods have received a non-sterilizing treatment (Hotchkiss and Banco, 1992; Marth, 1998). Other active packaging techniques such as antimicrobial packaging, incorporating antimicrobial agents into packaging systems, also provide better packaging for non-thermally processed foods. These combinations must not reduce food safety while they improve food quality, which is demanded by consumers.

References

Ahn, D. U., Olson, D. G., Lee, J. I., Jo, C., Wu, C. and Chen, X. (1998). Packaging and irradiation effects on lipid oxidation and volatiles in pork patties. *J. Food Sci.* **63(1),** 15–19.

Ancos, B., Cano, M. P. and Gomez, R. (2000). Characteristics of stirred low-fat yoghurt as affected by high pressure. *Intl Dairy J.* **10,** 105–111.

Ancos, B., Sgroppo, S., Plaza, L. and Pilar-Cano, M. (2002). Possible nutritional and health-related value promotion in orange juice preserved by high-pressure treatment. *J. Sci. Food Agric.* **82(8),** 790–796.

Ayhan, Z., Yeom, H. W., Zhang, Q. H. and Min, D. B. (2001). Flavor, color, and vitamin C retention of pulsed electric field processed orange juice in different packaging materials. *J. Agric. Food Chem.* **49,** 669–674.

Baner, A. L. and Piringer, O. (1999). Preservation of quality through packaging. In: *Plastic Packaging Materials for Food* (O. G. Piringer and A. L. Baner, eds), pp. 1–3. Wiley-VCH, New York, NY.

Baner, A. L., Kalyankar, V. and Shoun, L. H. (1991). Aroma sorption evaluation of aseptic packaging. *J Food Sci.* **56(4),** 1051–1054.

Barbosa-Canovas, G. V. and Rodriguez, J. J. (2002). Update on nonthermal food processing technologies: pulsed electric field, high hydrostatic pressure, irradiation and ultrasound. *Food Australia* **54(11),** 513–520.

Barbosa-Canovas, G. V., Pothakamury, U. R., Palou, E. and Swanson, B. G. (1998). *Nonthermal Preservation of Foods*, pp. 9–52, 139–213. Marcel Dekker, New York, NY.

Barbosa-Canovas, G. V., Gongora-Nieto, M. M., Pothakamury, U. R. and Swanson, B. G. (1999). *Preservation of Foods with Pulsed Electric Fields*, pp. 1–171. Academic Press, San Diego, CA.

Benz, R. and Zimmermann, U. (1980). Pulse-length dependence of the electrical breakdown in lipid bilayer membranes. *Biochim. Biophys. Acta* **597,** 637–642.

Bloomfield, L. (1993). Food irradiation and vitamin A deficiency: Public health implications. *Food Policy* **18(1),** 64–72.

Brody, A. (1996). Integrating aseptic and modified atmosphere packaging to fulfill a vision of tomorrow. *Food Technol.* **50(4),** 56–66.

Brown, W. E. (1992). Properties of plastics used in food packaging. In: *Plastics in Food Packaging*, pp. 103–138. Marcel Dekker, New York, NY.

Buchalla, R., Schuttler, C. and Bogl, K. W. (1993). Effects of ionizing radiation on plastic food packaging materials: a review. *J. Food Protect.* **56,** 991–997.

Calenberg, S. V., Philips, B., Mondelaers, W., Cleemput, O. V. and Huyghebaert, A. J. (1999). Effect of irradiation, packaging, and postirradiation cooking on the thiamin content of chicken meat. *J. Food Protect.* **62(11),** 1303–1307.

Caner, C., Hernandez, R. J., Pascall, M. A. and Buchanan, J. (2000). Effect of high-pressure processing on the permeance of selected high-barrier laminated films. *Packag. Technol. Sci.* **13(5),** 183–195.

Carpi, G., Squarcina, N., Gola, S., Rovere, P., Pedrielli, P. and Bergamaschi, M. (1999). Application of high pressure treatment to extend the refrigerated shelf-life of sliced cooked ham. *Industria Conserve* **74(4),** 327–339.

Chang, D. C. and Reese, T. S. (1990). Changes in membrane structure induced by electroporation as revealed by rapid-freezing electron microscopy. *Biophys. J.* **58,** 1–12.

Cruz, C., Moueffac, A. E., Antoine, M. *et al.* (2003). Preservation of fatty duck liver by high pressure treatment. *Intl J. Food Sci. Technol.* **38,** 267–272.

Dagbjartsson, B. and Solberg, M. (1973). Textural changes in precooked lobster (Homarus americanus) meat resulting from radurization followed by refrigerated storage. *J. Food Sci.* **38(1),** 165–167.

Dickson, J. S. (2001). In: *Food Irradiation* (R. A. Molins, ed.), pp. 23–35. John Wiley & Sons, New York, NY.

Dunn, J. (1996). Pulsed light and pulsed electric field for foods and eggs. *Poultry Sci.* **75,** 1133–1136.

Dunn, J. E., Clark, R. W., Asmus, J. F. *et al.* (1991). US Patent 5,034,235.

El Makhzoumi, Z. (1994). Effect of irradiation of polymeric packaging material on the formation of volatile compounds. In: *Food Packaging and Preservation* (M. Mathlouthi, ed.), pp. 88–99. Blackie Academic & Professional, London, UK.

Evrendilek, G. A., Jin, Z. T., Ruhlman, K. T., Qiu, X., Zhang, Q. H. and Richter, E. R. (2000). Microbial safety and shelf-life of apple juice and cider processed by bench and pilot scale PEF systems. *Innov. Food Sci. Emerg. Technol.* **1,** 77–86.

Evrendilek, G. A., Dantzer, W. R., Streaker, C. B., Ratanatriwong, P. and Zhang, Q. H. (2001). Shelf-life evaluations of liquid foods treated by pilot plant pulsed electric field system. *J. Food Process. Preserv.* **25(4),** 283–297.

Farr, D. (1990). High pressure technology in the food industry. *Trends Food Sci. Technol.* **1(7),** 14–16.

Fiala, A., Wouters, P. C., van den Bosch, E. and Creyghton, Y. L. M. (2001). Coupled electrical-fluid model of pulsed electric field treatment in a model food system. *Innov. Food Sci. Emerg. Technol.* **2(4),** 229–238.

Garcia, A. F., Butz, P., Bognar, A. and Tauscher, B. (2001). Antioxidative capacity, nutrient content and sensory quality of orange juice and an orange-lemon-carrot juice product after high pressure treatment and storage in different packaging. *Eur. Food Res. Technol.* **213,** 290–296.

Goodner, J. K., Braddock, R. J., Parish, M. E. and Sims, C. A. (1999). Cloud stabilization of orange juice by high pressure processing. *J. Food Sci.* **64(4),** 699–700.

Grecz, N., Rowley, D. B. and Matsuyama, A. (1983). The action of radiation on bacteria and viruses. In: *Preservation of Foods by Ionizing Radiation*, Vol. 2 (E. S. Josephson and M. S. Peterson, eds), pp. 167–218. CRC Press, Boca Raton, FL.

Harrison, S. L., Barbosa-Canovas, G. V. and Swanson, B. G. (1997). Saccharomyces cerevisiae structural changes induced by pulsed electric field treatment. *Lebensm. Wiss. u. Technol.* **30,** 236–240.

Hoover, D. G., Metrick, C., Popineau, A. M, Farkes, D. F. and Knorr, D. (1989). Biological effects of high hydrostatic pressure on food microorganisms. *Food Technol.* **3,** 99–107.

Hoover, D. H. (1997). Minimally processed fruits and vegetables: Reducing microbial load by nonthermal physical treatments. *Food Technol.* **51,** 66–71.

Hotchkiss, J. H. (1995). Safety considerations in active packaging. In: *Active Food Packaging* (M. L. Rooney, ed.), pp. 238–255. Blackie Academic & Professional, London, UK.

Hotchkiss, J. H. and Banco, M. J. (1992). Influence of new packaging technologies on the growth of microorganisms in produce. *J. Food Protect.* **55(10),** 815–820.

Ikegami, T., Nagashima, K., Shimoda, M., Tanaka, Y. and Osajima, Y. (1991). Sorption of volatile compounds in aqueous solution by ethylene-vinyl alcohol copolymer films. *J. Food Sci.* **56(2),** 500–509.

Jaenicke, R. (1981). Enzymes under extremes of physical conditions. *Ann. Rev. Biophys. Bioeng.* **10,** 1.

Jin, Z. T. and Zhang, Q. H. (1999). Pulsed electric field inactivation of microorganisms and preservation of quality of cranberry juice. *J. Food Process. Eng.* **23,** 481–497.

Jo, C., Jin, S. K. and Ahn, D. U. (2000). Color changes in irradiated cooked pork sausage with different fat sources and packaging during storage. *Meat Sci.* **55(1),** 107–113.

Katz, F. (1999). Smart packaging adds a dimension to safety. *Food Technol.* **53(11),** 106.

Kilcast, D. (1990). Irradiation of packaged food. In: *Food Irradiation and the Chemist* (D. E. Johnson and M. H. Stevenson, eds), pp. 140–152. Pub. No. 86. The Royal Society of Chemistry, Cambridge, UK.

Kim, K. H. and Gilbert, S. G. (1991). *Packag. Technol. Sci.* **4,** 35–48.

Kim, Y. H., Nam, K. C. and Ahn, D. U. (2002). Irradiation of packaged food. *Meat Sci.* **61(3),** 257–265.

Kinosita, K., and Tsong, T. Y. (1977). Formation and resealing of pores of controlled sizes in human erythrocyte membrane. *Nature* **268(4),** 438–440.

Knorr, D. (1995). Hydrostatic pressure treatment of food: microbiology. In: *New Methods of Food Preservation* (G. W. Gould, ed.), pp. 159–175. Blackie Academic & Professional, London, UK.

Kubel, J., Ludwig, H., Marx, H. and Tauscher, B. (1996). Diffusion of aroma compounds into packaging films under high pressure. *Packag. Technol. Sci.* **9,** 143–152.

Lambert, Y., Demazeau, G., Largeteau, A., Bouvier, J. M., Laborde-Croubit, S. and Cabannes, M. (2000). Packaging for high-pressure treatments in the food industry. *Packag. Technol. Sci.* **13,** 63–71.

Landau, J. V. (1967). Induction, transcription and translation in Escherichia coli: a hydrostatic pressure study. *Biochim. Biophys. Acta* **149,** 506–512.

Landois-Garza, J. and Hotchkiss, J. H. (1987). Plastic packaging can cause aroma sorption. *Food Eng.* **4,** 39–42.

Lee, J. I., Kang, S., Ahn, D. U. and Lee, M. (2001). Formation of cholesterol oxides in irradiated raw and cooked chicken meat during storage. *Poultry Sci.* **80(1),** 105–108.

Lee, M., Sebranek, J. G., Olson, D. G. and Dickson, J. S. (1996). Irradiation and packaging of fresh meat and poultry. *J. Food Protect.* **59,** 62–72.

Lopez-Gonzalez, V., Murano, P. S., Brennan, R. E. and Murano, E. A. (2000). Sensory evaluation of ground beef patties irradiated by gamma rays versus electron beam under various packaging conditions. *J. Food Qual.* **23,** 195–204.

Marque, D., Feigenbaum, A. and Riquet, A. M. (1995). Consequences of polypropylene film ionization on food/packaging interactions. *J. Polym. Eng.* **15,** 101–115.

Marth, E. H. (1998). Extended shelf life refrigerated foods: Microbiological quality and safety. *Food Technol.* **52(2),** 57–62.

Masuda, M., Saito, Y., Iwanami, T. and Hirai, Y. (1992). Effects of hydrostatic pressure on packaging materials for food. In: *High Pressure and Biotechnology* (C. Balny, R. Hayashi, K. Heremans and P. Masson, eds), pp. 545–547. John Libbey Eurotext, London, UK.

Matsui, T., Inoue, M., Shimoda, M. and Osajama Y. (1991). Sorption of volatile compounds into electron beam irradiated EVA film in the vapour phase. *J. Sci. Food Agric.* **54,** 127–135.

Mertens, B. and Deplace, G. (1993). Engineering aspects of high-pressure technology in the food industry. *Food Technol.* **6,** 1964–1969.

Min, S., Jin, Z. T. and Zhang, Q. H. (2003a). Commercial scale pulsed electric field processing of tomato juice. *J. Agric. Food Chem.* **51,** 3338–3344.

Min, S., Jin, Z. T., Min, S. K., Yeom, H. and Zhang, Q. H. (2003b). Commercial scale pulsed electric field processing of orange juice. *J. Food Sci.* **68(4),** 1265–1271.

Murano, E. A., Murano, P. S., Brennan, R. E., Shenoy, K. and Moreira, R. G. (1999). Application of high hydrostatic pressure to eliminate *Listeria monocytogenes* from fresh pork sausage. *J. Food Protect.* **62(5),** 480–483.

Nakayama, A., Yano, Y., Kobayahi, S., Ishikawa, M. and Sakai, K. (1996). Comparison of pressure resistances of spores of six bacillus strains with their heat resistances. *Appl. Environ. Microbiol.* **62(10),** 3897–3900.

Nam, K. C. and Ahn, D. U. (2002). Mechanisms of pink color formation in irradiated pre-cooked turkey breast meat. *J. Food Sci.* **67(2),** 600–607.

Nielsen, T. J., Jagerstad, I. M., Oste, R. E. and Wesslen, B. O. (1992). Comparative absorption of low molecular aroma compounds into commonly used food packaging polymer films. *J. Food Sci.* **57(2),** 490–492.

Nienaber, U. and Shellhammer, T. H. (2001). High-pressure processing of orange juice: Combination treatments and a shelf life study. *J. Food Sci.* **66(2),** 332–336.

Olson, D. G. (1998). Irradiation of food. *Food Technol.* **52(1),** 56–62.

Ozen, B. F. and Floros, J. D. (2001). Effects of emerging food processing techniques on the packaging materials. *Trends Food Sci. Technol.* **12,** 60–67.

Palou, E., Hernandez-Salgado, C., Lopez-Malo, A., Barbosa-Canovas, G. V., Swanson, B. G. and Welti-Chanes, J. W. (2000). High pressure-processed guacamole. *Innov. Food Sci. Emerg. Technol.* **1,** 69–75.

Pascat, B. (1986). In: *Food Packaging and Preservation: Theory and Practice* (M. Mathlouthi, ed.), pp. 7–24. Elsevier Applied Science, London, UK.

Pentimalli, M., Capitani, D., Ferrando, A., Ferri, D., Ragni, P. and Segre, A. L. (2000). Gamma irradiation of food packaging materials: an NMR study. *Polymer* **41,** 2871–2881.

Pilette, L. (1990). Effects of ionizing treatments on packaging–food stimulant combinations. *Packag. Technol. Sci.* **3,** 17–20.

Qin, B. L., Chang, F. J., Barbosa-Canovas, G. V. and Swanson, B. G. (1995a). Nonthermal inactivation of *Saccharomyces cerevisiae* in apple juice using pulsed electric fields. *Lebensm. Wiss. u. Technol.* **28,** 564–568.

Qin, B. L., Pothakamury, U. R., Vega-Mercado, H., Martin, O., Barbosa-Canovas, G. V. and Swanson, B. G. (1995b). Food pasteurization using high intensity pulsed electric fields. *Food Technol.* **49(12),** 55–60.

Qiu, X., Sharma, S., Tuhela, L., Jia, M. and Zhang, Q. H. (1998). An integrated PEF pilot plant for continuous nonthermal pasteurization of fresh orange juice. *Trans. ASAE* **41,** 1069–1074.

Raghubeer, E. V., Dunne, C. P., Farkas, D. F. and Ting, E. Y. (2000). Evaluation of batch and semicontinuous application of high hydrostatic pressure on foodborne pathogens in salsa. *J. Food Protect.* **63(12),** 1713–1718.

Robertson, G. L. (1993). *Food Packaging: Principles and Practice*, pp. 9–62, 318–321. Marcel Dekker, New York, NY.

Rodriguez-Amaya, D. B. (1993). Stability of carotenoids during the storage of foods. In: *Developments in Food Science 33*, pp. 591–628. Elsevier, New York, NY.

Rojas De Gante, C. and Pascat, B. (1990). *Packag. Technol. Sci.* **3,** 97–105.

Ruan, S. (2000). Particle engineering of polyimide composites. MSc thesis, Ohio State University, Columbus, OH.

Sahasrabudhe, M. R. (1990). *J. Can. Diet. Assoc.* **51(2),** 329–334.

Sharma, S. K., Zhang, Q. H. and Chism, G. W. (1998). Development of a protein fortified fruit beverage and its quality when processed with pulsed electric field treatment. *J. Food Qual.* **21,** 459–473.

Sheung, S. K. (1995). Sorption of orange juice flavor compounds into polymeric packaging materials. MSc thesis, Ohio State University, Columbus, OH.

Singh, R. K. and Nelson P. E. (1992). *Advances in Aseptic Processing Technologies.* Elsevier Science, London, UK.

Smelt, J. P. P. M. (1998). Recent advances in the microbiology of high pressure processing. *Trends Food Sci. Technol.* **9,** 152–158.

Sohn, K.-H. and Lee, H.-J. (1998). Effects of high pressure treatment on the quality and storage of kimchi. *Intl J. Food Sci. Technol.* **33(4),** 359–365.

Takahashi, F., Pehrsson, P. E., Rovere, P. and Squarcina, N. (1998). High-pressure processing of fresh orange juice. *Industria Conserve* **73(4),** 363–368.

Thakur, B. R., Singh, R. K. and Nelson, P. E. (1996). Quality attributes of processed tomato products: a review. *Food Rev. Intl* **12,** 375–401.

Tsong, T. Y. (1991). Electroporation of cell membranes. *Biophys. J.* **60,** 297–306.

Tuboly, E., Lebovics, V. K., Gaal, O., Meszaros, L. and Farkas, J. (2003). Microbiological and lipid oxidation studies on mechanically deboned turkey meat treated by high hydrostatic pressure. *J. Food Eng.* **56(2/3),** 241–244.

Van Willige, R. W. G., Linssen, J. P. H. and Voragen, A. G. J. (2000). Influence of food matrix on absorption of flavour compounds by linear low-density polyethylene: proteins and carbohydrates. *J. Sci. Food Agric.* **80,** 1779–1789.

Varsanyi, I. (1975). Investigation into the permeability of polymer membranes of food packaging quality to gases and water vapour after radiation treatment with radurizing doses. *Acta Aliment.* **4,** 251–269.

Varsanyi, I. (1986). In: *Food Packaging and Preservation: Theory and Practice* (M. Mathlouthi, ed.), pp. 25–38. Elsevier Applied Science, London, UK.

Yam, K. L. and Lee, D. S. (1995). Design of modified atmosphere packaging for fresh produce. In: *Active Food Packaging* (M. L. Rooney, ed.), pp. 55–73. Blackie Academic & Professional, London, UK.

Yeom, H. W., Streaker, C. B., Zhang, Q. H. and Min, D. B. (2000). Effects of pulsed electric fields on the quality of orange juice and comparison with heat pasteurization. *J. Agric. Food Chem.* **48(10),** 4597–4605.

Zhang, Q., Barbosa-Canovas, G. V. and Swanson, B. G. (1995). Engineering aspects of pulsed electric field pasteurization. *J. Food Eng.* **25,** 261–281.

Zhang, Q. H., Qiu, X. and Sharma, S. K. (1997). In: *New Technologies Yearbook* (D. I. Chandrana, ed.), pp. 31–42. National Food Processors Association, Washington, DC.

Ziegler, R., Schanderl, S. H. and Markakis, P. (1968). Gamma irradiation and enriched CO_2 atmosphere storage effects on the light-induced greening of potatoes. *J. Food Sci.* **33(5),** 533–535.

Zimmermann, U. (1986). Electrical breakdown, electropermeabilization and electrofusion. *Rev. Physiol. Biochem. Pharmacol.* **105,** 175–256.

Index

Food Science and Technology
International Series

Maynard A. Amerine, Rose Marie Pangborn, and Edward B. Roessler, *Principles of Sensory Evaluation of Food*. 1965.

Martin Glicksman, *Gum Technology in the Food Industry*. 1970.

Maynard A. Joslyn, *Methods in Food Analysis*, second edition. 1970.

C. R. Stumbo, *Thermobacteriology in Food Processing*, second edition. 1973.

Aaron M. Altschul (ed.), *New Protein Foods:* Volume 1, *Technology, Part A*—1974. Volume 2, *Technology, Part B*—1976. Volume 3, *Animal Protein Supplies, Part A*—1978. Volume 4, *Animal Protein Supplies, Part B*—1981. Volume 5, *Seed Storage Proteins*—1985.

S. A. Goldblith, L. Rey, and W. W. Rothmayr, *Freeze Drying and Advanced Food Technology*. 1975.

R. B. Duckworth (ed.), *Water Relations of Food*. 1975.

John A. Troller and J. H. B. Christian, *Water Activity and Food*. 1978.

A. E. Bender, *Food Processing and Nutrition*. 1978.

D. R. Osborne and P. Voogt, *The Analysis of Nutrients in Foods*. 1978.

Marcel Loncin and R. L. Merson, *Food Engineering: Principles and Selected Applications*. 1979.

J. G. Vaughan (ed.), *Food Microscopy*. 1979.

J. R. A. Pollock (ed.), *Brewing Science*, Volume 1—1979. Volume 2—1980. Volume 3—1987.

J. Christopher Bauernfeind (ed.), *Carotenoids as Colorants and Vitamin A Precursors: Technological and Nutritional Applications*. 1981.

Pericles Markakis (ed.), *Anthocyanins as Food Colors*. 1982.

George F. Stewart and Maynard A. Amerine (eds.), *Introduction to Food Science and Technology*, second edition. 1982.

Malcolm C. Bourne, *Food Texture and Viscosity: Concept and Measurement*. 1982.

Hector A. Iglesias and Jorge Chirife, *Handbook of Food Isotherms: Water Sorption Parameters for Food and Food Components*. 1982.

Colin Dennis (ed.), *Post-Harvest Pathology of Fruits and Vegetables*. 1983.

P. J. Barnes (ed.), *Lipids in Cereal Technology*. 1983.

David Pimentel and Carl W. Hall (eds.), *Food and Energy Resources*. 1984.

Joe M. Regenstein and Carrie E. Regenstein, *Food Protein Chemistry: An Introduction for Food Scientists*. 1984.

Maximo C. Gacula, Jr., and Jagbir Singh, *Statistical Methods in Food and Consumer Research*. 1984.

Fergus M. Clydesdale and Kathryn L. Wiemer (eds.), *Iron Fortification of Foods*. 1985.

Robert V. Decareau, *Microwaves in the Food Processing Industry*. 1985.

S. M. Herschdoerfer (ed.), *Quality Control in the Food Industry*, second edition. Volume 1—1985. Volume 2—1985. Volume 3—1986. Volume 4—1987.

F. E. Cunningham and N. A. Cox (eds.), *Microbiology of Poultry Meat Products*. 1987.

Walter M. Urbain, *Food Irradiation*. 1986.

Peter J. Bechtel, *Muscle as Food*. 1986.

H. W.-S. Chan, *Autoxidation of Unsaturated Lipids*. 1986.

Chester O. McCorkle, Jr., *Economics of Food Processing in the United States*. 1987.

Jethro Japtiani, Harvey T. Chan, Jr., and William S. Sakai, *Tropical Fruit Processing*. 1987.

J. Solms, D. A. Booth, R. M. Dangborn, and O. Raunhardt, *Food Acceptance and Nutrition*. 1987.

R. Macrae, *HPLC in Food Analysis*, second edition. 1988.

A. M. Pearson and R. B. Young, *Muscle and Meat Biochemistry*. 1989.

Marjorie P. Penfield and Ada Marie Campbell, *Experimental Food Science*, third edition. 1990.

Leroy C. Blankenship, *Colonization Control of Human Bacterial Enteropathogens in Poultry*. 1991.

Yeshajahu Pomeranz, *Functional Properties of Food Components*, second edition. 1991.

Reginald H. Walter, *The Chemistry and Technology of Pectin*. 1991.

Herbert Stone and Joel L. Sidel, *Sensory Evaluation Practices*, second edition. 1993.

Robert L. Shewfelt and Stanley E. Prussia, *Postharvest Handling: A Systems Approach*. 1993.

R. Paul Singh and Dennis R. Heldman, *Introduction to Food Engineering*, second edition. 1993.

Tilak Nagodawithana and Gerald Reed, *Enzymes in Food Processing*, third edition. 1993.

Dallas G. Hoover and Larry R. Steenson, *Bacteriocins*. 1993.

Takayaki Shibamoto and Leonard Bjeldanes, *Introduction to Food Toxicology*. 1993.

John A. Troller, *Sanitation in Food Processing*, second edition. 1993.

Ronald S. Jackson, *Wine Science: Principles and Applications*. 1994.

Harold D. Hafs and Robert G. Zimbelman, *Low-fat Meats*. 1994.

Lance G. Phillips, Dana M. Whitehead, and John Kinsella, *Structure-Function Properties of Food Proteins*. 1994.

Robert G. Jensen, *Handbook of Milk Composition*. 1995.

Yrjö H. Roos, *Phase Transitions in Foods*. 1995.

Reginald H. Walter, *Polysaccharide Dispersions*. 1997.

Gustavo V. Barbosa-Cánovas, M. Marcela Góngora-Nieto, Usha R. Pothakamury, and Barry G. Swanson, *Preservation of Foods with Pulsed Electric Fields*. 1999.

Ronald S. Jackson, *Wine Science: Principles, Practice, Perception*, second edition. 2000.

R. Paul Singh and Dennis R. Heldman, *Introduction to Food Engineering*, third edition. 2001.

Ronald S. Jackson, *Wine Tasting: A Professional Handbook*. 2002.

Malcolm C. Bourne, *Food Texture and Viscosity: Concept and Measurement*, second edition. 2002.

Benjamin Caballero and Barry M. Popkin (eds), *The Nutrition Transition: Diet and Disease in the Developing World*. 2002.

Dean O. Cliver and Hans P. Riemann (eds), *Foodborne Diseases*, second edition. 2002.

Martin Kohlmeier, *Nutrient Metabolism*. 2003.

Herbert Stone and Joel L. Sidel, *Sensory Evaluation Practices*, third edition. 2004.

Jung H. Han, *Innovations in Food Packaging*. 2005.

Da-Wen Sun, *Emerging Food Technologies*. 2005.